2013 IEEE 22nd Conference on Electrical Performance of Electronic Packaging and Systems

(EPEPS 2013)

San Jose, California, USA
27-30 October 2013

IEEE Catalog Number:	CFP13EPP-POD
ISBN:	978-1-4799-0708-3

Copyright © 2013 by the Institute of Electrical and Electronic Engineers, Inc
All Rights Reserved

Copyright and Reprint Permissions: Abstracting is permitted with credit to the source. Libraries are permitted to photocopy beyond the limit of U.S. copyright law for private use of patrons those articles in this volume that carry a code at the bottom of the first page, provided the per-copy fee indicated in the code is paid through Copyright Clearance Center, 222 Rosewood Drive, Danvers, MA 01923.

For other copying, reprint or republication permission, write to IEEE Copyrights Manager, IEEE Service Center, 445 Hoes Lane, Piscataway, NJ 08854. All rights reserved.

***This publication is a representation of what appears in the IEEE Digital Libraries. Some format issues inherent in the e-media version may also appear in this print version.**

IEEE Catalog Number: CFP13EPP-POD
ISBN 13:   978-1-4799-0708-3

Additional Copies of This Publication Are Available From:

Curran Associates, Inc
57 Morehouse Lane
Red Hook, NY 12571 USA
Phone: (845) 758-0400
Fax: (845) 758-2633
E-mail: curran@proceedings.com
Web: www.proceedings.com

2013 IEEE 22nd Conference on Electrical Performance of Electronic Packaging and Systems

October 27-30, 2013

DoubleTree by Hilton Hotel
San Jose, California

Sponsored by

The IEEE Microwave Theory and Techniques Society

and

The IEEE Components, Packaging and Manufacturing Technology Society

Copyright and Reprint Permission: Abstracting is permitted with credit to the source. Libraries are permitted to photocopy beyond the limit of U.S. copyright law for private use of patrons those articles in this volume that carry a code at the bottom of the first page, provided the per-copy fee indicated in the code is paid through Copyright Clearance Center, 222 Rosewood Drive, Danvers, MA 01923.

For other copying, reprint or republication permission, write to IEEE Copyrights Manager, IEEE Operations Center, 445 Hoes Lane, P.O. Box 1331, Piscataway, NJ 08855-1331. All rights reserved. Copyright 2009 by the Institute of Electrical and Electronics Engineers

IEEE 22nd Conference on Electrical Performance of Electronic Packaging and Systems

Welcome Message
by the
General Chair
EPEPS-2013, San Jose

Greetings EPEPS Attendees,

Welcome to the 22nd EPEPS conference in beautiful San Jose, CA, USA! The Technical Program Committee (TCP) is proud to present a diverse technical program of 35oral presentations in our traditional single-track format and 22 poster papers covering the latest advances and emerging technologies in signal and power integrity. We have excellent educational opportunities with three tutorials. New to the conference this year are Industry Spotlight presentations with one on Sunday and two on Monday. Meeting old friends and making new ones is also very important. We have arranged bus transportation on Tuesday evening for you to enjoy Santana Row, a 1.5 million square foot mixed-use development known as the Silicon Valley's premier destination for shopping, dining, living, working and nightlife. Surrounded by landscaped gardens, parks and plazas, Santana Row features over 70 retail shops and more than two dozen acclaimed restaurants for you to enjoy a relaxed evening for dinner on your own as well as an opportunity to come away a winner by participating in our Santana Row Scavenger Hunt. More information on Santana Row can be found at www.santanrow.com. More information is also available at our conference registration area. In addition to the EPEPS conference, we invite you to explore the Intel Museum nearby. http://www.intel.com/content/www/us/en/history/museum-visiting-intel.html. Although not one of our official conference scheduled activities, the museum is open from 9am-6pm and has many interactive displays regarding silicon chip fabrication and technology

EPEPS continues to be a highly diverse and international conference with authors from over ten countries. A strength of EPEPS has always been the good mix of industry and academic work, and this year continues that tradition. The oral papers are organized into sessions spanning novel signaling, system design, modeling, electromagnetics and CAD. The ever popular poster session continues as a highly technical interaction with a social flair.

The TPC has invited a number of eminent speakers to provide educational background and forward-looking technological insights. We are fortunate to have as our Keynote Speaker Todd Takken, Manager of Systems Power, Packaging and Cooling at IBM Research. His presentation *The Future of Optics in Supercomputers* will give us a glimpse in to next generation "green electronics".

Corporate sponsorship enables the TPC to provide you with a better EPEPS in many ways and hope many fruitful relationships are initiated. Please drop by the sponsor tables and see what they have to offer. We greatly appreciate the support of our Platinum Sponsors, Ansys and Cadence; Gold Sponsor Rambus; Silver Sponsors IBM, IdemWorks, CST, Agilent and Nimbic.

EPEPS is a volunteer effort requiring hours of dedicated effort by many individuals. For their efforts, the TPC thanks the invited and tutorial speakers, authors, presenters, and session chairs. Special thanks go to Mandy Wisehart of the University of Illinois for handling all of the logistical details and for putting her own personal touch on numerous visible aspects of EPEPS. Finally, we thank the IEEE Societies MTT-S and CPMT for their continuing support.

Relax and enjoy three days of the best the world has to offer in signal and power integrity presentations and professional interactions at EPEPS 2013.

Kathleen L. Melde
EPEPS 2013 General Co-Chair
University of Arizona

Tutorial Speaker - I (Sunday)

Title: NanoCarbon for Next-Generation Green Electronics: Status and Prospects

Speaker: Kaustav Banerjee, UC Santa Barbara

Time: Sunday, 3:00pm – 4:30pm

Abstract: NanoCarbons constitute the low-dimensional allotropes of carbon including 1-dimensional carbon nanotubes (CNTs) and 2-dimensional (2D) graphene. These nanomaterials have extraordinary physical properties that can be exploited to bring forward their exciting prospects for a variety of applications. This tutorial will highlight and discuss the unique prospects of NanoCarbon materials (especially graphene and CNTs) for designing next generation low-power, low-loss and ultra-energy-efficient active and passive devices targeted for designing next-generation "green electronics". The discovery of graphene has also opened up a new era for a wide range of 2D crystals and their unprecedented electronic applications. This tutorial will also provide a brief overview of such materials and related opportunities, especially in the electronics domain.

Bio: Kaustav Banerjee is Professor of Electrical and Computer Engineering and Director of the Nanoelectronics Research Lab at the University of California, Santa Barbara (UCSB). Initially trained as a physicist, he received the Ph.D. degree in Electrical Engineering and Computer Sciences from the University of California, Berkeley, in 1999. Prior to joining the UCSB Faculty in 2002, he was a Research Associate at the Center for Integrated Systems in Stanford University during 1999-2001.

His research interests include nanometer scale issues in CMOS VLSI as well as emerging nanoelectronics. Prof. Banerjee's ideas and innovations chronicled in over 275 publications have not only received thousands of citations but also have played a decisive role in steering worldwide research. He was elected a Fellow of IEEE in Fall 2011, and has served as a Distinguished Lecturer of the IEEE Electron Devices Society since 2008. Prof. Banerjee is one of five engineers worldwide to receive the Friedrich Wilhelm Bessel Research Award from Alexander von Humboldt Foundation, Germany, in 2011 for his outstanding contributions in nanoelectronics. More information about him and his research can be found at: http://nrl.ece.ucsb.edu/

Keynote Speech (Monday)

Title: The Future of Optics in Supercomputers

Speaker: Dr. Todd Takken, Manager of Systems Power, Packaging and Cooling at IBM Research

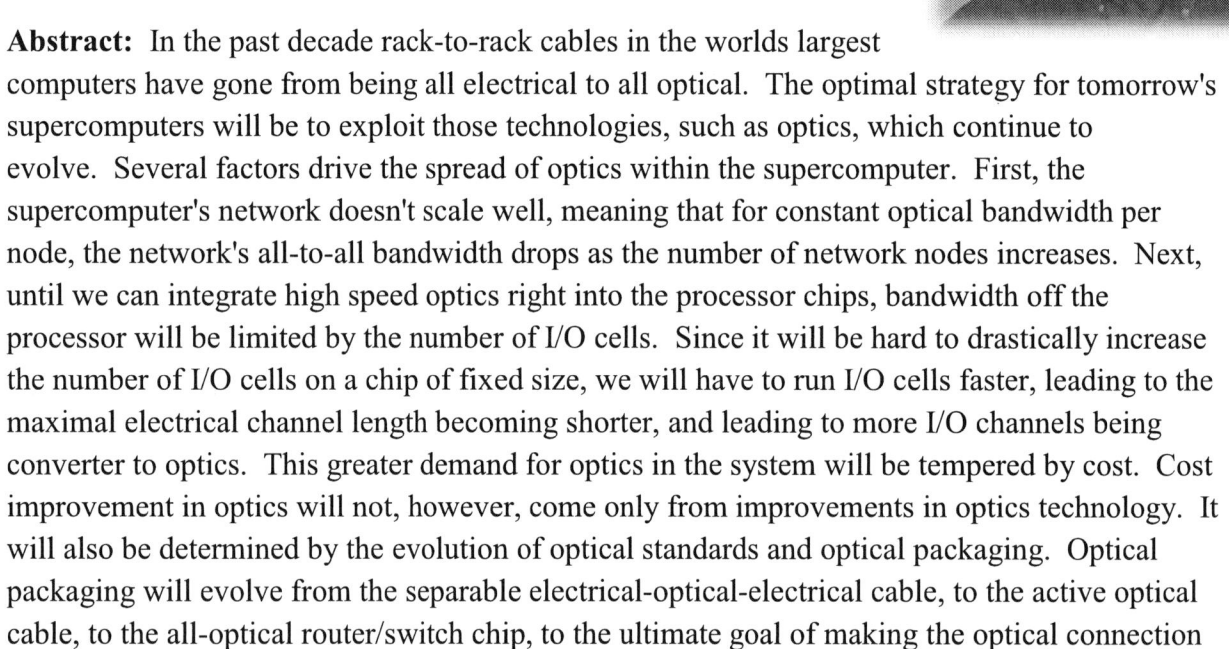

Time: Monday, 8:45am – 9:30am

Abstract: In the past decade rack-to-rack cables in the worlds largest computers have gone from being all electrical to all optical. The optimal strategy for tomorrow's supercomputers will be to exploit those technologies, such as optics, which continue to evolve. Several factors drive the spread of optics within the supercomputer. First, the supercomputer's network doesn't scale well, meaning that for constant optical bandwidth per node, the network's all-to-all bandwidth drops as the number of network nodes increases. Next, until we can integrate high speed optics right into the processor chips, bandwidth off the processor will be limited by the number of I/O cells. Since it will be hard to drastically increase the number of I/O cells on a chip of fixed size, we will have to run I/O cells faster, leading to the maximal electrical channel length becoming shorter, and leading to more I/O channels being converter to optics. This greater demand for optics in the system will be tempered by cost. Cost improvement in optics will not, however, come only from improvements in optics technology. It will also be determined by the evolution of optical standards and optical packaging. Optical packaging will evolve from the separable electrical-optical-electrical cable, to the active optical cable, to the all-optical router/switch chip, to the ultimate goal of making the optical connection at the processor chip. Many a supercomputer system is designed around the network and the connectors, and a 1st-level module to waveguide connector, coupled with an optical connector at the tailstock, should be the most effective combination for improving network performance of future systems.

Bio: Dr. Todd Takken manages the Systems Power, Packaging and Cooling group at IBM's T.J. Watson Research Center. Since receiving his Ph.D. in Electrical Engineering from Stanford University in 1997, Dr. Takken has researched new architecture, networking, high speed signaling, signal integrity, power delivery, packaging and cooling solutions for IBM servers and systems. Starting with the inception of the Blue Gene program in 1999, Dr. Takken has been one of the primary engineers responsible for the system packaging of IBM's largest parallel supercomputers. With over 30 patents in his field, Dr. Takken continues to investigate the technologies which will be needed for the ExaFlop generation of systems.

Tutorial Speaker - II (Tuesday)

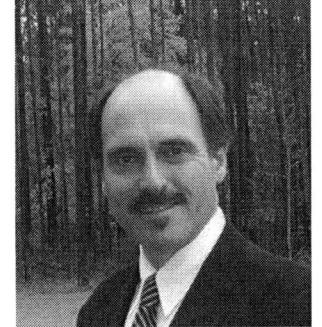

Title: Design and Test of 2.5D and 3D stacked ICs

Speaker: Paul Franzon, North Carolina State

Time: Tuesday, 8:30am – 10:10am

Abstract: Three dimensional chips stacked using Through Silicon Via (TSV) technology has been under consideration and the subject of intensive research for several years now. Soon the technologies will become available through standard fabs. Will the technology be an instant hit, a niche, or a flop? What is needed to ensure it reaches hit status? What are the basic manufacturing steps and flows? This tutorial will discuss these question mainly in the context of the opportunities and challenges that face the designer. What are the significant opportunities presented by 3DIC? What problems will the designer face that will need clever solutions? What are the potential solution paths?

Bio: Paul D. Franzon is currently a Distinguished Alumni Professor of Electrical and Computer Engineering at North Carolina State University. He earned his Ph.D. from the University of Adelaide, Adelaide, Australia in 1988. He has also worked at AT&T Bell Laboratories, DSTO Australia, Australia Telecom and two companies he cofounded, Communica and LightSpin Technologies. His current interests center on the technology and design of complex microsystems incorporating VLSI, MEMS, advanced packaging and nano-electronics. He has lead several major efforts and published over 200 papers in these areas. In 1993 he received an NSF Young Investigators Award, in 2001 was selected to join the NCSU Academy of Outstanding Teachers, in 2003, selected as a Distinguished Alumni Professor, and received the Alcoa Research Award in 2005. He served with the Australian Army Reserve for 13 years as an Infantry Solider and Officer. He is a Fellow of the IEEE.

Tutorial Speaker - III (Wednesday)

Title: On-Chip Measurement and Characterization for High –Speed Links

Speaker: Chris Madden and Hai Lan, Rambus

Time: Wednesday, 8:30am – 9:10am

Chris Madden

Hai Lan

Abstract: Today's high-speed interface designers are facing increasingly complex challenges imposed by the requirements of high performance, optimal power efficiency, and often by tight 3-D integration, such as in package-on-package (PoP) systems. Answers to many difficult, yet critical questions, need to be found by measurement inside the high-speed interfaces. What is the signal quality after the package pin and equalization circuitry? What is the relative jitter impact with clock-data-recovery (CDR) on versus off? What is the actual dynamic supply noise experienced by the various sections of the on-chip circuitry? What is the power delivery network (PDN) impedance seen from the silicon? Conventional off-chip probing at the board- or package-pin-level is neither sufficient nor relevant to answer these questions. On-chip signal integrity (SI) and power integrity (PI) characterization techniques are therefore indispensable. This tutorial first gives an overview of on-chip SI/PI measurement requirements. It then reviews several key on-chip measurement concepts and techniques to allow true on-chip SI/PI characterization:

- *eScope* for measuring operating voltage and timing margin of the entire link
- *eWave* for capturing equivalent-time signal waveform as seen by the actual data sampler
- *nScope* for monitoring on-chip power supply noise
- *zScope* for characterizing on-chip self and mutual PDN impedances

Application examples for both parallel and serial interfaces will be discussed, including low-power memory interfaces and Serializer/Deserializer (SerDes) links.

Bios: Chris Madden received the Ph.D. degree in electrical engineering from Stanford University, Stanford, CA, in 1990. He is a Senior Signal Integrity Principal Engineer with Rambus Inc., Sunnyvale, CA, where he has worked since March 2003 on link modeling and device characterization methodologies for multi-gigabit CMOS interfaces. Prior to Rambus, he worked at several companies developing modules for fiber-optic communications, notably Agilent and Finisar. Also, he was at HP Laboratories for nine years where he did mm-wave circuit design and characterization for instrumentation and wireless application. He currently holds six U.S. patents.

Hai Lan is a Senior Principal Engineer in signal and power integrity at Rambus Inc. He received the Ph.D. degree from Stanford University in 2006, the M.S. degree from Oregon State University in 2001, and the B.S. degree from Tsinghua University in 1999, all in Electrical Engineering. His professional interests include power integrity and signal integrity, mixed-signal integrated circuits design and noise analysis, high-speed on-chip interconnects, and advanced silicon effects such as substrate coupling noise in SoC and mixed-signal IC. Dr. Lan has published more than thirty papers in journals and conferences and authored four chapters in two books. He received the Best Paper Award at DesignCon 2012. He is an IEEE Member.

Technical Program Committee (TPC)

Co-chair
Vikram Jandhyala
University of Washington; Nimbic, Inc.

Conference chair
Kathleen Melde
University of Arizona

Co-chair
Dale Becker
IBM

Peter Aaen
University of Surrey

Ramachandra Achar
Carleton University

Wendem Beyene
Rambus

Henning Braunisch
Intel Corporation

Flavio Canavero
Politecnico di Torino

Paul Franzon
North Carolina State University

Kevin (Xiaoxiong) Gu
IBM

Joungho Kim
KAIST

Michael Lamson
Texas Instruments (retired)

Michel Nakhla
Carleton University

Dan (Kyung Suk) Oh
Altera

Albert Ruehli
Emeritus IBM, Missouri U. of Science and Tech.

Jose Schutt-Aine
University of Illinois

Andreas Weisshaar
Oregon State University

Thomas-Michael Winkel
IBM

Every product is a promise

For all its sophisticated attributes, today's modern product is, at its core, a promise.

A promise that it will perform properly, not fail unexpectedly, and maybe even exceed the expectations of its designers and users. ANSYS helps power these promises with the most robust, accurate and flexible stimulation platform available.

To help you see every possibility and keep every promise.

Realize Your Product Promise®

For more information, visit: **www.ansys.com/promise**

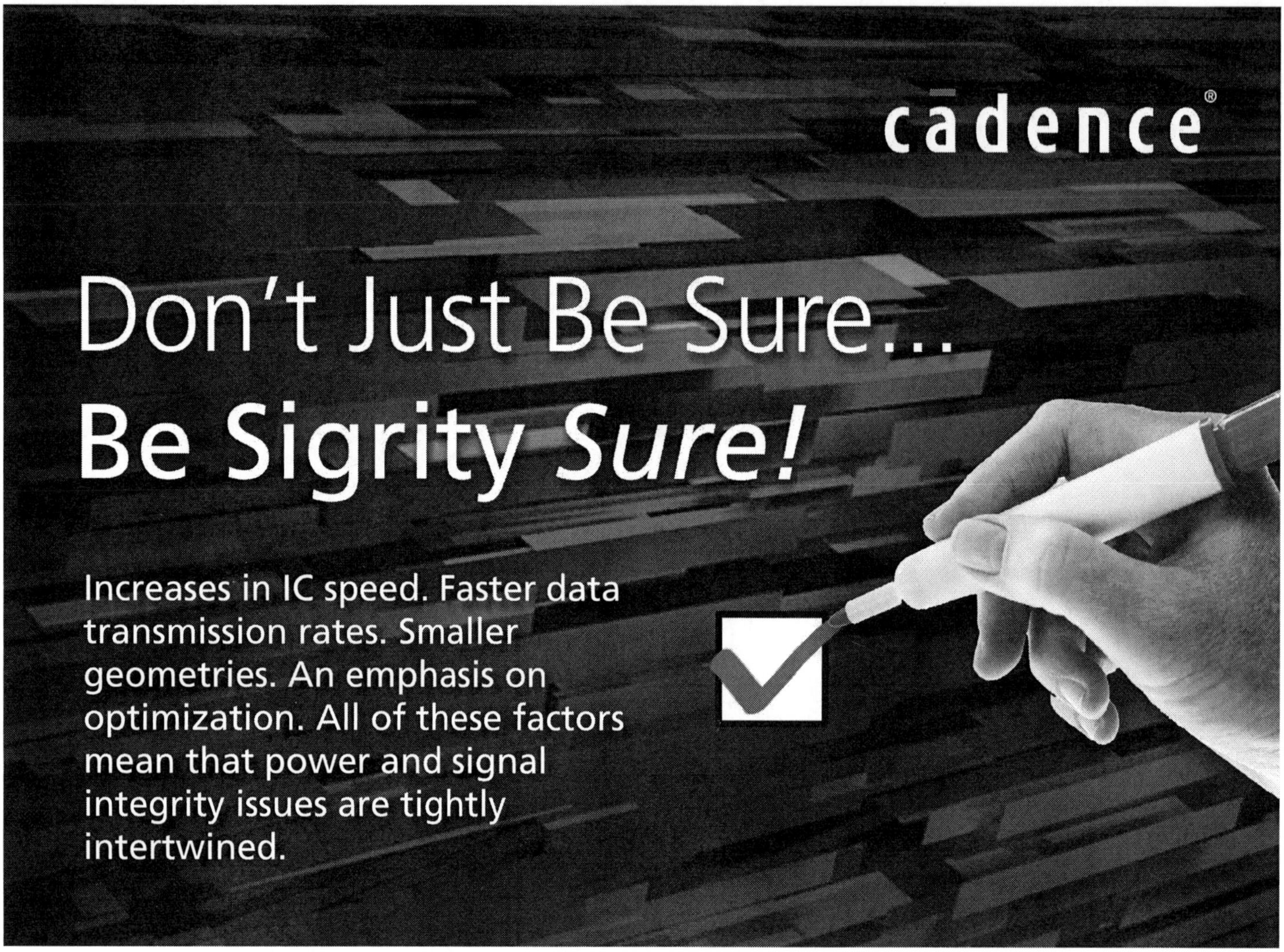

cādence®

Don't Just Be Sure...
Be Sigrity *Sure!*

Increases in IC speed. Faster data transmission rates. Smaller geometries. An emphasis on optimization. All of these factors mean that power and signal integrity issues are tightly intertwined.

With Cadence® Sigrity® signal integrity and power analysis solutions for system-level verification and interface compliance, you'll have what you need to sign off with confidence and be Sigrity Sure!

Our power-aware signal integrity tools integrate our Allegro® and Sigrity technologies. You get signoff-level, accurate signal integrity analysis for PCBs and IC packages. Our Sigrity technology-based power integrity tools deliver signoff-level accuracy for AC and DC power analysis of PCBs and IC packages.

Be sure to visit Cadence at EPEPS to learn more about our newest products. We look forward to meeting you and chatting about your signal integrity and power analysis challenges!

Learn more today at:
http://www.cadence.com/cadence/ads/sigrity/Pages/default.aspx

© Cadence Design Systems, Inc. All rights reserved. Cadence, the Cadence logo, and Sigrity are registered trademarks of Cadence Design Systems, Inc.

Rambus

Bringing **invention** to market

R+ Enhanced
Standard Solutions

Industry-compatible memory and serial link interface IP developed with a **system-aware design methodology** for **ease of integration** and **robust manufacturability**

Standards supported include:

- DDR4/3/2
- LPDDR3/2
- PCI-Express3/2/1
- 10GBASE-KR
- CEI-11G, CEI-28G
- SATA1/2/3

rambus.com/rplus

/R

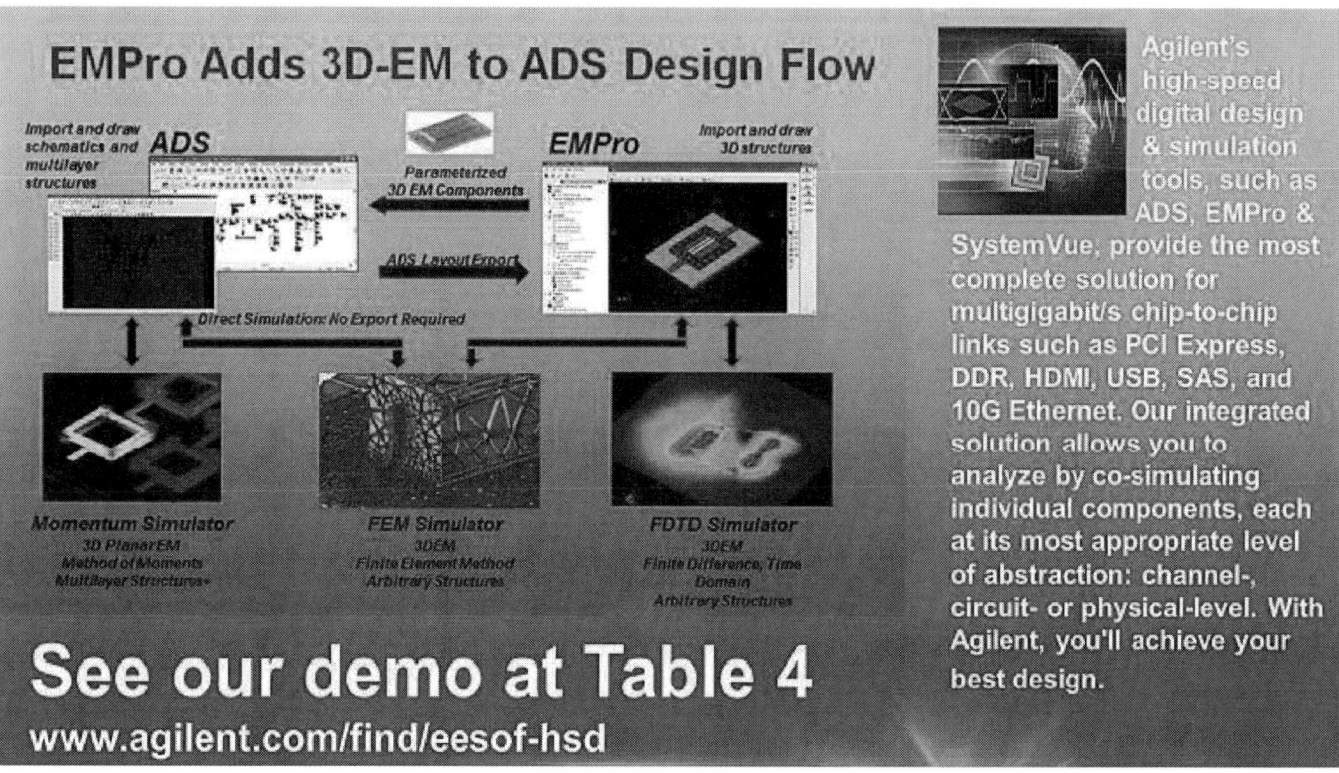

Make the Connection

Find the simple way through complex EM systems with CST STUDIO SUITE

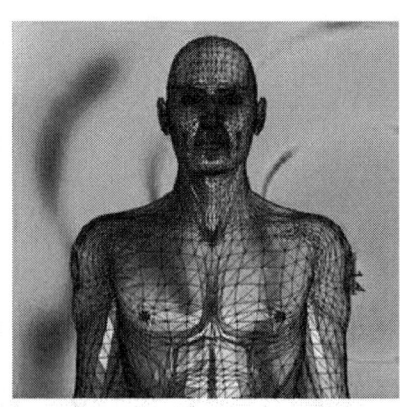

Components don't exist in electromagnetic isolation. They influence their neighbors' performance. They are affected by the enclosure or structure around them. They are susceptible to outside influences. With System Assembly and Modeling, CST STUDIO SUITE helps optimize component and system performance.

Involved in antenna development? You can read about how CST technology is used to simulate antenna performance at www.cst.com/antenna.

If you're more interested in filters, couplers, planar and multilayer structures, we've a wide variety of worked application examples live on our website at www.cst.com/apps.

Get the big picture of what's really going on. Ensure your product and components perform in the toughest of environments.

Choose CST STUDIO SUITE – Complete Technology for 3D EM.

IBM
Research

www.research.ibm.com

From the invention of DRAM to innovations in nanotechnology and silicon, IBM has been at the forefront of Science and Technology breakthroughs in materials, processes and devices. Through our expertise in the physical sciences, silicon technology, and communication and computation subsystems we're helping to create the next generation of IBM systems, servers and storage products. We've also applied our research to a wide range of client needs, including projects to develop polymers that can destroy antibiotic resistant bacteria, polymers to more efficiently filter and desalinate water, and materials, devices and systems that convert sunlight into electricity.

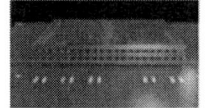     

3D Chip   Holey Optochip   Carbon Nanotube   Spintronics   Quantum Computing   Silicon Photonics

IdemWorks is an industry leading provider of modeling tools and services for fast EMC, Signal and Power Integrity, system verification. IdemWorks tools allow faster product development, reducing risk, time-to-market and costs.

IdEM is aimed at model extraction for Electrical Interconnects and Packages, providing broadband computational models and/or equivalent netlists that can be used in any circuit-based simulation environment for transient or AC analyses. An advanced multiprocessing module is also available.

IdEM MORE performs Model Order Reduction of large circuits obtained from parasitic extraction.

HPM is the first High Performance Macromodeling service, providing on-demand interactive access to parallel model extraction and qualification algorithms via an intuitive secure web interface.

www.idemworks.com

Seeking Signal Integrity Sign-off?

Now, you can benefit from electromagnetic simulation as an integral part of your production design flow. And, get a full package and PCB signal integrity, power integrity and EMI sign-off.

Contact us today for a no-obligation demo. We'll show you how customers world-wide are benefiting from our solutions and how we can help you as well.

A representative sample of our global customers:

www.nimbic.com/offer

TABLE OF CONTENTS

Millimeter Wave and RF Technologies I

Electrical Characterization of Low-Cost glass epoxy laminates at Millimeter Wave Frequencies 7
A Method for Modeling the Impact of Conductor Surface Roughness on Waveguiding Properties of Interconnects 11
A Perturbation Technique to Analyze the Influence of Fiber Weave Effects on Differential Signaling 15

Millimeter Wave and RF Technologies II

Insertion Loss Characterization of Tightly Spaced Interconnects with an Embedded Patterned Layer 21
Modeling I/O Buffers Using X-Parameters 25
Design and Fabrication of Geometrically Complicated Multiband Microwave Devices Using A Novel Integrated 3D Printing Technique 29

EM I

Parallel Processing Improvements for Full-Wave Electromagnetic Solvers 35
A Generalized Modeling Method for Signal/Power Integrity Analysis of 3D Coupled Interconnects In Finite Cavity Based on 1D Technology 39
New Simulation Procedure for Accurate Package Modeling Considering Chip-Package Interaction 43
Optimum Implementation of a Locally Implicit Leapfrog Scheme for Fast Simulation of Inhomogeneously-Meshed Plane Structures 47
Memory Efficient Laguerre-FDTD Scheme for Dispersive Media 51

EM II

Modeling Methodologies for Multi Level PCB- Package Co-Simulation & Co-Design 57

EM III

Accelerate High Speed IO Design Closure with Distributed Chip IO Interconnect Model 61

3-D Packaging

Applying a Physics-Based Via Model for the Simulation of Through Silicon Vias 65
Timing Analysis for Thermally Robust Clock Distribution Network Design for 3D ICs 69
Mitigation of TSV-Substrate Noise Coupling in 3-D CMOS SOI Technology 73

Power Integrity and Wireless Power Transfer

Power Integrity Analysis for Core Logic Blocks 79
Power Integrity of a 4.8Gbps-per-link Low-swing Single-ended-I/O Server Memory Interface 83

Power Distribution Network Design Optimization with On-Die Voltage-Dependent
Leakage Path 87
Design, Implementaion and Measurement of Board-to-Board Wireless Power
Transfer (WPT) for Low Voltage Applications..91

Advanced CAD Techniques ....

SPICE-Based Statistical Assessment of Interconnects Terminated by Nonlinear Loads
with Polynomial Characteristics 99
Cost Effective Modeling Methodologies and Evaluating Electrical Interaction in
FCBGA Packages 103
Electrical-Thermal Co-Simulation for DC IR-Drop Analysis of Large-Scale
Integrated Circuits..107
Reliable Detection of Causality Violations in Tabulated Scattering Parameters through
Filtered Dispersion Relations..111
A Parallel, Adaptive, Multi-Point Model Order Reduction Algorithm 115
Multiple and Non-Existent Barnes-Hut Center-of-Charge (CoC) Solution in Lossy
Layered Substrates: Half Space Analytic and Numerical Study 119

Macromodeling...

An iterative reweighting process for macromodel extraction of power distribution
networks..125
Noise compliant macromodel synthesis for RF and Mixed-Signal applications....................129
Connecting vector fitting to barycentric interpolation and the Loewner matrix 133
A Novel Algorithm for Optimum Order Estimation of Nonlinear Reduced Macromodels........137
Fixed-Order Parametric Macromodeling of Interconnects from S-parameter Data
using Loewner Matrix based Method 141

High Speed Links I...

On-die Supply-inducecd Jitter Behavioral Modeling 147
Mitigating the Impact of Sinusoidal Jitter and Duty Cycle Distortion on Random
Jitter estimation by Tailfit Algorithm 151
A Novel Flexible On-Die Decoupling Scheme Using Package Interconnects.........................155
Robust PoP Probing Solutions for High-Performance Application Processor Developments....159

High Speed Links II..

High-speed DIMM-in-a-Package (DIAP) Memory Module 165
Peak Distortion Analysis of Nonlinear Links 169
A Low-Frequency Enhanced S-Parameter Handling Scheme for Time Domain
Simulation of High Speed Interconnects 173

Poster Session ....

A package-level implementation of traveling-wave switch using PIN-diodes............................179

Implementation of a RF Front-end Module by Embedding ICs in Molding Package.................183

Accurate Characterization of Lossy Interconnects from TDR Waveforms............................187

Per-Unit-Length Parameter Extraction for Lossy Multi-conductor Power Cables....................191

Differential Through-Silicon-Vias Modeling and Design Optimization to Benefit
3D IC Performance 195

On-line Real-time Temperature and Power Estimation of an IC Using Time-domain
Thermal Filters..199

Modeling Broadband Equivalent Circuit of Interconnects with Full Wave
Electromagnetic Solver..203

Efficient performance evaluation of high-speed differential interconnect lines with via
discontinuities 207

In-Depth Analysis of Power Noise Coupling Between 211

Analysis and Verification of Board Power Delivery Network Impact on DDR3L Memory
Interface in ARM SoC Application 215

Minimal-Order Circuit Model Based Fast Electromagnetic Simulation 219

Characterization and Analysis of Vertical Coupling Impact on Receiver Performance in
High Speed Serial Interface 223

Some Internal Crosstalk Reduction Schemes 227

Crosstalk Mitigation in Dense Microstrip Wiring Using Stubby Lines 231

Simulation of the TSV-to-Device Coupling in 3D ICs for Short-Channel Strained
Silicon Transistors 235

Efficient Adaptive Mesh Refinement for MoM-based Package-Board 3D Full-wave
Extraction..239

A Novel EBG Structure with Super-Wideband Suppression of Simultaneous Switching
Noise in High Speed Circuits..243

Application of Qualitative Imaging Methods to Electrical Performance-Aware Package
Board Design 247

Characterization of TSVs by Cascaded Daisy Chains..251

Design and Verification of SMT MMIC Package using a 20 GHz LNA, a 40 GHz LNA
and a 40GHz Digital Attenuator 255

A Novel Common-Mode Filter for Multiple Differential Pairs with Low Crosstalk and
Low Mode Conversion Level 259

Tests for Time Domain EM Solvers for Stability and Towards Passivity........................263

22nd Conference on Electrical Performance of Electronic Packaging and Systems

978-1-4799-0708-3/13 $31.00 © 2013 IEEE

978-1-4799-0708-3/13 $31.00 © 2013 IEEE

Samsung Industry Spotlight

SoC Power Integrity from Early Estimation to Design Signoff

Melinda (Ling) Yang*, Anil Gundurao, Eileen You and Harpreet Gill
SoC Bay Area R&D
Samsung Semiconductor Inc.
601 McCarthy Blvd, Milpitas, CA 95035
*melinda.yang@ssi.samsung.com

Power integrity has become a major integration challenge even for low power ARM-based SoC design. As performance requirements demand higher speed, more advanced process nodes reduce transistor level voltage margin, and low power consumption dictate aggressive power saving schemes such as more frequent power mode shifts and power gating, it's challenging to satisfy the power integrity requirements. Following descriptions are some important aspects for SoC chip design.

First, SoC chips integrate many different functional blocks on the same chip, thus the chip's capacity may reach hundreds of millions of gates and tens of supply domains. In addition, a chip has to survive under all functional modes and operating conditions due to many different applications. For example, the ARM-based SoC chip that we taped out is designed for low power high performance applications. It is running on Samsung 28nm LP process to take advantage of the low power features. It has 50+ clock domains, 500+ macro cells, and places 20million+ instances. It has multiple power supply domains and applies power gate features.

Second, SoC chips have to be able to sustain huge system variation, such as process corner differences, different voltage supply and different thermal conditions. More advanced process only exacerbates this problem.

Third, SoC chips like any chips have to meet system constraints. This SoC chip resides on a 6-layer flip-chip thin core package with u-via technology. Advanced embedding decap technology is also experimented. All the supply domains compete for real estate and closeness to decoupling capacitors. As the simple method is proved to be good approximation, further tradeoff and optimization can be performed to provide design guideline for system design. One example of that is to select package number of balls when number of cores is doubled.

There are complicated design EDA tools to address the above issues. However, EDA tools put many constraints on design data input, often strict requirements, and many of which will only be available at very late stage. Also, it's not rare to take days of turn-around time. Thus relying on sign off tools to analyze and make decisions of the power integrity will be lame duck.

In this article, a simple yet comprehensive power integrity analysis method is used. This analysis is based on the hierarchy of the power delivery system, and estimations from each hierarchical level's physical implementation, thus it can perform analysis at a flashing speed and at very early stage. It also achieves very high accuracy, comparing to EDA sign off tools, a maximum difference of 15% is achieved.

This article provides five design examples implementing this system PI method. It explains that due to frequency contents, power supply noise is very different at different operation modes. It explores the on-chip decap amount, and points out increasing the decap area efficiency is the best way of suppressing noise. It also explores different board and package options, and indicates sign off analysis needs to include both board and package models. Then, this article predicts that under a hypothetical situation when mode shift activity, like high power – low power – high power is possible, previous benign operation mode could become very problematic. Finally, this article applies this method to guide a new generation design, and predict package feasibility for new chip configurations.

978-1-4799-0708-3/13 $31.00 © 2013 IEEE

Millimeter Wave and RF Technologies I

978-1-4799-0708-3/13 $31.00 © 2013 IEEE

Electrical Characterization of Low-Cost Glass / Epoxy Laminates at Millimeter Wave Frequencies

Noam Kaminski, Evgeny Shumakher, Danny Elad
IBM Research - Haifa,
Haifa University Campus, Carmel Mountain,
Haifa, Israel
E-mail: noamka@il.ibm.com

Keishi Okamoto, Kazushige Toriyama, Hiroyuki Mori
IBM Research - Tokyo
7-7, Shin-Kawasaki, Saiwai-ku, Kawasaki-shi
Kanagawa, Japan
E-mail: keishi@jp.ibm.com

Abstract - We describe the characterization of low-cost glass/epoxy laminates at millimeter-wave frequencies up to 110 GHz. We characterize the loss, dielectric constant and Fiber-Weave-Effect properties of 3 such laminates. We also demonstrate several test cases of transitions at E-band and D-band frequencies (up to 130 GHz) designed with these laminates, showing excellent performance.

Keywords—Printed circuits, millimeter wave circuits, dielectric substrates, dielectric measurement.

I. INTRODUCTION

The field of mmWave (millimeter-wave) technology is gaining a lot of interest lately driven by a growing number of applications such as high resolution radar, ultra-wide band communication and passive imaging using black-body radiation sensing. Such applications can be realized with IBM silicon-germanium BiCMOS 8HP technology that enables low-cost and high performance integrated circuits at the mmWave. To maintain the low-cost paradigm on a system level, inexpensive mmWave packages are required for integration of these chips.

The mmWave packages require a very tight tolerance due to the extremely short wavelength, which is commonly achieved with thin film technology (usually with ceramic substrates). This technology is however relatively expensive and is mostly limited to a 2-sided circuit board and is not suitable for inner layers of a multi-layers board. Another alternative allowing for tight tolerance is the sequential build-up (also referred to as High Density Interconnection). The two main drawbacks of this otherwise excellent technology are the high cost and the high dielectric loss.

In order to make the mmWave products suitable for high volume production, we look for a low-cost technology that exhibits low loss and reasonably tight tolerances. The specifications of the packaging materials were derived from the packaging requirements of the E-band point-to-point communication chipset [1]. For the reasons presented above, we chose the PCB (Printed Circuit Board) technology with laminates that can be fabricated using standard Glass/Epoxy (FR4) processes, and avoided the PTFE based laminates. We studied 3 different laminates from this technology, with dielectric properties reported by their manufacturers (see

TABLE I). The thickness of all the laminates was 0.1 mm. All the PCBs were fabricated by the same PCB fabricator.

A most prominent drawback of the chosen materials is the local variation in the ε_r (dielectric constant), resulting from the fiberglass weave pattern in the dielectric material, also called the FWE (Fiber Weave Effect). Usually, the FWE is measured through its impact on the skew and delay of long lines, but this information is hardly useful for the mmWave packaging, and therefore we measured the FWE with a more suitable technique for our usage, elaborated on in section II.

This paper is organized as follows: in section II we describe the FWE measurement technique and present the characterization results. Section III depicts the measured dielectric properties of the different laminates and section IV showcases the characterization results of several mmWave transitions, designed in these laminates.

II. FIBER WEAVE EFFECT MEASUREMENT

The FWE origin, measurement methods and influence on the printed circuits performance are well summarized in [2]. In our design, we used a 0.1 mm thick, double sided laminates, with microstrip Tline (transmission line). Since we use very short lines in the E-band package, the anisotropy could have little or no affect in the form of skew and delay, but rather in the form of localized effective ε_r variations since the Tline width required to form an 50 Ω impedance was found to be 160 um, while the weave unit size is 20 mil (~0.5 mm) [3].

In order to test the localized ε_r variations, we designed a band-stop filter composed of a single open stub (1.8 mm long) and measured its band-stop frequency, which has an estimated value (by finite-element simulation) of 74.8 GHz for a ε_r of 3.6. Five such filters were fabricated on the same coupon with the same orientation, and five more with a perpendicular orientation. The open stub location was moved 4mil between adjacent filters forcing the first and the fifth filter to experience the same ε_r. The photo of the filters and the measured $|S_{21}|$ are shown in Fig. 1. The results were compared with a finite-element simulation for different laminate ε_r. The comparison is listed in TABLE I, showing that the FWE is meaningful only with Laminate B, at the X orientation. Please note in Fig. 1(c) that the first and fifth filters show the same resonance frequency, as expected.

978-1-4799-0708-3/13 $31.00 © 2013 IEEE

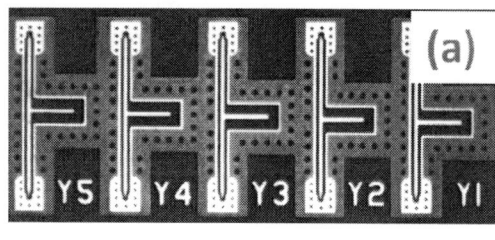

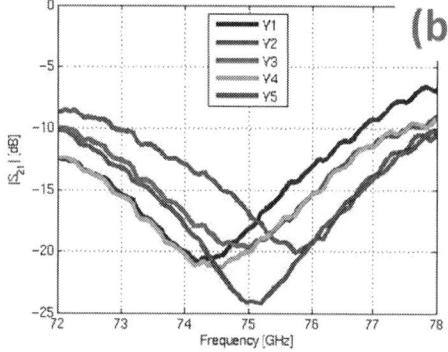

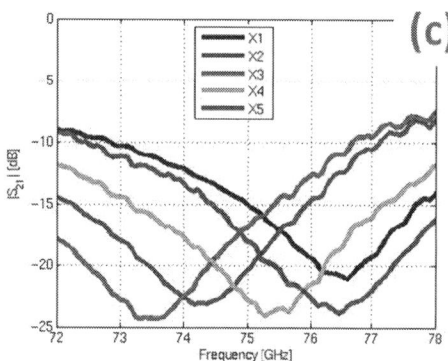

Fig. 1. (a) Photo of 5 filters in the Y orientation. The filters in the X orientation are not seen here, (b) $|S_{21}|$ of the 5 filters in the Y orientation (laminate B), (c) $|S_{21}|$ of the 5 filters in the X orientation (laminate B).

TABLE I - LIST OF ε_r DERIVED FROM THE RESONANCE FREQUENCIES

	Laminate A	Laminate B	Laminate C
ε_r / tanδ from the datasheets	3.66 / 0.0037 @10 GHz	3.62 / 0.006 @50 GHz	3.53 / 0.0012 @ 1GHz
Y orientation	3.7 – 3.8	3.5 – 3.7	3.15 – 3.25
X orientation	3.7 – 3.8	3.4 – 3.8	3 – 3.25

III. DIELECTRIC PROPERTIES OF THE 3 LAMINATES

Next we measured and compared the dielectric constant and the loss per mm of a microstrip on the 3 different laminates, based on S-parameters measurements of microstrip lines with different lengths (2.4 mm and 4.8 mm). The microstrips were aligned with the Y-orientation, that does not suffer from the FWE, as can be seen in TABLE I.

The probes were de-embedded using the well-known Through-Reflect-Line method. After calibration, the loss ($|S_{21}|$ [dB]) of the 4.8 mm long line, was exactly twice that of the 2.4 mm long line. Hence, the derivation of the microstrip line loss per mm is strait forward as seen in Fig. 2. The abrupt notch in laminates A and B curves at ~103 GHz results from a resonance due to the line length, which was difficult to de-embed. The same abrupt notch is shifted to above 110 GHz in laminate C due to the fact that its ε_r is lower (see TABLE I).

The dielectric constant of the 3 laminates was derived by comparing the phase accumulation per mm of the measured lines (phase of S_{21} after de-embedding) with a finite-element simulation of the lines. The measured ε_r in Fig. 3 is in agreement with the measured results in TABLE I.

Looking at Fig. 2 and Fig. 3, it is apparent that laminate C has the lowest ε_r as well as the lowest loss.

IV. MEASURED RESULTS OF MMWAVE TRANSITIONS TEST CASES

Apart from the characterization of the different laminates, we also designed several electrical transitions for advanced low-cost packages in the mmWave. Four such test cases are presented here: transition from microstrip to waveguide at the E-band 71-76 GHz and at the D-band 110-130 GHz, transition from a differential line to a waveguide at the E-band 71-76 GHz and a transition of a microstrip from one side of a 0.1 mm thick laminate to the other side.

A. Transition from microstrip to waveguide at the E-band, 71-76GHz.

In this transition, a microstrip probe enters a WR12 waveguide in the E-plane in a similar fashion to [4]. The measurement was performed in a back-to-back configuration by probing 2 microstrips, each terminated with a waveguide. The 2 waveguides then connect forming a "U-turn" shape. The cross section of the setup is seen in Fig. 4(a) and a photo of the PCB, the waveguides and the back-shorts are all shown in Fig. 4(b). The measurement results of a single transition are depicted in Fig. 4(c) and Fig. 4(d), showing excellent agreement with finite-element simulation predictions. The transition notch around 87 GHz, is a results of a high mode developing in the back-short.

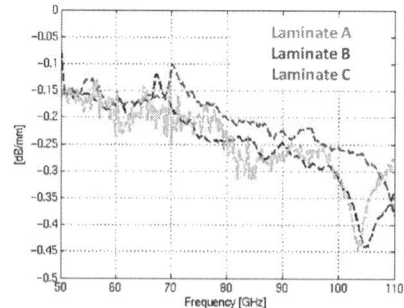

Fig. 2. Microstrip lines loss in [dB/mm].

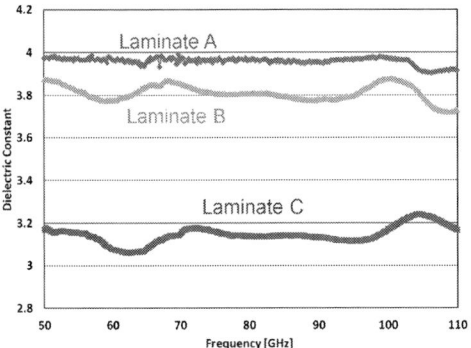

Fig. 3. Dielectric constant of the 3 laminates, measured by microstrip line method.

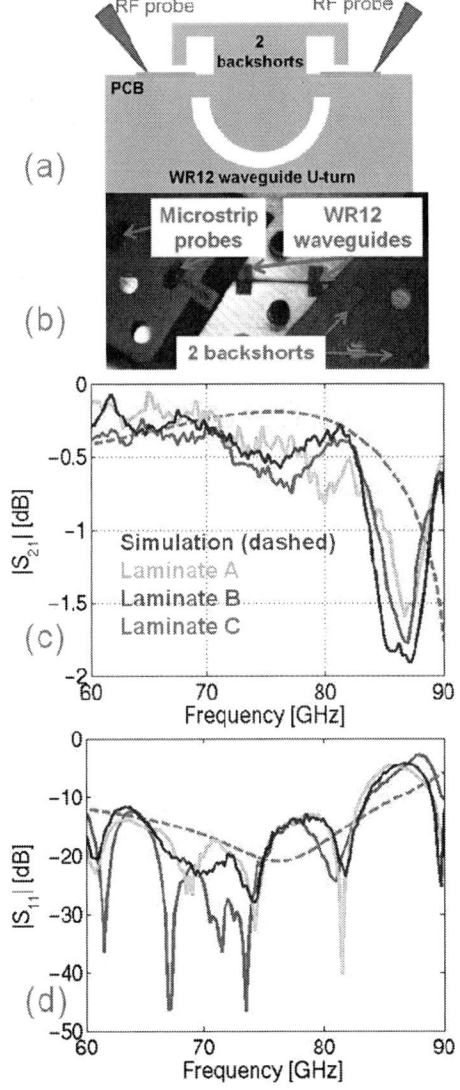

Fig. 4. Microstrip to WR12 transition. (a) Cross-section and (b) photo of the measurement set-up parts, (c) measured |S₂₁| and (d) measured |S₁₁| of the transition. Dashed line shows simulation predictions in both (c) and (d).

B. Transition from microstrip to waveguide at the D-band, 110-130 GHz.

This transition is similar to the E-band transition with a microstrip probe entering a WR6 waveguide in the E-plane. The measurement was performed in a back-to-back configuration with waveguide probes with a single 6 mm long microstrip connecting the two different WR6 waveguides. A photo of the PCB, the waveguides and the back-shorts appears in Fig. 5(a). Fig. 5(b) and Fig. 5(c) show the measured results of the total back-to-back set-up. There is a good agreement between the simulation and the measurement. The laminate C exhibits the lowest transition loss. It could be a result of the lower microstrip loss, which was not de-embedded in this measurement.

C. Transition from differential microstrip to wavegide at the E-band 71-76 GHz

This transition is based on [5] with some modification. In [5], the differential lines end with triangles that are shorted to the ground. We replaced the shunting element with λ/4 open stubs at the edges of the triangles. Fig. 6(a) shows a photo of the transition. Fig. 6(b) and Fig. 6(c) show the measurement of a single transition after de-embedding, with very good agreement to the simulation. Only laminate A transitions have been measured.

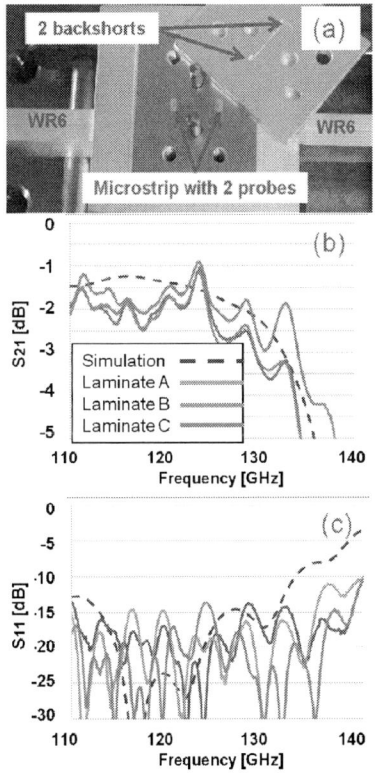

Fig. 5. A transition from a microstrip to a WR6 waveguide. (a) Photo of the measurement set-up, (b) measured |S₂₁| and (c) |S₁₁| of the entire back-to-back set-up, including the 6mm connecting microstrip.

978-1-4799-0708-3/13 $31.00 © 2013 IEEE

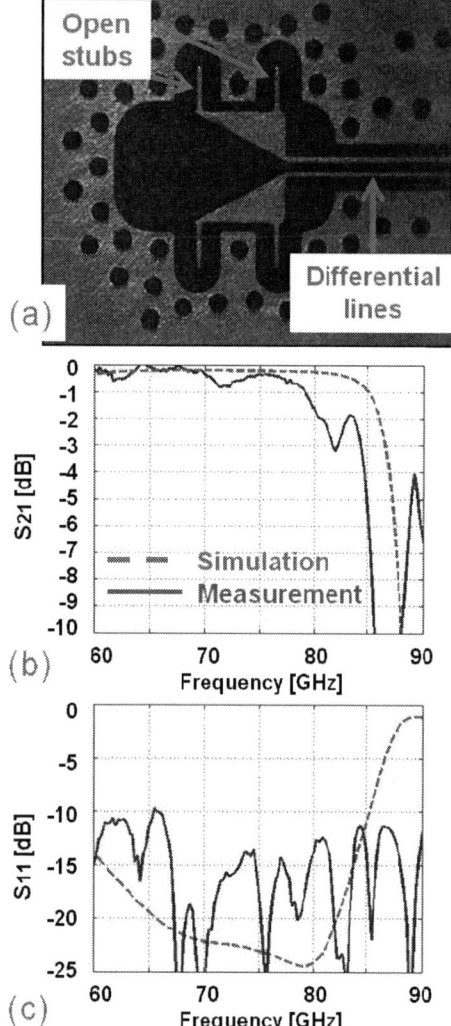

Fig. 6. A differential microstrip to waveguide transition. (a) Photo of the PCB, (b) |S₂₁| and (c) |S₁₁| measured results of laminate A and simulation of the transition.

D. Side to side transition

Here we show a transition of a microstrip from the top side of the 2-sided PCB, to a microstrip on the bottom side. We used six via for ground currents and shielding and a single via for the transition itself, as can be seen in Fig. 7(a). In this configuration, the simulated transition loss is less than 0.1 dB across the E-band frequency range, which is extremely difficult to measure. The measured results in Fig. 7(b) and Fig. 7(c) exhibit higher measurement errors than the simulated loss, and therefore the $|S_{21}|>0$ is visible in several frequencies.

V. CONCLUSIONS

We presented a complete electrical characterization of 3 low-cost Glass/Epoxy PCB laminates at frequencies up to 110 GHz. The laminates exhibit stable ε_r and low FEW, which makes them good candidates for mmWave packaging. We also designed and tested 4 transitions as test cases using these laminates at mmWave frequencies. The transitions show excellent performance and are in a very good agreement with finite-element simulation up to 130 GHz.

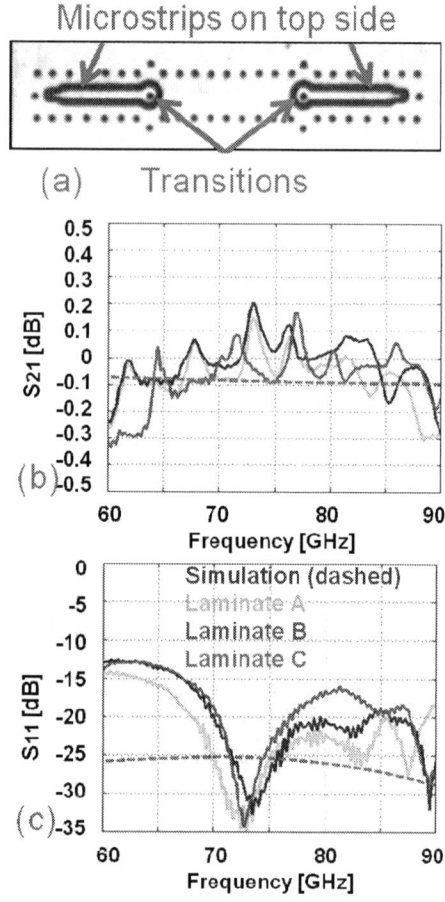

Fig. 7. Microstrip side-to-side transition. (a) Photo, (b) measured |S₂₁| and (c) measured |S₁₁| of the transition. The anticipated loss (according to simulation) is smaller than the measurement errors, therefore $|S_{21}|>0$ is also seen on several frequencies.

REFERENCES

[1] O. Katz, et al. "A Fully Integrated SiGe E-BAND Transceiver Chipset for Broadband Point-to-Point Communication", IEEE radio and wireless symposium , January 15-19, 2012, Santa Calra, CA, USA.

[2] Rautio, James C.; Rautio, B.J.; Arvas, S.; Horn, A.F.; Reynolds, J.W., "The effect of dielectric anisotropy and metal surface roughness," *Microwave Conference Proceedings (APMC), 2010 Asia-Pacific* , vol., no., pp.1777,1780, 7-10 Dec. 2010

[3] Heck, H.; Hall, S.; Horine, B.; Liang, T., "Modeling and mitigating AC common mode conversion in multi-Gb/s differential printed circuit boards," *Electrical Performance of Electronic Packaging, 2004. IEEE 13th Topical Meeting on* , vol., no., pp.29,32, 25-27 Oct. 2004

[4] Yoke-Choy Leong; Weinreb, S., "Full band waveguide-to-microstrip probe transitions," *Microwave Symposium Digest, 1999 IEEE MTT-S International* , vol.4, no., pp.1435,1438 vol.4, 13-19 June 1999

[5] Ortner, M.; Ziqiang Tong; Ostermann, T., "A millimeter-wave wide-band transition from a differential microstrip to a rectangular waveguide for 60 GHz applications," *Antennas and Propagation (EUCAP), Proceedings of the 5th European Conference on* , vol., no., pp.1946,1949, 11-15 April 2011

A Method for Modeling the Impact of Conductor Surface Roughness on Waveguiding Properties of Interconnects

Xiao Ma, Juan S. Ochoa, and Andreas C. Cangellaris
Department of Electrical and Computer Engineering
University of Illinois at Urbana-Champaign, Urbana, IL 61801
{xiaoma1, jsochoa2, cangella}@illinois.edu

Abstract—A methodology is proposed for accounting for the impact of random conductor surface roughness in the electromagnetic analysis of the propagation characteristics of high-speed interconnects. The proposed methodology replaces the rough-surface conductor cross-section with a compound one, where a thin outer layer, conforming to the conductor contour and of frequency-dependent thickness and conductivity is used in place of the surface roughness. The resulting compound conductor model facilitates extraction of the waveguiding properties of the interconnect using electromagnetic field solvers. The attributes of the method are demonstrated through the electromagnetic analysis of a microstrip transmission line structure and through correlations with measurement data.

Index Terms—Propagation losses, surface roughness, transmission lines.

I. INTRODUCTION

Substrates used for interconnects on printed circuit boards (PCBs) are often roughened to promote adhesion. The roughness of the substrate surface is transferred to the conductor surface during metallization. The existence of the conductor surface roughness disturbs current flow and influences the waveguiding properties of interconnects. Such effect is more significant at high frequency due to skin effect. At multi-GHz range, where the skin depth is comparable to the root-mean-square (rms) height of the rough surface, the effect of conductor surface roughness becomes a critical factor to consider when modeling high speed interconnects.

Over many years, analyses of rough surface effects were restricted to periodic structure models. Morgan [1] modeled the rough surface as 2D periodic grooves and solved for eddy current losses using finite difference method. Hammerstad and Bekkadal [2] proposed an empirical formula for calculating the power absorption enhancement factor based on Morgan's result. Holloway and Kuester [3] verified Morgan's result using finite element method with impedance boundary condition. Proekt and Cangellaris [4] used perturbation method to examine the effect of surface roughness and proposed the definition of effective conductivity to account for the impact of increased conductor resistance due to surface roughness. However, the above research used periodic rough surface and their results were subject to the specific geometry. Recently, Tsang et al. [5] proposed a random rough surface model, calculated the roughness-induced absorption by second order

small perturbation method (SPM2), and verified the results with method of moments (MoM). In a series of follow-up research Gu et al. extended the random rough surface model to 3D [6], Braunisch et al. [7], Gu et al. [8], Tsang et al. [9], and Ding et al. [10] used measured surface profile data to predict propagation loss in interconnects and waveguide structures.

In this paper we propose a methodology to account for the impact of random conductor surface roughness in the electromagnetic analysis of the propagation characteristics of high-speed interconnects. The methodology leads to a compound conductor model with the rough-surface conductor cross-section being replaced by a compound one, where a thin outer layer, conforming to the conductor contour and of frequency-dependent thickness and conductivity is used in place of the surface roughness. The compound conductor model facilitates extraction of the waveguiding properties of the interconnect using electromagnetic field solvers.

II. FORMULATION

A. The Proposed Compound Conductor Model

The idea of the compound conductor model is to replace the outer portion of the conductor cross section where surface roughness occurs with an effective, smooth conductor layer conforming to the conductor contour, while keeping the interior of the conductor cross section unaltered. This equivalent compound conductor model should be constructed in such a manner that the impact of conductor surface roughness on the waveguiding characteristics of the interconnect is accounted for in an accurate manner. This is shown pictorially in Fig. 1. As clearly shown in the figure, the thickness and conductivity of the outer layer, $h_e(f)$ and $\sigma_e(f)$, are frequency dependent, while the interior of the conductor cross section is assigned the electromagnetic properties of the original conductor. Thus, the development of the compound conductor model amounts to the calculation of the frequency-dependent conductivity and thickness of the outer layer. The way this is done is described next.

B. Extraction of outer layer thickness and conductivity

Our approach is guided by the ideas, analysis and results presented by Braunisch et al. [7]. It is assumed that data is

978-1-4799-0708-3/13 $31.00 © 2013 IEEE

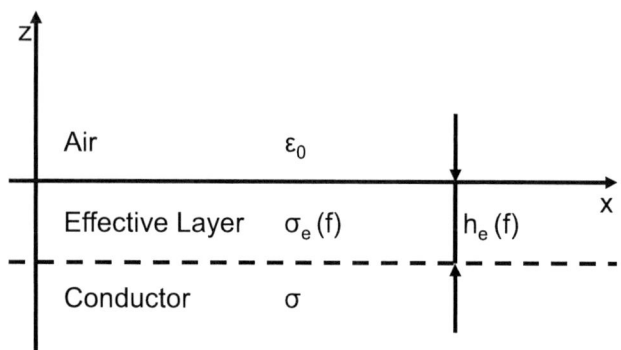

Fig. 1. The compound conductor model.

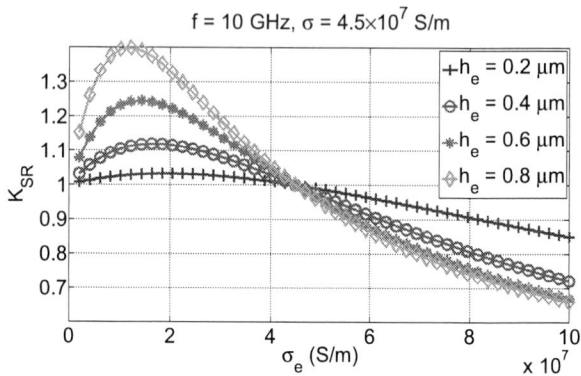

Fig. 2. K_{SR} vs σ_e curves for a family of h_e values.

available for the description of the conductor surface roughness. Rough surfaces are usually characterized by their power spectral density (PSD) [11]. As proposed in [6], the impact of conductor surface roughness on the frequency-dependent ohmic loss can be quantified in terms of the power absorption *enhancement* factor,

$$
\begin{aligned}
K_{SR} &= \frac{\langle P_{a,rough}\rangle}{P_{a,smooth}} \\
&= 1 + \frac{2h^2}{\delta^2} - \frac{2}{\delta}\int_{-\infty}^{\infty} dk_x \int_{-\infty}^{\infty} dk_y W\left(k_x, k_y\right) \\
&\quad \cdot \Re\left(\sqrt{-\frac{2j}{\delta^2} - k_x^2 - k_y^2}\right),
\end{aligned}
\tag{1}
$$

where h is the rms height of the rough surface, $\delta = \sqrt{\frac{1}{\pi\mu\sigma f}}$ is the skin depth in the conductor, $W\left(k_x, k_y\right)$ is the PSD of the rough surface profile, and k_x, k_y are angular spatial frequencies. The rms height of the rough surface is calculated based on the PSD as

$$
h^2 = \int_{-\infty}^{\infty}\int_{-\infty}^{\infty} dk_x dk_y W\left(k_x, k_y\right).
\tag{2}
$$

To extract the thickness and conductivity of the effective layer, we first calculate the power absorption enhancement factor for the compound conductor. The *smooth* surface case corresponds to the absence of the effective conductor layer in the compound conductor model. Assuming an air-conductor planar interface with the conductor occupying the bottom half space, for the case of a TM-polarized plane wave of magnitude $|\mathbf{H}_{t,0}|$ on the conductor surface, the time-average power absorbed in the conductor, assuming a conductor layer of infinite extent is calculated as

$$
P_{a,smooth} = \frac{1}{2}R_{s,smooth}\left|\mathbf{H}_{t,0}\right|^2 = \frac{1}{2\delta\sigma}\left|\mathbf{H}_{t,0}\right|^2,
\tag{3}
$$

where $Z_{s,smooth} = R_{s,smooth} + jX_{s,smooth} = \frac{1+j}{\sigma\delta}$ is the input impedance looking into the semi-infinite conductor, which, for the case of a homogeneous conductor equals its intrinsic impedance, and σ is its conductivity. Time-average power absorption in the presence of surface roughness using the proposed compound conductor model can be calculated in a

similar manner. However, in this case, the input impedance of the compound conductor is given by

$$
\begin{aligned}
Z_{s,rough} &= R_{s,rough} + jX_{s,rough} \\
&= \frac{1+j}{\sigma_e\delta_e}\frac{\sigma_e\delta_e + j\sigma\delta tan\left(k_{1z}h_e\right)}{\sigma\delta + j\sigma_e\delta_e tan\left(k_{1z}h_e\right)}.
\end{aligned}
$$

In the above equation, σ_e is the conductivity of the effective conducting layer, $\delta_e = \sqrt{\frac{1}{\pi\mu\sigma_e f}}$ is the skin dept in it, and $k_{1z} \approx \frac{1-j}{\delta_e}$ is the propagation constant inside it in the direction perpendicular to the air-conductor interface. Let $P_{a,rough,c}$ be the time-average power absorbed in the compound conductor model. Then, the power absorption enhancement factor obtained using the compound conductor model is,

$$
\begin{aligned}
K_{SR}^c &= \frac{P_{a,rough,c}}{P_{a,smooth}} \\
&= Re\left\{\frac{(1+j)\sigma\delta}{\sigma_e\delta_e}\frac{\sigma_e\delta_e + j\sigma\delta tan\left(k_{1z}h_e\right)}{\sigma\delta + j\sigma_e\delta_e tan\left(k_{1z}h_e\right)}\right\}.
\end{aligned}
\tag{4}
$$

The combination of values for h_e and σ_e that gives the desired K_{SR} value, as obtained by (1), at given frequency is not unique. Thus, a process must be defined, where additional considerations are used to choose these parameters. To begin with, we note that, since the effective conductor layer takes up the space that is originally occupied by the bulk conductor, and the bulk conductor does not always extend to infinity, the thickness of the effective layer should be constrained to be as small as possible. Therefore, we first find the smallest possible h_e and then determine the corresponding σ_e. Toward this, Fig. 2 suggests that there is a maximum K_{SR}, $K_{SR,max}$ that can be achieved at a given frequency if we vary only the value of σ_e. At each frequency point, interpolation is performed on the curve $K_{SR,max}$ vs h_e, as shown in Fig. 3 to determine the minimum h_e that makes it possible to obtain the desired K_{SR}. Once the optimal value for h_e, $h_{e,op}$, has been found, interpolation is performed on the curve K_{SR} vs σ_e for $h_{e,op}$ to determine the corresponding value for σ_e, $\sigma_{e,op}$, that generates the desired K_{SR} at a given frequency, as depicted in Fig. 4. This process leads to the desired result, namely, frequency-dependent values for effective layer thickness and conductivity.

978-1-4799-0708-3/13 $31.00 © 2013 IEEE

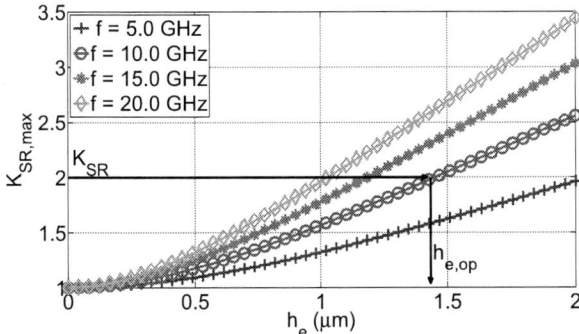

Fig. 3. $K_{SR,max}$ vs h_e curves for a family of f values, interpolation is performed on $f = 10GHz$ curve to decide the appropriate $h_{e,op}$ value.

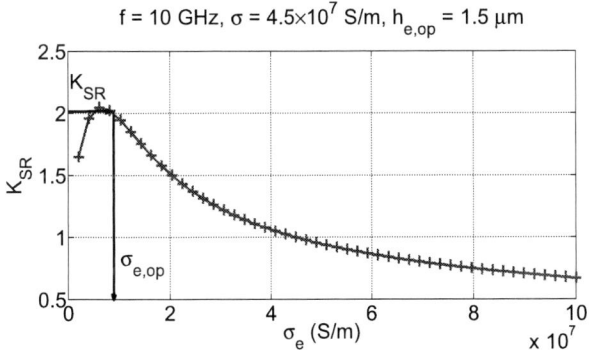

Fig. 4. K_{SR} vs σ_e curve for $h_{e,op}$, interpolation is performed to decide the appropriate $\sigma_{e,op}$ value.

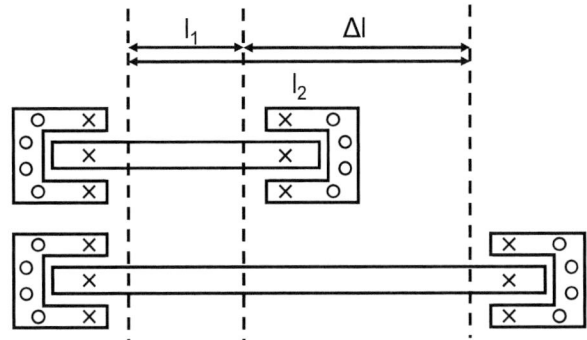

Fig. 5. Schematic of microstrip test structure. $l_1 = 12.3mm$, $l_2 = 24.3mm$, $\Delta l = 12mm$.

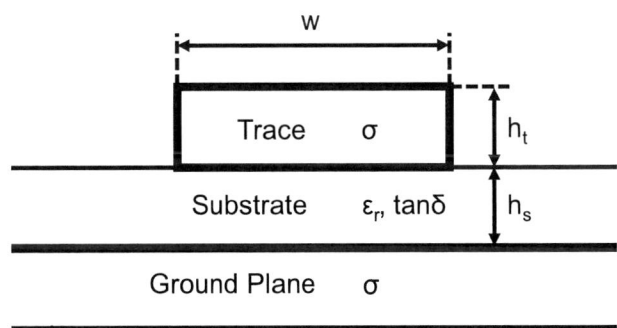

Fig. 6. Cross-section of the microstrip transmission line structure. Thick lines in the figure indicate rough conductor surfaces.

Once the frequency depend outer layer thickness and conductivity are extracted following the above steps, the compound conductor model could be used in electromagnetic field solvers to extract the waveguiding properties of interconnect structures.

III. NUMERICAL RESULTS

To illustrate the proposed methodology, the microstrip transmission line test structure used in [7] is considered. The test structure is shown schematically in Fig. 5, where a two-line method is used to characterize the transmission line. The cross-section of the test structure is shown in Fig. 6.

The cross-sectional geometry parameters are taken to be the average of measurement values $h_s = 30.6\mu m$, $h_t = 15.3\mu m$, $w = 65.5\mu m$. The relative permittivity and the loss tangent of the insulating dielectric are $\epsilon_r = 3.4$, $tan\delta = 0.017$, respectively, and the conductivity of the metallization is $\sigma = 4.5 \times 10^7 S/m$). K_{SR} is calculated using the measured PSD for the ground plane conductor in [7]. Use of the proposed methodology and the calculated K_{SR}, leads to the frequency-dependent thickness and conductivity of the outer conducting layer, as shown in Fig. 7. Note that the frequency dependent thickness and conductivity are calculated down to $0.1GHz$. As the frequency goes lower, the problem becomes ill-conditioned and interpolation couldn't be applied. For the extreme case corresponding to DC frequency, thickness and conductivity of

the effective layer could assume any value. All combinations of thickness and conductivity lead to the same desired K_{SR}, which is 1.

The surface roughness of the strip conductor is assumed to be the same with that for the ground plane. Hence, the effective outer layer of the same frequency-dependent thickness and conductivity is being used to develop compound conductor models for both the strip conductor and the ground plane, as depicted in Fig. 8.

The generated compound model is subsequently used in ANSYS® Q3D Extractor® [12] to obtain the per-unit-length resistance and inductance parameters used in the transmission-line modeling of the microstrip line. The extracted per-unit-length parameters are depicted in Fig. 9. The attenuation constant of the microstrip transmission line structure is extracted and compared to the results from measurement in [7]. As depicted in Fig. 10, very good agreement is observed.

IV. CONCLUSION

In this paper, a methodology has been proposed and demonstrated for the convenient and accurate handling of surface roughness of interconnects in their electromagnetic characterization using popular electromagnetic field solvers. The key idea of the proposed method is the replacement of the thin surface layer of the conductor within which the roughness

978-1-4799-0708-3/13 $31.00 © 2013 IEEE   13

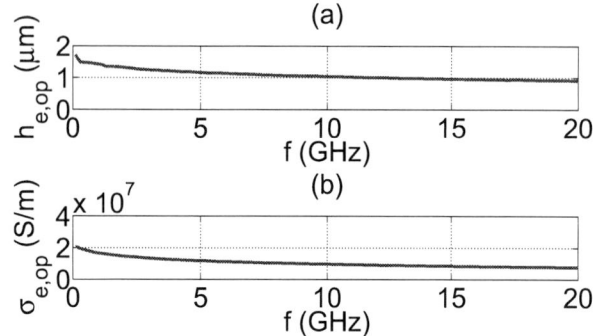

Fig. 7. Frequency dependence of the thickness (a) and the conductivity (b) of the effective outer conducting layer.

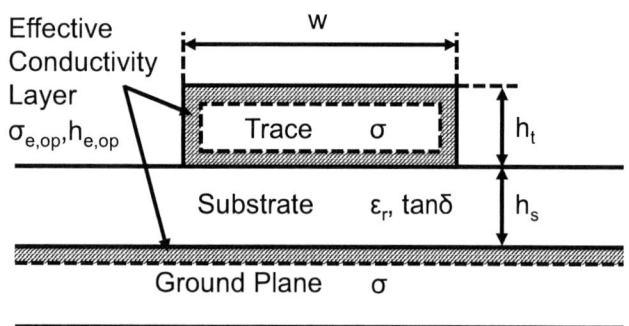

Fig. 8. Cross-section of the compound conductor model. The shaded layers indicate the effective outer layers used in place of the conductor surface roughness.

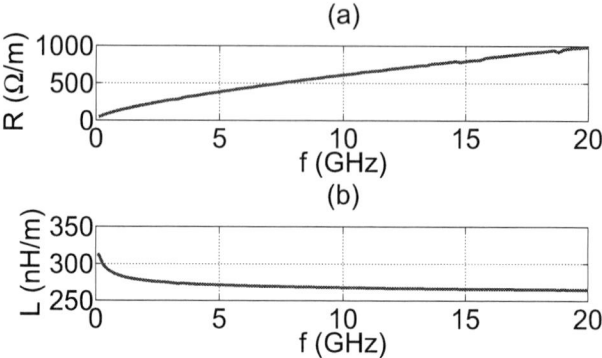

Fig. 9. Per-unit-length parameters. (a) Resistance. (b) Inductance.

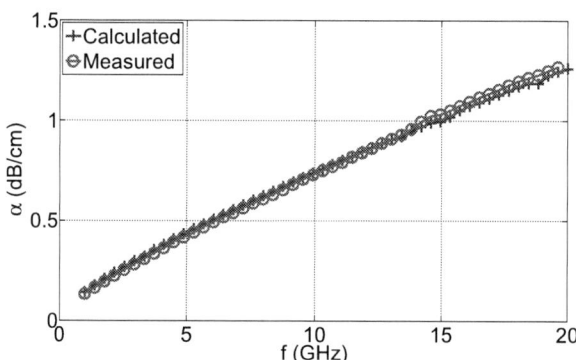

Fig. 10. Comparison of attenuation constant calculated using the proposed method with measured data from [7].

occurs with an smooth layer conformal to the conductor surface. The thickness and conductivity of this conformal thin outer layer are frequency dependent and their values are obtained through a systematic process that requires that the two surfaces, the one with roughness and the one with the conformal thin outer layer exhibit the same frequency-dependent time-average power absorption. The accuracy of the proposed approach was demonstrated through its application to the calculation of the attenuation constant of a microstrip structure, measured results for which were available in the open literature. As demonstrated through this validation study, use of the proposed methodology results in the replacement of conductors exhibiting surface roughness with smooth, compound conductors, which can be handled easily by popular electromagnetic field solvers.

ACKNOWLEDGMENT

This material is based upon work supported in part by the US Army Research Laboratory and the US Army Research Office under grant number W911NF-10-1-0269.

REFERENCES

[1] S. P. Morgan, "Effect of surface roughness on eddy current losses at microwave frequencies," *J. Appl. Phys*, vol. 20, p. 352, 1949.

[2] E. Hammerstad, *Microstrip Handbook*, ser. ELAB report, F. Bekkadal, Ed. Trondheim, Norway: Norwegian Institute of Technology, 1975.

[3] C. Holloway and E. F. Kuester, "Power loss associated with conducting and superconducting rough interfaces," *Microwave Theory and Techniques, IEEE Transactions on*, vol. 48, no. 10, pp. 1601–1610, 2000.

[4] L. Proekt and A. Cangellaris, "Investigation of the impact of conductor surface roughness on interconnect frequency-dependent ohmic loss," in *Electronic Components and Technology Conference, 2003. Proceedings. 53rd*, 2003, pp. 1004–1010.

[5] L. Tsang, X. Gu, and H. Braunisch, "Effects of random rough surface on absorption by conductors at microwave frequencies," *Microwave and Wireless Components Letters, IEEE*, vol. 16, no. 4, pp. 221–223, 2006.

[6] X. Gu, L. Tsang, and H. Braunisch, "Modeling effects of random rough interface on power absorption between dielectric and conductive medium in 3-d problem," *Microwave Theory and Techniques, IEEE Transactions on*, vol. 55, no. 3, pp. 511–517, 2007.

[7] H. Braunisch, X. Gu, A. Camacho-Bragado, and L. Tsang, "Off-chip rough-metal-surface propagation loss modeling and correlation with measurements," in *Electronic Components and Technology Conference, 2007. ECTC '07. Proceedings. 57th*, 2007, pp. 785–791.

[8] X. Gu, L. Tsang, and H. Braunisch, "Estimation of roughness-induced power absorption from measured surface profile data," *Microwave and Wireless Components Letters, IEEE*, vol. 17, no. 7, pp. 486–488, 2007.

[9] L. Tsang, H. Braunisch, R. Ding, and X. Gu, "Random rough surface effects on wave propagation in interconnects," *Advanced Packaging, IEEE Transactions on*, vol. 33, no. 4, pp. 839–856, 2010.

[10] R. Ding, L. Tsang, H. Braunisch, and W. Chang, "Wave propagation in parallel plate metallic waveguide with finite conductivity and three dimensional roughness," *Antennas and Propagation, IEEE Transactions on*, vol. 60, no. 12, pp. 5867–5880, 2012.

[11] L. Tsang, J. Kong, and K. Ding, *Scattering of Electromagnetic Waves, Theories and Applications*, ser. A Wiley interscience publication. Wiley, 2000.

[12] ANSYS, "Q3d extractor," release 14.0. [Online]. Available: http://www.ansys.com/

A Perturbation Technique to Analyze the Influence of Fiber Weave Effects on Differential Signaling

Mykola Chernobryvko, Dries Vande Ginste and Daniël De Zutter

Electromagnetics Group, Dept. of Information Technology, Ghent University, Gent, Belgium

Email: mykola.chernobryvko@intec.ugent.be

Abstract—We study differential signaling via a pair of striplines in a substrate that is comprised of an epoxy/fiberglass woven composite structure. The transmission characteristics, which are deteriorated due to the presence of the fiber weave, are analyzed via an efficient modeling technique for nonuniform transmission lines. This technique is based on the solution of the pertinent differential equations using a perturbation approach. For a challenging application example, it is shown that the unavoidable phase errors can be controlled by subdividing electrically long lines into smaller pieces, as such increasing accuracy whilst maintaining efficiency.

Keywords—*Nonuniform transmission line, perturbation, fiber weave, mode conversion*

I. Introduction

Nonuniform transmission lines (NUTLs) are found in various applications, such as filters, impedance transformers, etc. The modeling of these interconnects has, however, always been a challenging problem. Due to the varying per-unit-length (p.u.l.) parameters along the NUTL, the differential equations describing them cannot be solved analytically, except for some very special cases. In [1], the authors of the present contribution presented a two-step perturbation approach to model single and differential pairs of NUTLs. The focus was on the theoretical description of the technique and on demonstrating its applicability and limitations. In this contribution, however, we extend the perturbation technique [1] in order to apply it to a very important, but challenging, application example. Leveraging the novel perturbation technique, we analyze differential signaling using a pair of striplines embedded in a substrate that is comprised of an epoxy/fiberglass woven composite structure [2], [3]. In such a commonly used substrate, it is very likely that one trace of the differential pair is located mainly in the epoxy resin with low dielectric constant, while the other trace is located close to the glass fiber with a high dielectric constant. As the two traces "see" a different permittivity, a differential skew between the lines is observed. This skew results in insertion loss suck-outs of the transmitted (differential) signal. Additionally, the imbalance leads to conversion from the differential mode to the common mode [4]. All this may prohibit the use of these substrates at very high frequencies, or differently put, it poses a limit on the maximum (electrical) length of the lines.

II. Perturbation solution for a differential line pair

We analyse nonuniform differential lines within the framework of the quasi-TM approach and in the frequency domain (with the $e^{j\omega t}$ dependency suppressed). Consider voltage and current column vectors $\mathcal{V} = [V_1\,V_2]^T$ and $\mathcal{I} = [I_1\,I_2]^T$, holding the two voltages and two currents along the lines. To simplify the notations we work with 2×2 complex p.u.l. inductance $\mathcal{L}$ and capacitance $\mathcal{C}$ matrices, i.e. the p.u.l. resistance $\mathcal{R}$ and conductance $\mathcal{G}$ are understood to be part of $\mathcal{L}$ and $\mathcal{C}$ ($\mathcal{L} = L + \frac{\mathcal{R}}{j\omega}$ and $\mathcal{C} = C + \frac{\mathcal{G}}{j\omega}$). Our starting point is the well-known Telegrapher's equations:

$$\frac{d\mathcal{V}(z)}{dz} = -j\omega\mathcal{L}(z)\mathcal{I}(z), \tag{1}$$

$$\frac{d\mathcal{I}(z)}{dz} = -j\omega\mathcal{C}(z)\mathcal{V}(z), \tag{2}$$

with z being the signal propagation direction. To perform a perturbation technique, the following expansions are introduced:

$$\begin{aligned}
\mathcal{V}(z) &= \tilde{\mathcal{V}}(z) + \Delta\mathcal{V}_1(z) + \Delta\mathcal{V}_2(z) + ..., \\
\mathcal{I}(z) &= \tilde{\mathcal{I}}(z) + \Delta\mathcal{I}_1(z) + \Delta\mathcal{I}_2(z) + ..., \\
\mathcal{C}(z) &= \tilde{\mathcal{C}} + \Delta\mathcal{C}(z), \\
\mathcal{L}(z) &= \tilde{\mathcal{L}} + \Delta\mathcal{L}(z).
\end{aligned} \tag{3}$$

The leading terms of the series expansions (3), i.e. the voltage $\tilde{\mathcal{V}}(z)$ and current $\tilde{\mathcal{I}}(z)$, are labeled as the *unperturbed* values. The remaining terms are perturbations of order one, two, etc. $\mathcal{C}(z)$ and $\mathcal{L}(z)$ in (3) are simply written as the sum of a constant part and a place-dependent part. Here, $\tilde{\mathcal{C}}$ and $\tilde{\mathcal{L}}$ are the unperturbed values. $\Delta\mathcal{C}(z)$ and $\Delta\mathcal{L}(z)$ are the variations of the capacitance and inductance along the line which remain after subtracting the constant martrices $\tilde{\mathcal{C}}$ and $\tilde{\mathcal{L}}$ from $\mathcal{C}(z)$ and $\mathcal{L}(z)$ respectively. Remark that $\tilde{\mathcal{C}}$ and $\tilde{\mathcal{L}}$ are not necessarily the mean values of $\mathcal{C}$ and $\mathcal{L}$ over the line. We only suppose that $\Delta\mathcal{C}(z)$ and $\Delta\mathcal{L}(z)$ are small enough with respect to $\tilde{\mathcal{C}}$ and $\tilde{\mathcal{L}}$. The unperturbed matrices can be written as:

$$\tilde{C} = \begin{pmatrix} C_a & -C_b \\ -C_b & C_a \end{pmatrix} \qquad \tilde{L} = \begin{pmatrix} L_a & L_b \\ L_b & L_a \end{pmatrix}. \tag{4}$$

Consequently, the unperturbed solution consists of an even and an odd mode contribution, i.e.:

$$\begin{aligned}
\tilde{V}_1(z) &= [\tilde{V}_e(z) + \tilde{V}_o(z)]/2, \\
\tilde{V}_2(z) &= [\tilde{V}_e(z) - \tilde{V}_o(z)]/2, \\
\tilde{I}_1(z) &= [\tilde{I}_e(z) + \tilde{I}_o(z)]/2, \\
\tilde{I}_2(z) &= [\tilde{I}_e(z) - \tilde{I}_o(z)]/2.
\end{aligned} \tag{5}$$

978-1-4799-0708-3/13 $31.00 © 2013 IEEE

Using (5), the unperturbed differential equations for the even and odd mode are easily found to be

$$\frac{d\tilde{V}_e(z)}{dz} = -j\omega(L_a + L_b)\tilde{I}_e(z),$$

$$\frac{d\tilde{I}_e(z)}{dz} = -j\omega(C_a - C_b)\tilde{V}_e(z),$$

$$\frac{d\tilde{V}_o(z)}{dz} = -j\omega(L_a - L_b)\tilde{I}_o(z),$$

$$\frac{d\tilde{I}_o(z)}{dz} = -j\omega(C_a + C_b)\tilde{V}_o(z). \quad (6)$$

Hence, the modal voltages become:

$$\tilde{V}_e = (A_1 e^{-jk_e z} + B_1 e^{+jk_e z}),$$

$$\tilde{V}_o = (A_2 e^{-jk_o z} + B_2 e^{+jk_o z}). \quad (7)$$

Even and odd mode wave numbers k_e and k_o are given by:

$$\frac{k_e}{\omega} = \sqrt{(L_a + L_b)(C_a - C_b)}, \frac{k_o}{\omega} = \sqrt{(L_a - L_b)(C_a + C_b)}. \quad (8)$$

The corresponding modal currents are

$$\tilde{I}_e = (A_e e^{-jk_e z} - B_e e^{+jk_e z})/Z_e,$$

$$\tilde{I}_o = (A_o e^{-jk_o z} - B_o e^{+jk_o z})/Z_o, \quad (9)$$

with the even and odd mode impedances given by

$$Z_e = \sqrt{\frac{L_a + L_b}{C_a - C_b}}, \quad Z_o = \sqrt{\frac{L_a - L_b}{C_a + C_b}}. \quad (10)$$

The unknown coefficients A_e, A_o, B_e and B_o are determined by enforcing the boundary conditions at $z = 0$ and $z = l$ in terms of even and odd mode voltages and currents (see [1] for detailed expressions for these boundary conditions). Before turning to the first-order perturbation, let us take a closer look at $\Delta\mathcal{C}$ and $\Delta\mathcal{L}$. $\Delta\mathcal{C}$ can be written as

$$\Delta\mathcal{C} = \begin{pmatrix} \Delta C_{a1} & -\Delta C_b \\ -\Delta C_b & \Delta C_{a2} \end{pmatrix}. \quad (11)$$

As $\tilde{\mathcal{C}} + \Delta\mathcal{C}$ must have all the properties of a proper capacitance matrix in each point along the line pair, it can be asserted that the above matrix is symmetric but the entries of the matrix can *either* be *positive* or *negative*. It is useful to rewrite (11) as:

$$\Delta\mathcal{C} = \begin{pmatrix} \frac{\Delta C_{a1} + \Delta C_{a2}}{2} & -\Delta C_b \\ -\Delta C_b & \frac{\Delta C_{a1} + \Delta C_{a2}}{2} \end{pmatrix}$$

$$+ \begin{pmatrix} \frac{\Delta C_{a1} - \Delta C_{a2}}{2} & 0 \\ 0 & -\frac{\Delta C_{a1} - \Delta C_{a2}}{2} \end{pmatrix} \quad (12)$$

and

$$\Delta\mathcal{L} = \begin{pmatrix} \frac{\Delta L_{a1} + \Delta L_{a2}}{2} & \Delta L_b \\ \Delta L_b & \frac{\Delta L_{a1} + \Delta L_{a2}}{2} \end{pmatrix}$$

$$+ \begin{pmatrix} \frac{\Delta L_{a1} - \Delta L_{a2}}{2} & 0 \\ 0 & -\frac{\Delta L_{a1} - \Delta L_{a2}}{2} \end{pmatrix}. \quad (13)$$

With (12) and (13), the differential equations for the even and odd mode *first-order perturbation*, become

$$\frac{d\Delta V_{1e}}{dz} = -j\omega(L_a + L_b)\Delta I_{1e} - j\omega(l_a + l_b)\tilde{I}_e - j\omega l\tilde{I}_o,$$

$$\frac{d\Delta I_{1e}}{dz} = -j\omega(C_a - C_b)\Delta V_{1e} - j\omega(c_a - c_b)\tilde{V}_e - j\omega c\tilde{V}_o,$$

$$\frac{d\Delta V_{1o}}{dz} = -j\omega(L_a - L_b)\Delta I_{1o} - j\omega(l_a - l_b)\tilde{I}_o - j\omega l\tilde{I}_e,$$

$$\frac{d\Delta I_{1o}}{dz} = -j\omega(C_a + C_b)\Delta V_{1o} - j\omega(c_a + c_b)\tilde{V}_o - j\omega c\tilde{V}_e, \quad (14)$$

with

$$c_a = \frac{\Delta C_{a1} + \Delta C_{a2}}{2}, \quad c_b = \Delta C_b, \quad c = \frac{\Delta C_{a1} - \Delta C_{a2}}{2}$$

$$l_a = \frac{\Delta L_{a1} + \Delta L_{a2}}{2}, \quad l_b = \Delta L_b, \quad l = \frac{\Delta L_{a1} - \Delta L_{a2}}{2} \quad (15)$$

In (14) we have a separate set of equations for the two modes: the even mode comes with the $(C_a - C_b, L_a + L_b)$ p.u.l. set; the odd mode with the $(C_a + C_b, L_a - L_b)$ p.u.l. set. These equations still looks like Telegrapher's equations, but now, with additional distributed source terms. The source terms are responsible for *mode coupling*. By rewriting $\Delta\mathcal{C}$ and $\Delta\mathcal{L}$ as in (12) and (13), it becomes clear which part of the variation of the capacitance and inductance along the line is responsible for perturbation with and without mode coupling. The solution of (14) can be derived with standard mathematical techniques, as described in [1] for the single line case, but now for even- and odd-mode voltages and currents.

Following the same procedure as just outlined above, a second perturbation step may now be introduced, yielding equations similar to (14). As shown in [1], this second perturbation leads to a substantial gain in accuracy.

III. FIBER WEAVE APPLICATION EXAMPLE

A. Description of the example

To demonstrate our technique, consider the transmission of a differential signal over two copper ($\sigma = 5.8 \cdot 10^7$ S/m) stripline tracks embedded in a substrate. This substrate is nonhomogeneous due to the presence of fiber weave, as detailed below. The stripline pair is depicted in Fig. 1. The conductor thickness is 35 μm. The tracks are 180 μm wide with a distance of 630 μm separating them. The distance between top and bottom plate is 420 μm. These dimensions are such that at 10 GHz, a single line has an impedance of 50 Ω when a homogeneous background medium with $\varepsilon_r = 3.4$ is considered. However, here we consider a type 1080 fiber weave substrate, the top view of which is depicted in Fig. 2. To clearly illustrate the effect of fiber weave, we have opted to put the left line (line 1) on top of a glass bundle while the right line (line 2) mainly "sees" epoxy prepreg. Consequently, the tracks — which are running in the warp direction — are embedded in a periodically changing background medium. To model this background medium we consider two different cross-sections, indicated as cross-sections a and b in Fig. 2. These two cross-sections and all relevant dimensions are detailed in Figs. 3 and 4 respectively. The dielectric constant and the loss tangent of the glass and the epoxy prepreg are described by

978-1-4799-0708-3/13 $31.00 © 2013 IEEE

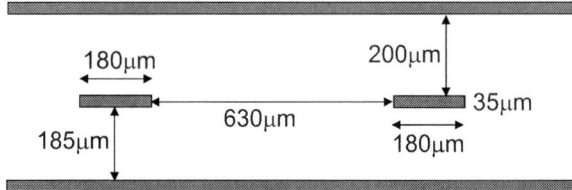

Fig. 1. Geometry of the differential stripline pair.

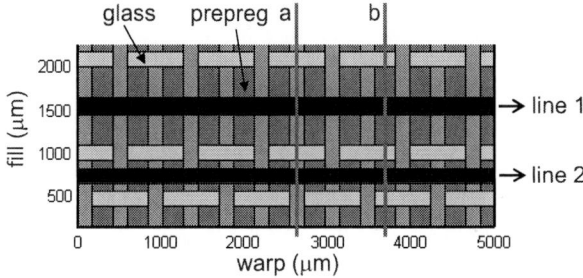

Fig. 2. Top view of the positioning of the lines w.r.t. the fiber weave.

a Debye model. For the glass, Fig. 5 depicts the real part of the relative dielectric constant, i.e. ε_r', and the loss tangent, $\tan\delta$, as a function of frequency. At 1 GHz, $\varepsilon_r' = 6$ and $\tan\delta = 0.015$. For the epoxy a similar model is used with the same loss tangent but with $\varepsilon_r' = 3$ at 1 GHz. The RLGC-parameters in each cross-section are modeled using an integral equation for the equivalent polarization charges and for the equivalent differential surface currents [5].

In the propagation direction z, the stripline pair is now modeled as the concatenation of alternating sections a and b, i.e. $a - b - a - b - \ldots$ Section a has a length of 171 μm; section b has a length of 253 μm. In this contribution we investigate such a line with a total length of 25.4 cm = $10''$, i.e. sections a and b are alternately repeated 600 times.

B. Numerical results

Such a model, where the p.u.l. parameters vary in a piecewise constant manner, allows using the chain matrix approach [6] as a reference technique. In this approach, the chain matrices of sections a and b are computed and the overall result is obtained by alternately concatenating the sections, i.e. by multiplication of the $2 \times 600 = 1200$ chain matrices.

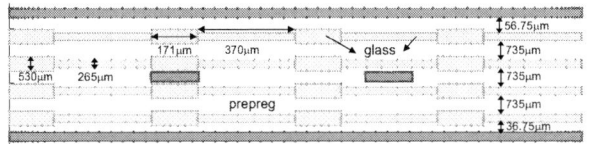

Fig. 3. Detail of cross-section a as defined in Fig. 2.

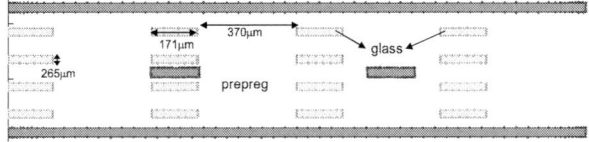

Fig. 4. Detail of cross-section b as defined in Fig. 2.

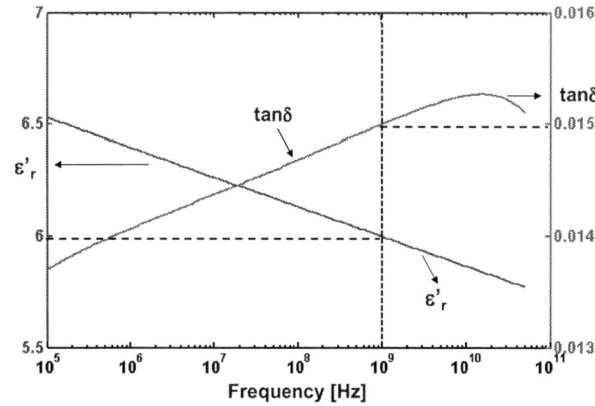

Fig. 5. Variation of the real part of the relative dielectric permittivity (left) and the loss tangent (right) of the fiber weave glass as a function of frequency.

In contrast to the chain matrix approach, the perturbation technique proposed in this contribution can also handle continuously varying p.u.l. parameters. Due care has to be taken, however, to ensure a high precision for electrically (very) long lines. Indeed, as reported in [1], when the line becomes electrically long, say about five to ten wavelengths, phase errors start to accumulate. Unfortunately, the fiber weave effect as described in the introduction, is best visible for very long lines, such as the one we analyze here. (Indeed, assuming an average ϵ_r of 3.4, a line length of $10''$ corresponds to approximately 80 wavelengths at 50 GHz!) To improve the accuracy of the method at very high frequencies, we need to subdivide the long line into shorter sections, model these sections separately with the perturbation technique and concatenate the models. Nonetheless, the number of sections may remain limited, making the perturbation technique still much faster than the reference chain matrix technique, as shown below.

First, we focus on the accuracy of the perturbation technique by presenting mixed-mode S-parameters w.r.t. 50 Ω reference impedances [4]. In particular, we study the magnitude of the differential transmission coefficient S_{dd21} and the differential-to-common-mode conversion S_{cd21} in the frequency range from DC to 50 GHz, as shown in Figs. 6(a) and (b). To perform the perturbation analysis up to 50 GHz, the line was subdivided into 20 sections. As can be seen, the results of the perturbation approach are very accurate in comparison to the reference technique. To assess the influence of the fiber weave, the results for a *uniform* interconnect with constant (unperturbed) p.u.l. C- and L-matrices as specified in (4), are also shown. It is clear that, next to mode conversion, the presence of fiber weave leads to insertion loss suck-outs. Additionally, Fig. 7 is shown to demonstrate the evolution of the accuracy of the perturbation technique when we vary the number of concatenated sections. When modeling the entire line of $10''$ as one section, the perturbation technique captures the correct behavior accurately up to 5 GHz (ca. 8 wavelengths). Modeling five sections of $2''$ each, allows to improve the results up to the first insertion loss suck-out peak at ca. 13 GHz. A section of $1''$ can be accurately modeled up to 35 GHz and to obtain accurate results up to 50 GHz, 20 sections are needed, as was already presented in Fig. 6(a).

978-1-4799-0708-3/13 $31.00 © 2013 IEEE

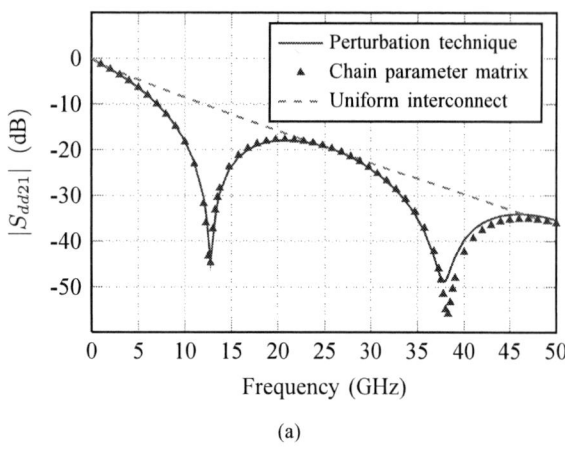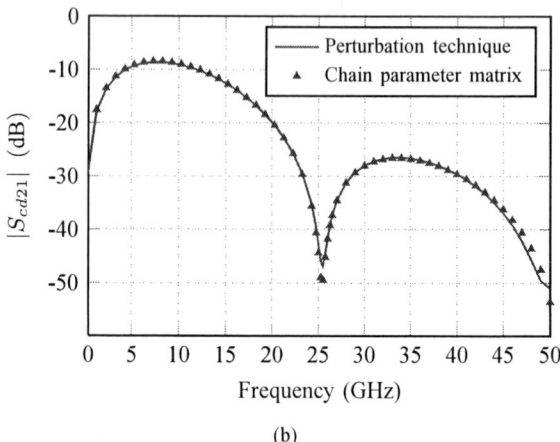

(a)　　　　　　　　　　　　　　　　　　　　(b)

Fig. 6. (a) Differential mode transmission coefficient and (b) forward differential-to-common mode conversion of the pair of coupled lines embedded in the fiber weave substrate. For the perturbation technique, the line was divided into 20 sections that were modeled separately and then concatenated. The chain matrix approach relies on a concatenation of all 1200 sections. To illustrate the fiber weave effect, the differential mode transmission coefficient for a uniform interconnect is also shown.

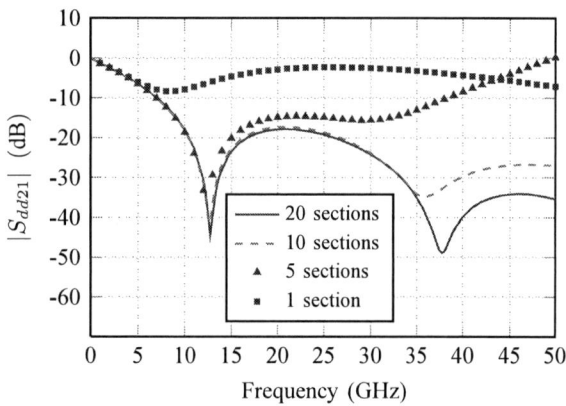

Fig. 7. Differential mode transmission coefficient of the entire line obtained with the perturbation technique when subdividing the line into a varying number of sections.

TABLE I.　EFFICIENCY OF THE PERTURBATION TECHNIQUE.
THE CPU TIME NEEDED WITH THE REFERENCE APPROACH IS 156.3 S.

Number of sections	CPU time	Speed-up factor
1	10.6 s	14.7
5	15.8 s	9.9
10	22.8 s	6.9
20	35.6 s	4.4

Second, to demonstrate the efficiency of our novel technique, we consider the computation time of the code in Matlab R2009a. All calculations were performed on a computer with an Intel(R) Core(TM) Quad CPU Q9650 and 8 GB of installed memory (RAM) for 201 frequency samples (linearly spaced between DC and 50 GHz). Table I shows the computation time of the perturbation method for a varying number of sections. The speed-up factor is calculated w.r.t. the CPU time of 156.3 s needed by the reference technique. As can be seen, even when subdividing the line into 20 sections, we still obtain a speed-up factor of 4.4.

IV. CONCLUSION

A perturbation technique to model NUTLs, in particular differential lines embedded in a substrate composed of woven glass fibers, was presented. The fiber weave effect causes a differential skew between the two traces, leading to insertion loss suck-outs and mode conversion. By subdividing electrically very long lines into a limited number of shorter sections of about 5 to 10 wavelengths, these fiber weave effects are precisely captured by the perturbation technique. Compared to a standard chain matrix approach, excellent accuracy and improved efficiency was obtained. Additionally, it is worth mentioning that the reference chain matrix approach does not allow the modeling of NUTLs with continuously varying p.u.l. parameters. Such NUTLs will however be further investigated, leveraging the novel perturbation technique.

ACKNOWLEDGMENT

The authors like to thank Stefaan Sercu and Jan De Geest of FCI for providing a model of the fiber weave cross-sections.

REFERENCES

[1] M. Chernobryvko, D. Vande Ginste, and D. De Zutter, "A two-step perturbation technique for nonuniform single and differential lines," *IEEE Trans. Microw. Theory Tech.*, vol. 61, no. 5, pp. 1758 – 1767, May 2013.

[2] S. McMorrow and C. Heard, "The impact of PCB laminate weave on the electrical performance of differential signaling at multi-gigabit data rates," in *Proc. DesignCon*, 2005.

[3] E.-P. Li, X.-C. Wei, A. C. Cangellaris, E.-X. Liu, Y.-J. Zhang, M. D'Amore, J. Kim, and T. Sudo, "Progress review of electromagnetic compatibility analysis technologies for packages, printed circuit boards, and novel interconnects," *IEEE Trans. Electromagn. Compat.*, vol. 52, no. 2, pp. 248–265, May 2010.

[4] C. Gazda, D. Vande Ginste, H. Rogier, R.-B. Wu, and D. De Zutter, "A wideband common-mode suppression filter for bend discontinuities in differential signaling using tightly coupled microstrips," *IEEE Trans. Adv. Packag.*, vol. 33, no. 4, pp. 969–978, Nov. 2010.

[5] T. Demeester and D. De Zutter, "Quasi-TM transmission line parameters of coupled lossy lines based on the Dirichlet to Neumann boundary operator," *IEEE Trans. Microw. Theory Tech.*, vol. 56, no. 7, pp. 1649–1660, Jul. 2008.

[6] C. R. Paul, *Analysis of Multiconductor Transmission Lines.* John Wiley & Sons, 1994.

978-1-4799-0708-3/13 $31.00 © 2013 IEEE

Millimeter Wave and RF Technologies II

978-1-4799-0708-3/13 $31.00 © 2013 IEEE

978-1-4799-0708-3/13 $31.00 © 2013 IEEE

Insertion Loss Characterization of Tightly Spaced Interconnects with an Embedded Patterned Layer

Marcos A. Vargas and Kathleen L. Melde

Department of Electrical and Computer Engineering
University of Arizona
Tucson, USA
melde@email.arizona.edu

Abstract—As smaller packaging footprints and faster data rates are pursued, signal integrity suffers as a result of interconnects routed in close proximity to one another. This paper focuses on two tightly spaced microstrips and highlights the use of an embedded patterned layer (EPL) of conductive elements to improve insertion loss and far end crosstalk. The frequency domain S-parameter performance is characterized with a commercial full wave solver and effective permittivity is extracted. The effect of relative permittivity on insertion loss is investigated. The largest improvement is seen for the permittivity of 10, with insertion loss improving from -6.3dB to -2.3dB at 67GHz. The same case shows a far end crosstalk improvement from −1.5dB to -4.8dB at 67GHz. However, a tradeoff with return loss and near end crosstalk is observed.

Keywords- Crosstalk, insertion loss, interconnects, metasurface, microstrip.

I. INTRODUCTION

High density electronic circuit packaging is a keystone technology in computing and communications. There is an ever increasing demand to create transmission lines with larger signal bandwidths to support multi-GHz data rate transfers in computing and ultra wideband (UWB) signals in communications. Although advances in chip scaling have led to increases in computing capabilities, interconnect scaling still remains a technical bottleneck. As edge rates increase, additional frequencies appear in the signal spectrum, further highlighting the need for broadband interconnects. Moreover, as the frequency range increases to support higher data rates, interconnect components such as wirebonds, pads, bends, and modal transitions become parasitic components that degrade performance. To recapture performance at high frequencies despite the hostile packaging and operation constraints, this paper considers an approach that can maintain high I/O count while overcoming some of the performance drawbacks of high density routing.

One emerging area in materials research involves the transformation or modulation of signal propagation using embedded, sub-wavelength elements: metamaterials. Metamaterials display negative permittivity and permeability resulting in backward traveling waves. Packaging interconnects using metamaterial properties have been demonstrated in [1-2]. Metasurfaces are derived from metamaterials by using a thin patterned layer that alters the signal properties in interconnects using periodic or aperiodic elements. While not displaying the backward travelling wave of metamaterials, metasurfaces

modulate the boundary conditions of material interfaces, thus, transforming a guided wave from one mode to another as in [3-4]. This work is inspired by the patterning of elements and the resulting modulated signal properties of metasurfaces to improve signal integrity in a common interconnect upwards of 30GHz.

Since signal integrity for interconnects at high data rates is so important, tightly coupled, single-ended microstrips are considered in this work. The use of an embedded patterned layer (EPL) is aimed at optimizing for insertion loss and reducing far end crosstalk while considering the tradeoff with return loss and near end crosstalk. Data is presented from simulations using the full-wave solver HFSS [5]. The models with and without the EPL are simulated from 30GHz to 67GHz.

II. MODELING THE EPL

The EPL consists of a patterned layer of elements. Realization of the EPL involves printed, flat, sub-wavelength conductors placed between the ground and signal layers of a RF transmission line. The EPL conductors in this work are flat disks. By implementing the patterned layer, the motivation is to mold the electromagnetic propagation as inspired by the molding of light in photonic crystal structures detailed in [6].

To investigate the effect of the EPL a structure consisting of two tightly spaced adjacent microstrips is modeled. Megtron 6 is chosen as the nominal dielectric with ε_r=3.5, tan δ=0.002 and a substrate height h=4mil. The thin substrate and tightly spaced microstrips create a challenging, tightly coupled microwave interconnect. The microstrips are a width w=6mil with a spacing s_1=4.5mil and have a ground plane at the bottom surface of the substrate that results in 59Ω single-ended characteristic impedance. A conductor thickness of 0.7mil is included to model lossy copper conductors with a conductivity σ_c=5.8x10^7 S/m. The substrate width and length is 200mil long while the microstrips span the same length. The HFSS simulation includes a 175mil tall airbox that surrounds the substrate with absorbing boundary conditions on all four sides and the top surface of the airbox. Fig. 1 illustrates the interconnect structure and dimensions.

When adding the EPL to the nominal model, disks were chosen with the following dimensions: diameter d=6.5mil, spacing s_2=4.5mil and thickness t=0.35mil. The EPL consists of 3 columns and 15 rows of disks that are placed parallel to microstrips along the x-axis and embedded in the substrate in

---

This work is supported by the Army Research Office (ARO) under Contract W911NF-12-1-015 and the National Science Foundation (NSF) under Grant ECCS-1231368.

978-1-4799-0708-3/13 $31.00 © 2013 IEEE

between the top conductors and the bottom ground plane. The z-position, *zpos*, between the top and bottom conductors will be varied to characterize the performance of this structure. The edge of the disks begin and end 20mil away from the HFSS waveports to avoid any reflections from the discontinuity back into the waveport. Terminal ports are used in HFSS due to the tight spacing of the microstrips, resulting in a terminal full wave solution. Both ports are renormalized to $Z_0=59\Omega$ and deembedded up to the location of the EPL.

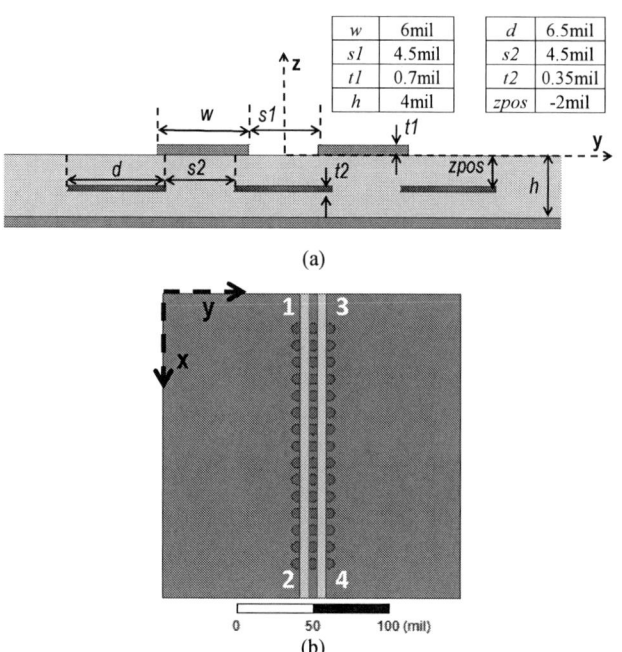

w	6mil	d	6.5mil
$s1$	4.5mil	$s2$	4.5mil
$t1$	0.7mil	$t2$	0.35mil
h	4mil	$zpos$	-2mil

(a)

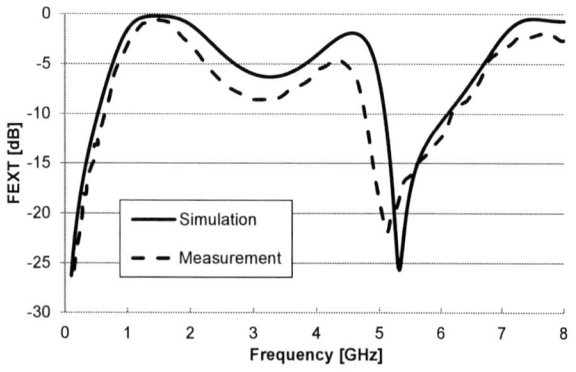

(b)

Fig. 1. (a) Cross section of the tightly coupled microstrips with the embedded patterned layer accompanied with tabled dimensions. (b) Top down view of the full model with port number assignments.

III. EPL PERFORMANCE

To verify the far end crosstalk (FEXT) simulation results on a similar structure with measurements, a comparison to results reported in [7] was conducted. Fig. 2 shows that the measured and simulated FEXT generally agree with one another.

Fig. 2. Simulation verification with documented measurement of far end crosstalk in [7].

After the simulations were validated with measurements, a set of parameter studies on coupled microstrips with EPLs were conducted. The first study includes a set of coupled microstrips ($\varepsilon_r=3.5$ and $\tan\delta=0.002$) with and without the EPL. The position of the EPL is considered. In this study, *zpos*, is placed at either 1mil, 2mil, or 3mils below the top surface. The overall result of including the EPL shows an improved insertion loss, S_{21}, as frequency increases. The following details characterize how much performance is recaptured. In Fig. 3a S_{21} shows a general improvement as the EPL moves closer to the top surface. At 67GHz a maximum 0.7dB improvement or 8% magnitude improvement is seen when the EPL is 1mil below the top surface compared to the nominal No-EPL case. While S_{21} improved, Fig. 3b shows that return loss S_{11} increased as the EPL approached the substrate's top surface. The increased return loss can be intuitively explained by the increased capacitances between the EPL and the adjacent microstrips as the EPL is positioned closer, resulting in a lower characteristic impedance of the EPL section and increasing the impedance mismatch with the load lines.

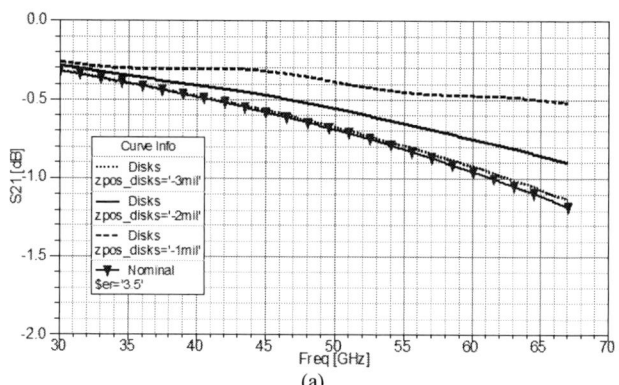

(a)

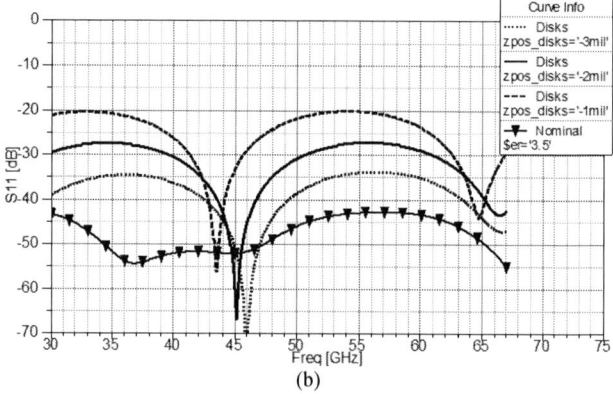

(b)

Fig. 3. For $\varepsilon_r=3.5$, (a) insertion loss S_{21} and (b) return loss S_{11} adjacent microstrips with EPL.

Since the microstrips are tightly spaced, crosstalk performance is also characterized. With the 6.5mil diameter disks and the z-position 1mil below the top surface, Fig. 4 shows far end crosstalk S_{41} (FEXT) improved across the simulation frequency range with an improvement of 5dB at the upper frequency limit. Also in Fig. 4 near end crosstalk S_{31} (NEXT) degraded to -26dB compared to the nominal No-EPL model, although a narrowband high-isolation region appears at 44.8GHz.

978-1-4799-0708-3/13 $31.00 © 2013 IEEE

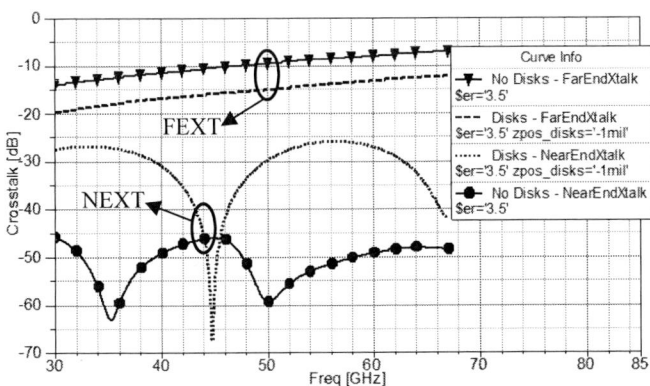

Fig. 4. For ε_r=3.5, near end (S_{31}) and far end (S_{41}) crosstalk comparison between the nominal No-EPL case and the EPL 1mil below the substrate top surface.

While the cases with ε_r=3.5, show that EPLs offer some improvement in S_{21} and FEXT, simulations on coupled microstrip lines with EPLs when ε_r=10 show even more performance improvements. The next study considered a permittivity of 10 and disk diameters fixed at 6.5mil. The position of the EPL is considered. In this study *zpos* is placed at either 1mil, 2mil, or 3mils below the top surface. Fig. 5a and 5b show the return loss and insertion loss as a function of frequency. Fig. 5a shows a general insertion loss improvement across frequency with a maximum reduction of 4dB or 58% magnitude improvement in S_{21} at 67GHz when *zpos* of the EPL

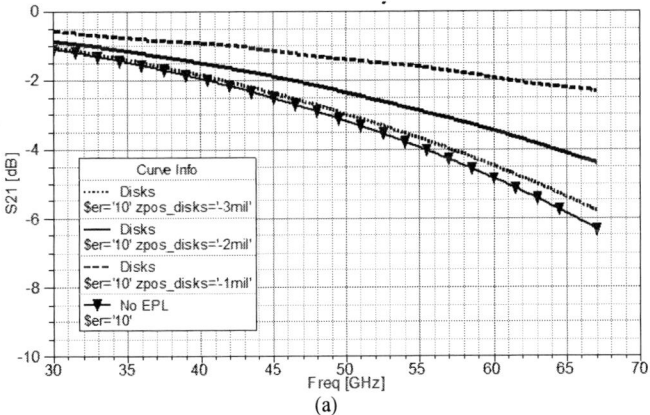

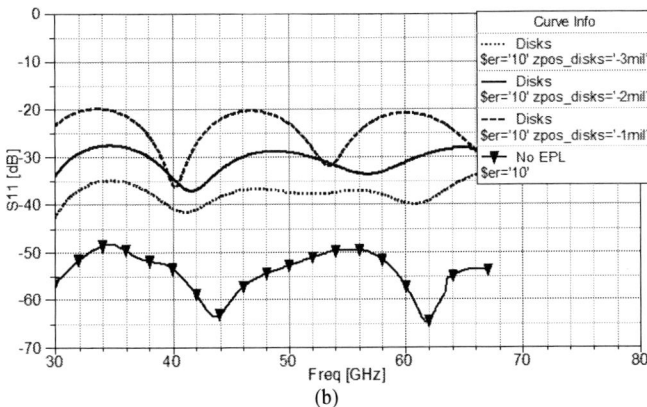

Fig. 5. For ε_r=10, (a) insertion loss S_{21} and (b) return loss S_{11} adjacent microstrips with EPL.

is 1mil below the top surface. Microstrip interconnects on higher permittivity material confine the electric fields more in the dielectric than in the air above. This means that the EPL has a greater impact on the field interactions between the adjacent microstrips and propagation performance. Fig. 5b shows S_{11} increasing as the EPL approaches the top surface. Return loss is -20dB at worst when the EPL is 1mil below the top surface.

Keeping with ε_r=10, the FEXT increases with increasing frequency. Applying the EPL improves FEXT by 3.3dB-4.2dB across the frequency range when compared to the No-EPL case as seen in Fig. 6. Also in Fig. 6, NEXT increased to a maximum -22dB, highlighting a degradation trend with increasing ε_r when compared to the prior ε_r=3.5. The resulting tradeoff is then insertion loss versus far end crosstalk when choosing a relative permittivity. A lower ε_r provides less insertion loss and better far end crosstalk than the higher ε_r. However, if a higher ε_r is needed for smaller distributed components, then the EPL makes higher permittivities viable at millimeter wave frequencies.

Consideration was taken to determine what would be the appropriate line spacing of a nominal two microstrip structure to produce performance comparable to the EPL structure. With the EPL performance characterized for ε_r=10, Fig. 7a shows the 9mil spacing needed for a nominal two microstrip structure to achieve the equivalent performance of two microstrips spaced 4.5mil apart with an EPL 1mil below the top surface. The insertion loss and far end crosstalk are plotted in Fig. 7b to highlight this equivalency.

To quantify the insertion loss trend across ε_r, Fig. 8 charts the increasing improvement of S21 with increasing ε_r. The permittivity is varied using values of packaging dielectrics: 3.5 (Megtron 6), 7 (LTCC), and 10 (HTCC) while the dielectric loss is held constant at the nominal tan δ=0.002. The z-position of the EPL 1 mil below the top surface had the most marked improvement in S_{21} for all cases.

The changes in S_{11} indicated changes in Z_o of the EPL interconnect. This indication implied that the effective permittivity, ε_{eff}, is changed from the nominal case when the EPL is present. To investigate this shift, the effective permittivity of a single ended microstrip line with and without the EPL was calculated. The ε_{eff} was extracted with (1) using group delay GD and the physical length of the embedded patterned layer

$$\varepsilon_{eff} = \left(\frac{c*GD}{l_{phys}} \right) \qquad (1)$$

Group delay is calculated using (2),

$$GD = \frac{1}{360°} \frac{d\varphi}{df} \qquad (2)$$

with dφ/df being the rate of phase change relative to the constant frequency interval of the simulation. The resulting ε_{eff} of the EPL has a higher value than the nominal ε_{eff} as is shown in Fig. 9. Also highlighted is the relatively more pronounced increase between the nominal No-EPL case and the embedded patterned layer at ε_r=10 than at ε_r=3.5. The engineered higher effective permittivity offers miniaturization of millimeter wave interconnects with higher ε_r materials.

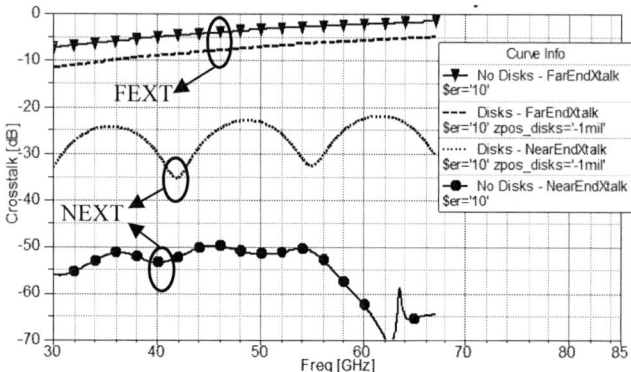

Fig. 6. For ε_r=10, near end (S_{31}) and far end (S_{41}) crosstalk comparison between the nominal No-EPL case and the EPL 1mil below the substrate top surface.

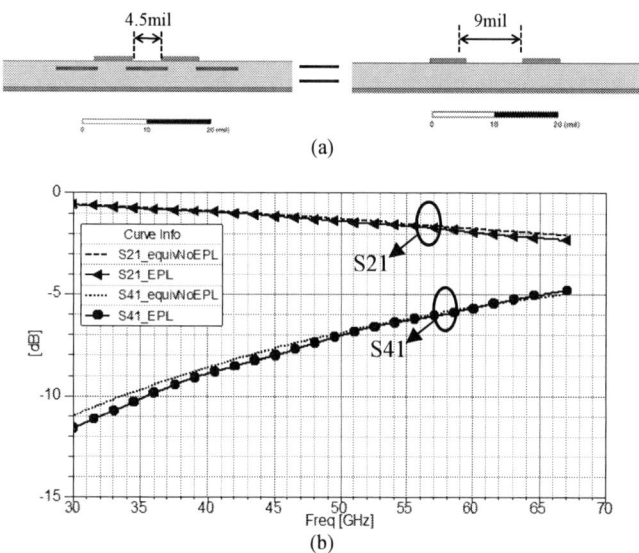

Fig. 7. For ε_r=10, (a) at left an EPL transmission line with the EPL 1mil below the top surface and 4.5mil microstrip spacing and the microstrip-only structure needed for (b) the equivalent insertion loss and FEXT of the EPL.

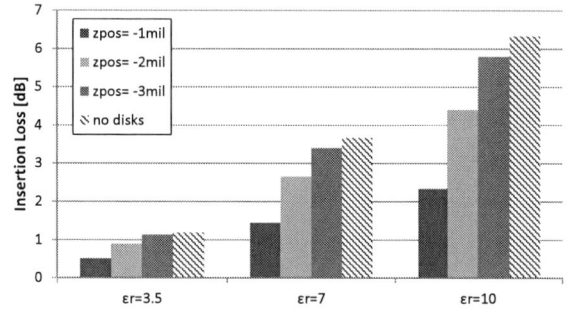

Fig. 8. Insertion loss, S_{21}, at 67GHz for different ε_r when tanδ is held constant at the nominal 0.002.

IV. CONCLUSION

With the electrical performance of adjacent interconnects bottlenecked by faster data rates and smaller form factors, this paper details a passive technique to improve S_{21} and FEXT when two microstrips are tightly spaced in millimeter wave

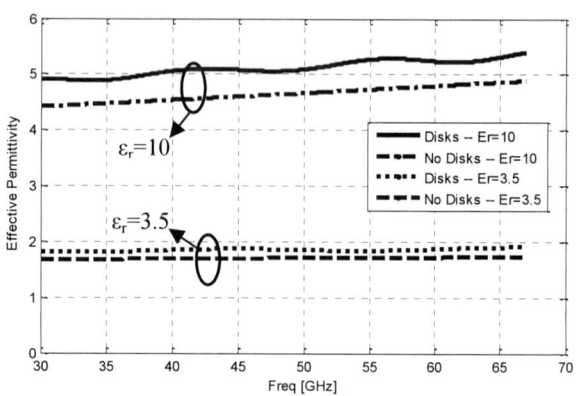

Fig. 9. Extracted effective permittivity for ε_r=3.5 and ε_r=10 compared to the nominal no-EPL case. Data corresponds to the model with disk diameters of 6.5mil and the z-position of the EPL 1mil below the surface.

frequencies. The use of an embedded patterned layer of conductive elements permits this improvement with S_{11} and NEXT being the performance tradeoff. This performance recapture is across different ε_R but shows the most marked effect in higher ε_R. Placement of the EPL as close to the top conductor layer provides the largest performance improvement. The extracted $\varepsilon_{\varepsilon FF}$ increases with the use of the EPL, thus, lowering the characteristic impedance of the microstrips while maintaining narrow conductor widths. Based on the findings in this paper, future work will consider aperiodic patterning of the EPL as an impedance control technique and the characterization of other element shapes for the EPL.

REFERENCES

[1] C. Caloz and T. Itoh, *Electromagnetic Metamaterials: Transmission Line Theory and Microwave Applications*. Hoboken, New Jersey: Wiley, 2006.

[2] A. Suntives, A. Khajooeizadeh, R. Abhari, "Charaterization of Metamaterial Interconnects," *Proc. Electrical Performance of Electronic Packaging and Systems Conf.*, Atlanta, GA, Oct. 29-31, 2007, pp. 211-214

[3] C. Holloway, et. al. "An Overview of the Theory and Application of Metasurfaces," *IEEE Antennas and Propagat. Magazine*, April 2012, pp. 10-35

[4] S. Maci, G. Minatti, M. Casaletti, and M. Bosiljevac, "Metasurfing: Addressing Waves on Impenetrable Metasurfaces," *IEEE Antennas Wireless Propag. Lett.*, 2011, vol.10, pp.1499-1502

[5] [Online] www.ansys.com, Ansys Corporation, High Frequency Structure Simulator, HFSS, Version 15.

[6] J. D. Joannopoulos et. al., *Photonic Crystals, Molding the Flow of Ligh t*, 2nd Edition, Princeton Univ. Press, 2007

[7] K. Hollaus, et al., "Simulation of Crosstalk on Printed Circuit Boards by FDTD, FEM, and a Circuit Model," *IEEE Trans. Magn.*, June 2008, vol. 44, no. 6

Modeling I/O Buffers Using X-Parameters

Thomas M. Comberiate and José E. Schutt-Ainé
Department of Electrical and Computer Engineering
University of Illinois at Urbana-Champaign
Urbana, Illinois 61801
{tcomber2, jesa}@illinois.edu

Abstract—**X-parameters have been shown to have a wide array of applications in the modeling of nonlinear devices and systems. This work analyzes the large-signal portion of an X-parameter model of a buffer to determine the extent of its nonlinearity and use for modeling input/output buffers.**

Index Terms—**X-parameters; nonlinearity; polyharmonic distortion (PHD); harmonic balance; I/O buffers.**

I. INTRODUCTION

With the increasing complexity and transmission speeds of input/output (I/O) buffers, circuit designers are in need of improved buffer models for simulation and design of high-speed links. Frequency-domain methods are becoming attractive for signal integrity purposes because they offer high accuracy, even at the very high frequencies at which buffers will be operating in the near-future. X-parameters*, a nonlinear frequency-domain measurement formalism, has been shown to have promise for modeling I/O buffers [1]. X-parameter models can be generated via simulation with either harmonic balance [2] or the latency insertion method [3] or via measurement with the use of a nonlinear vector network analyzer (NVNA) [4]. They are also blackbox models, thus they protect the intellectual property of the circuit for which they are used.

The X-parameter formalism is based on the Polyharmonic Distortion (PHD) model introduced by Verspecht and Root [5]. The basis of the PHD model is the linearization of a nonlinear scattering function with respect to a dynamic large signal incident wave [6]. This allows the scattering function to be decomposed into a large-signal response, a simple nonlinear function of the large-signal fundamental input with its DC bias, and a small-signal response, a nonanalytic linear superposition of small inputs that are harmonically related to the large-signal fundamental.

As the large-signal fundamental input shrinks, the simple nonlinear function becomes linear and approximates the traditional scattering parameter paradigm. The transient simulation technique developed in [1] used this insight to linearize the large-signal response while leaving the harmonic superposition intact. While this makes for a very convenient matrix representation of the nonlinear device, the linear approximation of the large-signal response might not be valid for a typical buffer, which switches from rail to rail.

*"X-parameters" is a registered trademark of Agilent Technologies.

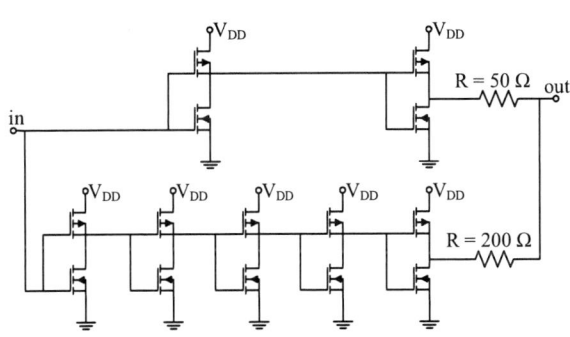

Fig. 1. Transmitter with equalization branch using a 2.5 V CMOS process. This circuit is used with and without the equalization branch in this work.

Although much work on X-parameters has been published, there is very little sense of what X-parameters, particularly the large-signal parameters, actually look like, particularly for an I/O buffer. The aim of this work is to investigate properties of the simple nonlinear function of the large signal and its DC bias in the context of using X-parameters to model the behavior of a simple I/O buffer. The example circuit analyzed in this work, shown in Fig. 1, is an equalized buffer consisting of a one-tap FIR filter implemented in a modified single-ended push-pull configuration using a typical 2.5 V CMOS process. The near-end signal driving the channel is obtained by voltage division, and the tap delay is implemented using a simple four-inverter chain. This circuit will also be analyzed without the equalization branch to provide additional intuition into the large-signal response of a buffer and its corresponding X-parameters.

The paper is organized as follows. Section II gives a brief overview of the PHD model and the X-parameter formalism. Section III demonstrates the effects of a DC voltage bias at the buffer input on the large-signal X-parameters. Section IV studies the degree of nonlinearity and frequency dependence on the large-signal response of an I/O buffer in the context of its normal operation. Section V provides a comparison between a transient simulation of the transistor-level model of the example buffer and a harmonic balance simulation made with its X-parameter model with a large-signal stimulus to show that the X-parameter model captures the nonlinear behavior of the buffer. Lastly, a conclusion and some insights

978-1-4799-0708-3/13 $31.00 © 2013 IEEE

into future work using X-parameters to model I/O buffers are provided in Section VI.

II. X-PARAMETER FORMALISM

The X-parameter formalism is based on the polyharmonic distortion (PHD) model, pioneered by Verspecht and Root [5]. The PHD model is a "black-box" behavioral model of a nonlinear component. The response of the nonlinear circuit is linearized around a large-signal dynamic operating point. The nonlinear behavior is taken into account by the reaction of the circuit to a large-amplitude input tone. It is then assumed that additional "small" perturbations due to signal components at harmonic frequencies respond linearly. Similar to the traditional scattering parameter formalism, the PHD model is defined in the frequency domain as a relation between dependent scattered waves and independent incident waves. Due to the nonanalytic property of the linearization around the large-signal dynamic operating point, the scattered waves in the PHD model also depend on the complex conjugate of the incident waves [5], accounting for an arbitrary phase difference between the large signal and the small signal. Consequently, the power wave relationship takes the form [7]:

$$B_{p,k} = X_{p,k}^{(FB)}(|A_{1,1}|, DC, f) \cdot P^k$$

$$+ \sum_{\substack{q=1 \\ l=1 \\ (q,l) \neq (1,1)}}^{\substack{q=N \\ l=K}} X_{p,k;q,l}^{(S)}(|A_{1,1}|, DC, f) \cdot A_{q,l} \cdot P^{k-l}$$

$$+ \sum_{\substack{q=1 \\ l=1 \\ (q,l) \neq (1,1)}}^{\substack{q=N \\ l=K}} X_{p,k;q,l}^{(T)}(|A_{1,1}|, DC, f) \cdot A_{q,l}^* \cdot P^{k+l} \qquad (1)$$

where $A_{q,l}$ is the contribution from the incident wave of the l-th harmonic at port q, $A_{q,l}^*$ is its complex conjugate, and $B_{p,k}$ is the contribution from the scattered wave of the k-th harmonic at port p. $X_{p,k}^{(FB)}$ is a scattering parameter of type FB that accounts for the contribution from the large-amplitude input tone, $A_{1,1}$, to the k-th harmonic of port p. This parameter has the same units as the scattered and incident waves. $X_{p,k;q,l}^{(S)}$ is a scattering parameter of type S that accounts for the contribution to the k-th harmonic of the scattered wave at port p from the l-th harmonic of the incident wave at port q. $X_{p,k;q,l}^{(T)}$ is a scattering parameter of type T that accounts for the contribution to the k-th harmonic of the scattered wave at port p from the l-th harmonic of the conjugate of the incident wave at port q. The S- and T-parameters are ratios of power waves and thus, like traditional scattering parameters, unitless. DC is the DC voltage or current biasing and f is the frequency. $P = e^{j \cdot \arg(A_{1,1})}$ is a pure phase term that compensates for the magnitude-only dependence of $A_{1,1}$ on the FB-, S-, and T-parameters to ensure the time-invariance of the model. N is the total number of ports and K is the total number of harmonics. The FB-, S-, and T-parameters together characterize the nonlinear dynamics of the network of interest with a large-signal operating point at a particular input fundamental frequency and power in

accordance with the PHD model. Note that because of the large-signal fundamental input, X-parameter measurements are inherently unidirectional.

The primary use of X-parameters thus far has been for the steady-state simulations of amplifiers and mixers using harmonic balance or circuit envelope simulation techniques for design or modeling purposes [8]. These simulation techniques construct a Volterra model for a device from X-parameter measurements with a variety of large-signal frequencies and powers. This allows the steady-state response of the circuit to be modeled for input modulation signals whose carrier frequency is close to the large-signal or any of its harmonics. In addition, the X-parameter formalism allows for an arbitrary number of large tones, which could be used to model circuit behavior for multiband inputs or multiple input ports, like a mixer [9].

There are several expansions to the X-parameter formalism that are relevant to the modeling of I/O buffers. One is the modeling of long-term memory effects [10], which can be used to describe phenomena like a buffer's sensitivity to temperature or bias voltage change. Another is the addition of impedance tuners to the NVNA measurement system that allow for measurement of non-50-Ω devices [11], potentially improving the characterization of buffers, many of which do not operate in 50-Ω environments. Last, X-parameters have also been used to generate IBIS models, which are the current standard for modeling I/O buffers [12].

III. LARGE-SIGNAL DC INPUT

The first nuance of using the X-parameter formalism for I/O buffers is the DC bias voltage on the input. Typical amplifiers might not need a DC bias at the input, but most I/O buffers are operated from rail-to-rail where the lower rail is the ground. As Fig. 2 shows, a bias voltage away from the middle of the rails produces data that is not useful because much of the input signal is outside of the rail-to-rail voltage range and is being clipped. Therefore, care must be taken to include this DC voltage while making X-parameter measurements of I/O buffers with an NVNA in the laboratory.

IV. LARGE-SIGNAL FUNDAMENTAL INPUT

The first term in (1) is the response to the large-signal fundamental input, $A_{1,1}$, at the k-th harmonic. For $k = 1$, this term contains the AM/AM and AM/PM terms often used to characterize amplifiers [13]. As the magnitude of $A_{1,1}$ shrinks, the linear approximation of $X_{2,1}^{(FB)} = S_{2,1;1,1} \cdot A_{1,1}$, where $S_{2,1;1,1}$ is a constant, can be used. This approximation was leveraged in the transient simulation technique in [1]. In doing so, that work essentially assumed that all of the nonlinear behavior is modeled in the existence of harmonics. By then separating the real and imaginary portions of the incident and scattered waves for each harmonic, the relationship between the scattered and incident waves can be written as a linear transformation. Because the large-signal fundamental behavior is assumed to be linear, the S- and T-parameters that normally vary with respect to $|A_{1,1}|$ can instead be treated as constants.

978-1-4799-0708-3/13 $31.00 © 2013 IEEE

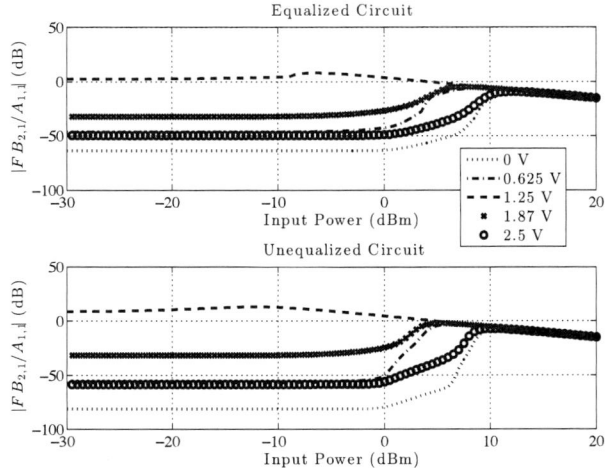

Fig. 2. Magnitude of $FB_{2,1}/A_{1,1}$ for the test circuit with and without equalization for multiple values of input DC voltage.

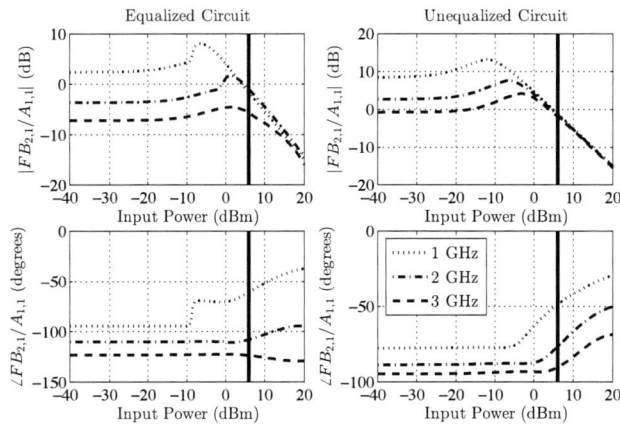

Fig. 3. Magnitude and phase of gain = $X_{2,1}^{(FB)}/A_{1,1}$ for the test circuit with and without equalization at frequencies equal to 1, 2, and 3 GHz and a DC voltage bias of 1.25 V. The thick black line marks the input power level where the input voltage waveform amplitude is equal to 2.5 V, the rail-to-rail voltage.

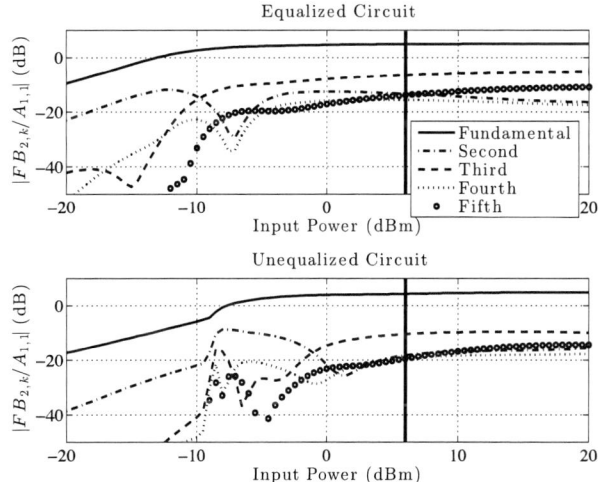

Fig. 4. Magnitude of $X_{2,k}^{(FB)}/A_{1,1}$ for $k = 1$ to 5 for the test circuit with and without equalization at 1 GHz and a DC voltage bias of 1.25 V. The thick black line marks the input power level where the input voltage waveform amplitude is equal to 2.5 V, the rail-to-rail voltage.

While this matrix representation is an incredibly desirable form, the linear approximation of the large-signal response is often not valid for the typical buffer that switches from rail to rail at its input. Figure 3 shows a plot of the simulated $X_{2,1}^{(FB)}$ term versus input power for the test circuit with and without equalization operating at 1, 2, and 3 GHz. The vertical line marks the input power equivalent to a input voltage with magnitude equal to the rail-to-rail voltage of the circuit. This line was determined from

$$V_1 = V_{DC1} + 2\sqrt{Z_0}\,(A_1 + B_1)$$
$$\approx V_{DC1} + 2\sqrt{Z_0}\left(|A_{1,1}| + \sum_{k=1}^{K}\left|X_{1,k}^{(FB)}\right|\right) \quad (2)$$

where A_1 and B_1 are the total incident and scattered waves at the input port. These waves can be decomposed into K harmonic wave components and the scattered wave harmonic components are equal to the FB-parameters assuming that there are no small-signal incident waves on either port. This is approximately true when the ports are well-matched, which was shown to be a reasonable assumption through comparison to a harmonic balance simulation. In addition, a DC voltage bias, V_{DC1}, of one half of the rail-to-rail voltage was provided so as to minimize the clipping of the large-signal input $A_{1,1}$ as discussed in the previous section.

As can be seen, the typical operating input amplitude is well into the nonlinear region for both the unequalized and the equalized circuits. There is a short region of gain expansion before the device enters saturation, more exaggerated for the equalized circuit. The large-signal term of the equalized circuit also exhibits stronger frequency-dependence than the unequalized circuit. This is appropriate because the equalization is a short-term memory effect. These types of memory effects manifest themselves in frequency-dependence of the AM/AM and AM/PM curves, which are equivalent to the magnitude and phase of $X_{2,1}^{(FB)}$ respectively [14].

While the AM/AM and AM/PM curves of a device can tell a great deal about a device's nonlinear distortion, especially in a well-matched environment, the large signal also produces a response at each harmonic k, characterized at port p by $X_{p,k}^{(FB)}$. The large-signal return losses at port 1, the input for each harmonic k, denoted by $X_{1,k}^{(FB)}$, are dominated by the fundamental, $X_{1,1}^{(FB)}$, which is at least 40 dB larger than that for every other harmonic at an input equal to the rail-to-rail voltage. Figure 4 shows the large-signal response at port 2, the output port, denoted by $X_{2,k}^{(FB)}$ for $k = 1$ to 5. As can be seen, $X_{2,3}^{(FB)}$ is less than 15 dB smaller than $X_{2,1}^{(FB)}$ for both the equalized and unequalized circuits. The second, fourth, and fifth harmonics are about 20 dB smaller than the fundamental. Thus, these harmonics have significant contributions on the large-signal output and cannot be neglected.

978-1-4799-0708-3/13 $31.00 © 2013 IEEE

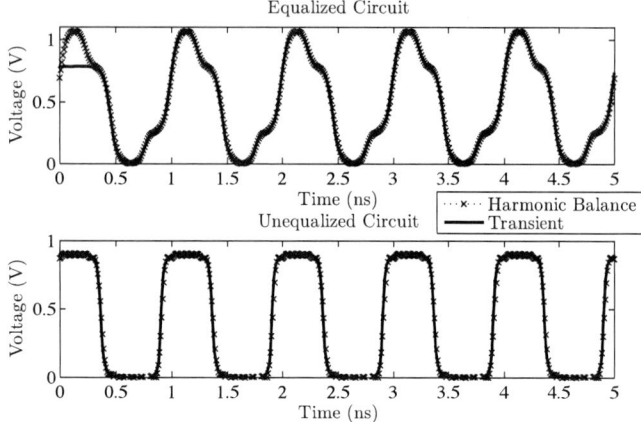

Fig. 5. Comparison of harmonic balance simulation of X-parameter model of the equalized and unequalized circuits with V_{DC1} = 1.25 V and input fundamental spanning the rail-to-rail voltage versus transient simulation of the transistor-level model with similar excitation. The model is terminated on both ends with 50 Ω in addition to the excitation at the input. The plots match very closely except in the beginning where the transient simulation has not reached its operating state.

V. SIMULATION OF LARGE-SIGNAL RESPONSE

Combining the fundamental and DC portions completes the large-signal stimulus, which spans the rail-to-rail voltage of the input. As shown above, this is well into the saturation region of both the equalized and unequalized circuits, so harmonics will be produced at the output. Figure 5 shows a comparison between a harmonic balance simulation performed with ADS [15] using the X-parameter models of both circuits with only the large-signal stimulus using 11 harmonics to a transient simulation of the transistor-level model of the circuit, also performed in ADS. The nonlinear behavior is matched very well for both the equalized and unequalized circuits, save for some Gibbs phenomenon ripple in the unequalized case. The DC-to-DC response was calculated separately and added into the solution. Thus, there is potential for accurate modeling of the nonlinear behavior of a buffer with X-parameters.

VI. CONCLUSION AND FUTURE WORK

This work has provided some intuition into the behavior of the large-signal response of an I/O buffer in the context of the X-parameter formalism. The example buffer has both significant nonlinear distortion and harmonic generation at a power level equivalent to an input spanning the rail-to-rail voltage. Thus, it is appropriate to model the large-signal response as nonlinear. As has been shown, X-parameters can capture the nonlinear behavior seen in I/O buffers. They remain a very exciting platform for the modeling of I/O buffers because of this, their mathematical robustness, and their IP protection.

Future work into modeling I/O buffers with X-parameters should leverage the insights of nonlinear modeling found in the power amplifier community, but should use the multitone X-parameter extension to emulate the wideband inputs to an I/O buffer. Most of the signals that drive a buffer can be modeled

with only a few large tones and harmonic superposition for the very high-frequency components. From these, an arbitrary bit stimulus with noise and jitter can be generated and used to construct an eye diagram for bit error rate estimation.

ACKNOWLEDGMENT

This research was made possible with United States Government support under and awarded by DoD, Air Force Office of Scientific Research, National Defense Science and Engineering Graduate (NDSEG) Fellowship, 32 CFR 168a and the National Science Foundation.

The authors also thank Agilent Technologies Inc. for their financial support of this work and for providing the ADS X-parameter generation platform. We especially thank Eric Iverson, Loren Betts, Steve Fulwider, and David Root from Agilent for their fruitful discussions, insightful comments and helpful suggestions. Lastly, the authors thank Xu Chen, Drew Newell, and Rishi Ratan of the Signal Integrity Group at the University of Illinois at Urbana-Champaign for their useful insights and contributions to this work.

REFERENCES

[1] J. E. Schutt-Aine, P. Milosevic, and W. T. Beyene, "Modeling and simulation of high speed I/O links using X parameters," in *2010 IEEE 19th Conference on Electrical Performance of Electronic Packaging and Systems, EPEPS 2010*, 2011.

[2] http://www.agilent.com/find/eesof-x-parameters.

[3] T. M. Comberiate and J. E. Schutt-Ainé, "Using the latency insertion method (LIM) to generate X parameters," in *2012 IEEE 21st Conference on Electrical Performance of Electronic Packaging and Systems, EPEPS-2012*, 2012.

[4] Agilent Technologies, *Agilent Nonlinear Vector Network Analyzer (NVNA)*, 2011.

[5] J. Verspecht and D. E. Root, "Polyharmonic distortion modeling," *IEEE Microwave Magazine*, vol. 7, no. 3, pp. 44 – 57, 2006.

[6] J. Verspecht, D. Williams, D. Schreurs, K. Remley, and M. McKinley, "Linearization of large-signal scattering functions," *Microwave Theory and Techniques, IEEE Transactions on*, vol. 53, no. 4, pp. 1369–1376, 2005.

[7] D. Root, J. Xu, J. Horn, M. Iwamoto, and G. Simpson, "Device modeling with NVNAs and X-parameters," in *2010 Workshop on Integrated Nonlinear Microwave and Millimeter-Wave Circuits (INMMIC)*, 2010.

[8] A. Soury and E. Ngoya, "Implementation of X-parameter models in harmonic-balance simulators," in *IEEE MTT-S International Microwave Symposium Digest*, 2012.

[9] Agilent Technologies, Inc. (2010, August) Notation and equations for multiport / multi-tone X-parameters. [Online]. Available: http://na.tm.agilent.com/pna/nvna/NVNAWebHelp/NotationMultiPortEquations.pdf

[10] J. Verspecht, D. Root, and T. Nielsen, "Dynamic X-parameters*: Behavioral modeling in the presence of long term memory effects," in *Microwave Conference (GeMiC), 2012 The 7th German*, 2012, pp. 1–4.

[11] G. Simpson, J. Horn, D. Gunyan, and David, "Load pull + NVNA = enhanced X-parameters for PA designs with high mismatch and technology-independent large-signal device models," Maury Microwave Corporation, Tech. Rep., 2009.

[12] T. M. Comberiate and J. E. Schutt-Aine, "Using X-parameters to generate IBIS models," in *Signal and Power Integrity (SPI), 2013 17th IEEE Workshop on*, 2013.

[13] A. Saleh, "Frequency-independent and frequency-dependent nonlinear models of TWT amplifiers," *Communications, IEEE Transactions on*, vol. 29, no. 11, pp. 1715–1720, 1981.

[14] P. Singerl and G. Kubin, "Constructing memory-polynomial models from frequency-dependent AM/AM and AM/PM measurements," in *Circuits and Systems, 2007. MWSCAS 2007. 50th Midwest Symposium on*, 2007, pp. 321–324.

[15] *Agilent Advanced Design System*, Version 2013.06. Copyright (c) 1983-2013, Agilent Technologies.

978-1-4799-0708-3/13 $31.00 © 2013 IEEE

Design and Fabrication of Geometrically Complicated Multiband Microwave Devices Using A Novel Integrated 3D Printing Technique

Majid Ahmadloo, *Member, IEEE*
Telecommunication Research Laboratories; TRTech (former TRLabs)
Edmonton, Canada
mahmadlo@alumni.uwo.ca

Abstract— **In this work a novel fast and efficient three dimensional (3D) printing technique is presented for prototyping and fabrication of 3D microwave structures to print both the conductive nanoparticle ink together with the dielectric material in an integrated process. This facilitates the fabrication of complicated 3D electromagnetic (EM) structures such as wide range of different multiband antennas and microwave devices. The process includes characterization of conductive ink and polymer based substrate to ensure proper RF, electrical, thermal and mechanical performances of both substrate and the ink. As most of the printed devices are relatively simple single band systems, in this work in order to demonstrate the performance of the proposed 3D printing technique, a geometrically complicated multiband meander line 3D dipole antenna is printed, tested and measured results are compared with simulations to verify the accuracy of the fabrication technique. Good agreement between measured and simulated results shows the efficiency and accuracy of the proposed fabrication technique. This not only provides a low cost, environmentally friendly integrated fabrication process but also enables us to use the technique on geometrically complicated multiband EM devices.**

Keywords—fabrication of 3D microwave structures, integrated 3D printing, nano-particle conductive ink, multiband printed meander line dipole antenna

I. INTRODUCTION

Due to ever increasing demand for the integrations of different communication services such as GSM bands, DSC, PCS, UMTS, Bluetooth/WLAN, WiMAX and other additional required bands in a single device, multiband antennas in small sizes are required at the front-end of such systems to transmit and receive the signals. As a result of multi-frequency nature of such antennas multiple resonance paths should be included in the design. Many researches have been conducted in this field to develop such antennas [1]-[6] Some of the conventional designs for multiband systems are in the form of planar inverted-F antennas [2]-[3], which may include parasitic elements in order to enhance and widen the original bandwidth of the antenna. Meander line dipole antennas are also other options to add multiple resonance modes for multiband purposes. In all such scenarios adding parasitic elements or increasing the length in the 3D structures make the antenna geometry complicated and many compromises should be made

to maintain minimum performance for the antenna. As a result despite all these advantages in achieving wider bandwidths and multiple resonance modes, such structures suffer from a major drawback as their accurate fabrication is costly and time consuming. To date mechanical machining and micromachining techniques such as CNC have been the dominant methods of fabrication of such complicated antennas and other EM devices. Application of these techniques is not only costly but also requires highly trained personnel to program the machine and operate it.

Printing technologies are an alternative approach to address such fabrication complexities. Printed electronics is relatively a new concept compare to other common technologies in the area of electronic circuit fabrication. The major advantage of this technology is to avoid costly and time consuming steps of traditional methods such as photolithography including etching, masking and plating which is also requires the use of bio-hazardous material. Unlike traditional machining techniques which create the object by drilling or etching to remove and shape materials, 3D printing can also generate three dimensional objects from a computer model by laying down successive layers of material. On the other hand due to recent advancements in the field of nanotechnology,

Fig.1. Ultimaker open source 3D printing machine

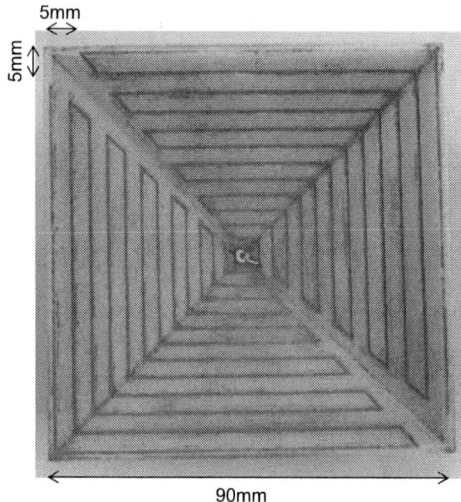

Fig. 2. 3D Printed multiband meander line dipole antenna on a 3D substrate.

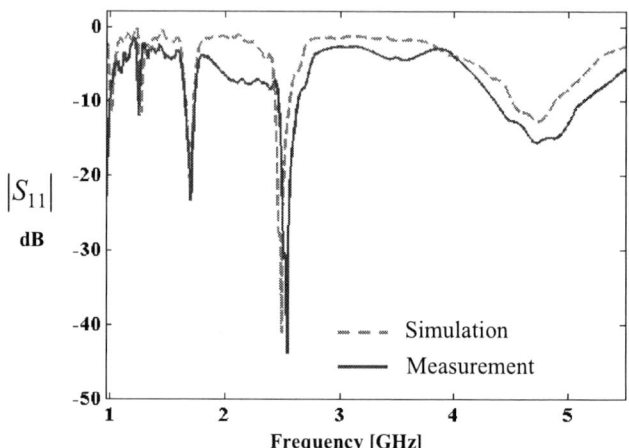

Fig. 3. S_{11} of the 3D Printed meander line dipole antenna.

nanoparticle conductive inks are playing an important role as a cost effective and environmentally friendly solution for the fabrication of printed circuits. Traditionally conductive ink printing has been mostly applied on planar surfaces [7]–[13]. However using proper printers enables us not only to print planar circuits but also to print over 3D structures separately fabricated to form non-planar antennas [14]. This however may bring difficulties to achieve proper alignment and guarantee high quality results. In this work based on the experience gained in the design, fabrication and testing of a meander lined V-shaped dipole antenna [15]–[16X], the integrated 3D printing technique is extended to fabricate a further complicated and more electromagnetically versatile multiband structure. In this integrated process conductive traces of a multiband meander dipole are printed using nanoparticle based conductive inks as well as the dielectric substrates to mechanically support antenna structure. As the complexities of the ink clogging and alignment issues have been addressed, endless opportunities for cost effective and efficient fabrication of wide range of different 3D EM structures are provided at relatively low temperatures. In the result section return loss and radiation pattern of the designed antenna are presented and simulated and measured data are compared to show the quality

and accuracy of the proposed fabrication technique and to demonstrate its potential for fast microwave device prototyping.

II. SUBSTRATE MATERIAL AND NANO-PARTICLE CONDUCTIVE INK PREPARATION AND PRINTING

Selection and preparation of conductive ink and appropriate substrate are the major steps in the integrated 3D printing process. Each of these steps requires careful investigation among available options to have proper curing temperatures and appropriate surface adhesion between the ink and polymer substrate. One of the available low cost and open source 3D printers is Ultimaker [17] shown in Fig. 1. This machine is a low cost open source table-top printer which has been modified for this work to have the ability of printing ink as well as traditional successive layers of polymer and perform integrated printing scenarios. Wide range of different polymers with different thermal characteristics and melting temperatures below 270°C can be directly used in this machine. Most common choices for 3D fabrication are ABS (Acrylonitrile Butadiene Styrene) and PLA (Poly Lactic Acid). However for better RF performance over higher frequencies other options such as Rexolite can be used. This requires reaching higher temperatures (more than 300°C) to melt the polymer and some modifications in the heating element of the extruder are required to achieve it. Like other conventional 3D printers in this class, printing is performed by laying successive layers of molten polymer to build up the 3D structure. In this work we utilized the same extruder to dispense ink with the nuzzle tip size of 400 micron and standard temperature of 240°C.

Regarding the selection of conductive ink, in spite of the availability of many commercial conductive inks in market, many research groups prefer to synthesize their own formulations based on the specific applications. This is to ensure proper electrical conductivity as well as thermal and mechanical characteristics during the continuous printing process. One common issue in working with conductive inks is the clogging of the nuzzle, which usually happens during the

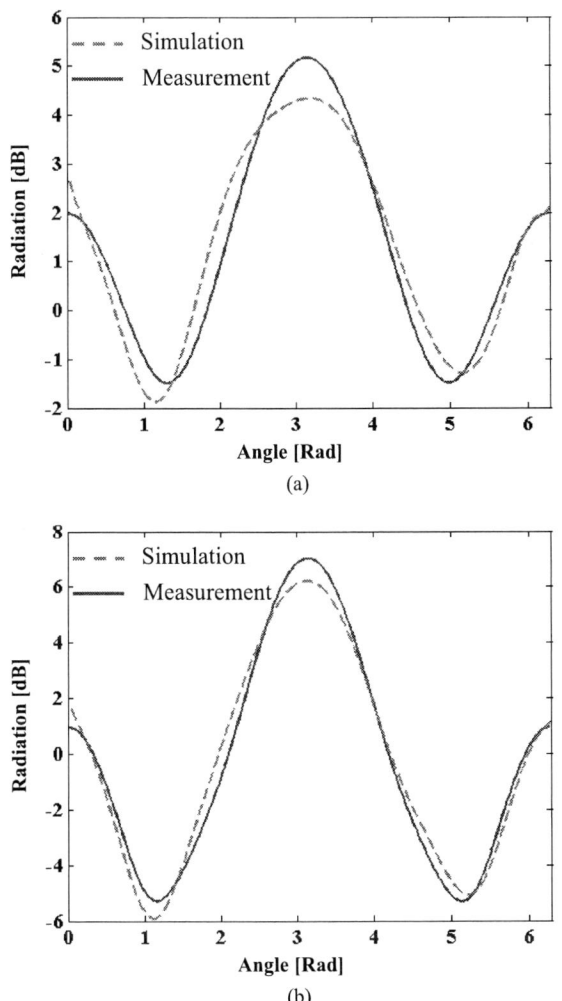

Fig. 4. Radiation pattern of the 3D Printed multiband meander line dipole antenna (a) 1GHz, (b) 1.8GHz.

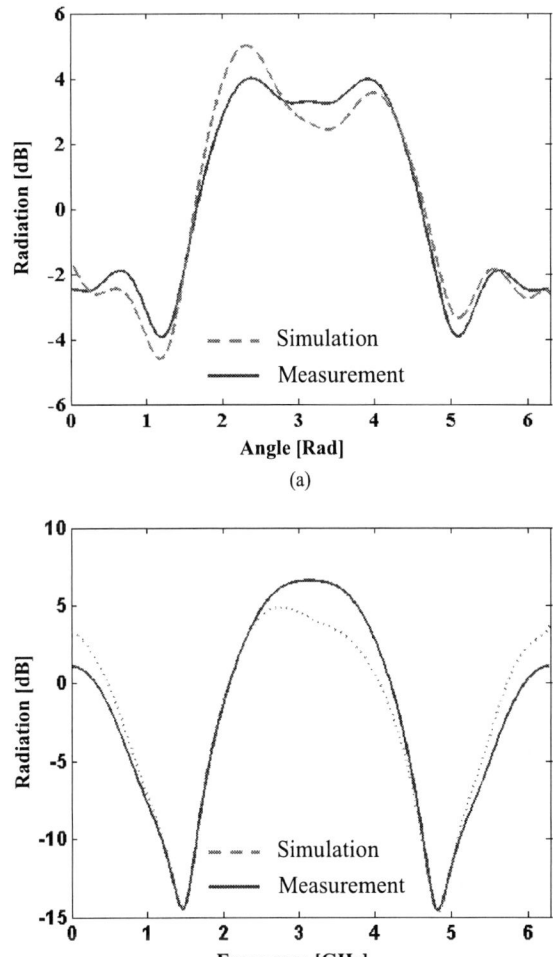

Fig. 5. Radiation pattern of the 3D Printed multiband meander line dipole antenna (a) 2.45GHz, (b) 3.5GHz.

longer printing times. To address this issue home-made recipes with variety of different viscosities, concentration of conductive particles per unit volume of ink and some mechanical adjustments in printing setups have been developed. Such recipes in this work were developed based on altering commercially available inks from IIMAK family including SSC, SSX and FSX [18], due to the ease of access and cost effectiveness. As each of these inks features different nanoparticle size, solvent agents, per unit resistivity and viscosity, the working solution has been developed and characterized by diluting them at different ratios, curing them at different temperatures for different times in order to get thermal and surface adhesion required in printing process and avoid nuzzle clogging issue which happens in all three standard inks. Solvent agents Methyl Ethyl Ketone and Glycol Ether PM are applied to dilute the original inks at different rates to achieve suitable viscosity and RF properties for consistent printing. SSC samples of 2.8 gr/ml, 1gr/ml and 0.5gr/ml, SSX samples of 9.5gr/ml, 4.5 gr/ml and 2 gr/ml and 4gr/ml and 3

gr/ml samples of FSX ink are made, printed on polymer test surfaces, and each one cured at 70°C, 85°C and 110°C for 10, 15 and 20 minutes. Resistivity of printed lines and the overall printing quality are tested and measured. Overall 72 samples printed and tested to characterize the ink and Lucas Labs S-302-4 four point probes is used to measure electric resistivity of printed lines. Amongst all the samples, 4.5 gr/ml SSX ink cured at 85°C for 15 minutes proved to be the best printable option with acceptable sintering temperature without deforming the substrate during the curing process in the oven, proper mechanical performance, good adhesion to the surface and very low resistivity of about 7×10^{-7} ohm/cm at its best. This customized ink is applied to print 3D meander line dipole antenna arms and form the multiband antenna. Characterization of the integrated printing process also involved wide range of different test prints to consistently have best printing speed, melting temperature, and ink adhesion to the polymer surface without nuzzle clogging.

III. MULTIBAND MEANDER LINE 3D DIPOLE ANTENNA PRINT AND MEASUREMENTS

3D printed meander dipole antenna fabricated in an integrated process has been previously investigated [15]; however such structure is relatively simple geometrically and only performs in a single frequency band. In order to demonstrate the performance of the novel 3D printing technique in fabrication of more complicated EM structures, a 3D multiband meander line dipole antenna is designed and fabricated. As it is shown in Fig. 2 antenna structure is in the form of a pyramid, a thin shell with 1mm thickness on which the arms of the dipole antenna are printed. The width of the printed traces is 0.8mm and each meandered arm is printed on the two adjacent faces of the pyramid as it is shown in Fig. 2 using the modified conductive ink formulation together with the PLA substrate. Dielectric constant of PLA is about 3 with the loss tangent of 0.02. The antenna doesn't require a substrate to radiate however in order to provide a base for printing the conductive ink, thin walls of the pyramid with 1mm thickness are printed to also provide structural integrity during the measurements. In order to reduce the effect of lossy substrate, antenna design doesn't include a ground plate. However this is not a limitation of the integrated printing methodology as it can be easily incorporated in the printing process using different substrates. Radiation pattern and return loss of the antenna are shown in Fig. 3, Fig. 4 and Fig. 5. As it is shown in these diagrams there is a good agreement between simulated result and measured data proving the accuracy of printing process as a reliable fabrication technique. The slight shift between simulation and measured results is due to the slight deformation of the plastic substrate and as a result the ink traces mostly during the thermal curing stage. Using more temperature tolerable plastics and inks with lower curing temperatures even as low as room temperature is undergoing to improve this process further more.

IV. CONCLUSIONS

This work presented the design and fabrication of a multiband antenna using a novel 3D printing process applicable on wide range of EM structures in an integrated nano-particle conductive inks and polymer based substrates simultaneous printing. Efficient substrate performance, low sintering temperatures and strong adhesion of the conductive ink are carefully investigated and applied to fabricate a 3D multiband meander line antenna. Numerical simulations and measured data demonstrated the accuracy and efficiency of such integrated 3D printing process and its potential in packaging, prototyping and fabrication of wide range of microwave structures and antennas. Further modifications on the preparation of ink, new sintering techniques, and materials can widen the scope of this work further into new micro structure and applications.

REFERENCES

[1] .-L. Wong, Y. C. Lin, and B. Chen, "Internal patch antenna with a thin air-layer substrate for GSM/DCS operation in a PDAphone," *IEEE Trans. Antennas Propag.*, vol. 55, pp. 1165–1172, Apr. 2007.

[2] K. L. Wong, Y. C. Lin, and T. C. Tseng, "Thin internal GSM/DCS patch antenna for a portable mobile terminal," *IEEE Trans. Antennas Propag.*, vol. 54, pp. 238–242, Jan. 2006.

[3] C. W. Chiu and Y. J. Chi, "Planar hexa-band inverted-F antenna for mobile device applications," *IEEE AntennaWireless Propag. Lett.*, vol. 8, pp. 1099–1102, 2009.

[4] J.-Y. Jan and L.-C. Tseng, "Small planar monopole antenna with a shorted parasitic inverted-L wire for wireless communications in the 2.4, 5.2, and 5.8-GHz bands," *IEEE Trans. Antennas Propag.*, vol. 52, no. 7, pp. 1903–1905, July 2004.

[5] M. Ali, G. J. Hayes, H.-S. Hwang, and R. A. Sadler, "Design of a multiband internal antenna for third generation mobile phone handsets," *IEEE Trans. Antennas Propag.*, vol. 51, pp. 1452–1461, Jul. 2003.

[6] K. L. Wong, W. J. Chen, L. C. Chou, and M. R. Hsu, "Bandwidth enhancement of the small-size internal laptop computer antenna using a parasitic open slot for the penta-bandWWAN operation," *IEEE Trans. Antennas Propag.*, vol. 58, pp. 3431–3435, Oct. 2010.

[7] A. Russo, B.Y. Ahn, J.J. Adams, E. Duoss, J. Bernhard, and J.A. Lewis, "Pen-on-Paper Flexible Electronics", *Advanced Materials, 23, pp. 3426-30, 2011*.

[8] J. J. Adams, E. J. Duoss, T. Malkowski, M. Motala, B. Y. Ahn, R. G. Nuzzo, J. T. Bernhard, and J. A. Lewis, "Conformal Printing of Electrically Small Antennas on Three-Dimensional Surfaces", *Advanced Materials, 23 [11], pp. 1304-1413, 2011*.

[9] S. B. Walker and J. A. Lewis, "Reactive Silver Inks for Pattering High-Conductivity Features at Mild Temperatures", *Journal of the American Chemical Society, 120105082029000, 2012*.

[10] G. Shaker, L. Ho-Seon, S. Safavi-Naeini, M. Tentzeris, "Printed electronics for next generation wireless devices", *IEEE Antenna and propagation conference (LAPC), pp. 1-5, November 2011*.

[11] J. Mei, M. Lovell, and M. Mickle, "Formulation and processing of novel conductive solution inks in continuous inkjet printing of 3-D electric circuits," *IEEE Transactions on Electronics Packaging Manufacturing*, vol. 28, no. 3, pp. 265- 273, July 2005.

[12] G. Shaker, S. Safavi-Naeini, N. Sangary, N.; M. Tentzeris, "Inkjet Printing of Ultrawideband (UWB) Antennas on Paper-Based Substrates", *Antennas and Wireless Propagation Letters, IEEE, Vol. 10, pp. 111-114, 2011*.

[13] J. Siden, M. Fein, A. Koptyug, and H. Nilsson, "Printed antennas antennas with variable conductive ink layer thickness", *Microw., Ant. Propag.*, vol. 1, no. 2, pp. 401–407, April 2007.

[14] J. Mei, M. Lovell, and M. Mickle, "Formulation and processing of novel conductive solution inks in continuous inkjet printing of 3-D electric circuits," IEEE Trans. on Electronics Packaging, vol. 28, no. 3, pp. 265-273, July 2005.

[15] M. Ahmadloo, and P. Mousavi, "A Novel Integrated Dielectric-and-Conductive Ink 3D Printing Technique for Fabrication of Microwave Devices", IMS 2013, Seattle, USA, June 2-7, 2013.

[16] M. Ahmadloo, and P. Mousavi, "Application of Novel Integrated Dielectric and Conductive Ink 3D Printing Technique for Fabrication of Conical Spiral Antennas", APS/USNC-URSI 2013, Orlando, Florida, USA, July 7-13, 2013.

[17] www.ultimaker.com

[18] www.iimak.com

EM I

978-1-4799-0708-3/13 $31.00 © 2013 IEEE

978-1-4799-0708-3/13 $31.00 © 2013 IEEE

Parallel Processing Improvements for Full-Wave Electromagnetic Solvers

Seung-Cheol Lee, Sergey Polstyanko, Denis Soldo, Matthew Commens, Prakash Vennam, and Steven G. Pytel Jr.

ANSYS, Inc.
275 Technology Drive
Canonsburg, PA 15317

Abstract—As computing resources become increasingly parallel for both shared and distributed memory systems, computational electromagnetic methods need to take full advantage of new architectures in order to reduce simulation times. Several different aspects of the solution process make it difficult for traditional finite element field solvers to effectively utilize distributed and shared memory resources. This paper will provide insight into recent advancements made in the finite element method matrix solve (shared memory) and frequency sweep (distributed memory) that drastically reduce simulation times for signal and power integrity applications.

Keywords—finite element method; HPC; high performance computing; signal integrity; power integrity

I. INTRODUCTION

Compute clusters utilizing engineering software with schedule management tools such as IBM Platform LSF, Altair PBS Professional, Windows HPC, Open Grid engine, and many others are becoming increasingly more common within IT departments. In addition, several commercial cloud networks such as Amazon EC2 and Microsoft Azure have been deployed in recent years. As Moore's law reaches its limit and these systems become more readily available, it becomes necessary to develop new numerical algorithms to take full advantage of these parallel systems in order to reduce simulation time and increase simulation capacity. However, a naive implementation may not get much benefit due to the trade-off between using more distributed nodes and keeping data local. To see performance gains, one has to place a significant amount of effort in optimizing the algorithms for parallelism and, in some instances, redesign the algorithms from scratch.

In this paper, we will review three different commercially available full-wave electromagnetic solvers and will focus on recent High Performance Computing (HPC) advancements made to accommodate large scale high performance shared and distributed memory systems. Throughout this paper, *shared memory* will be used to describe processes where multi-threaded cores within a single compute node share RAM during a single process and *distributed memory* refers to processes that can be deployed simultaneously over multiple compute nodes to speed up the solution process.

II. NUMERICAL METHODS

Among the various numerical methods, the finite element method (FEM) is one of the most versatile and accurate techniques for the analysis of arbitrary complex 3D structures. When combined with proper meshing and error control techniques it has proven to produce highly reliable results [1]. In this paper, we will review the application of the FEM for analyzing signal integrity (SI) and power integrity (PI) effects in planar, layered geometries encountered in modern package and printed circuit board (PCB) designs, and we present three different full-wave, FEM based electromagnetic field solvers that can be used for these applications. The differences between these approaches will be highlighted both in terms of performance and applicability.

When dealing with arbitrary 3D structures, tetrahedral meshes are best suited for accurate geometry representation. They make no geometry assumptions or simplifications and minimize geometry representation error resulting from initial discretization. As a result, they have no problems handling arbitrary bondwire shapes, plane etching, 3D solderball shapes and other complex shapes typically present in PCB and package structures. Furthermore, tetrahedral meshes can be easily adapted or refined in order to improve overall solution accuracy. HFSS is a good example of a 3D full-wave high frequency solver based on this technology [5].

Since package and PCB geometries are primarily planar, one can argue that prismatic mesh elements should be used instead of tetrahedral elements; this can result in smaller meshes and reduced run times without any loss of accuracy. This method works very well for planar structures that do not possess any 3D components such as connectors and wire bonds. Sentinel PSI is a good example of a full-wave high frequency solver based on this technology [7].

Finally, one can use a specialized formulation for this specific application by realizing that the total electromagnetic fields inside these multilayered structures can be decomposed into parallel-plate and transmission-line modes. Moreover, only the TEM mode needs to be considered due to the high cut-off frequency associated with the higher order modes. With those assumptions, the overall finite element formulation can be restricted to 2D and coupled with a localized field solution [2]. As a result, 2D triangular meshes can be used, which leads to dramatic improvement of computation time and memory usage with slight degradation of accuracy. This approach allows the

978-1-4799-0708-3/13 $31.00 © 2013 IEEE

simulation of large-scale industrial problems to be practical. SIwave is an example of a hybrid full-wave high frequency solver based on this technology [6].

For a given frequency, the solution process of any FEM based solver can be broken into several general steps: geometry meshing, unknown initialization, sparse matrix assembly, and matrix solution. If a frequency sweep is requested by a user, then the sparse matrix assembly and matrix solution steps are repeated for each frequency. Of all the steps, matrix solution is by far the most dominant one and can account for 90 percent or more of total solution time. In the remainder of this paper, we focus on recent HPC advancements made to speed up the matrix solution step both for single frequency point using shared memory parallel techniques within a workstation and for the frequency sweep using distributed memory methods within a workstation and across a cluster.

III. PARALLEL PROCESSING ALGORITHMS

A. Shared Memory Multi-threading

One common approach to speed up the finite element analysis is to enhance the performance of the matrix solution process by using multiple cores. Direct methods are often used to solve sparse linear systems of matrix equations coming from FEM analysis of packages and PCBs due to the ill-conditioned nature of these matrices. Therefore, this section will focus on the multi-threaded direct sparse matrix solution process for shared memory machines.

Direct methods for solving sparse matrix equations have long been studied, and one of the main areas of study is parallel processing. The shared memory matrix solver implementation discussed within this paper is based on the multi-frontal approach Parallelization can be achieved by exploring fine grain parallelism occurring in dense matrix kernels generated during the matrix solution process [8]. Another possibility comes from the tree-like structure of the factorization procedure, in which sub-trees can be processed independently of each other [3]. However, the best parallel performance is often achieved when both schemes are combined, using both tree-level parallelization and dense kernel parallelization concurrently. It's worth noting, that the overall parallel efficiency for either algorithm depends on several factors such as the sparsity of the matrix, shape of the elimination tree [8], number of right hand side vectors, problem size, memory bus speed, and processor architecture. As a result, one has to run a large number of examples in order to understand which method performs the best for a given class of problems.

B. Spectral Decomposition Method (SDM)

When a range of frequency responses needs to be computed, which is often the case for SI/PI simulations, the simplest and perhaps the most cost effective way to extract this data is to distribute the individual frequency points or different frequency bands across multiple compute nodes. Solving each frequency point or band is essentially an independent task, which provides a great benefit for parallel algorithm design. Once all frequency points are solved, the overall solution for the entire sweep can be reassembled. This parallelization is beneficial even for interpolating or curve fitting algorithms.

The process dramatically decreases the overall simulation time required to obtain highly accurate broadband full-wave results, and, at the same time, reduces the data transfer requirements between the computational nodes. One drawback of this approach however is that the memory usage of the overall computation increases linearly as the number of concurrent frequency solves is increased. Naturally, this frequency level parallelism is well suited for distributed memory environments when different frequency solves can be distributed between different compute nodes.

Furthermore, the recent trend of efficient utilization of shared and distributed memory architectures allows for the computational algorithms to become hybridized. This means that to achieve the greatest performance improvements the software must deploy both shared and distributed memory parallelization simultaneously. A simple way to achieve the goal for the problems of interest in this paper is to leverage orthogonalized multi-core matrix level and multi-node frequency level parallelization. Namely, each frequency or frequency band can be processed in different compute nodes or machines, and each frequency solve is multi-threaded through the shared memory parallel matrix solver. Processing multiple frequencies even within each compute node is also a possible option as long as the memory capacity of each node is large enough to hold the parallel frequency factorization. Software that generates smaller sized matrices in terms of memory is an ideal candidate for the latter scheme.

IV. PERFORMANCE BENEFITS

In the following section, several multi-layer packages and PCBs are solved on shared and distributed memory systems and performance data for different FEM solvers is collected to show scalability with respect to the number of cores. Up to 128 cores were used during the solution process to demonstrate the applicability of these methods for large scale compute clusters.

A. Multi-threading Results

In the first study, the performance of different shared memory parallel algorithms for the direct matrix solver has been compared when solving large, complex, symmetric, sparse, and indefinite matrices coming out of the hybrid solver. First, only the dense kernel parallel algorithm has been used to speed up the matrix solution step during a frequency sweep for 38 different package and PCB structures on a Dell Precision 490 system with an Intel Xeon X5450 quad-core processor running at 3.0 GHz. All parallel runs were done using four cores. Then, only tree-level parallel algorithm has been used to speed up the matrix solution step and all 38 projects were run again. In all runs, total run times have been collected, which includes meshing, initialization, and frequency sweep. Results were then compared with reference performance data for each project, obtained when using only one core.

Based in the results presented in Figure 1 and Figure 2, where only one shared memory parallel algorithm is used at a time, one can conclude that tree level parallelization provides better run time performance, but tends to use a bit more memory than dense kernel parallelization scheme. Keep in mind however, that even with a modest memory increase, all 38 projects solved in-core on a Dell system with 16 GB of RAM.

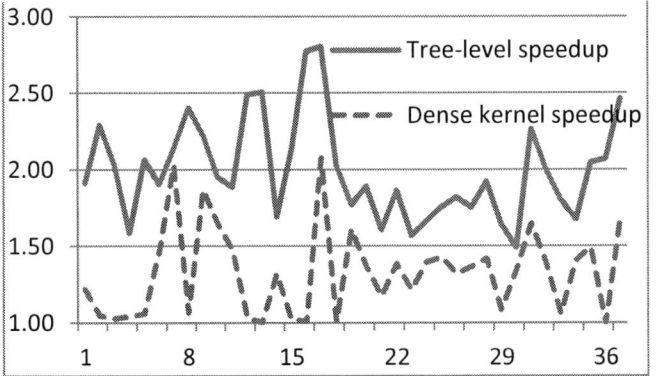

Figure 1: Speed up across 38 packages and PCBs when using tree level and dense kernel parallel implementations. Performance for single core run was used as a reference. Total solution time is used when calculating the speed up factors.

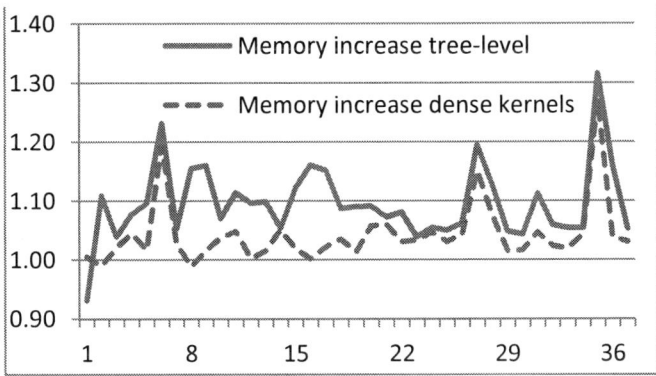

Figure 2: Memory increase across 38 projects when using both parallel schemes. Single core run was used as a reference.

The tree parallelism method for the matrix solver was also applied to a 3D full-wave FEM field solver [4],[5]. A 35% (19min vs. 14 min) speed up versus dense kernel level parallelism has been measured on 8 cores for the geometry shown in Figure 3. This is a six layer electronic package connected to a 12 layer PCB. The extracted geometry consists of two differential pairs that transition through multiple layers, with spherical solderballs mounting the package to the PCB.

Figure 3: Extracted geometry on eight cores that shows a 35% speed up in the matrix solver time. These results are for a 3D

full-wave FEM field solver comparing tree-level parallelism within matrix solver vs. dense kernels parallelism.

B. SDM Results for Interpolating Frequency Sweeps

The first set of results for SDM utilizes an interpolating sweep for the 3D full-wave FEM field solver [5]. The results were obtained using a Linux cluster that utilized the Open Grid Engine scheduler. The cluster consists of one head node, one post processing node and 16 execution nodes. Each execution node is a dual six-core Xeon X5760 blade with 96GB of shared memory. This equates to an HPC cluster that totals 192 cores and 1,536 GB of distributable memory.

SDM results for the geometry shown in Figure 3 show significant speed improvements due to the parallelization of frequency points across compute nodes. The project shown in Figure 3 contained 987,195 tetrahedral mesh elements and utilized ~ 35GB RAM/frequency point. A substantial reduction in solve time (~19x faster) was achieved due to the distributed memory solution of SDM. The project initially solved in approximately 3 days using eight cores of shared memory; combining shared and distributed memory schemes the solve time was reduced to 3 hours and 50 minutes.

Total Cores	Configuration	Runtime	Speed Up
8	1 nodes 8 cores per node	74hr 28min	Baseline
32	4 nodes 8 cores per node	14hr 48min	~5x
128	8 nodes 16 cores per node	3hr 50min	~19x

Table 1: For the geometry shown in Figure 3 a comparison in speed up for SDM is shown for a 3D full-wave FEM field solver. Frequency was swept from DC to 20GHz with a 50 MHz step size. Each frequency point used approximately 35 GB of RAM.

C. SDM Utilizing MPI with Discrete Frequency Points

An implementation of the SDM has been developed for the hybrid full-wave field solver utilizing LSF and Intel MPI [6]. This solution uses an embarrassingly parallel architecture for distribution of discrete frequency points as opposed to the adaptive interpolating frequency sweep discussed in IV. B. above. The hybrid solver implementation takes advantage of the fact that its memory footprint is considerably smaller than that of a 3D FEM solver. The number of parallel frequencies in each node/machine is set to the number of cores in the node/machine. In essence, we are trying to maximize the number of frequencies being solved in parallel as long as available memory allows it.

The results were obtained using a Windows cluster that utilized the Windows Server 2008 HPC Edition scheduler. The cluster consists of one head node, two post processing nodes and 16 execution nodes. Each execution node is a dual eight-core Xeon E5-2680 blade with 128GB of shared memory. This equates to an HPC cluster that totals 256 cores and 2,048 GB of distributable memory.

978-1-4799-0708-3/13 $31.00 © 2013 IEEE

The electromagnetic extraction consisted of 37 electrically connected nets (74 ports) within a 25 metal layer PCB and 500 frequency points. The final 2D mesh contained 390,000 triangles and the resultant matrix had 2.9 million unknowns. This required ~10GB RAM per frequency point. Please note that since each compute only had 128GB of RAM, only 10 or so points could be solved in parallel, with some of the points using more than 1 core to solve the matrix. A substantial reduction in solve time (~61x faster) was achieved due to the distributed memory solution of SDM. The results are shown in Table 2. The project initially solved in approximately 4 days using a single core; combining shared and distributed memory schemes the solve time was reduced to 1 hour and 31 minutes.

Total Cores	Configuration	Runtime	Speed Up
1	1 node 1 cores	92hr 39 min (~ 4 days)	Baseline
16	1 node 16 cores	16hr 18min	~5.7x
32	2 nodes 16 cores per node	5hr 28min	~17.0x
64	4 nodes 16 cores per node	2hr 50min	~32.7x
128	8 nodes 16 cores per node	1hr 31min	~61.1x

Table 2: Hybrid solver discrete SDM results for a 25 layer PCB. 500 discrete frequency points were solved from DC to 20 GHz.

Discrete SDM solution has also been extended to another specialized 3D full-wave solver Sentinel PSI [7], which utilizes prism element for the domain discretization. The test structure for the shared and distributed memory project is a 14 layer FBGA package with one power, one ground and 14 signal nets (29 ports).

The simulation frequency range of 1Hz to 10GHz has been partitioned based on the number of compute nodes. For example, a 3 node simulation partitions the frequency range into 3 bands. As shown in Table 3, the run time scales almost linearly with the number of nodes (or machines). These results show that in a shared memory environment, increasing the number of cores to 16 does not provide significant benefit compared to a distributed memory environment where two machines with 8 cores each solved the same project. This summarizes the advantages that SDM has for frequency point distribution compared to shared memory multi-threading.

Total Cores	Configuration	Runtime	Speed Up
16	1 node 16 cores	18hr 2min	Baseline
16	2 nodes 8 cores per node	7hr 32min	~2.3x
24	3 nodes 8 cores per node	5hr 27min	~3.3x

Table 3: Results from a 14 layer package using the specialized 3D FEM solver. The results show that greater speed up is obtained by distributed memory compared to shared memory solves.

V. CONCLUSION

This paper provides an overview of the different methods being utilized in three different commercial FEM field solvers to take advantage of HPC clusters. The paper distinguishes between shared memory (multi-threading) and distributed memory (SDM) processes to achieve peak performance.

During the matrix solution process the best parallel performance is achieved when both tree-level parallelization and dense kernel parallelization are used concurrently. A slight increase in memory usage is the tradeoff and is shown in Figure 1 and Figure 2 for SIwave. An improvement of 35% was shown for HFSS.

Finally we show the greatest performance benefits for FEM field solvers are obtained utilizing distributed memory when frequency sweeps are required (SDM). Table 1, Table 2, and Table 3 show significant performance improvements on compute clusters for HFSS, SIwave, and Sentinel PSI respectively.

Finally the greatest overall performance gains are achieved by hybridizing the memory allocation for different processes during the solve process. In other words, maximizing the distribution of frequency points on a HPC cluster while simultaneously using multi-threading for the matrix solution yields the overall best performance.

ACKNOWLEDGMENT

The authors would like to thank Eric Bracken for his invaluable comments and suggestions for improvements and implementation of this work.

REFERENCES

[1] J. Jin, The Finite Element Method in Electromagnetics, John Wiley & Sons, Inc, New York, 1993.

[2] J. E. Bracken, S. Polstyanko, S. Raman, Z. J. Cendes, "Efficient full-wave simulation of highspeed printed circuit boards and electronic packages", *Proc. 2004 IEEE Symposium on Antennas and Propagation*, vol. 3, pp. 3301-3304, June 2004.

[3] O. Schenk, K. Gartner, "Two-level dynamic scheduling in PARDISO: Improved scalability on shared memory multiprocessing systems," *Parallel Computing.*, vol. 28, pp. 187-197, 2002.

[4] S.G. Pytel, S.C. McMorrow, T. Dagostino, S. Polstyanko, W. Thiel, and R. Hall, "Successful Practices for the Modeling of Printed Circuit Boards and Substrates Using Electromagnetic Field Solvers", *43rd International Symposium on Microelectronics*, October 31 – November 4, 2010.

[5] HFSS version 15, ANSYS Inc., 225 West Station Square Dr., Pittsburgh, PA 15219, http://www.ansys.com.

[6] SIwave version 8, ANSYS Inc., 225 West Station Square Dr., Pittsburgh, PA 15219, http://www.ansys.com.

[7] Sentinel PSI version 13.1.1, Apache Design, Inc A Subsidiary of ANSYS, Inc., 2645 Zanker Rd., San Jose, CA 95134, http://www.apache-da.com/

[8] J.W.H. Liu, "The Multifrontal Method for Sparse Matrix Solution: Theory and Practice", *SIAM Review*, vol. 34, no. 1, pp. 82-109, March 1992.

A Generalized Modeling Method for Signal/Power Integrity Analysis of 3D Coupled Interconnects In Finite Cavity Based on 1D Technology

Xin Chang and Leung Tsang
Department of Electrical Engineering
University of Washington
Seattle, USA
changx@uw.edu, tsang1@uw.edu

Abstract— **In this paper, an innovative 1D technology is used to model complex vias structures in 3D IC and package system, which can include the layout of both eccentric single-ended and differential signaling types. The problem under investigation is an essential basic unit which can be applied to build and model complex multilayer structures. We first obtain the impedance matrix for finite cavity, which includes the reflection feartures of the cavity boundaries. Then the scattered field from a single via and T matrix in the presence of walls for a single via are derived. The Foldy-Lax multiple scatteing equations are furthermore listed based on the scattering mechanism. For the incident field, we calculate the exciting and scattering field coefficients on the vias in the arbitrarily shaped antipad based on 1D technology. Then the coupling among vertical interconnects are solved by applying reformulated Foldy-Lax multiple scattering equation. Finally, the scattering matrix of coupling among vias is calculated. Numerical results for the method are in good agreement with a commercial full wave numerical tool up to 50 GHz.**

Index Terms— **interconnects, Foldy-Lax equations, signal/ power integrity.**

I. INTRODUCTION

Signal/power integrity (SI/PI) effect is one of the crucial technologies in the future for the microelectronic industry. High performance computing systems for 3DI in which the electrical I/O bandwidth and density have been the bottleneck at different packaging levels (PCB, package, interposer, 3D chip stack, wafer and die, etc). Detailed trade-off analysis requires fast and accurate simulation tools of massively coupled vertical interconnects.

Different modeling strategies can be found in the literature [1-5]. However, there are some assumptions and limitations in the most of them. For example, the high order mode effects are usually ignored in the method of physics-based circuit model [1-3], which leads to an inaccuracy in high frequency range or when vias layout are complex, e.g., vias are placed closely to each other in differential signaling, vias are in irregular shaped antipad or the substrate are finite, etc. The RLGC equivalent circuit concept actually ceases to be effective above 1GHz in these complex layouts. Instead, rigorous electromagnetic analysis must be given. On the other hand, most of the current fast electromagnetic analysis methods for modeling vias are mainly developed based on the Foldy-Lax multiple scattering equations method [4,6-8], which is a mostly analytic method of

full wave solutions of multiple scattering among massive number of 3D vertical interconnects. In the initial work on the Foldy Lax approach [6], the vias layout are regular, which means the antipad is assumed to be circular and the via is concentric with the antipad.

In the recent work on the Foldy-Lax approach, for the case of multiple vias in antipad of arbitrary shape in infinite planar waveguide [7], we showed that the incident fields on the vias can be calculated by 1-dimensional line integrals of the surface charges on the vias and on the ground planes. Thus the 2-dimensional surface integration is reduced to 1-dimensional line integration which is much simpler. This also avoids the complication of surface integration that needs to be very accurate particularly for closely spaced vias. For the case of regular vias residing in circular cavity, T matrix for a via in the presence of the cavity walls has been derived and incorporated to Foldy-Lax equations, for capturing the scattering effects among vias and cavity boundaries [8].

In this paper, we study the approach of modeling vias with arbitrarily shaped antipad in finite cavity, including the cases of one via with one eccentric antipad or single-ended eccentric signaling layout, and two vias sharing one antipad or differential signaling layout, both in finite substrate filled cavity. The approach used in this paper is based on mostly analytic reformulated Foldy-Lax equations and 1D numerical technology. The 1D technology used in this paper means all integrals are 1-dimensional integrals, and all integral equations are also 1-dimensional. Notations in this paper follow [6-8].

II. T-MATRIX FOR VIA IN THE PRESENCE OF THE CAVITY WALL

Consider the problem of multiple vias residing in finite cavity as shown in figure 1. The boundary walls of the cavity will be treated as perfect magnetic conductors and will cause reflections. In the multiple scattering using Foldy-Lax equations, we consider the TM modes with $k_{zl} = \frac{l\pi}{d}$ and $k^2 - k_{zl}^2 = k_{\rho l}^2$, where $l = 0,1,2.....$, d is the separation between the two plates, and k is the wavenumber. Consider an arbitrary shape cavity. Since $kd \ll 1$, the modes with $l \neq 0$ are evanescent and will not propagate to the wall. The $l \neq 0$ modes have negligible reflection from the boundary walls. To incorporate the finite cavity boundary effect, we consider only the $l = 0$ mode. For TM_0 mode, the electric field only has the

978-1-4799-0708-3/13 $31.00 © 2013 IEEE

E_z component to represent the interactions. The "z" subscript will be suppressed. For the $l = 0$ mode, $k_{zl} = 0$ and $k_{\rho l} = 0$. Since there is no variation with z for the $l = 0$ mode, we use only the 2-dimensional ∇_t opertaor and the two dimensional position vector $\bar{\rho}$.

Conside an incident electrc field $E_{inc}(\bar{\rho})$ incident on the boundary wall. Let $E(\bar{\rho})$ be the electric field on the boundary. We use the t coordinate to describe the line contour of the boundary. In the MoM formulation, 1-dimensional discretiation of the boundary wall is used. Let there be N_t segments. We use pulse basis functions and point matching for the t_v coordinates with $v = 1,2...N_t$ and the length of segment v is Δt_v.

Using the discretized surface field, the scattered field from the wall $E_s^W(\bar{\rho})$ is

$$E_s^W(\bar{\rho}) = -\sum_{v=1}^{N_t}[\hat{n}' \cdot \nabla_t' g(\bar{\rho} - \bar{\rho}')\Delta t_v]_{\bar{\rho}' = \bar{\rho}(t_v)}(\bar{\bar{Z}}^{-1}\bar{E}_{inc})_v \quad (1)$$

where the impedance matrix elements are

$$Z_{\mu v} = \begin{cases} [\hat{n}' \cdot \nabla_t' g(\bar{\rho}, \bar{\rho}')\Delta t_v]_{\bar{\rho} = \bar{\rho}(t_\mu), \bar{\rho}' = \bar{\rho}(t_v)} & \mu \neq v \\ \frac{1}{2} & \mu = v \end{cases} \quad (2)$$

for $\mu, v = 1,2...N_t$, and $g(\bar{\rho}) = \frac{1}{4j}H_0^{(2)}(k\rho)$, $\bar{E}_{inc}$ and $\bar{E}$ are $N_t \times 1$ column vectors with $(\bar{E})_\mu$, $(\bar{E}_{inc})_\mu$, the values of the fields at the μ point. The impedance matrix $\bar{\bar{Z}}$ is of dimension $N_t \times N_t$. More details can be found in [9].

The difference of scattering from a via for finite cavity is the addtional reflection from the wall.

Consider the qth via centered at $\bar{\rho}_q$. In general, we include harmonics of $m = 0, \pm 1, ... \pm M$. Harmonics of $m \neq 0$ give rise to anisotropic effects. The outgoing cylindrical wave of the mth harmonic of the electric field is $E_m(\bar{\rho} - \bar{\rho}_q) = kH_m^{(2)}(k|\bar{\rho} - \bar{\rho}_q|)e^{-jm\phi_{\bar{\rho}\bar{\rho}_q}}$. The electric field is incident on the wall, and the scattered field from the wall is,

$$E_s^{Wqm}(\bar{\rho}) = -\sum_{v=1}^{N_t}[\hat{n}' \cdot \nabla_t' g(\bar{\rho}, \bar{\rho}')\Delta t_v]_{\bar{\rho}' = \bar{\rho}(t_v)}(\bar{\bar{Z}}^{-1}\bar{E}_m^q)_v \quad (3)$$

where $\bar{E}_m^q$ is of dimension $N_t \times 1$, with $(\bar{E}_m^q)_\mu = (E_m(\bar{\rho}(t_\mu) - \bar{\rho}_q))$.

Next we use vector addition theorem to expand $E_s(\bar{\rho})$ about via q. By using the addition theorem as we followed notations in refernce [7], let $[\bar{E}(\bar{\rho})]_n = E_n(\bar{\rho})$ and

$$[\bar{\bar{\alpha}}_{qp}^+]_{nm} = H_{n-m}^{(2)}(k|\bar{\rho}_p - \bar{\rho}_q|)e^{j(n-m)\phi_{\bar{\rho}_p\bar{\rho}_q}} \quad (4)$$

Define

$$[\bar{\bar{\gamma}}^{(qW)}]_{nv} = -\frac{1}{4jk}[\hat{n}' \cdot \nabla_t'[\bar{\bar{\alpha}}_{q\bar{\rho}'}^+]_{n0}\Delta t_v]_{\bar{\rho}' = \bar{\rho}(t_v)} \quad (5a)$$

$$[\bar{\bar{Q}}^{(Wq)}]_{\mu n'} = E_{n'}(\bar{\rho}(t_\mu) - \bar{\rho}_q) = kH_{n'}^{(2)}(k|\bar{\rho}(t_\mu) - \bar{\rho}_q|)e^{-jn'\phi_{\bar{\rho}(t_\mu)\bar{\rho}_q}} \quad (5b)$$

Then the wave from via q to via q through wall reflection for the mth harmonic incident on the wall is $E_s^{Wqqm}(\bar{\rho})$

$$E_s^{Wqqm}(\bar{\rho}) = [Rg\bar{E}(\bar{\rho} - \bar{\rho}_q)]^t \bar{\bar{\gamma}}^{(qW)}\bar{\bar{Z}}^{-1}\bar{E}_m^q \quad (6)$$

The scattered field from via q needs to include $E_s^{Wqqm}(\bar{\rho})$

$$E_s^{qqm} = E_m(\bar{\rho} - \bar{\rho}_q) + E_s^{Wqqm}(\bar{\rho}) \quad (7)$$

is the total scattered field from via q to q for the mth harmonic. After some derivations, the T-matrix of via q in the presence of the wall can be obtained as

$$\bar{\bar{\tau}}^{(q)} = [\bar{\bar{I}} - \bar{\bar{T}}^q \bar{\bar{\gamma}}^{(qW)}\bar{\bar{Z}}^{-1}\bar{\bar{Q}}^{(Wq)}]^{-1}\bar{\bar{T}}^q \quad (8)$$

where the T matrix, $\bar{\bar{T}}^q$, is a $(2M + 1) \times (2M + 1)$ diagonal matrix, with diagonal element equal to $T_m = -\frac{J_m(ka)}{H_m^{(2)}(ka)}$.

III. FOLDY-LAX MULTIPLE SCATTERING EQUATIONS

Consider multiple vias scattering problem. Let the scattered field from via p, $p \neq q$, is $E_s^p = \sum_m A_m^p E_s^{pm}$. Using the $\bar{\bar{\tau}}^{(q)}$, T matrix of via in the presence of the wall and the scattered field coefficients $\bar{A}^q = \bar{\bar{\tau}}^{(q)}\bar{w}^q$ of the scattered field that includes wall reflection, consider multiple vias, $q = 1,2,3,....N$ (Figure 1). Then the Foldy Lax equations can be written as

$$\bar{w}^q = \bar{a}^{q,inc} + \sum_{\substack{p=1 \\ p \neq q}}^{N}(\bar{\bar{\alpha}}_{qp}^+ + \bar{\bar{\gamma}}^{(qW)}\bar{\bar{Z}}^{-1}\bar{\bar{Q}}^{(Wp)})\bar{\bar{\tau}}^{(p)}\bar{w}^p \quad (9)$$

where $\bar{a}^{q,inc}$ is the incident filed coefficients for via q and $\bar{w}$ is the final exciting field coefficients.

The derivations about the Foldy-Lax equations are based on as follows. The exciting field on via q has three contributions: (i) incident field from source via, (ii) incident field from source via that is reflected from the wall, and (iii) field from other vias which is the sum of the direct and the reflection from the wall. More details are shown in [9].

IV. VIA CURRENTS

After the Foldy Lax equations are solved, the final exciting field coefficient $\bar{w}(q), q = 1,2...N$ are obained. The total field at via q is the sum of the excited field E_{ex}^q and the scattered field E_s^q

$$E_{ex}^q + E_s^q = \sum_m w_m^q RgE_m(\bar{\rho} - \bar{\rho}_q) + \sum_m A_m^q E_m(\bar{\rho} - \bar{\rho}_q) + \sum_m RgE_m(\bar{\rho} - \bar{\rho}_q)(\bar{\bar{X}}^{Wqq}\bar{A}^q)_m \quad (10)$$

where

$$\bar{\bar{X}}^{Wqp} = \bar{\bar{\gamma}}^{(qW)}\bar{\bar{Z}}^{-1}\bar{\bar{Q}}^{(Wp)} \quad (11)$$

Then the magnetic field in the $\hat{\phi}$ direction can be obtained. The surface currents J_s^q is in the z direction and is calculated from the magnetic field on the surface of the via

$$J_s^q = \frac{k^2}{j\omega\mu}\left(\sum_m w_m^q J_m'(ka)e^{-jm\phi_{\bar{\rho}\bar{\rho}_q}} + \sum_m A_m^q H_m^{(2)'}(ka)e^{-jm\phi_{\bar{\rho}\bar{\rho}_q}} + \sum_m J_m'(ka)e^{-jm\phi_{\bar{\rho}\bar{\rho}_q}}(\bar{\bar{X}}^{Wqq}\bar{A}^q)_m\right) \quad (12)$$

Current I_s^q is integration over the surface of the via, so that only the $m = 0$ in the above contributes

$$I_s^q = a\int_0^{2\pi} d\phi_{\bar{\rho}\bar{\rho}_q}J_s^q = \frac{-j2\pi ak}{\eta}\left\{\left(H_0^{(2)'}(ka)\bar{\bar{\tau}}^{(q)} + J_0'(ka)(\bar{\bar{I}} + \bar{\bar{X}}^{Wqq}\bar{\bar{\tau}}^{(q)})\right)\bar{w}^q\right\}_0 \quad (13)$$

where $\bar{\bar{I}}$ is $(2M + 1) \times (2M + 1)$ unit matrix and $\{\ \}_0$ represents the 0th harmonic of the column vector. Note that in the above, we calculate the current of the $l = 0$ mode. The $l \neq 0$ modes are also calculated but without considering the wall effects. The total current is the sum of currents of all l. Once the currents are determined, the calcuation of the scatteirng matrix proceed in manners as described in previous papers [6-8].

V. NUMERICAL RESULTS AND DISCUSSIONS

In this section we illustrate the results and compare the results with Ansoft's HFSS version 12.

All single-ended S-parameters provided here are referenced to 50 Ω, and all mixed-mode S-parameters are referenced to

100 Ω for differential mode. The configurations of PC used in the simulations are: Intel(R) Core(TM)2 Duo E7300 2.66GHz processor, 3GB RAM and Windows Vista 32-bit operating system. The average CPU times per frequency for simulation of single-ended S parameters by using Foldy-Lax method and HFSS are compared and shown in Table I. In the case specifications, h stands for waveguide thickness, and t stands for the plane thickness in the specifications of both cases.

A. 10 Vias Array for Each Via Going Through One Antipad Eccentrically (Single-Ended Signaling Case) in A Rectangular Cavity

Figure 1 shows top view of structure about 10 vias geometry with 10 signal vias, and the right one is the zoom in plot for the detail of via-antipad region. Each signal via goes through one antipad eccentrically. The specifications are: $\varepsilon_r = 4.0$ (silicon dioxide), $R_{via} = 5\ mil$, $R_{antipad} = 10\ mil$, eccentric pitch $e = 2\ mil$, via pitch $p_x = 150\ mil$, $p_y = 100\ mil$, $h = 50\ mil$, $t = 1.3\ mil$, plane width $W = 500\ mil$, plane length $L = 800\ mil$. The 10 signal vias locate at $(102,200)\ mil$, $(252,200)\ mil$, $(402,200)\ mil$, $(552,200)\ mil$, $(702,200)\ mil$, $(102,300)\ mil$, $(252,300)\ mil$, $(402,300)\ mil$, $(552,300)\ mil$, and $(702,300)\ mil$.

In figure 2 and 3, we compared both the insertion loss and return loss between bottom corner via and bottom center via, up to 20GHz. The simulation results obtained from the Foldy-Lax/1D technology agreed well with HFSS. From the figures, we can see that the resonances happened for the left bottom corner via are always earlier than the bottom center via. This can be explained by the physical layout of the vias. The resonances are due to the reflections of the multiple scattering among vias and the cavity walls. The bottom center via locates at a wider local region than the corner one, which leads to early resonances in frequency domain.

B. 2 Vias Sharing Same Antipad (Differential Signaling Case) in A Square Cavity

Figure 4 shows top view of structure about 2 vias sharing one antipad which is called differential vias. The specifications are: $\varepsilon_r = 4.4$, $tan\delta = 0.02$, (FR4_eproxy), $R_{via} = 15\ mil$, $R_{antipad} = 30\ mil$, pitch $= 40\ mil$, $h = 20\ mil$, $t = 1.3\ mil$. The cavity is square and the length is $L = W = 500\ mil$. Two vias are at $(200,300)\ mil$ and $(240,300)\ mil$ repectively.

Figure 5 and 6 show the insertion loss and return loss for both differential mode and common mode respectively, up to 50GHz, in order to observe the resonances. Good agreements are obtained between the results obtained from Foldy-Lax/1D technology and HFSS simulation. One can see that the differential mode usually has less loss than the common mode for the insertion loss, due to its high immunity to the noise and high tolerance to link path discontinuities.

VI. CONCLUSIONS

With the incorporation of the 1D technology into the reformulated Foldy-Lax multiple scattering equations method, both accuracy and computation speed are greatly improved for signal/power integrity analysis of high speed vertical interconnects modeling in 3D ICs and packaging system.

TABLE I. CPU RUN TIME PER FREQUENCY COMPARISONS FOR CASE A-B

	Foldy-Lax	HFSS v12
A	2.5 sec	46 sec
B	0.8 sec	3.2 sec

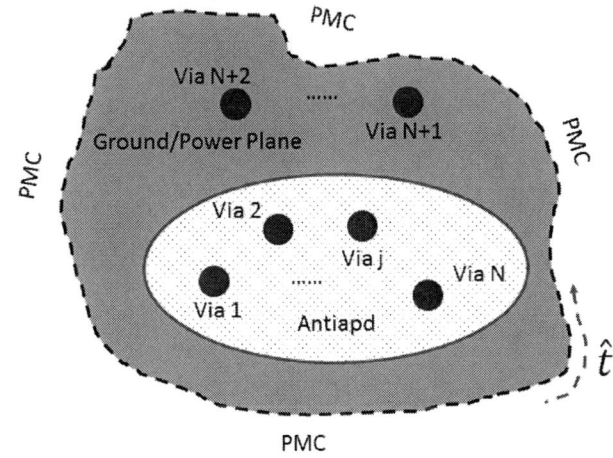

Fig. 1. Top view of N vias sharing the same antipad with arbitrarily shaped power/ground planes.

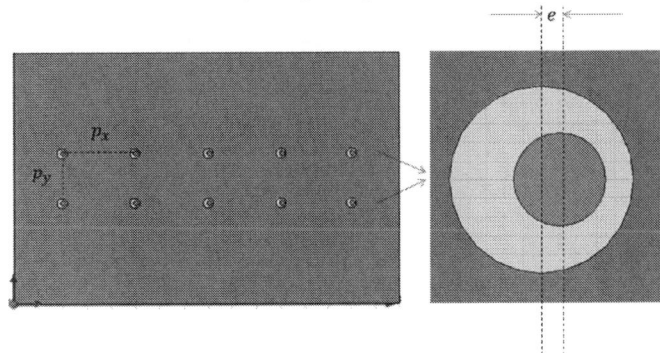

Fig. 2. (Left) Top view of 10 eccentric vias layout for case A. (right) Zoom in for one eccentric via in anitpad.

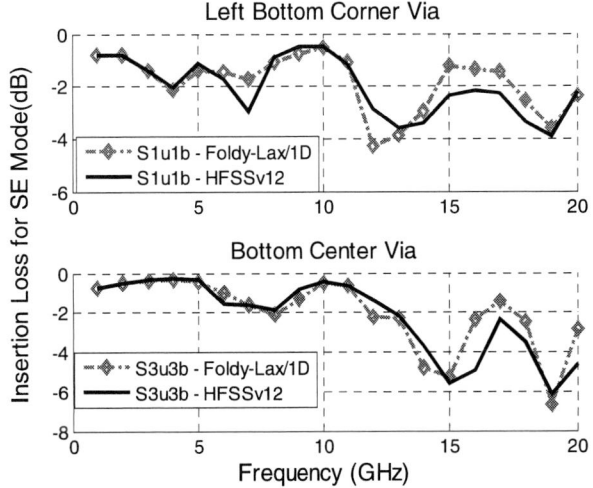

Fig. 3. Insertion loss comparisons of left corner bottom via and center bottom via for case A.

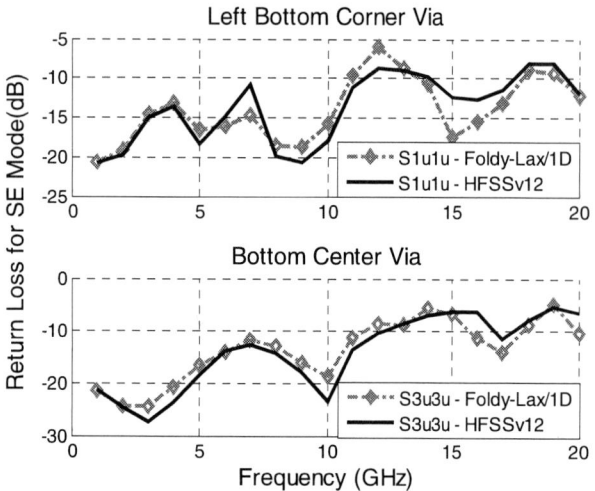

Fig. 4. Return loss comparisons of left corner bottom via and center bottom via for case A

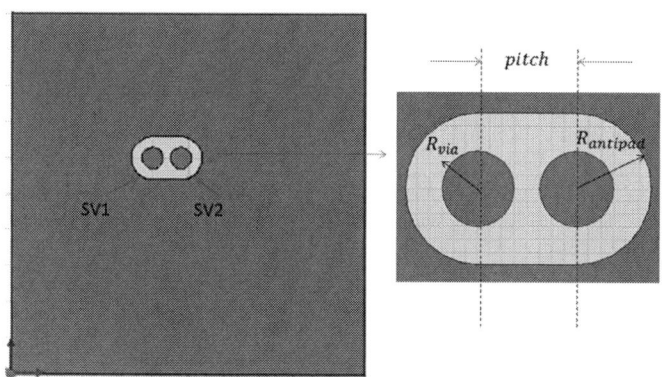

Fig. 5. Top view of 2 vias sharing one antipad in a square cavity for case B.

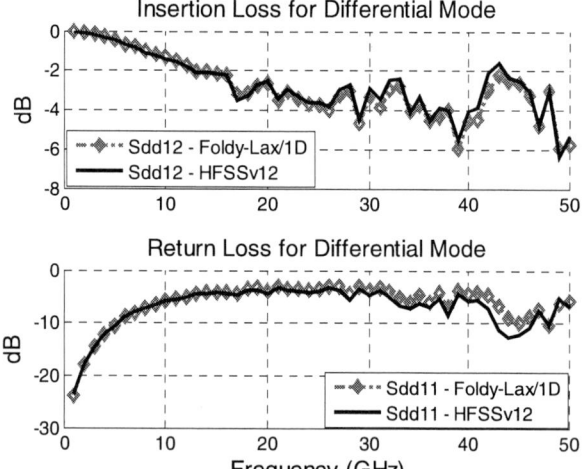

Fig. 6. Insertion/Return loss of differential mode for case B.

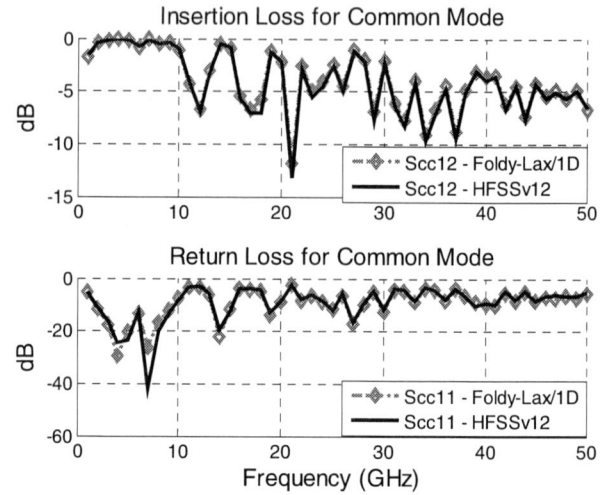

Fig. 7. Insertion/Return loss of common mode for case B.

REFERENCES

[1] X. Chang, B. Archambeault, M. Cocchini, F. De Paulis, V. Sivarajan, et al., "Return via connections for extending signal link path bandwidth of via transitions," *in Proc. Int. Symp. Electromagn. Compat. Eur.*, Hamburg, Germany, pp. 1-6, Sep. 2008.

[2] J. Kim, L. Ren, and J. Fan, "Physics-based inductance extraction for via arrays in parallel planes for power distribution network design," *IEEE Trans. Microw. Theory Tech.*, vol. 58, no. 9, pp. 2434–2447, Sep. 2010.

[3] S. Wu, X. Chang, C. Schuster, X. Gu, and J. Fan, "Eliminating Via-Plane Coupling Using Ground Vias for High-Speed Signal Transitions", 2008 *Electrical Performance of Electronic Packaging International Symposium*, Santa Rose, CA, USA, 2008.

[4] Y. Zhang, G. Feng, and J. Fan, "A novel impedance definition of a parallel-plate pair for an intrinsic via circuit model," *IEEE Trans. Microwave Theory Tech.*, vol. 58, no. 12, pp. 3780-3789, Aug. 2010.

[5] S. Muller, F. Happ, X. Duan, Rimolo-Donadio, R., H. Bruns, and C. Schuster, "Complete modeling of large via constellations in multilayer printed circuit boards," *IEEE Trans. Compon. Packag. Manuf.*, vol. 3, no.3, pp. 489-499, March. 2013.

[6] L. Tsang, H. Chen, C.-C. Huang, and V. Jandhyala, "Modeling of multiple scattering among vias in planar waveguides using Foldy-Lax equations," *Microwave Optical Technol. Lett.* vol. 31, pp. 201-208, Nov. 2001.

[7] L. Tsang and X. Chang, "Modeling of vias sharing the same antipad in planar waveguide with boundary integral equation and group T matrix method," *IEEE Trans. Comp.Packag. Manuf. Technol.*, vol. 3, pp. 315–327, Feb. 2013.

[8] L. Tsang and D. Miller, "Coupling of vias in electronic packaging and printed circuit board structures with finite ground plane", *IEEE Trans. Advanced Packaging*, vol. 26, pp. 375-384, Nov. 2003.

[9] X. Chang and L. Tsang, "Fast and broadband modeling method for multiple vias with irregular antipad in arbitrarily shaped power/ground planes in 3D IC and packaging based on generalized Foldy-Lax equations", *submitted to IEEE Trans. Compon. Packag. Manuf.*.

[10] X. Chang and L. Tsang, "A new efficient method for modeling dense via arrays with 1D discretization in 2D method of moment and group T matrix", *2012 Electrical Performance of Electronic Packaging International Symposium*, pp. 163-166, Tempe, AZ, USA, Oct. 2012.

[11] X. Chang and L. Tsang, "Modeling Multiple Scattering among Vertical Interconnects for SIW Structures and 3D ICs in Arbitrarily Shaped Waveguide", *2013 IEEE International Symposium on Antennas and Propagation and USNC-URSI National Radio Science Meeting*, Orlando, FL, USA, July, 2013.

New Simulation Procedure for Accurate Package Modeling Considering Chip-Package Interaction

E. Seler[1], M. Wojnowski[2], R. Weigel[1], A. Hagelauer[1]

[1]University of Erlangen-Nuremberg, Cauerstr. 9, 91058 Erlangen, Germany
[2]Infineon Technologies AG, Am Campeon 1-12, 85579 Neubiberg, Germany
E-mail: Ernst.Seler@Infineon.com

Abstract—**This paper presents a novel simulation procedure for full-wave electromagnetic simulations for generating package models taking into account chip-package interaction. If a simulated structure is split into two sub simulations, the accuracy of the overall simulation decreases. This is mainly the case if no clean separation of a composition into sub simulations is possible due to interactions between the sub domains. We show by means of a simple example of an embedded Wafer Level Ball grid array (eWLB) package that the simulation results of a cascade of an eWLB package and a chip are different to the simulation of the complete composition. The proposed simulation procedure delivers a package model with the package-chip interactions included and therefore cascading with integrated circuit model is possible with a higher accuracy. We show that the simulation results of a cascade of a package model obtained with the new procedure and a chip are virtually equal to a simulation of the whole composition. Furthermore we show that the created model of the package obtained from the proposed procedure is independent of the electrical circuit on the chip. This proves that the separation of the package from the chip with the proposed procedure is valid.**

Keywords — 3D EM simulation; package modeling; simulation accuracy; eWLB;

I. INTRODUCTION

During the development process of new radio-frequency (RF) products and applications 3D electromagnetic (EM) field simulations are used to predict and analyze the electrical performance at an early point of time regarding the overall development process. Using 3D EM field simulations during the development process is the fastest way to achieve information about the design and the achievable performance. This leads also to cost effectiveness. Therefore, much research is ongoing to improve the accuracy of 3D EM simulations [1].

To simulate the overall behavior of a complex electrical system, these 3D EM field simulation results have to be combined with simulation results of other simulators. The cascading of sub simulations is performed on a SPICE level. For instance, a system can contain an integrated circuit (IC), a package for the IC, a board and an antenna. Than this system is divided into four sub systems (Fig.1). Due to the increase of integration density the definition of the interface between two sub systems becomes more crucial. In an embedded Wafer Level Ball grid array (eWLB) package for instance it is not possible to define a clear interface between the chip and the package because there exists no clear transverse electromagnetic (TEM) wave at the physical chip-package interface. To achieve a better accuracy of the simulation this issue has to be considered. High accuracy is important especially for RF chip-package-board transitions and in the field of integrated passive component technology [2]. To enable innovation in packaging such as complex 3D System-in-Package (SiP) products and comply with the need of decreasing the interconnection gap between semiconductor and packaging [3][4][5] this increase in accuracy is required. Therefore we present a new de-embedding method to increase the accuracy of models generated with 3D EM field simulators in a decisive way. This EM models for packages can be used for circuit simulations.

II. PROBLEM FORMULATION

A. eWLB Technology

The eWLB technology is an innovative package concept based on an embedded device technology with fan-out redistribution. The fan-out area extends from the chip edge to the package edge, whereas the fan-in area is limited by the chip size. In the additional fan-out area around the chip the redistribution of the signals is realized with thin-film redistribution layer (RDL). The use of the fan-out area allows increased input/output density. The eWLB technology has demonstrated outstanding electrical capabilities for RF and mm-wave applications [5][6][7].

B. Example of cascading Chip with Package (eWLB)

The presented example illustrates the problem of cascading different simulation solutions with respect to accuracy. As

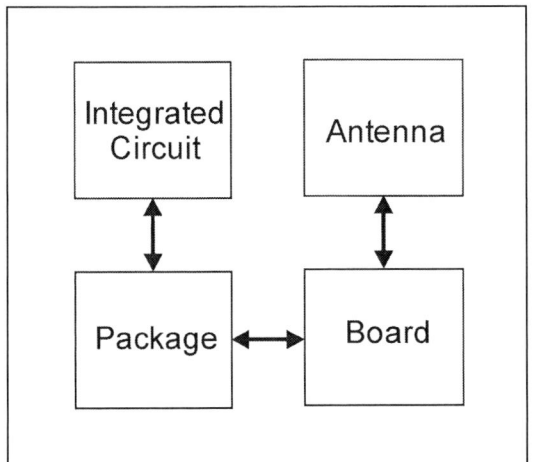

Fig. 1. Example of an electrical system divided into sub simulations.

978-1-4799-0708-3/13 $31.00 © 2013 IEEE

already mentioned above, cascading of simulation models is a problem if a model boundary at a position is required where no TEM wave exists.

To simulate the electrical performance of the complete chip with the complete package, a 3D EM field simulation of the package has to be combined with a circuit simulation of the chip. To explain the problem of cascading we restrict the problem to the chip-to-package transition. Fig. 2 shows the transition from an eWLB package to a chip [6][7]. The conventional approach is to divide the composition into two sub simulations. One simulation covers the electrical behavior of the integrated circuit on the chip, the other one the electrical influence of the package described with S-parameters. In our example we represent this procedure by making two separate simulations. One simulation represents the characteristics of a short interconnection in the RDL to the vias located at the chip pads. The other simulation represents the characteristics of a short interconnection on the chip to the chip pads. Fig. 3 and Fig. 4 show both simulation models. To achieve accurate results, one has to subtract the inductances of the lumped ports. These two separate simulations have to be compared to one simulation of the complete structure (Fig. 2). Then a statement regarding the accuracy of the cascading is possible.

Fig. 5 and Fig. 6 show the simulation results of the shown example. The insertion loss is shown in Fig. 5. The dashed curve is the insertion loss of the complete structure. The

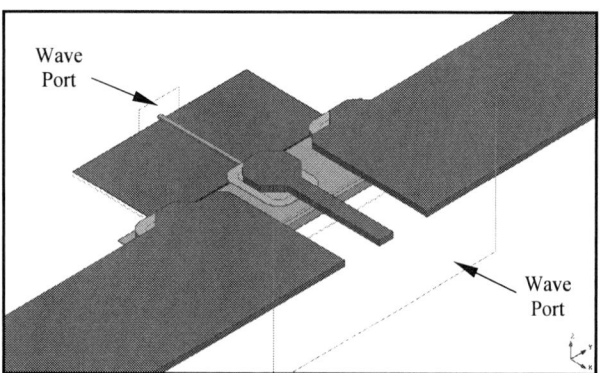

Fig. 2. Simulation model of the complete structure (part of chip and part of package). We use wave ports on both sides of the structure.

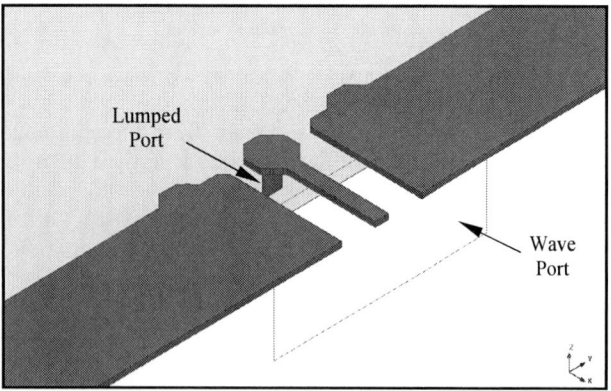

Fig. 3. Simulation model of the package part. We use a lumped port on the chip side and a wave port on the package side. The parasitic inductance of the lumped port is subtracted from the simulation.

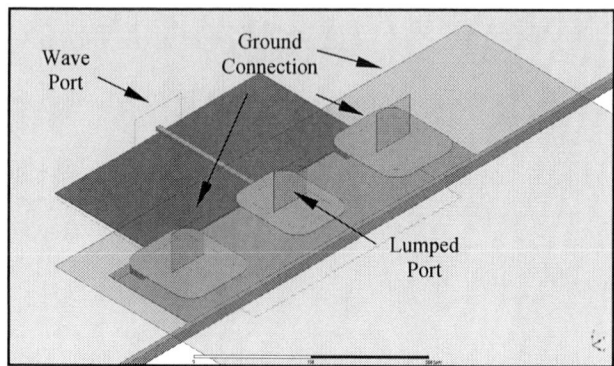

Fig. 4. Simulation model of the chip part. We use a wave port on the chip side and a lumped port on the package side. The parasitic inductance of the lumped port is subtracted from the simulation.

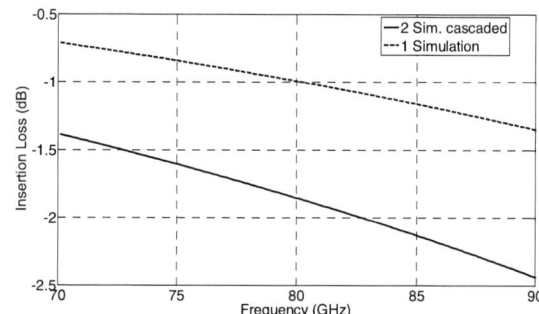

Fig. 5. Insertion loss of the complete structure and of the cascading.

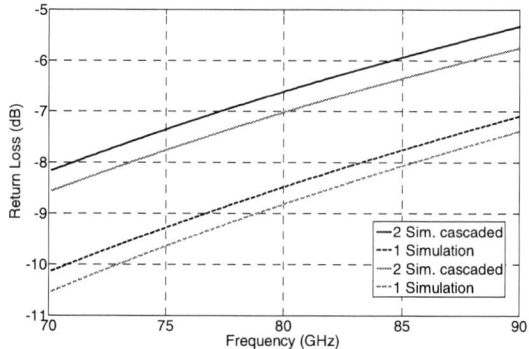

Fig. 6. Return loss of the complete structure and of the cascading.

continuous curve shows the insertion loss of the cascaded simulation. The difference of both approaches is about 0.8 dB. Fig. 6 shows the corresponding return loss. The difference between both approaches is about 1.8 dB.

C. Interpretation

The reason for the difference of both approaches is the missing capacitance between the RDL and the chip in the separate simulation models. In Fig. 7 the area of the chip-package interaction is visualized. The capacitance C_1 is the chip-pad capacitance and has no influence on the package characteristics. The capacitances C_2, C_3 and C_4 do influence the package characteristics. The capacitance C_2 represents the capacitance between the RDL and the chip pad, C_3 represents the capacitance between the RDL and some copper parts in the

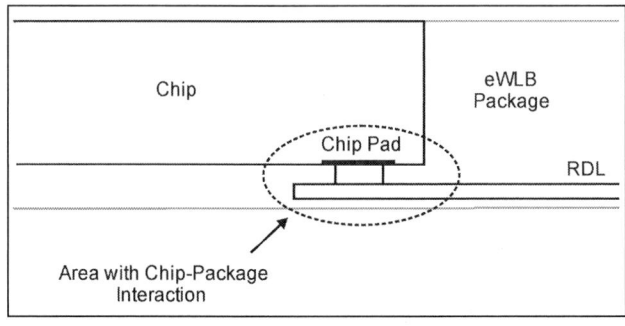

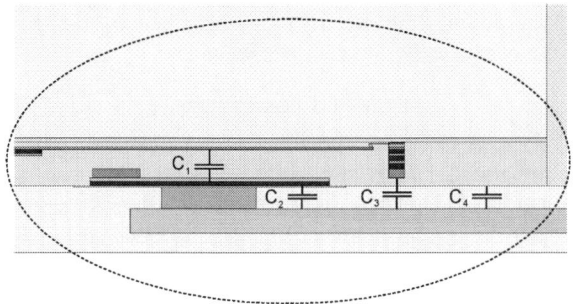

Fig. 7. Schematic visualization of the area with the chip package interaction and the corresponding capacitances.

chip and C_4 illustrates the capacitance between the RDL and silicon substrate. Therefore the goal is to provide simulation results of the package model with the capacitive influence of the chip included but without the electrical characteristics of the chip. The EM wave travels from the RDL through the vias and the chip pad into the chip. The marked area in Fig. 7 is the area where the interfaces of the integrated circuit simulations and the package simulations have to be defined.

To define accurate boundaries and ports in simulation models, it is required to have a good knowledge of the EM field in this region. This knowledge is not available if a package is modeled and no chip is included. But otherwise if a chip is included, the port definition in this region is not possible without modifying the simulation model. This problem can be resolved by using the following proposed simulation procedure.

III. NEW SIMULATION PROCEDURE

In this section we present the new simulation procedure. The method can be outlined as shown in Fig. 8.

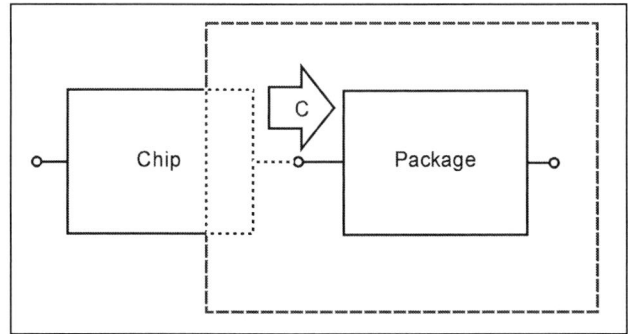

Fig. 8. Schematic visualization of the proposed simulation procedure.

As described above, the electrical simulation of a complete system needs to be divided into sub simulations. If one sub simulation is influenced electrically by its neighboring sub simulation, this influencing part has to be included (Fig. 8, dashed box). Therefore the relevant part of the chip (Fig. 8, dotted box) has to be included in the 3D EM simulation. In the second step, the electrical characteristics of the part that does not belong to the package need to be simulated alone (Fig. 8, dotted box). Finally the electrical characteristics of the chip (dotted box) need to be removed from the first described simulation (dashed box). As consequence the simulation results contain only the electrical performance of the package but including the influence of the chip (Fig. 8, indicated with an arrow). This influence is mainly a capacitance, therefore indicated as C in Fig. 8. In the next section this method is illustrated and verified on the basis of the previous eWLB example.

IV. METHOD VERIFICATION

The simulation procedure can be used to obtain a more accurate model of a chip package like eWLB. To verify the simulation procedure, two different chip models have to be used. In Fig. 9 the second chip model is shown. The chip model in Fig. 4 is also used. The verification follows the formula:

$$\mathbf{DP} = \mathbf{C} \qquad (1)$$

$\mathbf{D}$ represents the simulated ABCD transmission matrix [8] of the chip. $\mathbf{P}$ represents the transmission matrix of the package. Therefore $\mathbf{C}$ represents the transmission matrix of the chip and the package. To get the electrical characteristics of the package, the characteristics of the chip have to be de-embedded from $\mathbf{C}$:

$$\mathbf{D}_1^{-1}\mathbf{C}_1 = \mathbf{P}_1 \qquad (2)$$

$$\mathbf{D}_2^{-1}\mathbf{C}_2 = \mathbf{P}_2 \qquad (3)$$

$\mathbf{C}_1$ represents the transmission matrix of the package with a part of chip 1 (Fig. 4) included. $\mathbf{C}_2$ represents transmission matrix of the package and the chip 2 (Fig. 9) included. $\mathbf{D}_1$ and $\mathbf{D}_2$ represent the transmission matrix of chip 1 and chip 2. $\mathbf{P}_1$ is the transmission matrix of the package obtained from (2). $\mathbf{P}_2$ is the transmission matrix of the package obtained from (3). If $\mathbf{P}_1$ equals $\mathbf{P}_2$ we can state that the model of the package $\mathbf{P}$ is not influenced by the chip used for de-embedding:

$$\mathbf{P} = \mathbf{P}_1 = \mathbf{P}_2 \qquad (4)$$

Thus, the use of two different chips shows that the resulting package characteristics are universal and not depending on the electrical circuit on the chip.

The simulations of the chip need to be done with high accuracy. Therefore the port definition has to be considered in a precise way. In the proposed simulation procedure a wave port is used from chip side. Because the electrical characteristics of the circuit do not influence the obtained package model, it is useful to choose a location of the port where a proper TEM wave exists. The port definition at the chip model from package side at the chip pads is even more important. A decrease in accuracy of the simulation results in this area of the model would decrease the accuracy of the

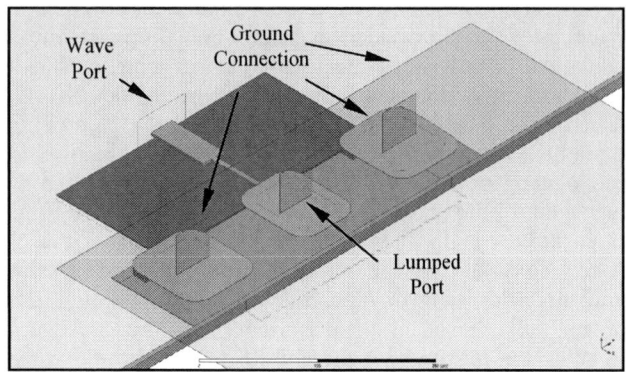

Fig. 9. Simulation model of the second chip part for the verification of the procedure.

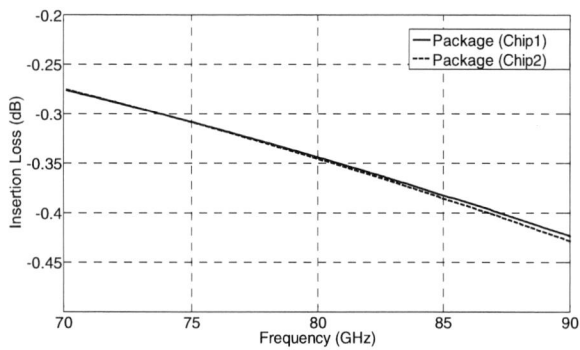

Fig. 10. Insertion loss of the package model obtained from the simulation with chip 1 and chip 2.

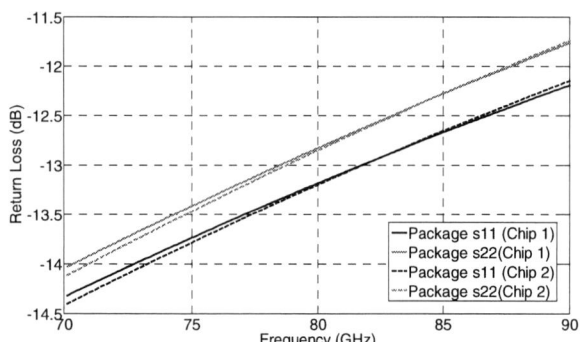

Fig. 11 Return loss of the package model obtained from the simualation with chip 1 and chip 2.

obtained package model. This is the area where the increase of accuracy by the proposed simulation procedure takes place. Compared to the wave port defined in the chip, the port at the chip pads is not included in the overall simulation. Thus a small error in this area of the model will not be de-embedded like it is the case at the wave port defined in the chip. In the shown example the used lumped port has an inductance of about 30 pH. Before de-embedding the chip part from the overall simulation, this inductance has to be subtracted. This procedure has also been verified.

Fig. 10 shows the insertion loss of the package models obtained by the proposed procedure. The continuous curve

shows the insertion loss of the package obtained from the model with chip 1. The dashed curve shows the insertion loss of the package obtained from the model with chip 2. The difference between both curves is virtually zero. Fig. 11 shows the corresponding return loss. The continuous curves are the return losses from chip 1, the dashed curves are the return losses from chip 2. The black curves are the return losses from chip side, the grey curves are the return losses from package side. Again the differences of the related curves are virtually zero. This proves that the obtained package models are not depending on the used chip model. This proves that the obtained package models are containing a general valid model which represents the electrical performance of the package with the package chip interactions included.

V. CONCLUSION

We showed that the cascading of 3D EM package models with integrated circuit simulations has less accuracy if the electrical chip-package interaction at the chip-package transition is not considered. Especially for system-in-package applications and the integrated passive component technologies high simulation accuracy is required. We proposed a new simulation procedure to include the chip-package interaction in the package model. To verify the procedure we showed that the obtained package model is not dependent on the used chip. Therefore the obtained model contains a general valid package model with the chip package interactions included.

ACKNOWLEDGMENT

This work has been supported by the German Bundes-ministerium für Bildung und Forschung (BMBF) under contract 01M3191A. A part of the work was also been performed in the European CATRENE project 3DIM3.

[1] A.C. Cangellaris, "Electrical modeling and simulation challenges in chip-package codesign," IEEE Micro, vol.18, pp.50-59, Jul., 1998

[2] Richard K. Ulrich, Leonard W. Schaper, "Integrated Passive Component Technology," First Edition, ISBN 0-471-24431-7, John Wiley & Sons, Inc, New York, 2003

[3] S. Wane and D. Bajon, "Partition-Recomposition Methodology for Accurate Electromagnetic Analysis of SIP Passive Circuitry," EUROCON 2007, pp. 15-23, Sep. 2007

[4] "The International Technology Roadmap for Semiconductors (ITRS), 2012 Update Overview," Available on-line: https://www.itrs.net.

[5] K. Pressel, G. Beer, T. Meyer, M. Wojnowski, M. Fink, G. Ofner, B. Römer, "Embedded Wafer Level Ball Grid Array (eWLB) Technology for System Integration," in Proc. International Symposium on Components, Packaging, and Manufacturing Technology (ICCSJ 2010), Tokyo, Aug. 2010

[6] M. Brunnbauer, T. Meyer, G. Ofner, K. Müller, R. Hagen, "Empedded Wafer Level Ball Grid Array (eWLB)," in Proc. 33rd Electronic Manufacturing Technology Symposium (IEMT 2008), Penang, Nov. 2008

[7] M. Wojnowski, R. Lachner, J. Böck, C. Wagner, F. Starzer, G. Sommer, K. Pressel, R. Weigel, "Embedded Wafer Level Ball Grid Array (eWLB) Technology for Millimeter-Wave Applications," in Proc. 13th Electronic Packaging Technology Conference (EPTC 2011), Singapore, Dec. 2011

[8] David M. Pozar, University of Massachusetts at Amherst, "Microwave Engineering," Second Edition, ISBN 0-471-17096-8, John Wiley & Sons, Inc, Crawfordsville, 1998

Optimum Implementation of a Locally Implicit Leapfrog Scheme for Fast Simulation of Inhomogeneously-Meshed Plane Structures

Tadatoshi Sekine
Dept. of Mechanical Eng.,
Shizuoka University
3–5–1 Johoku, Naka-ku, Hamamatsu-shi, 432–8561 Japan
Telephone: +81–53–478–1237
Email: sekine@tzasai7.sys.eng.shizuoka.ac.jp

Hideki Asai
Nanovision Research Division,
Research Institute of Electronics, Shizuoka University
3–5–1 Johoku, Naka-ku, Hamamatsu-shi, 432–8561 Japan
Telephone: +81–53–478–1237
Email: hideasai@rie.shizuoka.ac.jp

Abstract—**This paper describes an optimum implementation technique of a locally implicit leapfrog scheme for fast simulation of inhomogeneously-meshed conductor planes. A time step size is determined to optimize the computational efficiency of the scheme with consideration for the numerical stability, accuracy, and parallelization performance. The proposed technique is applied to transient simulations of two types of power/ground planes to evaluate the adequacy. Numerical results show that our implementation can improve the efficiency of the locally implicit scheme and enables it to be much faster than a conventional SPICE-like simulator and an existing fully-explicit scheme.**

I. Introduction

For power integrity in high-speed design, verification of power/ground planes has become a major problem because voltage fluctuations caused by power-supply noises directly affect the performance of an electronic circuit. Therefore, efficient modeling and simulation techniques for plane structures are important in the early stage of the design.

Triangular meshes are commonly used to model arbitrary shapes of conductor planes, and a circuit model obtained from the triangular meshes generally becomes a large-scale RLGC network [1]. Recently, a fully explicit leapfrog scheme gets much attention to analyze such a large network because it is matrix-free and cost-effective compared with conventional SPICE-like simulators [2], [3]. However, the explicit method has a strict numerical stability condition, by which a time step size is forced to be small if there exist small inductance and capacitance in the circuit. Since those small reactances are extracted from small meshes, the explicit method is not suitable for simulation of fine structure.

A locally implicit leapfrog scheme can overcome this drawback by combining the leapfrog scheme with the numerically-stable implicit method [4]. By applying the implicit method to local regions that require small meshes, a larger time step size than that of the explicit method can be used. However, although the method can reduce the number of time points, the total simulation cost may increase because the calculation cost per time step related to the implicit region is relatively large. In this work, we propose an optimum implementation technique

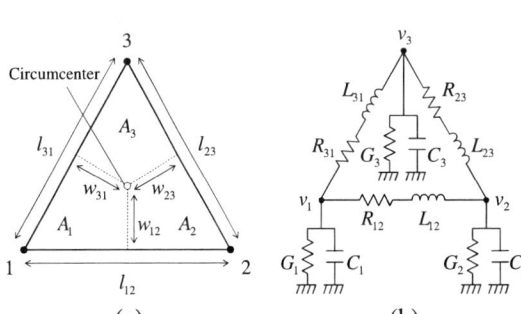

Fig. 1. The Delaunay-Voronoi modeling. (a) A Delaunay triangle. The dotted lines denote the Voronoi edges. (b) The corresponding equivalent circuit.

to minimize the total computational cost of the locally implicit scheme in simulations of conductor planes segmentalized by the inhomogeneous triangular meshes.

II. Related Modeling and Analysis Techniques

A. Circuit Modeling Using the Triangular Mesh

Parallel conductor planes and a dielectric between them constitute typical power/ground planes of a printed circuit board. For the equivalent circuit modeling of the plane structure, the Delaunay triangulation, which is dual to the Voronoi diagram, is commonly used to generate the inhomogeneous triangular meshes [1]. Fig. 1 illustrates the triangular mesh and the corresponding equivalent circuit, of which the element values per triangle are calculated as

$$R_{ab} = \frac{1}{\sigma} \frac{l_{ab}}{t w_{ab}}, \ L_{ab} = \mu \frac{h l_{ab}}{w_{ab}}, \ G_a = \frac{1}{\rho} \frac{A_a}{h}, \ C_a = \varepsilon \frac{A_a}{h} \quad (1)$$

where $a, b = 1, 2, 3$ are the vertex numbers of the triangle, l_{ab} is the length of the edge ab, w_{ab} is the shortest distance between the circumcenter of the triangle and the edge ab, t is the thickness of the conductor, h is the height of the dielectric, A_a is the intersection area of the triangle and the Voronoi cell around a, σ is the conductivity of the planes, and μ, ε, and ρ are the permeability, permittivity, and resistivity of the

978-1-4799-0708-3/13 $31.00 © 2013 IEEE

dielectric. In Fig. 1, an edge of the triangle is modeled as the RL branch, and a vertex corresponds to the node that has the GC path to the ground in the circuit.

B. Locally Implicit Leapfrog Scheme

The time step size Δt used in the explicit leapfrog scheme is limited by the numerical stability condition

$$\Delta t \leq \sqrt{2} \min_{a=1}^{N^{\mathrm{n}}} \left(\sqrt{\frac{C_a}{N_a^{\mathrm{b}}} \min_{m=1}^{N_a^{\mathrm{b}}} (L_{am})} \right) \qquad (2)$$

where N^{n} is the total number of the nodes, N_a^{b} is the number of the branches connected to the node a [3]. Clearly, small reactance elements force the time step size to be small. Because a small triangle leads to small reactance elements as seen in (1), it is not preferable to apply the explicit method to regions that require small meshes.

In the locally implicit leapfrog scheme, a numerically-stable implicit difference method is applied to such regions to alleviate the time step size limitation. First, the time step size to be used is selected. Next, the circuit to be analyzed is separated into low reactance parts (LRPs) and high reactance parts (HRPs). LRP is defined as a subcircuit in which (2) is satisfied, and HRP is a subcircuit other than LRPs. Then, node voltages and branch currents in HRPs are updated by using the following explicit updating formulas

$$v_a^{n+\frac{1}{2}} = \frac{C_a}{C_a + \Delta t G_a} v_a^{n-\frac{1}{2}} - \frac{\Delta t}{C_a + \Delta t G_a} \sum_{m=1}^{N_a^{\mathrm{b}}} i_{am}^n \qquad (3)$$

$$i_{ab}^{n+1} = \frac{L_{ab}}{L_{ab} + \Delta t R_{ab}} i_{ab}^n + \frac{\Delta t}{L_{ab} + \Delta t R_{ab}} \left(v_a^{n+\frac{1}{2}} - v_b^{n+\frac{1}{2}} \right) \quad (4)$$

where n is the index of the time step. On the other hand, the node voltages in LRP are calculated by using the implicit updating formula in a matrix-vector form

$$\left(\frac{1}{\Delta} \mathbf{Y}_{\mathrm{L}} + \Delta t \hat{\mathbf{Z}}_{\mathrm{L}} \right) \mathbf{v}_{\mathrm{L}}^{n+\frac{1}{2}} = \frac{1}{\Delta t} \mathbf{C}_{\mathrm{L}} \mathbf{v}_{\mathrm{L}}^{n-\frac{1}{2}} + \mathbf{b}_{\mathrm{L}}^{n-\frac{1}{2}} \qquad (5)$$

which is derived by assembling all equations associated with the nodes inside LRP in the form of

$$\left\{ \frac{Y_a}{\Delta t} + \sum_{p=1}^{N_{a,\mathrm{L}}^{\mathrm{b}}} \frac{\Delta t}{Z_{ap}} \right\} v_a^{n+\frac{1}{2}} - \sum_{p=1}^{N_{a,\mathrm{L}}^{\mathrm{b}}} \left(\frac{\Delta t}{Z_{ap}} v_p^{n+\frac{1}{2}} \right)$$

$$= \frac{C_a}{\Delta t} v_a^{n-\frac{1}{2}} - \left\{ \sum_{p=1}^{N_{a,\mathrm{L}}^{\mathrm{b}}} \left(\frac{L_{ap}}{Z_{ap}} i_{ap}^{n-\frac{1}{2}} \right) + \sum_{q=1}^{N_{a,\mathrm{H}}^{\mathrm{b}}} i_{aq}^n \right\} \quad (6)$$

where

$$Y_a = C_a + \Delta t G_a, \quad Z_{ap} = L_{ap} + \Delta t R_{ap} \qquad (7)$$

$N_{a,\mathrm{L}}^{\mathrm{b}}$ and $N_{a,\mathrm{H}}^{\mathrm{b}}$ are the numbers of the branches between the node a and other nodes inside LRP and HRP, respectively, and $N_a^{\mathrm{b}} = N_{a,\mathrm{L}}^{\mathrm{b}} + N_{a,\mathrm{H}}^{\mathrm{b}}$. Note that $N_{a,\mathrm{H}}^{\mathrm{b}} = 0$ if the node is completely inside LRP and not on the boundary of LRP. In (5), $\mathbf{Y}_{\mathrm{L}}$ is the diagonal matrix of which each element is Y_a, $\hat{\mathbf{Z}}_{\mathrm{L}}$

TABLE I
COMPUTATIONAL COST PER TIME STEP

Updating process		Eq. #	# of eqs.	# of ops. per variable	
Explicit	Voltage	(3)	$N_{\mathrm{H}}^{\mathrm{n}}$	M/D	2
				A/S	N_a^{b}
	Current	(4)	$N_{\mathrm{H}}^{\mathrm{b}}$	M/D	2
				A/S	2
Implicit	Voltage	(5)	$N_{\mathrm{L}}^{\mathrm{n}}$	M/D	$(N_{\mathrm{r}} + 1)(N_{a,\mathrm{L}}^{\mathrm{b}} + 1)$
				A/S	$(N_{\mathrm{r}} + 1)N_{a,\mathrm{L}}^{\mathrm{b}} + N_{a,\mathrm{H}}^{\mathrm{b}}$
	Current	(8)	$N_{\mathrm{L}}^{\mathrm{b}}$	M/D	2
				A/S	2

is the symmetric matrix constructed by stamping $1/Z_{ap}$ to the associated positions, $\mathbf{C}_{\mathrm{L}}$ is the diagonal matrix of which each element is C_a, $\mathbf{v}_{\mathrm{L}}$ is the unknown voltage vector containing the node voltages inside LRP, and $\mathbf{b}_{\mathrm{L}}$ is the known vector calculated from the past variables. Additionally, the implicit updating formula of a branch current in LRP is

$$i_{ap}^{n+\frac{1}{2}} = \frac{L_{ap}}{L_{ap} + \Delta t R_{ap}} i_{ap}^{n-\frac{1}{2}} + \frac{\Delta t}{L_{ap} + \Delta t R_{ap}} \tilde{v}_{ap}^{n+\frac{1}{2}}. \quad (8)$$

Equation (8) can be calculated without matrix operations by substituting the node voltages obtained from (5). Since there may be more than one LRP in the circuit, we derive and solve the associated implicit updating formula (5) for each of LRPs individually.

III. OPTIMIZING THE LOCALLY IMPLICIT SCHEME

Although the locally implicit scheme can use a relatively-large time step size and reduce the total number of the time steps, the computational cost to solve (5) is larger than those of the other updating formulas. Therefore, if the regions of LRPs is too large, the total amount of the simulation cost increases even if the large time step size is used. For this reason, we propose the methodology to determine the optimum time step size to maximize the efficiency of the locally implicit scheme.

In the proposed technique, we adopt the Gauss-Jacobi (GJ) method to solve the updating formula (5) in contrast to the original method proposed in [4], which uses the LU decomposition method. The calculation cost of the GJ method is generally less than that of the LU decomposition method, and it induces no fill-in. In addition, the GJ method is easy to estimate its computational cost and suitable for parallelization. Moreover, as reported in [5], a few iterations are sufficient to converge numerical solutions in the simulation of an RLGC circuit such as the equivalent circuit described in Section II-A.

A. Estimation of the computational cost

The computational cost is estimated by using the numbers of multiplications/divisions (M/D) and additions/subtractions (A/S) in a transient analysis. The numbers of M/D and A/S operations are determined by counting the numbers of the arithmetic operations required to update the variables in the locally implicit scheme. Those numbers per time step are summarized in Table I. In Table I, $N_{\mathrm{H}}^{\mathrm{n}}$ and $N_{\mathrm{H}}^{\mathrm{b}}$ are the total numbers of the nodes and branches in HRPs, $N_{\mathrm{L}}^{\mathrm{n}}$ and $N_{\mathrm{L}}^{\mathrm{b}}$ are

978-1-4799-0708-3/13 $31.00 © 2013 IEEE

those of the nodes and branches in LRPs, and N_r is the number of the iterations of the GJ method. In this case, we use a single fixed number for N_r in all LRPs. Then, the computational cost $\tilde{F}$ of the whole circuit per time step is calculated by

$$\tilde{F} = \sum_{a=1}^{N_H^n} \left(N_a^b + 2\right) + 4N_H^b + \sum_{m=1}^{N_L} N_{L,m} \quad (9)$$

where

$$N_{L,m} = \sum_{a=1}^{N_{L,m}^n} \left\{ (N_r+1)(2N_{a,L}^b + 1) + N_{a,H}^b \right\} + 4N_{L,m}^b \quad (10)$$

N_L is the number of LRPs, and $N_{L,m}^n$ is the number of the nodes in the m-th LRP. Finally, the total computational cost F for the transient analysis becomes $F = \lceil T/\Delta t \rceil \tilde{F}$, where T is the finish time of the transient analysis.

B. Optimization procedure

We find the optimum time step size Δt_{opt} which minimizes F subject to $\Delta t_{min} \leq \Delta t_{opt} \leq \Delta t_{max}$, where Δt_{min} is equal to the right-hand side of (2), $\Delta t_{max} = 1/(20 f_{max})$, and f_{max} is the maximum frequency of interest. In order to solve this optimization problem in terms of the cost function F, the simulated annealing (SA) is used [6]. The optimization process is as follows: First, the initial computational cost F is calculated by using the initial time step size $\Delta t = \Delta t_{max}$, which is also the best solution at this time. Then, Δt is slightly changed to perturb the regions of LRPs, and the cost is calculated again. If the new cost is smaller than the previous one, Δt in this case is regarded as the best solution as well as the solution adopted to determine the next Δt. Along the SA algorithm, the new solution with a larger cost can be the solution adopted to determine the next Δt with a probability, but can not be the best one. After repeating the above processes several times, the probability is reduced slightly. These processes are repeated until the solution converges or the probability becomes sufficiently small.

C. Parallelization

In this work, we focus on a shared-memory programming model based on OpenMP [7]. We can parallelize the processes in each of four sequential phases of the locally implicit scheme: the explicit updating processes of the voltages and currents in HRPs, the implicit updating process based on the GJ method for the voltages in LRPs, and the updating process of the currents in LRPs. Because the leapfrog-based updating calculations are performed for each variable individually, it is easy to parallelize [8]. Additionally, the GJ-based algorithm is inherently parallelizable. As for the currents in LRPs, because their updating calculations are similar to those of the currents in HRPs, there is no difficulty to parallelize.

IV. NUMERICAL RESULTS

The triangular meshes of two example plane structures are illustrated in Fig. 2. The plane for Example A in Fig. 2(a) is 100 mm × 100 mm and has 40 randomly-placed holes, each of

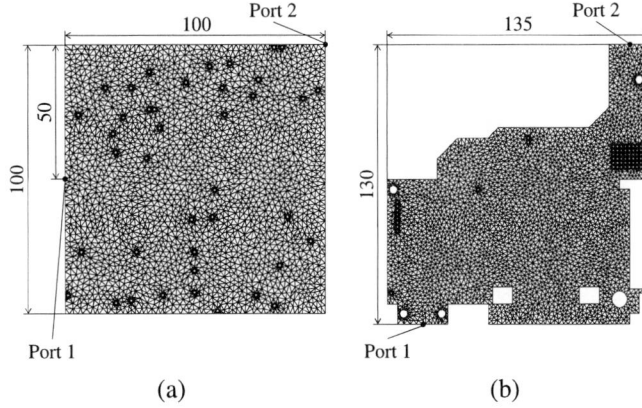

Fig. 2. The triangular meshes for Examples (a) A and (b) B.

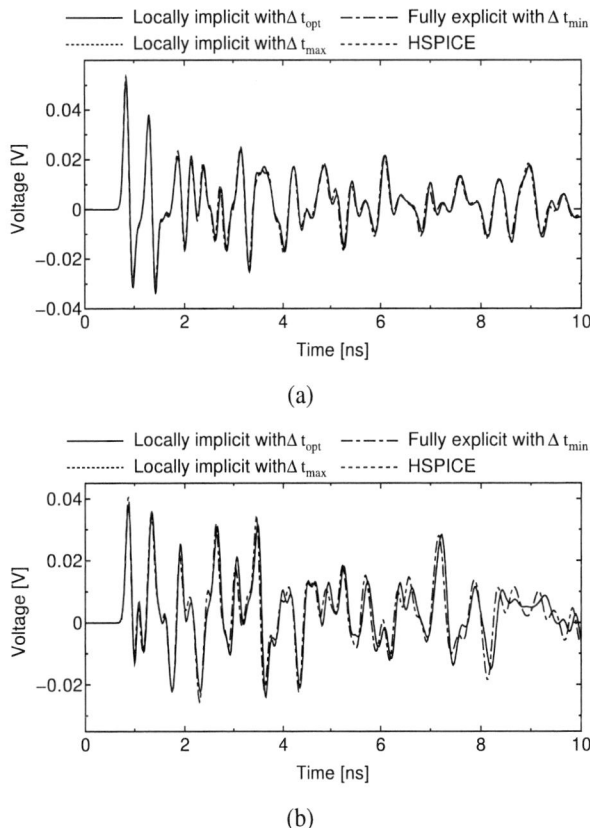

Fig. 3. The waveform results in the cases of Examples (a) A and (b) B.

which is 1 mm in diameter. On the other hand, the plane in Fig. 2(b) used for Example B is 130 mm × 135 mm and has the small holes and the nonorthogonal contours. The conductivity of each plane is 5.8×10^7 S/m, the relative permittivity of each dielectric between the planes is 4.0, and the thickness of the conductor and the height of the dielectric are 30 μm and 0.1 mm, respectively. The numbers of the circuit elements in Examples A and B are 34 626 and 44 626, respectively. In each example, a current source with the internal resistance 10 Ω is appended to Port 1 as an excitation, and voltage waveforms at Port 2 are observed. The excitation source is the triangular

978-1-4799-0708-3/13 $31.00 © 2013 IEEE

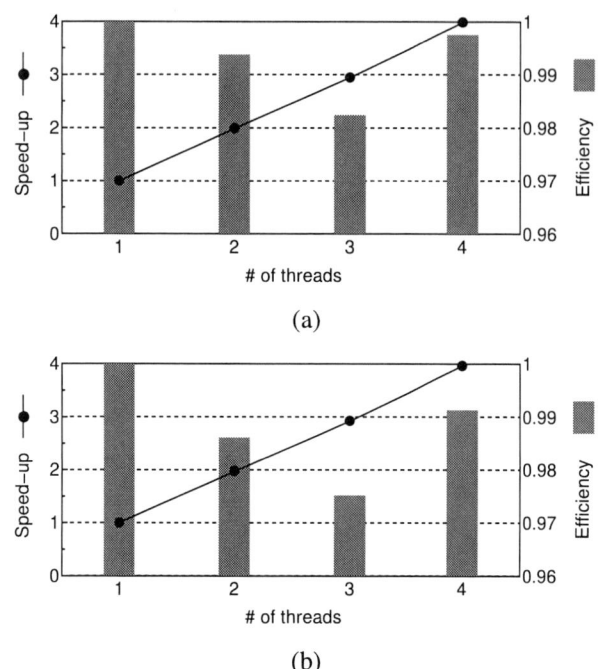

(a)

(b)

Fig. 4. Speed-up (solid line with circles, left scale) and parallelization efficiency (histogram, right scale) of the OpenMP implementation in Examples (a) A and (b) B.

TABLE II
CPU TIME AND SPEED-UP

Solver	Example A		Example B	
	CPU time (s)	Speed-up	CPU time (s)	Speed-up
Locally implicit with Δt_{opt}	11.03	82.61	13.92	89.05
Locally implicit with Δt_{max}	16.65	54.73	25.75	48.14
Fully explicit with Δt_{min}	59.46	15.32	240.65	5.15
HSPICE	911.21	1.0	1239.61	1.0

pulse of which the amplitude is 0.1 A, and the rise/fall times are 0.1 ns.

In these examples, we assume that the maximum frequency of interest is 5 GHz, and therefore, $\Delta t_{max} = 10$ ps. In addition, according to (2), $\Delta t_{min} = 0.66$ ps in Example A and $\Delta t_{min} = 0.23$ ps in Example B. By using the proposed optimization procedure with $N_r = 3$, the optimum time step sizes are calculated to be $\Delta t_{opt} = 7.83$ ps in Example A and $\Delta t_{opt} = 8.74$ ps in Example B. As a result, the numbers of the nodes in LRPs in the cases of Δt_{max} and Δt_{opt} are 3 553 and 1 458 in Example A and 5 065 and 1 906 in Example B.

We perform the transient simulations from 0 ns to 10 ns by using the locally implicit scheme with Δt_{opt}, locally implicit scheme with Δt_{max}, fully explicit leapfrog scheme with Δt_{min}, and HSPICE. The waveform results are plotted in Fig. 3. We can see that the waveforms obtained from the optimized method agrees well with those of the other solvers in both example simulations. Table II shows the CPU times and speed-

up ratios and indicates that the locally implicit scheme with the proposed implementation is more than 80 times faster than HSPICE in both examples. Additionally, the locally implicit scheme with Δt_{opt} is about 1.5 and 1.9 times faster than the scheme with Δt_{max} in Examples A and B, respectively, and therefore, the proposed implementation technique is adequate to improve the efficiency of the locally implicit scheme.

The two example simulations are also performed by using the parallelized programs based on OpenMP on Intel Core i7-2600 processor running at 3.4 GHz with four L2 caches of 256 KB and L3 cache of 8 MB. The processor has four cores and therefore the algorithm is parallelized using four threads. Fig. 4 shows the speed-up and the parallelization efficiency of the parallelized locally implicit leapfrog scheme. It is confirmed that our parallel implementations for both examples achieve nearly-ideal speed-up compared with a single-threaded implementation.

V. CONCLUSION

In this paper, we have proposed the optimization technique to improve the computational performance of the locally implicit leapfrog scheme. The criterion for determining the time step size was based on the numerical stability condition, accuracy, and computational cost. We adopted the GJ method to solve the locally implicit subcircuits because it was useful to estimate the computational cost and easy to parallelize. In the simulations of two types of the example plane structures, the optimized locally implicit scheme was more than 80 times faster than HSPICE without degrading the accuracy and improved the efficiency of the locally implicit scheme. Furthermore, it has been confirmed that the proposed scheme can be parallelized with the nearly-ideal efficiency by using OpenMP.

ACKNOWLEDGEMENT

This work is partially supported by Japan Society for the Promotion of Science (JSPS).

REFERENCES

[1] K.-B. Wu, G.-H. Shiue, W.-D. Guo, C.-M. Lin, and R.-B. Wu, "Delaunay-Voronoi modeling of power-ground planes with source port correction," *IEEE Trans. Adv. Packag.*, vol. 31, pp. 303–310, May 2008.

[2] J. E. Schutt-Ainé, "Latency insertion method (LIM) for the fast transient simulation of large networks," *IEEE Trans. Circuits Syst. I*, vol. 48, pp. 81–89, Jan. 2001.

[3] S. N. Lalgudi, M. Swaminathan, and Y. Kretchmer, "On-chip power-grid simulation using latency insertion method," *IEEE Trans. Circuits Syst. I*, vol. 55, pp. 914–931, Apr. 2008.

[4] H. Kurobe, T. Sekine, and H. Asai, "Locally implicit LIM for the simulation of PDN modeled by triangular meshes," *IEEE Microw. Wireless Compon. Lett.*, vol. 22, pp. 291–293, Jun. 2012.

[5] Y. Tanji, T. Watanabe, H. Kubota, and H. Asai, "Transient analysis of large scale interconnect networks using relaxation methods," *IEICE Trans. Electron. (Japanese Edition)*, vol. J89-C, pp. 809–816, Nov. 2006.

[6] S. M. Sait and H. Youssef, *Iterative Computer Algorithms with Applications in Engineering: Solving Combinatorial Optimization Problems*, 1st ed. Hoboken, NJ: Wiley-IEEE Computer Society, Feb. 2000.

[7] O. A. R. Board, "OpenMP C and C++ application program interface - version 2.0," Mar. 2002. [Online]. Available: http://openmp.org/

[8] Y. Inoue, T. Sekine, and H. Asai, "Parallel-distributed block-LIM for transient simulation of tightly coupled transmission lines," *IEEE Trans. Compon., Packag., Manuf. Technol.*, vol. 3, pp. 670–677, Apr. 2013.

978-1-4799-0708-3/13 $31.00 © 2013 IEEE

Memory Efficient Laguerre-FDTD Scheme for Dispersive Media

Ming Yi, Madhavan Swaminathan
Interconnect and Packaging Center
School of Electrical and Computer Engineering
Georgia Institute of Technology
Atlanta, GA 30332
myi9@gatech.edu,
madhavan.swaminathan@ece.gatech.edu

Myunghyun Ha, Zhiguo Qian, and Alaeddin Aydiner
Intel Corporation
Santa Clara, CA 95054
Chandler, AZ 85226,
and Hillsboro, OR 97124,
myunghyun.ha@intel.com, zhiguo.qian@intel.com, and
alaeddin.a.aydiner@intel.com

Abstract—**The unconditionally stable Laguerre-FDTD method is suitable for simulating 3-D structures with large time step. In this work, memory efficient Laguerre-FDTD scheme for dispersive materials is proposed to ensure accurate modeling and less memory consumption compared to standard procedures. The memory efficient scheme is realized by representing the Laguerre domain expression of electric susceptibility in a recursive manner. Formulations have been derived for both Debye and Lorentz media. Numerical results show that the proposed Laguerre-FDTD method exhibits significant peak memory usage reduction and equivalent calculation accuracy of dispersive material involved transient simulation.**

Keywords-Laguerre-FDTD; memory efficient; dispersive material ; Debye and Lorentz media

I. INTRODUCTION

The finite-difference time-domain (FDTD) method has been widely used to solve transient electromagnetic problems for decades. To overcome the intrinsic stability issue due to the Courant-Friedrichs-Lewy (CFL) stability condition, semi-implicit and implicit FDTD schemes have been studied extensively. The alternating direction implicit FDTD (ADI-FDTD) method has been introduced which is shown to be unconditionally stable [1]. More recently, the locally-one dimensional FDTD (LOD-FDTD) method has been proposed with reduction of arithmetic operations and increased computational efficiency compared to ADI-FDTD method [2].

One of the major challenges in time domain methods is to rigorously and efficiently model the material dispersion. In [3], the frequency-dependent material is modeled by incorporating a discrete time domain convolution in conventional FDTD and is efficiently evaluated using recursion. However, this method still suffers from CFL condition which makes it computational inefficient to simulate multiscale structures. To ensure unconditional stability, the ADI-FDTD method has been extended to be able to simulate dispersive material in [4]. Also, the frequency-dependent implementation of LOD-FDTD has been reported in [2] which showed reduction of simulation time compared to explicit FDTD method.

In recent years, the unconditionally stable Laguerre-FDTD method has been proposed and extended with algorithm modifications [5]-[7]. By transforming the time domain problem to Laguerre domain using temporal Galerkin's testing procedure, the transient solution is independent of time discretization. Thus, Laguerre-FDTD has the advantage of less numerical dispersion error when larger time step is used compared to ADI-FDTD. A Laguerre-FDTD formulation for frequency-dependent materials has been proposed in [7]. However, the formulation has a drawback of using all orders of solutions of Laguerre coefficients which requires considerable memory consumption to incorporate dielectric dispersion.

In this paper, memory efficient Laguerre-FDTD scheme for dispersive materials is proposed. To be specific, the electric susceptibility in time domain is obtained with Laguerre domain transformation which represents the Laguerre coefficient of susceptibility by the product of order-dependent and order-independent parts. By utilizing the unique mathematical properties of Debye and Lorentz models, the general Laguerre-FDTD formulations for dispersive materials are further rewritten into a recursive form which significantly reduces the memory storage in the simulation.

This paper is organized in the following manner: In Section II, the memory efficient Laguerre-FDTD schemes for Debye and Lorentz media is introduced with formulations and derivation. In Section III, the proposed method is verified with numerical examples which show that structures with Debye and Lorentz media can be analyzed efficiently using Laguerre-FDTD method. In Section IV, we summarize some conclusions.

II. PROPOSED SCHEME

A. General Formulations for Dispersive Materials

In Laguerre-FDTD method, time domain electric field components can be represented as a sum of infinite Laguerre basis functions $\varphi^q(\bar{t})$ scaled by Laguerre basis coefficient $\bar{E}^q$

$$\bar{E}(t) = \sum_{q=0}^{\infty} \bar{E}^q \varphi^q(\bar{t}) \qquad (1)$$

This work was supported in part by the Semiconductor Research Corporation under Project 2146 where $\bar{t} = t \cdot s$, s is the time scaling factor and t is time. Superscript q denotes the Laguerre coefficient of order q. The Laguerre basis functions $\varphi^q(\bar{t})$ can be expressed as

$$\varphi^q(\bar{t}) = e^{-\bar{t}/2} L^q(\bar{t}) \tag{2}$$

where $L^q(\bar{t})$ is the Laguerre polynomial which is defined recursively as

$$L^0(\bar{t}) = 1 \tag{3}$$

$$L^1(\bar{t}) = 1 - \bar{t} \tag{4}$$

$$qL^q(\bar{t}) = (2q - 1 - \bar{t})L^{q-1}(\bar{t}) - (q-1)L^{q-2}(\bar{t}), q \geq 2. \tag{5}$$

Assuming an isotropic, dispersive, lossy media, the wave equation can be written as

$$\nabla \times \nabla \times \bar{E} = -\mu \frac{\partial^2 \bar{D}}{\partial t^2} - \mu \frac{\partial(\bar{J} + \sigma \bar{E})}{\partial t} \tag{6}$$

where μ is the magnetic permeability, σ is the electric conductivity. $\bar{D}$ is the electric flux density which can be expressed as

$$\bar{D}(t) = \varepsilon_\infty \varepsilon_0 \bar{E}(t) + \varepsilon_0 \int_0^t \bar{E}(t - \tau)\chi(\tau)d\tau \tag{7}$$

where χ, ε_0 and ε_∞ are the electric susceptibility, electric permittivity of free space and infinite frequency relative permittivity, respectively.

Discretizing the differential equation (6) in Laguerre domain using temporal testing procedure yields

$$\nabla \times \nabla \times \bar{E}^q = -\mu s^2 \left[\frac{1}{4}\bar{D}^q + \sum_{n=0,q>0}^{q-1} (q-n)\bar{D}^n \right]$$
$$- \mu s \left[\frac{1}{2}\left(\bar{J}^q + \sigma \bar{E}^q\right) + \sum_{n=0,q>0}^{q-1} \left(\bar{J}^n + \sigma \bar{E}^n\right) \right] \tag{8}$$

where the Laguerre coefficient of electric flux density is given by [7] as

$$\bar{D}^q = \varepsilon_\infty \varepsilon_0 \bar{E}^q + \varepsilon_0 \left(\sum_{n=0}^{q} \bar{E}^n \chi^{q-n} - \sum_{n=0}^{q-1} \bar{E}^n \chi^{q-1-n} \right). \tag{9}$$

Using (9), the frequency-dependent dispersion can be incorporated in the Laguerre-FDTD scheme. However, due to the form of the transformed convolution term, all previous solutions are required to calculate $\bar{D}^q$. Therefore, as the order of Laguerre coefficient increases, significant amount of memory is required to store all the solution of previous orders. Such large memory consumption is undesirable for solving practical 3-D problems.

B. Debye Media

The frequency-dependent Debye model of order n can be written as

$$\varepsilon(\omega) = \varepsilon_\infty + (\varepsilon_s - \varepsilon_\infty) \sum_{i=1}^{n} \frac{a_i}{1 + j\omega\tau_i} \tag{10}$$

where a_i and τ_i denote the strength and time constant of various relaxation processes, ε_s is the static permittivity. For simplicity, considering only the case for $n = 1$. The method can be easily extended to $n > 1$ cases in a similar manner. Thus, the frequency dependent susceptibility function is given by

$$\chi(\omega) = (\varepsilon_s - \varepsilon_\infty)\frac{a}{1 + j\omega\tau} \tag{11}$$

where a and τ are the strength and time constant of the first-order Debye relaxation process. Performing Fourier transform of (11), the time domain expression for susceptibility is

$$\chi(t) = \frac{a(\varepsilon_s - \varepsilon_\infty)}{\tau} e^{-\frac{t}{\tau}}. \tag{12}$$

Using the definition in (1) and applied partial integration, with some manipulations, the Laguerre-domain expression for susceptibility is

$$\chi^n = \alpha_D \beta_D^2 \tag{13}$$

where

$$\alpha_D = \frac{2a(\varepsilon_s - \varepsilon_\infty)}{2 + s\tau} \tag{14}$$

$$\beta_D = \frac{2 - s\tau}{2 + s\tau}. \tag{15}$$

Rewriting (9) as

$$\bar{D}^q = \varepsilon_\infty \varepsilon_0 \bar{E}^q + \varepsilon_0 \bar{G}^q \tag{16}$$

results in

$$\bar{G}^q = \sum_{n=0}^{q} \bar{E}^n \chi^{q-n} - \sum_{n=0}^{q-1} \bar{E}^n \chi^{q-1-n} \tag{17}$$

and inserting (13) into (17) results in

$$\bar{G}^q = \alpha_D \left(\sum_{n=0}^{q} \bar{E}^n \beta^{q-n} - \sum_{n=0}^{q-1} \bar{E}^n \beta^{q-1-n} \right). \tag{18}$$

Subtracting $\beta_D \bar{G}^{q-1}$ from $\bar{G}^q$, we have

$$\begin{aligned} \bar{G}^q &- \beta_D \bar{G}^{q-1} \\ &= \alpha_D \left(\sum_{n=0}^{q} \bar{E}^n \beta_D^{q-n} - \sum_{n=0}^{q-1} \bar{E}^n \beta_D^{q-1-n} \right) \\ &\quad - \alpha_D \left(\sum_{n=0}^{q-1} \bar{E}^n \beta_D^{q-n} - \sum_{n=0}^{q-2} \bar{E}^n \beta_D^{q-1-n} \right) \\ &= \alpha_D \left(\bar{E}^q - \bar{E}^{q-1} \right) \end{aligned} \tag{19}$$

Therefore, $\bar{G}^q$ can be calculated recursively as

$$\bar{G}^q = \beta_D \bar{G}^{q-1} + \alpha_D \left(\bar{E}^q - \bar{E}^{q-1} \right). \tag{20}$$

For zero order

$$\bar{G}^0 = \alpha_D \bar{E}^0. \tag{21}$$

Inserting (20) and (21) into (16), the Laguerre coefficient for electric flux density $\bar{D}^q$ can be calculated recursively which requires only one previous order $(q - 1)$ solution.

978-1-4799-0708-3/13 $31.00 © 2013 IEEE

C. Lorentz Media

The frequency-dependent Lorentz model of order n can be written as

$$\varepsilon(\omega) = \varepsilon_\infty + (\varepsilon_s - \varepsilon_\infty) \sum_{i=1}^{n} \frac{a_i \omega_i}{\omega_i^2 + 2j\omega\delta_i - \omega^2} \quad (22)$$

where a_i, ω_i and τ_i represent the pole amplitude, the pole location and the damping factor. Again, only the case for $n = 1$ is discussed here. The frequency dependent susceptibility function is given by

$$\chi(\omega) = (\varepsilon_s - \varepsilon_\infty) \frac{a\omega_0^2}{a\omega_0^2 + 2j\omega\delta - \omega^2}. \quad (23)$$

The time domain expression of susceptibility can be obtained as

$$\chi(t) = \frac{a\omega_0^2(\varepsilon_s - \varepsilon_\infty)}{\sqrt{\omega_0^2 - \delta^2}} e^{-\delta t} \sin\left(\sqrt{\omega_0^2 - \delta^2}\right). \quad (24)$$

Introducing complex numbers, (24) can be rewritten into

$$\chi(t) = \mathrm{Im}\left[\frac{a\omega_0^2(\varepsilon_s - \varepsilon_\infty)}{\sqrt{\omega_0^2 - \delta^2}} e^{-\left(\delta - j\sqrt{\omega_0^2 - \delta^2}\right)t} \right]. \quad (25)$$

Transforming (25) into Laguerre domain yields

$$\chi^n = \mathrm{Im}\left(\alpha_L \beta_L^n\right) \quad (26)$$

where

$$\alpha_L = \frac{a\omega_0^2(\varepsilon_s - \varepsilon_\infty)}{\sqrt{\omega_0^2 - \delta^2}} \frac{2}{2\left(\delta - j\sqrt{\omega_0^2 - \delta^2}\right) + s} \quad (27)$$

$$\beta_L = \frac{2\left(\delta - j\sqrt{\omega_0^2 - \delta^2}\right) - s}{2\left(\delta - j\sqrt{\omega_0^2 - \delta^2}\right) + s}. \quad (28)$$

Performing the similar procedures as for Debye model, the electric flux density of the Lorentz model can be calculated recursively as

$$\bar{D}^q = \varepsilon_\infty \varepsilon_0 \bar{E}^q + \varepsilon_0 \mathrm{Im}\left(\bar{G}^q\right) \quad (29)$$

where

$$\bar{G}^q = \beta_L \bar{G}^{q-1} + \alpha_L \left(\bar{E}^q - \bar{E}^{q-1}\right) \quad (30)$$

and for zero order

$$\bar{G}^0 = \alpha_L \bar{E}^0. \quad (31)$$

III. NUMERICAL RESULTS

To validate the memory usage improvement using the proposed scheme, a simple microstrip line shown in Fig. 1 is simulated and analyzed. The structure has a dielectric substrate with width and thickness of $s = 30mm$ and $d = 0.305mm$. The dielectric material is FR-4 and is assumed to be dispersive and can be approximated with first order Debye model. Parameters for Debye model are $\varepsilon_s = 4.530$, $\varepsilon_\infty = 4.398$, $a = 1$, $\tau = 57.22ps$ given in [8]. The metal strip is considered

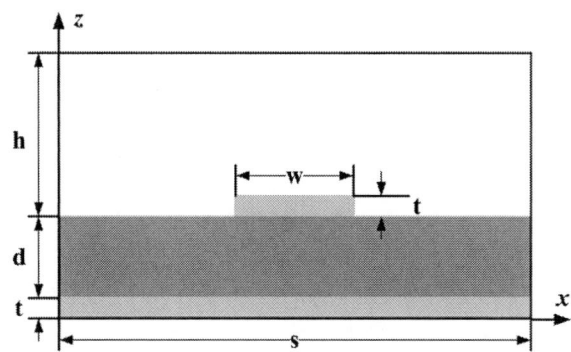

Fig. 1. Cross sectional view of the simulated microstrip line

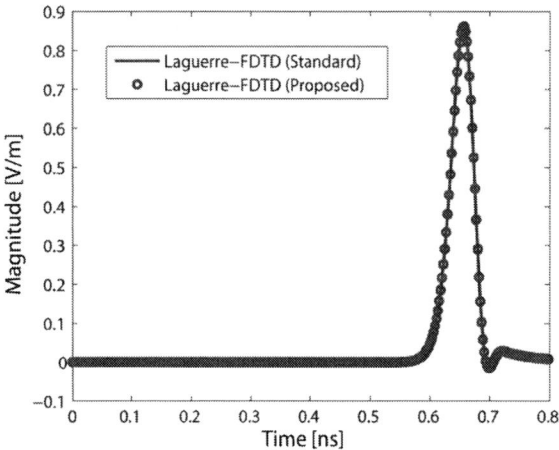

Fig. 2. Time domain response of the observation point of micrstrip line using (a) Standard Laguerre-FDTD (b) proposed memory efficient Laguerre-FDTD

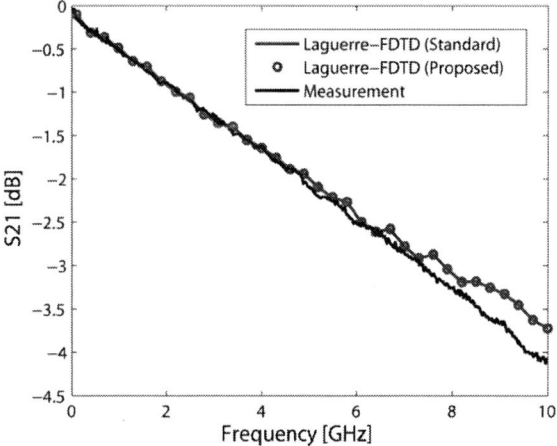

Fig. 3. Insertion loss of the simulated microstrip line with (a) Standard Laguerre-FDTD (b) proposed memory efficient Laguerre-FDTD and (c) measurement

as copper (conductivity $\sigma = 5.8 \times 10^7 S/m$) with length, width and thickness of $l = 93.5mm$, $w = 0.51mm$ and $t = 0.03mm$, respectively. The simulated structure is surrounded by an box

978-1-4799-0708-3/13 $31.00 © 2013 IEEE

TABLE I
COMPARISON OF MEMORY CONSUMPTION FOR DIFFERENT SCHEMES

Test Case[*]	Method	Memory	Improvement
1	Standard Laguerre-FDTD	0.92GB	-
	Proposed Laguerre-FDTD	0.48GB	52.2%
2	Standard Laguerre-FDTD	1.85GB	-
	Proposed Laguerre-FDTD	0.95GB	51.4%

[*] Test case 1: Microstrip line; 2: Patch antenna

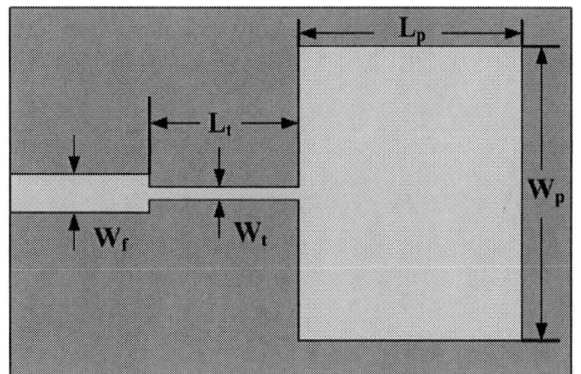

Fig. 4. Top view of the simulated microstrip patch antenna

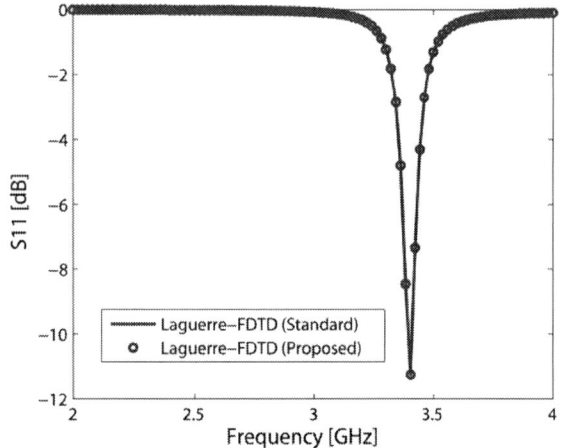

Fig. 5. Return loss of the simulated patch antenna with (a) Standard Laguerre-FDTD (b) proposed memory efficient Laguerre-FDTD

with first order ABC boundary with height $h = 1mm$. Two ports are set at each end of the metal strip.

Fig. 2 shows the time domain response of the observation point at one port of the microstrip line using the standard Laguerre-FDTD method and the proposed memory efficient Laguerre-FDTD method. Good agreement can be observed. Fig. 3 shows the comparison of insertion loss of the microstrip line using both methods along with the measurement. Good correlation with the measurement can be observed for both methods. The peak memory consumption using both methods are compared in Table I. It can be observed form Table I that under the same level of accuracy, the memory consumption for standard Laguerre-FDTD method is 0.92GB whereas the counterpart for memory efficient Laguerre-FDTD is 0.48GB. A 52.2% of improvement is achieved.

Fig. 4 shows the test case of a microstrip patch antenna. The structure has a dielectric substrate with thickness of $d = 0.8mm$. The dielectric material is FR-4 and is assumed to be dispersive. The Debye parameter of the material is the same as in the previous example. The metal strip is considered as copper whose conductivity is $\sigma = 5.8 \times 10^7 S/m$. The feature sizes shown in Fig. 4 are $W_f = 1.53mm$, $W_t = 0.4mm$, $W_p = 30mm$, $L_t = 12mm$, and $L_p = 20mm$, respectively. The simulated structure is surrounded by ABC boundary with one port assigned at the end of the feed line.

Fig. 5 shows the comparison of return loss of the patch antenna using the standard Laguerre-FDTD and the memory efficient Laguerre-FDTD. Good agreement can be observed. Table I shows the comparison of peak memory consumption for both methods. It can be observed that the memory consumption for standard Laguerre-FDTD and proposed Laguerre-FDTD are 1.85GB and 0.95GB, respectively. A 51.4% improvement is achieved.

IV. CONCLUSION

The memory efficient Laguerre-FDTD scheme for dispersive material is proposed. Formulations for both Debye and Lorentz media are derived in recursive manner. The proposed method is verified with results of structures with dispersive dielectric materials. Significant reduction of the peak memory consumption can be achieved using the proposed Laguerre-FDTD scheme.

REFERENCES

[1] F. Zheng, Z. Chen, and J. Zhang, "Toward the development of a three-demensional unconditionally stable finite-difference time-domain method," *IEEE Trans. Microw. Theory Tech.*, vol. 48, no. 49, pp. 1550-1558, Sept. 2000.

[2] J. Shibayama, R. Takahashi, J. Yamauchi, and H. Nakano, "Frequency-dependent LOD-FDTD implementations for dispersive media," *Electron. Lett.*, vol. 42, no. 19, pp. 1084-1085, Sept. 2006.

[3] R. Luebbers, F. Hunsberger, K. Kunz, R. Standler, and M. Schneider, "A frequency-dependent finite-difference time-domain formulation for dispersive materials," *IEEE Trans. Electromagn. Compat.*, vol. 32, no. 3, pp. 222-227, Aug. 1990.

[4] S. Garcia, R. Rubio, A. Bretones, and R. Martin, "Extension of the ADI-FDTD method to Debye media," *IEEE Trans. Antennas Propag.*, vol. 51, no. 11, pp. 3183-3186, Nov. 2003.

[5] M. Ha, K. Srinivasan, and M. Swaminathan, "Transient chip-package co-simulation using the Laguerre-FDTD scheme," *IEEE Trans. Adv. Packag.*, vol. 32, no. 4, pp. 816-830, Nov. 2009.

[6] M. Yi, M. Swaminathan, "Skin effect modeling of interconnects using the Laguerre-FDTD scheme," in *IEEE Electr. Performance Electron. Packag. Systems Conf.*, Oct. 2012, pp. 236-239.

[7] M. Ha, and M. Swaminathan, "A Laguerre-FDTD formulation for frequency-dependent dispersive materials," *IEEE Microw. Wireless Compon. Lett.*, vol. 21, no. 5, pp. 225-227, May. 2011.

[8] J. Zhang, M. Koledintseva, D. Pommerenke, and J. Drewniak, "Extraction of dispersive material parameters using vector network analyzers and genetic algorithms," *IEEE Instrumentation and Measurement Tech. Conf.*, pp. 462-467, 2006.

EM II

978-1-4799-0708-3/13 $31.00 © 2013 IEEE

978-1-4799-0708-3/13 $31.00 © 2013 IEEE

Modeling Methodologies for Multi Level PCB-Package Co-Simulation & Co-Design

Antonio Ciccomancini Scogna[#], ChunTong Chiang[*], Linus Lau[+]

[#]CST of America Inc, 492 Old Connecticut Path, #505 01701, Framingham, MA - antonio.ciccomancini@cst.com

[*]CST of South East Asia - ct.chiang@cst.com
[+]CST of Malaysia - linus.lau@cst-malaysia.com

LianKheng Teoh, HsuenYen Lee, eASIC of Malaysia, Level 6 Menara KWSP, 38 Jalan Sultan Admad Shah, 11900 Malaysia - lkteoh@easic.com , hylee@easic.com

Summary - The continuous demand on wide bandwidth and high-speed data rate is pushing the signal spectra to be considered in simulations at the stage of 50GHz and beyond. Data rate through backplane is developing at 15Gbps and the next generation is targeted at 25Gbps.

At such high data rates with extra wide frequency spectra for signals, lots of challenges are brought into the system level simulations including affordable simulation time, computational resources and required accuracy. Time to market is critical in industry due to the market share while products are outdated very quickly.

Minimum simulation time is always desired and appreciated. The minimum simulation time is related to the available computational resources, simulation tools and the requirement on accuracy. For a system with above given conditions, which are the most cases in product development in the industry, the simulation time is solely dependent on modeling methodology, which is the key in system level simulation.

Package and printed circuit board (PCB) co-design is common in system analysis for the prediction of high-speed channel performance. In recent times, chip companies have been closely looking into package and PCB co-simulations in order to guarantee the performances of these chips at the system level.

However, full-wave package and PCB co-simulation can be challenging: if the package and the PCB are modeled together, the co-simulation is limited by either available computational resource, due to the extremely large memory consumption, or by affordable simulation time. The current, widely adopted method in industry, for PCB and package co-simulation, is to model them separately.

There has been some investigation of where to separate/segment the PCB and package and build a reference plane; however, in most of the cases, only very simple models are used to verify the proposed methods.

It is known that the reference plane is critical to the simulation results of the entire channel and it is recognized that a transversal electromagnetic wave (TEM) wave is required at the reference plane so that the current and voltage are well defined. Unfortunately, this is sometimes challenging due to the complexity of the structure and the port definition in most of the commercial Electromagnetic (EM) tools.

This paper investigates these details on a realistic test structure and gives some general solution for PCB and package system co-simulation.

The model under test consists of a 10 layers PCB in standard FR4 material with $\varepsilon r=4.5$ and $tg\delta=0.035$ at 1GHz. The package is 8 layers flip chip technology, $\varepsilon r=3.32$ and $tg\delta=0.0175$ at 1GHz. Due to the complexity and the high density of the nets routing, waveguide ports cannot be used. Therefore, in order to mimic a port set up as close as possible to TEM wave propagation, a virtual reference plane is used and the port is defined by a small port source gap between the end of a bump and/or pin and the metal sheet which acts as the virtual ground plane.

Results of the proposed methodology applied to the partitioned model of PCB and package are compared with the 3D EM simulation of the full model and validated with measurements up to 40GHz.

Other two modeling techniques are also investigated. In the first one, the GND layers are slightly extended and the ports are located inside the ground planes. In the second one, the ports are defined at the ground plane edges and a small gap is added between the boundaries and the truncated area. Details will be provided in the presentation. Very good agreement can be observed when comparing the S-parameters of the cascaded models with the S-parameters of the full model.

The proposed techniques greatly simplified the co-design process and reduce the simulation time up to 5X, therefore they can be easily extended to study the influence of the PCB effects, such as PCB via length, trace routing, and escape area definition.

978-1-4799-0708-3/13 $31.00 © 2013 IEEE

978-1-4799-0708-3/13 $31.00 © 2013 IEEE

EM III

978-1-4799-0708-3/13 $31.00 © 2013 IEEE

978-1-4799-0708-3/13 $31.00 © 2013 IEEE

Accelerate High Speed IO Design Closure with Distributed Chip IO Interconnect Model

Yun Dai, Patrick Ho, Tiejun Yu, Jiayuan Fang
Sigrity R&D
Cadence Design System, Inc.
San Jose, CA 95134, USA
yund@cadence.com, patrickh@cadence.com, tyu@cadence.com, fangj@cadence.com

Abstract

This paper presents an overview of the applications of the distributed models representing chip IO power, ground and signal distribution systems from IO cells to die bumps. It provides a methodology of applying such models for on-die electrical performance assessment of IO power, ground and signal interconnects. It also demonstrates the die-to-die system-level IO SSO analysis with the chip interconnect model together with other models.

The conventional chip interconnect model extraction tools are designed either for STA (static timing analysis) or for voltage drop analysis. In the former, the extraction tools mainly focus on RC (resistance and capacitance) extraction of signal traces but are neither keen to power rail network nor the couplings between power rails and signals. In the later, the tools extract the power rail network RC only by completely ignoring the signal parasitics. In both cases, the size of parasitic netlists including chip power and ground networks could be too large to be simulated. As a result, chip engineers have to guard their designs with a large amount of safety margin and leave the final signal integrity verification and fixes to the system engineers. On the other hand, the system engineers often have to either omit the chip IO interconnect model completely by connecting IO buffers with the package model directly or include a very simplified lumped on-chip power/ground model across all the driver power and ground terminals. Hence, the system-level SSO simulation results become either unduly pessimistic or over optimistic.

A breakthrough extraction technology which generates a detailed model of chip IO power, ground and signal interconnects from bumps to IO circuits that fully represent the distributed nature of power, ground and signals as well as their electromagnetic coupling effects has been introduced by Cadence, which fills the gap between EDA tools and chip design needs for accurate on-die and system-level analysis of high-speed channels and buses .

The newly introduced chip IO interconnect model extraction takes chip layout data in GDSII or LEF/DEF formats, as well as the technology file for stackup process parameters and a user-specified configuration for net names, circuit definitions, etc. and then generates a comprehensive SPICE netlist that consists of a fully distributed IO power, ground and signal connections from IO cells to die bumps, including RDL and all the other metal layers, and bumps/ubumps. It accounts for all inductive and capacitive couplings between power, ground and signals on the chip. This extraction method offers both high spatial resolution and compact circuit size to ensure accuracy and efficiency. There is no practical limitation on the number of external nodes of the SPICE circuit netlist. By default, it generates a distinct external node for each die bump connected to off-chip structures and each pin connected to IO cells. An option is provided for a user to group bumps by region for accuracy and performance/capacity trade-off.

The on-die interconnect model thus extracted enables a quick assessment of on-chip power and ground quality along with signal performance at every IO cell. Intuitive graphical representation of the electrical performance at each cell and query functions help designers easily verify each IO channel characteristics, quickly identify weak or problematic physical areas and perform what-if analysis to rapidly improve the design.

Once the design meets the chip-level specification, the SI engineers can assemble the chip IO power, ground and signal interconnect model with other off-chip models through Cadence Model Connection Protocol (MCP) interface and perform the system-level IO SSO simulation. A typical die-to-die IO SSO simulation for a DDR memory interface, which includes power-aware IBIS (Input Output Buffer Information Specification) models for drivers and receivers, and distributed and coupled power/ground/signal models for chip, package and board interconnects, will be demonstrated.

978-1-4799-0708-3/13 $31.00 © 2013 IEEE

978-1-4799-0708-3/13 $31.00 © 2013 IEEE

3D Packaging

978-1-4799-0708-3/13 $31.00 © 2013 IEEE

978-1-4799-0708-3/13 $31.00 © 2013 IEEE

Applying a Physics-Based Via Model for the Simulation of Through Silicon Vias

David Dahl*, Xiaomin Duan*, Anne Beyreuther†, Ivan Ndip†, Klaus-Dieter Lang†, and Christian Schuster*

*Institut für Theoretische Elektrotechnik, Technische Universität Hamburg-Harburg (TUHH), Harburger Schloßstr. 20
21079 Hamburg, Germany, Email: david.dahl@tuhh.de, Phone: +494042878-2173

†Fraunhofer Institute for Reliability and Microintegration (IZM), Berlin, Germany and
School of Electrical Engineering and Computer Sciences, Technical University Berlin, Berlin, Germany

Abstract—This paper presents a first approach for the efficient modeling of Through Silicon Vias based on a Physics-Based Via model for application in silicon interposers with metallic boundaries.

Keywords—*Through Silicon Vias, 3D interconnects, Physics-Based Via Model*

I. INTRODUCTION

3D-integration of integrated circuits is of major interest because many advantages can be drawn from the vertical stacking of integrated circuits. One of the most important is the significantly reduced mean interconnect length [1] which results in reduced power and signal integrity issues. A key component of this technology is the vertical interconnect termed Through Silicon Via (TSV).

Through Silicon Vias are fabricated in silicon interposers where they allow for signal routing and/or power supply. Such an interposer is the considered structure of this paper. Here we restrict the analysis to the case where it consists of a thinned silicon plate that has metallizations on its top and bottom surfaces. These are electrically insulated by an intermittent silicon dioxide layer. For the frequency range under consideration (100 MHz to 100 GHz), this cavity can be regarded as a parallel plate structure for the propagation of waves along the horizontal direction. In the vertical direction, the via traverses this structure in form of a conducting barrel (typically made of metal or polysilicon) which is also electrically insulated with regard to the silicon layer by a silicon dioxide layer. In order to provide a well-defined current return path, a via carrying a signal needs to be accompanied by at least one sufficiently closely located ground via which connects at its designated position the top and bottom metal layers. The effects of ground vias can be deduced from the presented results by placing a short-circuit at both ports of a via. It is assumed here that the boundary of the layered structure can with sufficient accuracy be modeled as (perfectly) conducting layers thus representing the metallic traces of the redistribution layers and power planes which enclose the dielectric layers.

In the past Through Silicon Vias have been investigated by using general-purpose full-wave solvers or methods based on multi-scattering techniques as presented for example in [2]. The Physics-Based Via Model has already been applied to printed circuit board structures [3], [4] and in this paper the goal is to present first steps of the extension to the treatment of silicon interposer structures. The main objective is to show that the presented physics-based via model is a promising method

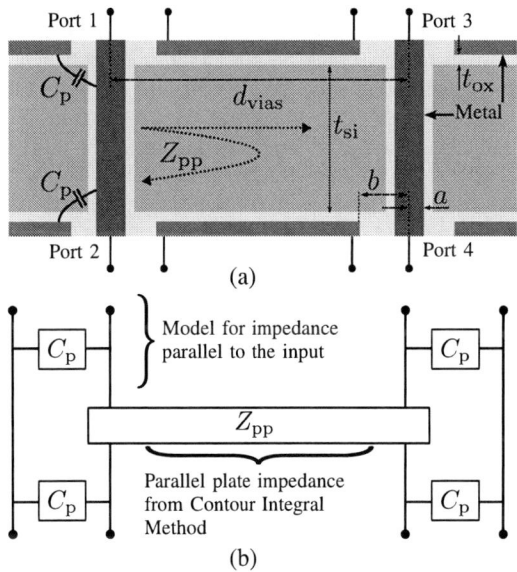

Fig. 1. Physics-based via model: Parallel plate impedance (Z_{pp}) and the parasitic capacitance (C_{p}) are obtained from a Contour Integral Method and an analytical model, respectively, and assembled for overall transmission and reflection properties. The applied dimensions of this investigations are: Silicon thickness $t_{\mathrm{si}} = 50...200\ \mu m$, oxide thickness $t_{\mathrm{ox}} = 1\ \mu m$, via barrel radius $a = 15\ \mu m$, antipad radius $b = 30\ \mu m$, and via pitch $d_{\mathrm{vias}} = 75...300\ \mu m$. The assumed conductivity is 1...10 S/m.

for the efficient simulation of the interposer and thus enables also the simulation of large via arrays and the execution of extensive parameter studies.

II. ADAPTATION OF A PHYSICS-BASED VIA MODEL

The basic approach of the physics-based via model is the representation of the electrical behavior of the TSV by a lumped impedance parallel to every top and bottom port and a parallel plate impedance as illustrated in Fig. 1. The former characterizes mainly the via to plane capacitance, the latter the interaction with the vias in the vicinity and the board edges. This interaction is of relevance because the vertical signal transmission is related to a horizontal wave propagation in the parallel plate structure where waves are scattered at the discontinuities represented by the via barrels and board edges. The parallel plate impedance of the structures which are significantly larger in the horizontal directions compared to the transverse direction can be computed in a numerically efficient way e.g. by using a Contour Integral Method (CIM) [5].

978-1-4799-0708-3/13 $31.00 © 2013 IEEE

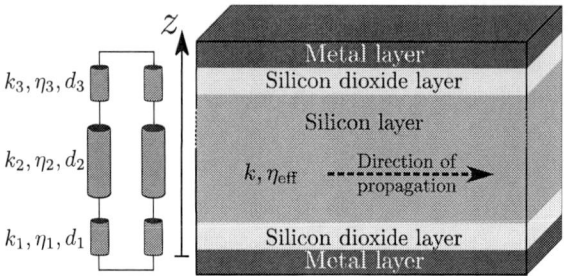

Fig. 2. Illustration of the equivalent circuit representation of the Transverse Resonance Method which is used for the determination of the wave number k of the parallel plate mode.

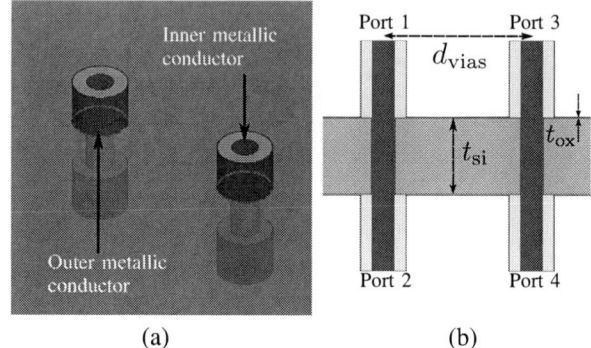

(a)   (b)

Fig. 3. (a) Detail view and (b) cross-sectional view of the simulated structure. When simulating the presented examples using a commercial FEM full-wave solver [8], coaxial extensions at the ports are added and deembedded. Relatively large planes need to be used in order to achieve good convergence in the lower frequency range. When choosing the plane size as 25 mm x 25 mm the typical CPU time on a 3.2 GHZ CPU is 32 minutes per frequency point. On the same CPU, the proposed method requires a CPU time of approximately 7 ms per frequency point of which approximately 6 ms are necessary for the TRM.

Here, the layered nature of the structure is accounted for by computing the complex propagation constant using the Transverse Resonance Method (TRM). To compute the overall transmission, reflection and crosstalk properties in terms of scattering parameters at the plane of the TSV terminals, the parallel plate impedance and the impedance parallel to the input are connected, most conveniently performed as admittance parameters, and converted to scattering parameters.

A. Calculation of wave number and effective wave impedance

In order to extend the physics-based via model to the treatment of stratified media the frequency dependent propagation constant k and the (effective) wave impedance η_{eff} are required for the computation of the parallel plate impedance and the via to plane capacitance. A detailed investigation is important because the propagation characteristics will in general be different from those of a single layer and more complicated when lossy materials are involved [6]. As illustrated in Fig. 2, the Transverse Resonance Method considers the problem of finding the modes which are supported by a stratified structure by considering an equivalent circuit for the stranding wave of the transverse direction which is connected to the propagating wave in the direction which is in the planes of layering. Of the two fundamental modes which are supported by this layered structure only the fundamental TM mode needs to be considered. More details and further references can be found in a preceding paper [7]. By solving the transcendental equation which arises from the matching of the tangential field components at the interfaces, the exact wave number of the structure is obtained. When considering single layers, the wave impedance of the transverse magnetic mode is related to the thus determined wave number by $\eta = k/(\omega\epsilon)$. For the layered structure a corresponding relation of the effective wave impedance to an effective permittivity is defined as

$$\eta_{\mathrm{eff}} = k/(\omega\epsilon_{\mathrm{eff}}). \tag{1}$$

In the following investigation a weighted average (by the relative thickness of the layers) of the complex, frequency dependent permittivities of the single layers to compute an effective permittivity

$$\epsilon_{\mathrm{eff}} = \frac{\sum_{i=1}^{3} t_i}{\sum_{i=1}^{3} t_i/\epsilon_i} \tag{2}$$

is used. Here t_i is the thickness of the i-th layer and ϵ_i is its complex permittivity. It is examined if the application of these definitions can be justified by agreement with full-wave reference results.

B. Integration into the physics-based via model

The results for wave number and effective wave impedance can now be used to extend the equations of the physics-based via model to the case of the stratified medium. The Contour Integral Method, presented in detail in [5], [9], allows for the efficient simulation of structures where two dimensions are significantly larger than the third, which is the case for most printed circuit boards and also for the interposer structures of interest here. For these 2D-structures only the inner and outer contours need to be discretized. The numerical effort can be further reduced by considering the case of infinite planes which will be valid for sufficiently large planes and finite silicon conductivities. The via barrels represent inner circular contours. The case of infinite planes and only circular inner contours is considered, for which analytical solutions can be used for the isotropic fields and a discretization of the contour becomes unnecessary. Only the fundamental mode is considered by using Eq. (4) of Ref. [9].

The impedance parallel to the input of our physics-based via model approach shows mainly capacitive behavior, at least for low substrate conductivities. For the dominant wave propagation in the horizontal direction of the planes, which is assumed here, it is typically composed of a coaxial capacitance for the region where the via barrel crosses the planes and of a via barrel to plane capacitance. The latter has been shown to be the manifestation of the non-propagating parallel plate modes [10] but also exists in the static case. The effect of the coaxial component is assumed to be negligible for this investigation and an analytical model for the via barrel to plane capacitance is used. The used reactance of this model, for the frequency range and geometrical dimensions of interest, is given by "B_a" of Eq. (37) in Ref. [11]. It is assumed that the properties of the higher-order, non-propagating modes which need to be considered here are also described with sufficient accuracy by the already computed wave number and effective wave impedance so these values can be used instead of those for the empty guide for which the equation referred to was originally derived.

978-1-4799-0708-3/13 $31.00 © 2013 IEEE

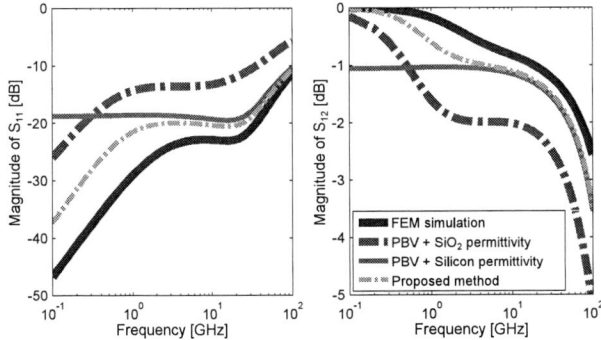

Fig. 4. Comparison of the results for transmission and reflection at one via when different effective permittivities are used for the calculation of the effective wave impedance of the interposer structure in the physics-based via model (PBV). Results of proposed method are compared to full-wave simulations based on FEM [8].

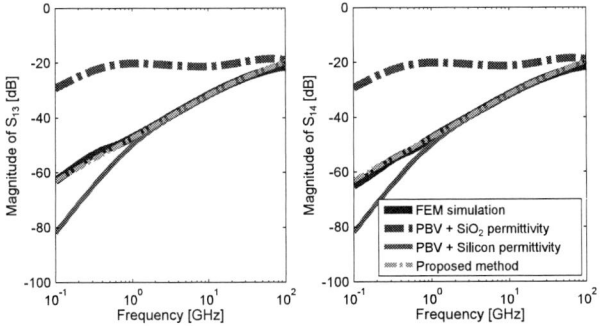

Fig. 5. Comparison of the results for near-end and far-end crosstalk when different effective permittivities are used for the calculation of the effective wave impedance of the interposer structure in the physics-based via model (PBV). Results of the proposed method are compared to full-wave simulations based on FEM [8].

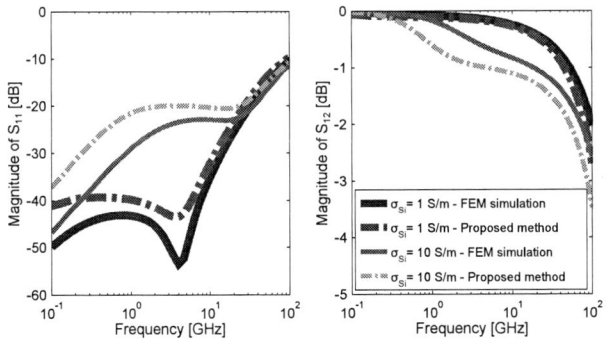

Fig. 6. Comparison of transmission and reflection results obtained with full-wave simulations based on FEM [8] and with the proposed method for two different silicon conductivities.

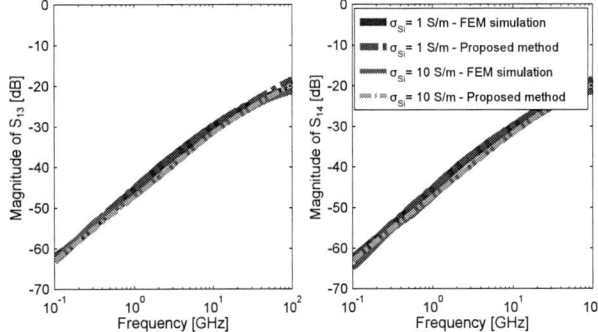

Fig. 7. Comparison of near-end and far-end crosstalk results obtained with full-wave simulations based on FEM [8] and with the proposed method for two different silicon conductivities.

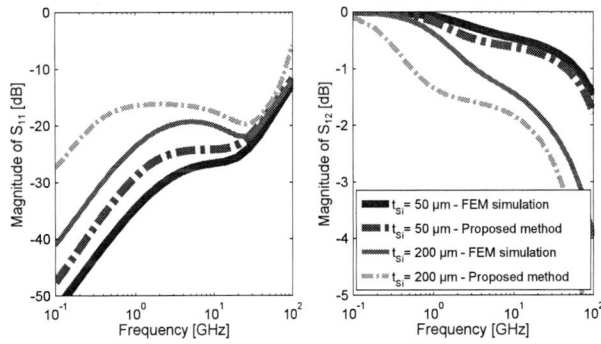

Fig. 8. Comparison of transmission and reflection results obtained with full-wave simulations based on FEM [8] and with the proposed method for two different silicon layer thicknesses.

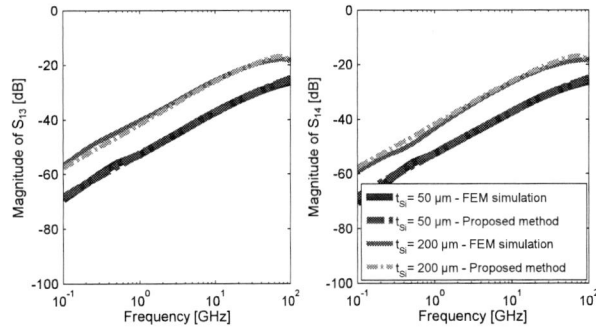

Fig. 9. Comparison of near-end and far-end crosstalk results obtained with full-wave simulations based on FEM [8] and with the proposed method for two different silicon layer thicknesses.

III. APPLICATION TO INTERPOSER STRUCTURES

The reference results presented in this paper are obtained using a FEM (finite element) based full-wave solver [8]. Various parameter studies are performed to ensure the reliability and stability of these reference results. The obtained scattering parameters are used to compare to the results of the proposed method on multiple levels. The segmentation approach together with the assumed model of the configuration allows to compare on the level of scattering parameters which are presented here,

but also on the levels of the parallel plate impedance and the extracted via to plane capacitance. A cross-sectional view and the port definitions are given in Fig. 3. The basic example considered here is the one of a differential pair of vias. Ground signal configurations can be considered by terminating ports with short-circuits. The via barrel radius is 15 μm, the radius of the antipad is 30 μm. The investigated case features an oxide thickness of 1 μm and perfectly conducting outer metal layers. The default thickness of the silicon layer is 100 μm, its conductivity 10 S/m and the default separation between the vias is 200 μm.

In order to motivate the choice made for the effective per-

978-1-4799-0708-3/13 $31.00 © 2013 IEEE

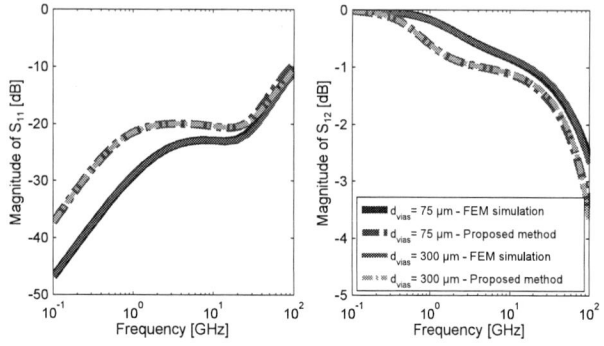

Fig. 10. Comparison of transmission and reflection results obtained with full-wave simulations based on FEM [8] and with the proposed method for two different values of center to center distance of the vias.

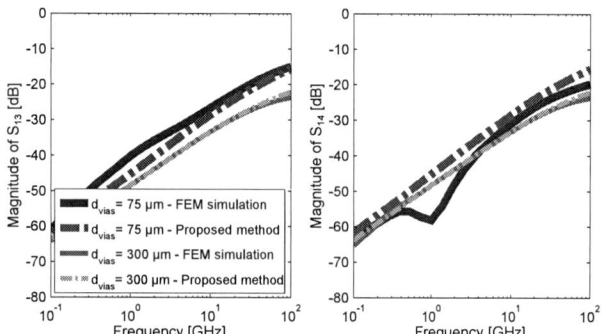

Fig. 11. Comparison of near-end and far-end crosstalk results obtained with full-wave simulations based on FEM [8] and with the proposed method for two different values of center to center distance of the vias.

mittivity, used when computing the effective wave impedance which is inserted into expressions for the via to plate capacitance and the parallel plate impedance, the results obtained for different choices of this permittivity are compared. As shown in Figs. 4 and 5, for the layered interposer structure this choice shows the best agreement over the complete considered frequency range when compared to two alternatives which are using either the (complex) permittivity of silicon or silicon dioxide alone to determine an effective wave impedance from the relation to the wave number (Eq. (1)).

When comparing the results for different conductivities of the silicon layer, the results presented in Figs. 6 and 7 are obtained. In the case of the lower conductivity of 1 S/m the agreement slightly is better for S_{11} and S_{12}. With respect to the cross-parameters, it is found that the agreement is good and almost independent of the conductivity value. The results for a varied silicon layer thickness, Figs. 8 and 9, show that thinner silicon layers lead to reduced reflections. Also for the thin silicon layer the agreement of the FEM simulation with the proposed method is better. When looking at the cross-parameters, it can be observed that thin silicon layers lead to reduced crosstalk. The agreement of the results if good for both thicknesses and the whole frequency range. Also, a variation of the distance between the vias has been performed, the results are shown in Figs. 10 and 11, and it is found that in the investigated range the transmission and reflection are only very slightly influenced by this variation. The agreement of these parameters is not affected by this variation. This is not true for the crosstalk

for a very close spacing of the vias. The observed effects are probably due to the fact that only the lowest order mode is taken into account in the proposed method.

IV. CONCLUSION

The investigations presented in this paper show that a physics-based via model which takes into account the modified parallel plate impedance and the modified via to plane capacitance already leads to good agreement of the crosstalk parameters with reference results provided that the vias are not too close to each other. The main expectation of achieving significant reduction in simulation time compared to FEM simulations is fulfilled. Future investigations should aim at further justifying the applicability and establishing an extended model that captures the electrical behavior even better and also considers the case without bounding metal layers.

ACKNOWLEDGMENT

This work was funded in part by the German Research Foundation (Deutsche Forschungsgemeinschaft, DFG).

REFERENCES

[1] J. Knickerbocker, P. Andry, E. Colgan, B. Dang, T. Dickson, X. Gu, C. Haymes, C. Jahnes, Y. Liu, J. Maria, R. Polastre, C. K. Tsang, L. Turlapati, B. Webb, L. Wiggins, and S. Wright, "2.5D and 3D technology challenges and test vehicle demonstrations," in *Electronic Components and Technology Conference (ECTC), 2012 IEEE 62nd*, 2012, pp. 1068–1076.

[2] X. Gu, B. Wu, M. Ritter, and L. Tsang, "Efficient full-wave modeling of high density TSVs for 3D integration," in *Electronic Components and Technology Conference (ECTC), 2010 Proceedings 60th*, Jun. 2010, pp. 663 –666.

[3] S. Muller, X. Duan, M. Kotzev, Y.-J. Zhang, J. Fan, X. Gu, Y. Kwark, R. Rimolo-Donadio, H.-D. Bruns, and C. Schuster, "Accuracy of physics-based via models for simulation of dense via arrays," *IEEE Transactions on Electromagnetic Compatibility*, vol. 54, no. 5, pp. 1125–1136, 2012.

[4] R. Rimolo-Donadio, X. Gu, Y. Kwark, M. Ritter, B. Archambeault, F. de Paulis, Y. Zhang, J. Fan, H. Bruns, and C. Schuster, "Physics-based via and trace models for efficient link simulation on multilayer structures up to 40 GHz," *IEEE Transactions on Microwave Theory and Techniques*, vol. 57, no. 8, pp. 2072–2083, 2009.

[5] T. Okoshi, *Planar circuits for microwaves and lightwaves*, ser. Springer series in electrophysics. Springer, 1985.

[6] H. Hasegawa, M. Furukawa, and H. Yanai, "Properties of microstrip line on Si-SiO2 system," *IEEE Transactions on Microwave Theory and Techniques*, vol. 19, no. 11, pp. 869 – 881, Nov. 1971.

[7] D. Dahl, X. Duan, A. Beyreuther, I. Ndip, K.-D. Lang, and C. Schuster, "Application of the transverse resonance method for efficient extraction of the dispersion relation of arbitrary layers in silicon interposers," in *IEEE Workshop on Signal and Power Integrity (SPI), 2013 Proceedings 17th*, 2013, pp. 145 –148.

[8] Ansys Inc. High Frequency Structure Simulator (HFSS) Ver. 13. Pittsburgh, USA. [Online]. Available: http://www.ansoft.com/products/hf/hfss/

[9] X. Duan, R. Rimolo-Donadio, H.-D. Bruns, and C. Schuster, "Circular ports in parallel-plate waveguide analysis with isotropic excitations," *IEEE Transactions on Electromagnetic Compatibility*, vol. 54, no. 3, pp. 603–612, 2012.

[10] Y. Zhang, J. Fan, G. Selli, M. Cocchini, and F. de Paulis, "Analytical evaluation of via-plate capacitance for multilayer printed circuit boards and packages," *IEEE Transactions on Microwave Theory and Techniques*, vol. 56, no. 9, pp. 2118–2128, 2008.

[11] A. G. Williamson, "Radial-line/coaxial-line junctions: analysis and equivalent circuits," *International Journal of Electronics*, vol. 58, no. 1, pp. 91–104, 1985.

978-1-4799-0708-3/13 $31.00 © 2013 IEEE

Timing Analysis for Thermally Robust Clock Distribution Network Design for 3D ICs

Sung Joo Park, Nitish Natu, and Madhavan Swaminathan
School of Electrical and Computer Engineering,
Interconnect & Packaging Center,
Georgia Institute of Technology,
Atlanta, GA
{sjoo, natu.nitish}@gatech.edu;
madhavan.swaminathan@ece.gatech.edu

Byunghyun Lee, Sang Min Lee, Woong Hwan Ryu, and Kee Sup Kim
Design Technology Team,
System LSI Business, Device Solutions
Samsung Electronics Co., Ltd,
Yongin, Korea
{byhy.lee, sm69.lee, woong.h.ryu, kee.sup.kim}@samsung.com

Abstract—**Three-dimensional Integrated Circuits provide a solution to overcome bottlenecks in performance and power management issues. However, the drawback arises in the form of increased thermal density that results in thermal gradients that affect signal integrity. Since, the clock signal is critical for ensuring the performance of synchronous digital systems, its design is very important. In this paper we analyze the effect of thermal gradient on the clock distribution networks in the context of 3D ICs. We also propose novel methods for compensating the thermal effects which have been validated through extensive simulations and preliminary hardware measurements.**

Keywords-3D IC, TSV (Throung Silicon Via), Temperature gradient, CDN (Clock Distribution Network), Propagation Delay.

I. INTRODUCTION

The semiconductor industry has accepted 3D integration as a possible solution to address speed and power management problems. Through Silicon Vias (TSVs) are popular in such 3D structures due to their small lengths and high densities. These 3D stacking techniques have proved to dramatically increase the density of transistors in digital and mixed-signal systems. However, heat management has proved to be a concern with TSV-based systems as they are prone to temperature gradients as much as 50 °C [1].

For instance, consider a 3D system shown in Figure 1 (a) comprising of two stacked dies with different power maps. They are stacked on top of a silicon interposer which is in turn mounted on a printed circuit board [2]. The hot spots in various parts of the system shown in Figure 1 (b, c) can affect active as well as passive circuits with changes in resistivity, mobility, and threshold voltages.

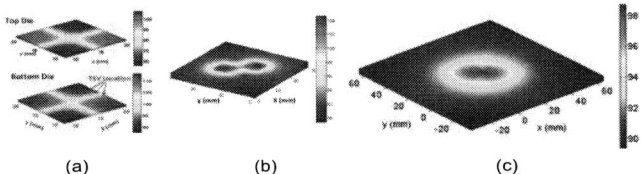

Figure 1.   Temperature Distribution in a 3D System containing TSVs.
(a) Dies. (b) Inteposer. (c) PCB.

Synchronous digital systems thrive on the reliability of the clock distribution network as it is responsible for integrity of data paths across the length and breadth of the IC. The clock distribution network (CDN) has some of the largest fanouts and longest distances. Adding the requirement for higher frequency operation makes the clock signals largely sensitive to temperature variations across the chip [3].

In this paper, we analyze the effects of temperature on the CDN and propose a combination of modifications of two techniques, adaptive voltage and controllable delay, to overcome them. This paper is organized as follows: the system configuration is described in Section II followed by the thermal analysis and the electrical analysis in Section III. Delay analysis and subsequent measurements have been presented in section IV. Section V lists the compensation methods followed by conclusions in Section VI.

II. SYSTEM CONFIGURATION

In this paper, we assume that a TSV-based 3D system comprising of 3 dies mounted on an interposer is to be designed as given in Figure 2(a). The CDN is sandwiched between two dies of logic and are integrated and connected to a PCB through the interposer by means of TSVs.

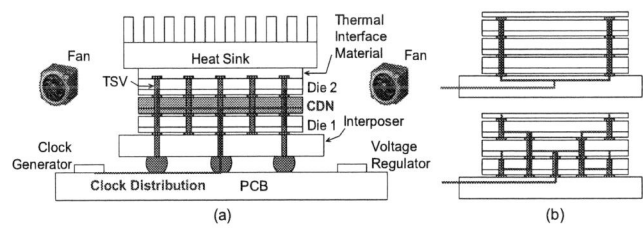

Figure 2.   System Configuration with CDN structure. (a) Centered CDN. (b) CDN on an interposer [4] and CDN with tree-structure TSV[5].

Some approaches to the CDN architecture in 3D systems have been presented in [4-5]. Clock was routed from the interposer, and distributed through the TSVs in [4] while a design using interconnects of CDN being present in the bottom die with multiple symmetrical TSVs distributing it on their way up has been shown in [5]. In this paper, we propose a system with CDN on the center die. A more detailed structure is presented with electrical modeling in Section III.

TABLE I.   SYSTEM CONDITION AND ASSUMPTION

Parameter	Unit	Value	Note
Die size	mm³	10 x 10 x 0.2	L x W x t
Interposer size	mm³	30 x 30 x 0.2	L x W x t
PCB size	mm³	100 x 100 x 1.27	L x W x t
Air convection	W/(m²K)	20	With fans
TIM conductivity	W/(m·K)	2	
T_A	°C	25	Heat sink
Underfill	W/(m·K)	4.3	

978-1-4799-0708-3/13 $31.00 © 2013 IEEE

Sizes of the chip, the interposer and the PCB are 10 mm x 10 mm, 30 mm x 30 mm, and 100 mm x 100 mm, respectively. Ratios of TSV diameters and heights are 30 µm / 100 µm for interposer and 5 µm / 50 µm for the die. We have also assumed the thermal environment and related parameters such as convection, ambient temperature, and thermal conductivities, as shown in Table I.

III. THERMAL AND ELECTRICAL CO-ANALYSIS

A. Thermal Modeling and Simulation

A finite volume formulation presented in [6] has been used for running thermal simulations. The solver can accurately capture voltage and current distributions with temperature distribution across any layer in the 3D stack with Joule heating. It takes the material parameters shown in Table I as input and considers them in correspondence to the 3D structure.

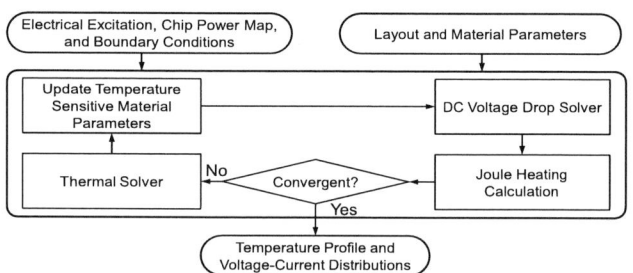

Figure 3.   Thermal Solver Operation [6].

The 3D IC configuration shown in Figure 2 was constructed and simulated using the solver. Power was randomly distributed in all the three layers for generating an initial temperature profile. The clock distribution network architecture was then considered (H-Tree) to build a power map for the center die (CDN). The solver was fed with the 3D structure and material parameters to generate temperature profiles for all the layers. The profile for the center die was evaluated further against expected heat-maps. The results show that the temperature gradient is significantly more in case of power being randomly distributed across the die as against an evenly distributed power map.

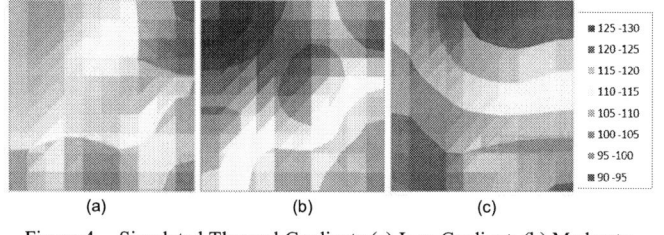

Figure 4.   Simulated Thermal Gradients (a) Low Gradient. (b) Moderate Gradient. (c) High Gradient.

The profiles were sorted by the temperature gradient across the die as it would have the maximum effect on the propagation delay of the clock signals. Three profiles of varying gradients were then selected to be superimposed on the electrical model for the final simulation as shown Figure 4.

B. Electrical Modeling and Simulation

The CDN was considered to be constructed on the center die and have an H-Tree architecture with inverters being used

as clock repeaters and placed 500µm along the trace with metal interconnects, as shown in Figure 5. The technology nodes had a significant impact on the electrical parasitics, which, along with higher resistivity, have a large effect on the clock signals.

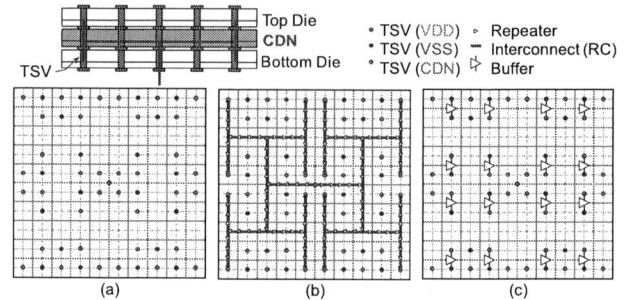

Figure 5.   Chip Configuration (a) Bottom Die. (b) CDN Die. (c) Top Die.

A BSIM4 CMOS model of 45nm technology from [7] was chosen, with clock repeaters that depict a buffer sizing profile described in [8]. The evaluation of the CDN was done with lumped RC values for interconnects from [9]. The unit buffer sizes (W/L) for PMOS and NMOS are 630 nm and 195 nm respectively. Note that the TSVs used for connecting the ends of the CDN to adjacent dies are only a subset of TSVs used for the complete 3D integration. A lumped TSV model from [10] was selected to complete the electrical model of the system.

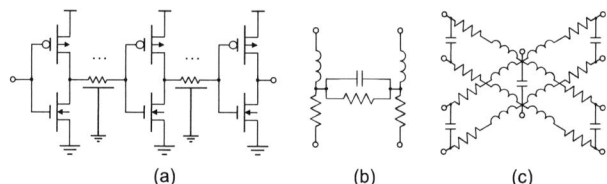

Figure 6.   Schematic of Simulation Model. (a) CDN. (b) TSV. (c) PDN.

A meshed on-chip power distribution network model in [11] has been added to estimate Power Distribution Network (PDN) effect. Data buffers for PDN noise sources were added along with their model for on-chip decoupling capacitors [12]. Schematics of the models can be seen in Figure 6. The resistance of the lumped resistor used for modeling the CDN and PDN interconnects as well as the TSVs contain a temperature coefficient which is directly responsible for the thermal dependency which is included in the BSIM4 model.

TABLE II.   GEOMETRAL PARAMETERS FOR ELECTRICAL PARISITICS

Component	Width/Diameter	Thickness/Height	Pitch/Space
CDN	1 um (w_{CDN})	1 um (t_{CDN})	N/A
TSV	5 um (d_{TSV})	50 um (h_{TSV})	N/A (p_{TSV})
PDN	10 um (w_{PDN})	50 um (t_{PDN})	50 um (s_{PDN})

Table II shows the geometral parameters for the CDN, the TSVs, and the PDN models while Table III lists the parasitics from the models with respect to the geometral parameters.

Electrical transient simulations were done using Agilent's ADS 2009 with the aforementioned BSIM4 model. The clock signal has amplitude of 1.1V with a frequency of 500MHz. A supply voltage (V_{DD}) of 1.1V is fed to the clock buffers by the voltage regulator through the PDN model.

TABLE III. GEOMETRAL PARAMETERS FOR ELECTRICAL PARISITICS

Component	R	L	C	Note
CDN	30 ohm	N/A	200 fF	per mm
TSV	61.4 mohm	29.4 pH	4.0 fF	per TSV
PDN	430 mohm	22.3 pH	1740 fF	per mm^2

Simulations show that the base condition of an ideal PDN (a) has a skew of 30.7 ps. The addition of PDN anomalies (b) increases the skew by 19.2 ps (62.5%). Addition of temperature effects to the ideal PDN gives a rise of 143.6 ps (467.8%). A further rise of 20.3 ps (68.2%) can then be seen when temperature gradient is superimposed on the PDN (d).

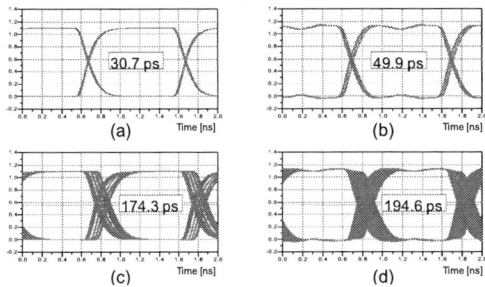

Figure 7. Simulated skew (a) Ideal PDN without temperature effects. (b) With PDN effects without temperature effects. (c) Ideal PDN with temperature effects. (d) With PDN and temperature effects.

IV. DELAY ANALYSIS

A. Analytical Analysis

In this paper, we have performed an analytical analysis for the propagation delay and the findings are reinforced using measurements from the test vehicle. The Elmore delay model has proved to be accurate for analyzing RC interconnects [13]. It defines propagation delay as given in equation (1).

$$t_p = 0.69R_{DR}C_W + 0.69R_{DR}(C_O + C_I) + 0.38R_W C_W + 0.69R_W C_I \quad (1)$$

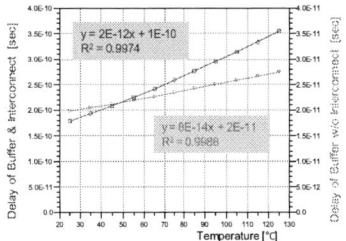

Figure 8. Four portion of propagation delay of an unit circuit. (a) Inverter driving wire cap. (b) Inverter. (c) Wire delay. (d) Wire driving an inverter.

The RC delay unit of the CDN has been divided into 4 parts. Each part can be defined as a combination of resistance and capacitance along with an inverter or a wire, as shown in Figure 8. The wire capacitance (C_W) and the driver resistance (R_{DR}) have the largest impact on the delay as can be seen through the calculations listed in Table IV.

TABLE IV. DELAY VALUES WITH BUFFERS AND WIRE

	Formula			Values			Delay	Note
	Network	R	C	Coeff.	R [Ω]	C [F]	[ps]	
(a)	Lumped	R_{DR}	C_W	0.69	1.25k	100f	86.3	82.7 %
(b)	Lumped	R_{DR}	C_O+C_I	0.69	1.25k	20f	17.3	16.6 %
(c)	Distributed	R_W	C_W	0.38	15	100f	0.6	0.6 %
(d)	Lumped	R_W	C_I	0.69	15	10f	0.2	0.2 %

The values for the delay are dependent on temperature. Figure 9 shows thermal dependency of each of the 4 parts of the RC Delay unit defined earlier. Resistance of copper is known to have a linear dependency on temperature with a coefficient of 0.0039 while the capacitance constructed on a silicon substrate or SiO_2 is considerably stable with a negligible temperature coefficient. It can be concluded from the figure that the RC delay has a linear relationship with temperature.

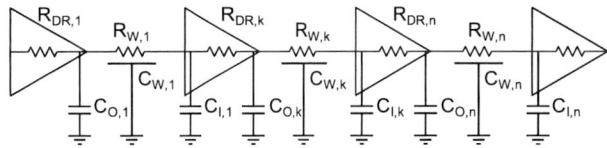

Figure 9. Temperature dependency of the delay.

A lumped electro-thermal model, accounting for thermal effects through temperature coefficient of the interconnect resistance was presented in [14]. Here, we accommodate the driver resistance as well as the buffer capacitance to extend the formula as given in equation (3).

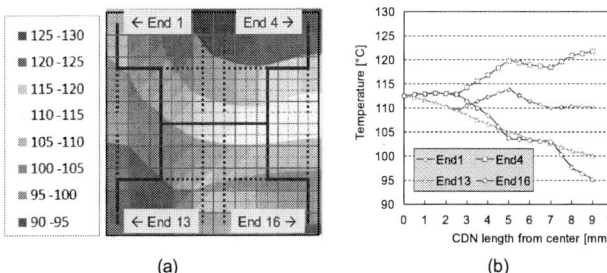

Figure 10. Cascaded buffer and interconnect.

Thus, equation (2) provides an accurate measure for delay (D) as it considers the effects of temperature on the buffers given in equation (3), as well as the effects of temperature on the interconnects given in equation (4). The constants for lumped (a = 0.69) and distributed (b = 0.38) networks can then be multiplied to obtain the delay.

$$D = a\sum_{i=1}^{n} R_{DR,i}\left(C_{O,i} + C_{W,i} + C_{I,i}\right) + \sum_{i=1}^{n} R_{W,i}\left(bC_{W,i} + aC_{I,i}\right) \quad (2)$$

$$R_{DR,k} = R_{DR,0}\left(1 + \beta_{DR}(T_k - T_0)\right) \quad (3)$$

$$R_{W,k} = R_{W,0}\left(1 + \beta_W(T_k - T_0)\right) \quad (4)$$

Figure 11. (a) Temp gradient. (b) Temp profile for delay calculation.

A delay profile simulated for all the 4 ends of the CDN, with the temperature profile superimposed on the electrical model is shown in Figure 11. It provides correlation between calculations and simulations as shown in Figure 12 (a).

B. Measurements

The analytical and simulated results have been confirmed using the measurements done through an FPGA-based test vehicle. A part of the center die (CDN) was reconstructed by structural Verilog coding on the Spartan®-6 Evaluation Board and the temperature gradient was simulated by an external input. Only a single buffer with the RC interconnect model was coded to establish correlation with the simulations. The Timing Analyzer tool was used for further evaluations on the module.

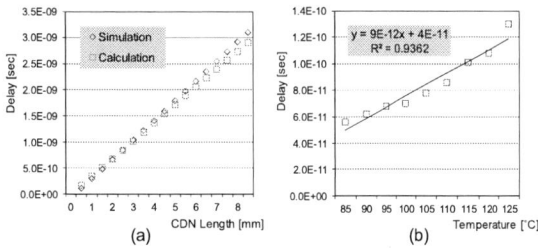

Figure 12. (a) Delay profile. (b) Measured unit delay from the test vehicle.

V. DELAY COMPENSATION

Delay variations caused by thermal variations can be compensated by adjusting buffer parameters. In this paper, we propose a combination of two methods to ensure compensation against delay variations: varying buffer strength and controlling interconnect delay. First approach makes use of the fact that temperature gradient affects threshold voltage and mobility, which can also be controlled through bias voltages, V_{DD} [15]. This would require temperature sensors and level converters. The other approach compensates by delaying faster signals using adjustable loads [16].

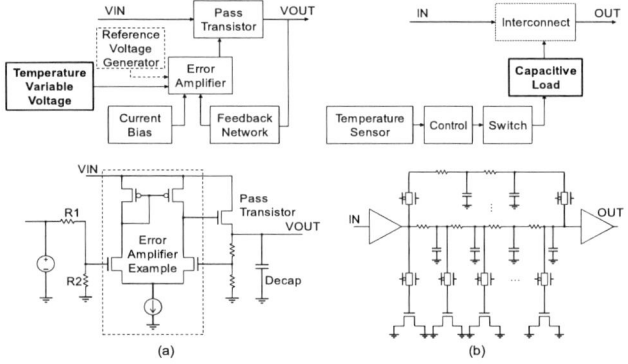

Figure 13. Block diagram and schematic of delay compensation. (a) Variable reference voltages for linear regulators. (b) Controllable delay for interconnect.

However, the values of additional tunable loading capacitors tend to cause problems as they are delay dependent. Here, we modify and investigate these methods as shown in Figure 13. We use variable output voltages of linear regulators obtained by adjusting their reference for the clock repeaters. The temperature coefficient of the resistance is responsible for implementing the thermal dependent reference voltage, thus eliminating the need for a temperature sensor. Larger variations are however left for the controllable delay mechanism to handle. Figure 14 shows the simulations for both the compensation methods. The values of skew are marked in the figure itself. A combination of these methods can be used depending on the amount of compensation required.

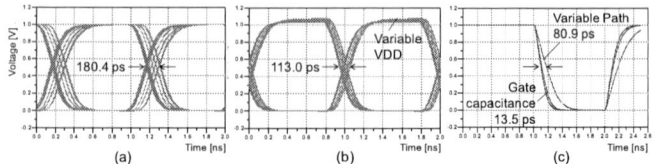

Figure 14. Delay compensation (a) Without compensation. (b) Compensated by variable VDD. (c) Variable unit delays by gate capacitances or paths.

VI. CONCLUSIONS

We have quantified the effects of temperature on the clock skew in TSV-based 3D ICs and systems in the 45nm CMOS technology with an on-chip PDN. The results clearly show that the temperature gradient affects the clock skew due to thermal dependent components in the PDN. A new equation for calculation of delay including effects on buffers as well as interconnects has been proposed. The analytical results are reinforced through measurements performed on the test vehicle. Lastly, the methods to compensate the temperature effects have been analyzed and modifications to the same have been proposed.

ACKNOWLDEGEMENT

This work was sponsored by Samsung Electronics Co., Ltd.

REFERENCES

[1] S. Borkar et al., "Parameter variations and impact on circuits and microarchitecture," Proc. of DAC, vol. 64, pp. 338-342, 2003.

[2] J. Xie, M. Swaminathan, "Electrical-thermal co-simulation of 3D integrated systems with micro-fluidic cooling and Joule heating effects," IEEE Trans. on CPMT, vol. 1, no. 2, pp. 234-246, 2011.

[3] E. G. Friedman, "Clock distribution networks in synchronous digital integrated circuits," Proc. of the IEEE, vol. 89, no. 5, pp. 665-692, 2001.

[4] D. Kim et al., "Distributed multi TSV 3D clock distribution network in TSV-based 3D IC," IEEE 20th Conference on EPEPS, pp. 87-90, 2011.

[5] D. Kim et al., "Vertical Tree 3-dimensional TSV Clock Distribution Network in 3D IC," IEEE 62nd ECTC, pp. 1945-1950, 2012.

[6] J. Xie and M. Swaminathan, "Fast electrical-thermal co-simulation using multigrid method for 3D integration," IEEE 62nd ECTC, pp. 651-657, 2012.

[7] 45nm NCSU FreePDK™, http://www.si2.org, 2012.

[8] PTM (Predictive Technology Model), http://ptm.asu.edu, 2012.

[9] ITRS Roadmap - Interconnect, http://www.itrs.net, 2011.

[10] J. Kim et al., "High-frequency scalable electrical model and analysis of a Through Silicon Via (TSV)," IEEE Trans. on CPMT, vol. 1, no. 2, pp. 181-195, 2011.

[11] J. Pak et al., "PDN impedance modeling and analysis of 3D TSV IC by using proposed P/G TSV array model based on separated P/G TSV and chip-PDN models," IEEE Trans. on CPMT, vol. 1, pp. 208-219, 2011.

[12] K. Kim et al., "Modeling and analysis of a Power Distribution Network in TSV-based 3-D memory IC including P/G TSVs, on-chip decoupling capacitors, and silicon substrate effects," IEEE Transactions on CPMT, vol. 2, no. 12, pp. 2057-2070, 2012.

[13] J. M. Rabaey et al, "Digital Integrated Circuits ," Prentice-Hall, 2003.

[14] N. Spennagallo et al., "Lumped electro-thermal model of on-chip interconnects," Proc. of 12th THERMINIC, pp. 220-224, 2006.

[15] K. Shakeri and J. D. Meindl, "Temperature variable supply voltage for power reduction," Proc. of ISVLSI, pp. 1-4, 2002.

[16] A. Chakraborty et Al., "Dynamic thermal clock skew compensation using tunable delay buffers," IEEE Trans. on VLSI Systems, vol. 16, no. 6, pp. 639-649, 2008.

Mitigation of TSV-Substrate Noise Coupling in 3-D CMOS SOI Technology

Xiaoxiong Gu and Keith Jenkins

IBM T. J. Watson Research Center, Yorktown Heights, NY, USA {xgu, jenkinsk} @ us.ibm.com

Abstract— **Substrate noise coupling in 3-D CMOS SOI technology is characterized using hardware measurement. Couplings between device contacts and through-silicon vias (TSVs) are measured in frequency domain. Time domain simulations based on the measured S-parameters are performed to assess the impact of TSV-induced noise coupling on active circuit performance. Equivalent circuits are constructed with good model-to-hardware correlation. The characterization results demonstrate a dominant noise-coupling path through N+ epi layer in the SOI substrate. The data also successfully validates our proposed noise mitigation technique of using CMOS process compatible buried interface contacts, e.g., achieving over 20-dB reduction for TSV-induced substrate noise coupling at 1 GHz.**

Keywords- 3-D integrated circuit (IC), 3-D integration, through silicon via (TSV), substrate noise.

I. INTRODUCTION

Three-dimensional (3-D) silicon integration technology, featuring thinned die-to-die bonding and through-silicon-via (TSV) interconnections, enables dense local chip-to-chip interconnect and holds promise for improved performance of integrated system by increasing interconnect bandwidth [1]. Integration of copper-based TSV with high-K/metal gate and embedded-DRAM to build functional 3D modules has been demonstrated [2, 3]. As the TSV pitch keeps scaling down for higher I/O density, concerns have been raised that TSV-induced substrate noise coupling can cause detrimental effects on nearby sensitive circuits. Different noise isolation techniques have been proposed including a guard ring design [4], grounded TSV shielding structures [5], and ground metal "plugs" which penetrate the substrate with a substantial depth [6]. Bulk silicon substrates have been considered in the analyses of these approaches.

We recently examined TSV-induced noise coupling mechanism for silicon-on-insulator (SOI) substrates, which demonstrate superior noise isolation compared to bulk silicon due to the buried oxide layer capacitance [7]. However, the buried oxide layer becomes transparent and offers little advantage above a few GHz, unless combined with underlying high resistivity silicon [8]. Our analysis took into account an additional highly doped N+ epi layer in SOI CMOS. Equivalent circuit models were extracted and used in time domain analysis to assess the impact of noise coupling on active circuit performance. The simulation results demonstrate that the N+ epi layer dominates the SOI substrate noise path due to its high doping profile.

In order to mitigate TSV-induced noise coupling, we introduced a technique of using a specific type of substrate contact, i.e., a buried metal-filled interface tie that connects the N+ epi layer to ground [9]. A low impedance ground return path can be easily formed with this technique to effectively mitigate TSV-to-device noise coupling without significantly affecting IC wiring density and flexibility. Fig. 1 (right) illustrates how a substrate contact penetrates the device layer and buried oxide layer and connects the N+ epi layer with ground through redistribution layer (RDL) or back-end-of-line (BEOL) layers. A device contact is shown on the left for noise coupling characterization.

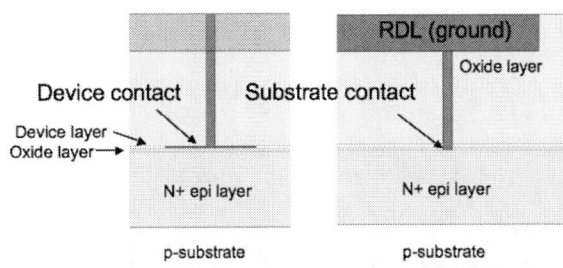

Figure 1. Illustration of cross-sections of device contact (left) and substrate contact (right) including buried oxide layer and N+ epi layer.

This paper presents the recent hardware measurement results for characterizing TSV-induced substrate noise coupling in advanced 3-D CMOS SOI technology. Frequency-domain S-parameter measurement data are presented up to 20 GHz for two-port test sites including contact-to-contact and TSV-to-contact couplings. First-order equivalent circuit models are extracted to correlate with the measurement data. Time-domain analysis based on the measured S-parameters is also conducted to assess the impact of noise coupling on active circuit performance. As will be shown in this work, applying substrate contacts to connect the N+ epi layer to ground can effectively build a low-impedance return path, which significantly mitigates the substrate noise coupling in 3-D CMOS SOI technology.

II. MEASUREMENT SET-UP AND TEST SITE DESCRIPTION

Fig. 2 illustrates a test macro for 2-port substrate noise coupling measurement. The macro was fabricated on a 300mm wafer using advanced CMOS SOI technology. Ground-signal-ground 100μm-pitch surface probe launch is designed for broadband S-parameter measurement. The measurement set-up is shown in Fig. 3. Five groups of test sites are implemented, including contact-to-contact, TSV-to-contact and TSV-to-TSV

978-1-4799-0708-3/13 $31.00 © 2013 IEEE

couplings. The nominal spacing between the source and the victim is 200μm (groups 1~4) except that the spacing is changed to 100μm and 300μm in the group 5.

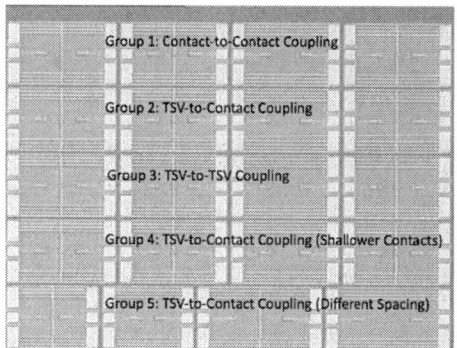

Figure 2. A microscropic view of on-wafer test sites for substrate noise coupling measurement.

Here, both TSV and contact are connected to the signal pad through wires on the top BEOL layer. The copper-based TSVs with silicon dioxide liner are immersed in a wafer before the thinning process. The 25μm × 25μm device contacts in groups 1, 2 and 5 go all the way through BEOL to the poly-Si layer and diffused regions, whereas the contacts in the group 4 go only to the bottom M1 layer. Furthermore, the effects of using substrate contact to mitigate noise coupling are investigated. In groups 1~4, in addition to the baseline test site, we also add a substrate contact in the middle which has 0, 4 or 100 buried interface ties which connect the N+ epi layer to ground.

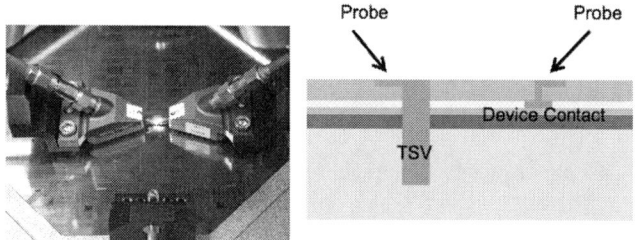

Figure 3. Two-port S-parameter measurement using 100μm-pitch microwave probes (left) and an illustration of TSV to device contact coupling measurement (right).

III. FREQUENCY DOMAIN MEASUREMENT RESULTS

Figs. 4 and 5 plot the measured noise coupling signals for contact-to-contact and TSV-to-contact coupling, respectively. In each case, four different configurations are compared: (1) without a middle substrate; (2) with a middle substrate contact which has no buried interface tie to connect to the N+ epi layer; (3) with a middle substrate contact which has 4 ties; and (4) with a middle substrate contact which has 100 ties. The measurement data are shown from 500 MHz to 20 GHz. Notice that two lowest frequency points (45 MHz and 295 MHz) are excluded here because the excessive probe coupling (i.e., -62 dB and -72 dB when lifted in air) overwhelms the measured substrate noise coupling at these frequencies.

For SOI substrates, it can be observed in Figs. 4 and 5 that the coupling level increases in the gigahertz range when N+ epi

layer is present. More importantly, the data demonstrates that adding a substrate contact with buried interface ties to the N+ epi layer effectively reduces substrate noise coupling over a wide frequency range. For example, over 10-dB and 20-dB noise coupling reduction are shown at 1 GHz using 4 ties and 100 ties in both cases.

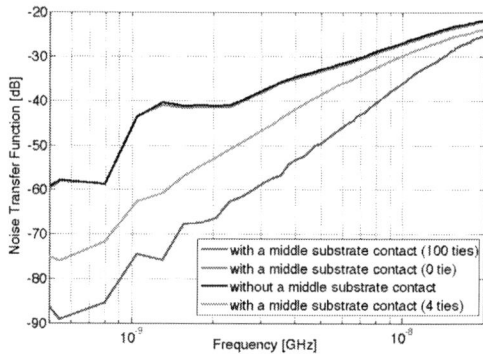

Figure 4. Measured contact-to-contact noise transfer function with and without buried interface ties for coupling mitigation.

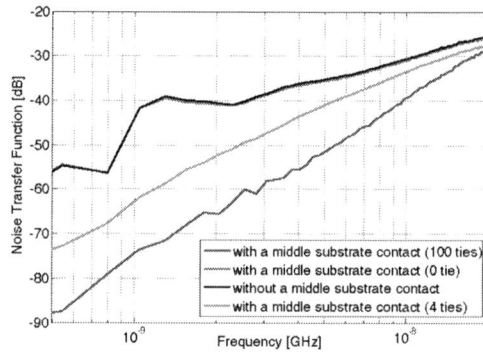

Figure 5. Measured TSV-to-contact noise transfer function with and without buried interface ties for coupling mitigation.

IV. EQUIVALENT CIRCUIT MODELS

First-order equivalent circuit models shown in Figs. 6 and 7 are constructed to shed light on different coupling mechanisms found in the measurement. Fig. 6 illustrates a circuit model for the contact-to-contact coupling. The probe launch parasitics are represented by C_{launch} (31.6 fF) and R_{launch} (74.8 mΩ). C_{BOX} (15 fF) is the buried oxide layer capacitance. The N+ epi layer is modeled by R_{N+} (800 Ω) due to a high doping profile. A small loop inductance (L=0.3 nH) is taken into account in the model. When a substrate contact with buried interface ties is applied, $R_{contact}$ (800 Ω) is used to model the connection to ground with 4 ties.

Similarly, Fig. 7 shows the equivalent circuit model for TSV-to-contact coupling. The TSV oxide liner capacitance is broken into two parts to capture the two coupling paths through the N+ epi layer and p-substrate respectively (C_{TSV1} = 222 fF and C_{TSV2} = 6.5 fF). C_{sub} and R_{sub} are the silicon substrate capacitance and resistance, respectively (C_{sub} = 1 fF, R_{sub} = 100 kΩ).

978-1-4799-0708-3/13 $31.00 © 2013 IEEE

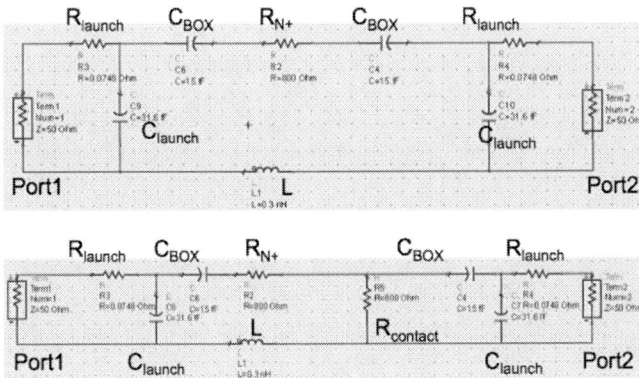

Figure 6. First-order equivalent circuit to model contact-to-contact coupling: (top) without and (bottom) with buried interface ties.

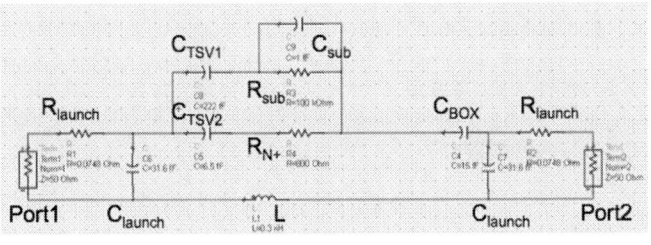

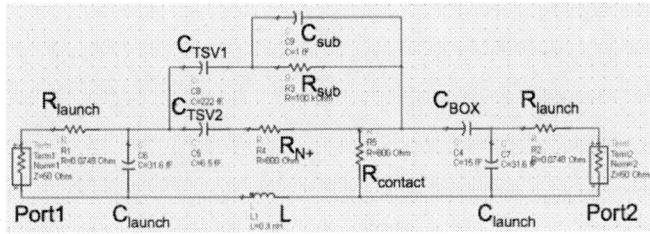

Figure 7. First-order equivalent circuit to model TSV-to-contact coupling: (top) without and (bottom) with buried interface ties.

Figs. 8 and 9 plot the model-to-hardware correlation for contact-to-contact and TSV-to-contact coupling, respectively. From a circuit point of view, because a majority of substrate noise coupling in terms of a displacement current occurs in the N+ epi layer, R_{N+} and C_{BOX} dominate the noise transfer function curves over the entire frequency range.

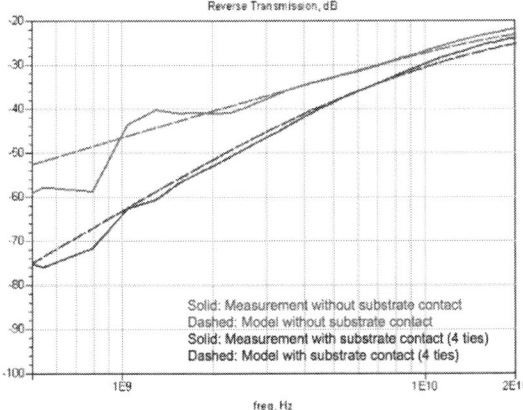

Figure 8. Model-to-hardware correlation for contact-to-contact coupling with and without buried interface ties.

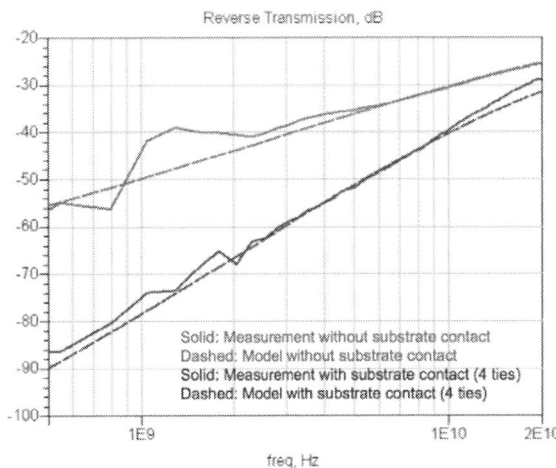

Figure 9. Model-to-hardware correlation for TSV-to-contact coupling with and without buried interface ties.

The effect of adding a substrate contact with buried interface ties on reducing noise coupling is also clearly visible. In the circuit simulation, the value of $R_{contact}$ can be tuned to account for different number of substrate contacts for noise mitigation.

V. TIME DOMAIN SIMULATION RESULTS

Time domain simulation using the measured TSV-to-contact S-parameters are used to analyze the TSV-induced noise amplitude that could appear at an active device. Fig. 10 shows the simulation setup, where a 1 volt 10 Gb/s pulse signal with 20 ps rise time is applied to the TSV at port 1. The 50-Ω resistance on the left side represents the termination of TSV. The load at the device end (port 2) is a resistance representing the on-resistance of a CMOS device in parallel with a capacitance emulating the total capacitance including p-n junction, wire and gate capacitances.

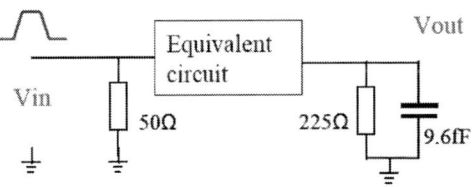

Figure 10. Time domain simulation setup based on TSV-to-contact measurement data in Fig. 5.

Simulation results are plotted in Fig. 11. The input pulse waveform is shown in Fig. 11-a as a reference. Fig. 11-b shows the output noise waveform at the device end for an SOI substrate with N+ epi layer. A noise peak at ~32 mV is observed if no substrate contact is used to mitigate noise coupling. Fig. 11-c and Fig. 11-d show the output noise waveform for the same substrate but using a substrate contact with 4 and 100 ties. The noise peaks are reduced to ~24 mV and ~12 mV, respectively. This validates the effectiveness of TSV-to-device coupling reduction with the proposed technique.

978-1-4799-0708-3/13 $31.00 © 2013 IEEE

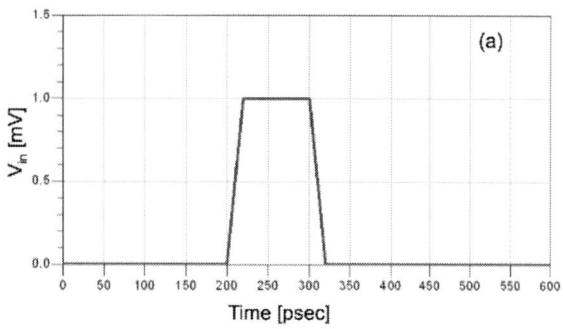

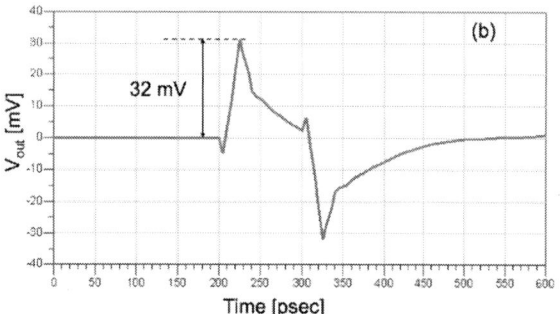

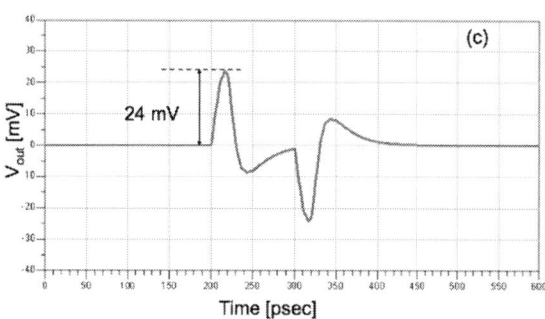

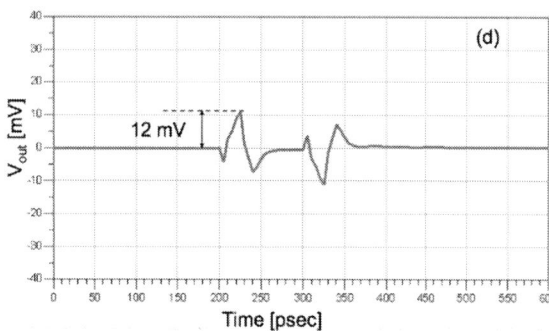

Figure 11. Time domain simulation results: (a) input pulse waveform; (b) output noise waveform for SOI substrate with N+ epi layer; (c) and (d) output noise waveform after including a substrate contact with 4 and 100 buried interface ties.

VI. CONCLUSION

Our previous frequency- and time- domain modeling and analysis demonstrated that TSV-induced substrate noise coupling to active area deteriorates for SOI substrates after taking into account a highly doped N+ epi layer. We proposed using buried interface contacts to mitigate the noise coupling. The new hardware characterization results in this paper successfully demonstrate that the substrate noise coupling in 3-D SOI substrates can be significantly reduced by connecting the N+ epi layer to ground to form a low impedance signal return path. The proposed substrate contacts are small in size, flexible for placement, compatible with advanced SOI CMOS fabrication processes and therefore offer a practical and effective solution for mitigating TSV-induced noise in 3D IC.

ACKNOWLEDGMENT

The authors would like to thank Joel Silberman, Shih-Hsien Lo, Yong Liu, Rachel Gordin, Michael Cranmer, John Safran at IBM for their discussions and technical support, as well as Christy Tyberg, Peilin Song and Clint Schow at IBM for their management support.

REFERENCES

[1] J. Knickerbocker et al., "3-D Silicon Integration and Silicon Packaging Technology Using Silicon Through-Vias," IEEE J. Solid-State Circuits, vol. 41, no. 8, pp. 1718-1725, Aug. 2006.

[2] G. Wang, et al., "Scaling Deep Trench Based eDRAM on SOI to 32nm and Beyond," Proceedings of IEEE Devices Meeting (IEDM), pp. 1-4, 2009.

[3] M. G. Farooq, et al., "3D Copper TSV Integration, Testing and Reliability," Proceedings of IEEE Devices Meeting (IEDM), pp. 7.1.1-7.1.4, 2011.

[4] J. Cho et al., "Modeling and Analysis of Through-Silicon Via (TSV) Noise Coupling and Suppression Using a Guard Ring," IEEE Trans. Components, Packaging and Manufacturing Technology, Vol. 1, Issue 2, pp. 220-233, 2011.

[5] J. Cho et al., "Through Silicon Via (TSV) Shielding Structures", Proceedings of IEEE 19th Conference on Electrical Performance of Electronic Packaging and Systems (EPEPS), pp. 269-272, 2010.

[6] N. H. Khan, S. M. Alam and S. Hassoun, "Mitigating TSV-induced Substrate Noise in 3D-IC using GND Plugs," Proceedings of 12th International Symposium on Quality Electronic Design, pp. 751-757, 2011.

[7] X. Duan, X. Gu, J. Cho and J. Kim, "A Through-Silicon-Via to Active Device Noise Coupling Study for CMOS SOI Technology," Proceedigns of IEEE 61st Electronic Components and Technology Conference (ECTC), pp. 1791-1795, 2011.

[8] K. A. Jenkins, "Substrate Coupling Noise Issues in Silicon Technology", Digest of Silicon Monolithic Integrated Circuits in RF Systems, pp. 91-94, 2004.

[9] X. Gu, J. Silberman, Y. Liu and X. Duan, "Mitigating TSV-induced Substrate Noise Coupling in 3-D IC Using Buried Interface Contacts," Proceedings of IEEE 21st Conference on Electrical Performance of Electronic Packaging and Systems (EPEPS), pp. 75-78, 2012.

978-1-4799-0708-3/13 $31.00 © 2013 IEEE

Power Integrity and Wireless Power Transfer

978-1-4799-0708-3/13 $31.00 © 2013 IEEE

978-1-4799-0708-3/13 $31.00 © 2013 IEEE

Power Integrity Analysis for Core Logic Blocks

Dan Oh, Alex Razmadze, and Karthik Chandrasekar

Altera Corporation
101 Innovation Drive, San Jose, CA 95134
doh@altera.com

Abstract— **In this paper, the new framework for power integrity analysis for core logic blocks is presented. Balancing on-chip timing budget becomes more challenging as both data and clock jitter increase due to large power noise. Conventional power integrity (PI) analysis focuses on reducing supply noise and does not provide timing jitter information. This paper proposes a general framework to model timing jitter due to supply noise. The proposed methodology can be used to define power distribution network (PDN) design requirements. This timing-based PDN design leads to significant reduction in the pessimism associated with either conventional target impedance concept, or static or dynamic power integrity analysis.**

Keywords-target impedance, core timing, digital timing, core jitter, power integrity, clock uncerntainty

I. INTRODUCTION

Target impedance concept [1] has been used to design power distribution network for more than a decade. The target impedance method limits AC voltage noise based on Ohm's law. Although this method warrants proper circuit operations, it can lead to significant overdesign. In modern IC design with a large current requirement, the target impedance can be very difficult to meet, and alternative or modified approaches have been used in practice. For instance, a heuristic approach based experiment is used to build an impedance border line in [2], whereas a transient switching current is used to define the target impedance in [3]. Most existing approaches focus on voltage noise and the impact on timing is not considered. Typical AC power noise significantly increases timing jitter; hence, it may lead to more detrimental impact to chip Fmax performance than noise. Also, the timing-based approach avoids the system overdesign as it is directly related to system performance.

Power supply noise induced jitter (PSIJ) has been extensively studied in recent years [4]. Much work has been focused on interface (IO) designs where timing jitter is a crucial part of the overall system performance. Proper modeling of PSIJ can lead to reduction in PDN requirements; for example, in clock forwarding architecture, the impact of jitter is drastically reduced due to jitter tracking between data and clock signals [5]. Unfortunately, no major work has been performed to analyze timing jitter for core logic blocks. The majority of work has been still focused on quantifying or limiting supply voltage noise, and the supply noise impact on core Fmax performance still remains to be unknown. Static timing analysis (STA) with a simple clock uncertainty term is typically used in digital core design for several decades to capture any jitter due to supply voltage noise.

The first serious attempt to analyze jitter impact to core performance was carried out by Intel [6], [7]. In the early 2000s, Intel was performing aggressive frequency scaling, reaching 4GHz core frequencies with large current consumption. To keep the supply noise under the traditional target impedance goal, a very large on-die decoupling capacitor (ODC) would have been needed. To quantify a true PDN impact to core performance, Intel fabricated the same CPU architecture with and without (extrinsic) ODC and performed extensive measurements [6]. Surprisingly, they observed minimum impact to core Fmax performance in spite of large voltage noise increases caused by the removal of extrinsic ODC. Although a few potential root causes were suggested, no clear explanations were given by the authors. In a few years later, a more meaningful explanation based on jitter tracking between data and clock signals was given by Intel [7]. This work became a baseline for later work on optimizing clock circuitries [8], [9].

In this paper, we present a rigorous formulation of PSIJ for core logic blocks. The proposed formulation accurately predicts core performance loss due to supply noise by accounting potential data and clock jitter tracking. We illustrate that the formulation described in [7] is suitable for traditional IO interface but may not be applicable for on-chip logic timing analysis. We also note that, in order to avoid double counting of some of timing errors, PSIJ calculation must account for any static delay variation modeled in STA timing analysis.

II. IMPACT OF ODC ON CORE FMAX PERFORMANCE

Intel's experiment in the early 2000s [6] was surprising to many signal and power integrity engineers. Power integrity simulation was predicting an increase of 15% supply noise when extrinsic ODC was removed from their Pentium™ processor. Significant performance degradation due to this power noise was expected, but very minimal performance degradation (1%) was observed. To get further insights on this phenomenon, similar experiment was performed by Altera. Thirty four percent of the total ODC was removed from one of Altera device and the performance of the test chip was compared with the normal production device.

Table I summarizes findings. The register-to-register performance is measured as shown in Figure 1. Noise is generated by an internal noise generator using different excitation patterns. Detail procedure is described in [10]. The measured voltage noise is increased significantly depending on different excitation patterns (30%~85%). Period jitter has been also increased by 20% to 50%. However, similar to the Intel case, the Fmax performance loss was relatively small (a few percent). The difference from Intel's experiment is that we had a simple register transaction and our noise was manually generated using a noise source instead of real applications.

978-1-4799-0708-3/13 $31.00 © 2013 IEEE

Table I. Performance comparison of the production device and test chip device with ODC removed

Noise Pattern	Noise (p-p, mV)			Period Jitter (p-p, psec)			Fmax Loss Variation (%)		
	Production	Test Chip	% inc	Production	Test Chip	% inc	Production	Test Chip	% diff
AC Steady State (20MHz)	62	115	85%	272	391	43%	-7.19	-9.29	-2.1%
AC Steady State (50MHz)	43	62	44%	243	293	21%	-2.88	-3.73	-1%
PRBS15 Burst (280Mb/s)	149	197	32%	303	449	48%	-12.5	-16.0	-3.5%

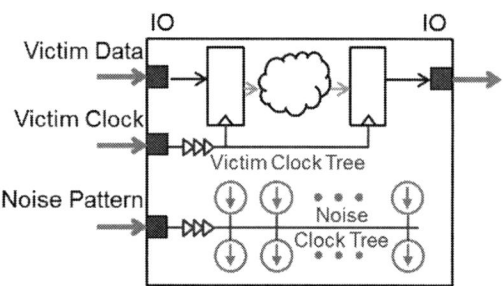

Figure 1. Register-to-register performance testing setup

A few speculative root causes were given by Intel such as wrong current estimation, multicycle critical path being not sensitive to noise, and jitter tracking between data and clock [6] [7]. Although from these studies the conclusion was made that jitter tracking is the major contribution for this minimal Fmax performance degradation, an alternative explanation is presented in this paper. In Section III the reason is shown why the jitter tracking is less likely to be the cause for this phenomenon although it somewhat reduces the jitter impact. Our explanation is based on the ramp time required to have a large surge current. Supply noise is a function of both the current amount and rate of change. When a large device goes through a big current surge, it would take a significant amount of time to ramp up. Figure 2 illustrates the difference between fast and slow ramp time. Most of power integrity engineers simulate the power noise with a fast step current as shown in Figure 2 (a). If ramp up and down times are sufficiently slow, the resonance noise disappears as shown in Figure 2(b).

In real applications, it is impossible to have an actual application to demand such large a current at once. It would take at least tens of clock cycles to ramp up large current if not hundreds. There are some corner cases such as clock and power gating which cause a sudden current change. In such cases, the performance is not critical and the operation can be staged sufficiently to reduce power noise or normal operation can wait until supply noise settles down. Figure 3 shows another experiment with partitioning our noise source into four smaller sources. These smaller sources are excited in stages. As shown in the figure, we can observe significant reduction in noise and period jitter (~80%). Fmax testing shows the performance degradation is reduced by ~44%.

Although this explanation clarifies Intel's results, it still does not explain our results in Table I as our experiment used the step surge current [10]. In the following section, we present a rigorous formulation of core jitter modeling and it is used to explain our results based on numerical examples.

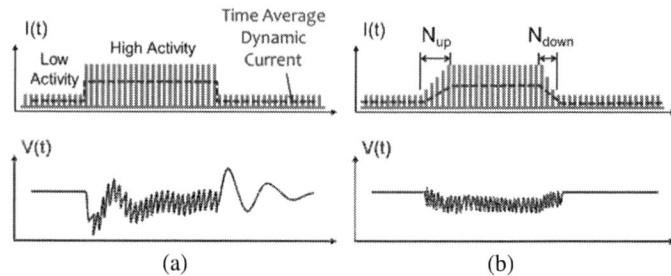

Figure 2. Supply noise due to (a) a sudden step surge current and (b) gradual ramp current

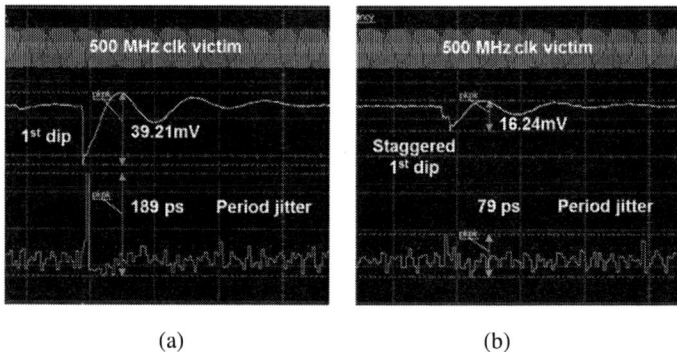

Figure 3. Measurements of supply noise and period jitter due to (a) a step surge current and (b) gradual ramp current with four-stages

III. FORMULATION OF CORE JITTER SENSITIVITY

Jitter tracking between data and clock paths is considered the main reason behind the minimal impact of supply noise on Fmax performance [7]. Although this is a convincing argument, the formulation used in [7] is only valid for general IO analysis and it may not be accurate for core logic timing analysis. This section presents the formulations for both IO interface and core logic blocks and we demonstrates that, although jitter tracking occurs in high-frequency regions, it is not so effective around the package resonance frequency where power noise dominates.

A. Traditional IO Case

Assuming multiple data bits are transferred along multiple clock cycles and data is captured using the rising edges of clock signals as shown in Figure 4(a), the noise-to-jitter sensitivity function, $H_{Net}(\omega)$, can be derived as follows:

$$H_{Net}(\omega) = H_d(\omega) - H_{ck}(\omega)(1 - e^{-j\omega T_{ck}}) \qquad (1)$$

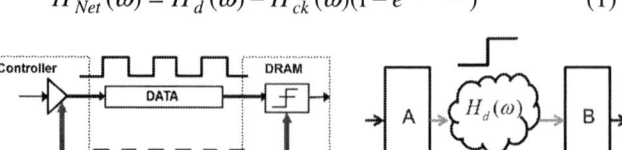

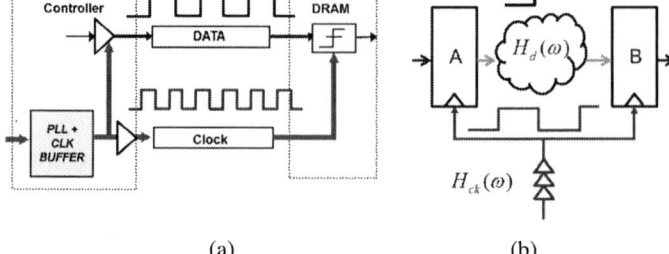

Figure 4. High-level representations of (a) traditional I/O interface and (b) digital logic transfer cases

where $H_d(\omega)$ and $H_{ck}(\omega)$ are the noise-to-jitter sensitivity of the data and clock paths, and T_{ck} is the clock period. Jitter is, then, calculated by multiplying voltage noise in frequency domain. For a typical buffer path, the sensitivity can be approximated as [7]:

$$H(f) = H_0 \sin(\pi f \tau_d)/(\pi f \tau_d) \qquad (2)$$

where $\tau_d(\omega)$ is the buffer delay and H_0 is the delay sensitivity at DC.

Example sensitivity curves using 1.25nsec of data path delay and 0.7nsec of clock path delay are shown in Figure 5. The net sensitivity function is also shown in the figure. Notice that the jitter tracking occurs at relatively high frequency regions. Typical core power has low PDN impedance and the resonance frequency is around 50MHz. Hence, the jitter due to the dominant core PDN noise would not be tracked out in this case. Also, at low frequency the net jitter would be dominated by data jitter. The high frequency jitter tracking is also a function of the phase and insertion delays of data and clock. For instance, if the insertion delay of clock is longer than data, the net sensitivity will be increased at the high frequency region as it will be demonstrated in the next section.

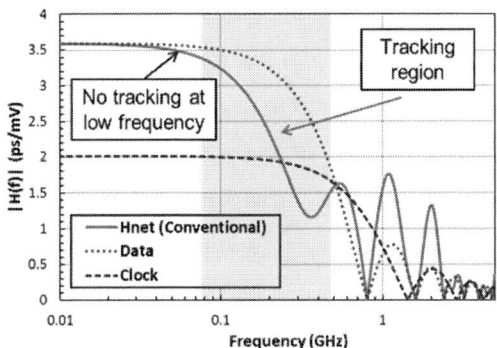

Figure 5. Noise-to-jitter sensitivity function of conventional I/O interface case

B. Digital Blocks

The previous formulation is not truly applicable to logic timing analysis for core or digital blocks. As shown in Figure 4(b), during the register transaction, there is only one data transition during a clock period. Each clock period, the data transition is regenerated from source and the jitter on data path can only occur during one clock period. The previous data edge does not impact the current edge transition; hence, only high frequency voltage noise can affect data jitter as illustrated in Figure 6. In typical IO interface, any one data edge can be affected by all the previous data edge transitions.

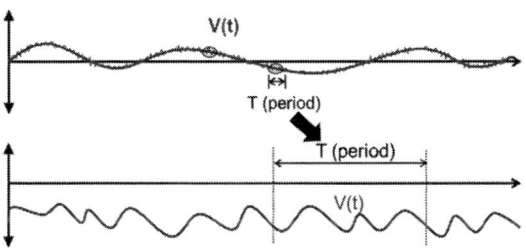

Figure 6. Noise frequency behavior over one period

To accurately model this phenomenon, we have to model the high and low frequency voltage noise separately. The sensitivity of the high frequency jitter over one clock period, $H_{OnePeriod}(\omega)$, can be modeled using the following expression:

$$H_{OnePeriod}(\omega) = (H_d(\omega) - H_{ck}(\omega))(1 - e^{-j\omega T_{ck}}) \qquad (3)$$

The low frequency jitter can be modeled by filtering the high frequency noise using the following expression:

$$V_{filtered}(\omega) = V(\omega)e^{-\omega T_{ck}} \qquad (4)$$

The high frequency jitter, $J_{HF}(\omega)$, is calculated by multiplying $H_{OnePeriod}(\omega)$ with the total voltage, $V(\omega)$. The low frequency jitter, $J_{LF}(\omega)$, is calculated by multiplying $H_d(\omega)$ by $V_{filtered}(\omega)$. Note that low frequency clock jitter does not impact the final performance. Then, $J_{LF}(\omega)$ and $J_{HF}(\omega)$ are converted to time domain and then binned to obtain jitter histograms. The final jitter distribution is obtained by convolving the two jitter histograms.

The one-period sensitivity function is plotted using the previous setting and compared with the convention IO formulation in Figure 7. The one-period sensitive function is a high pass filter as expected. Also, the tracking is better than the conventional case. To cover more general cases, the clock insertion delay of 2.1ns, which is larger than the data delay, is also considered. As shown in the figure, the sensitivity function of the conventional formulation increased compared to the 0.7ns case. On the other hand, the one-period sensitivity function did not make significant difference. The large difference in the high frequency region is not as relevant as power noise is small due to local on-chip decoupling capacitor.

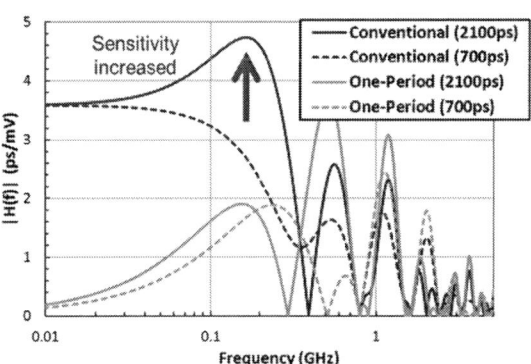

Figure 7. Noise-to-jitter sensitivity functions of (a) conventional I/O formulation and (b) one-period formulation

IV. NUMERICAL RESULTS

Timing jitter analysis is performed using a memory controller IP block. Noise profile is generated using on-chip current activity and shown in Figure 8. Both the original noise and filtered noise profiles in frequency and time domains are plotted. The previous clock and data paths with 0.7ns and 1.25ns insertion delays are used for the sensitivity functions. The corresponding jitter profiles for high-frequency and low frequency are shown in the figure for both frequency and time domains.

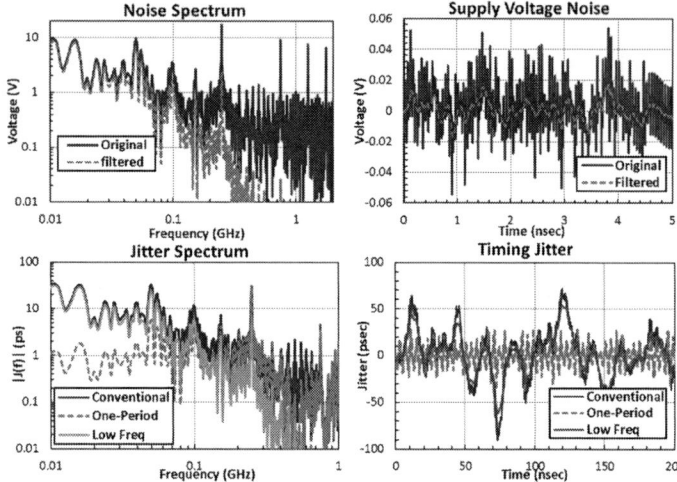

Figure 8. Simulation results of a memory controller block

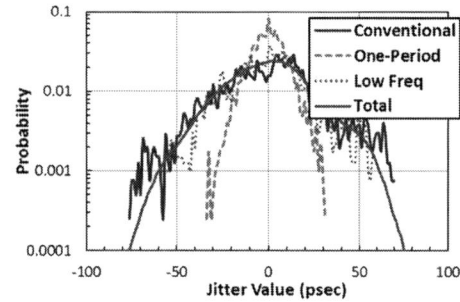

Figure 9. Jitter distributions of a memory controller case

The time domain jitter profile is binned to generate the histogram or equivalently the probability mass functions (PMFs). The resulting PMFs for high frequency (one-period) and low frequency jitter components are convolved to get the final PMF. The final results are compared to the conventional case in Figure 9 and summarized in Table II.

Three difference approaches can be used in modeling this supply noise induced timing jitter. Case A is based on pure DC sensitivity of data path and the peak-to-peak voltage noise. This approach results in 389ps. Since conventional static or dynamic timing analysis does not capture the frequency variation of the sensitivity function, it produces a pessimistic value. Case B is the conventional I/O interface formulation and it results in 166ps. Case C based on the proposed formulation results in 70ps of high frequency jitter and 122ps of low frequency jitter. The high frequency jitter can be modeled as a clock uncertainty term. The low frequency jitter term is typically modeled in static timing analysis and has no need to be explicitly budgeted.

V. CONCLUSIONS

A new formulation is proposed to model the impact of power noise for core logic blocks. The formulation accurately captures both low and high frequency impacts of voltage noise. This formulation can be applied to reduce the pessimism built in the traditional target impedance approach. It can be easily extended to include more complicated cases where some parts of data and clock paths use different power supply rails.

Table II. Summary of various of timing jitter modeling approaches

Noise Pattern	Case 1 (clk delay = 700ps)		Case 2 (clk delay = 2100ps)	
	Peak-Peak (ps)	RMS (ps)	Peak-Peak (ps)	RMS (ps)
One-Period	70 (C)	9.6	64	9.1
Conventional	166 (B)	26.5	180	27
Increase	x2.4	x2.8	x2.8	x3.0

Noise	Noise (p-p,mV)	Delay Variation
Filtered	34	122ps (C)
Original	108	389ps (A)
Increase	x3.2	

- Tperiod = 1.250ns
- Delay variation is computed based on DC sensitivity

ACKNOWLEDGMENT

The authors wish to thank Dr. Daniel Chow and Albert Zhang for helpful discussions and Dr. Yuri Tretiakov, Wern Shin Choo, and Dr. Shishuang Sun for developing measurement flow and data.

REFERENCES

[1] L. D. Smith, R. E. Anderson, D. W. Forehand, T. J. Pelc, and T. Roy, "Power distribution system design methodology and capacitor selection for modern CMOS technology," *IEEE Transactions on Advanced Packaging*, pp. 284-291, Aug. 1999.

[2] M. S. Tanaka, M. Toyama, H. Nakashima, J. Yamada, M. Haida, and I. Osshima, "Chip oriented target impedance for digital power distribution network design," *in Proceedings of IEEE Electrical Performance of Electronic Packaging and Systems Conference*, Oct 2012, pp. 220-223.

[3] J. Kim, Y. Takita, K. Araki, and F. Fan, "Improved target impedance for power distribution network design with power traces based on rigorous transient analysis in a handheld device," *IEEE Transactions on Components, Packaging and Manufacturing Technology*, pp. 1-10, Jan. 2013.

[4] D. Oh and C. Yuan, *High-Speed Signaling: Jitter Modeling, Analysis and Budgeting*, Prentice Hall, 2011.

[5] D. Oh, "System-level jitter characterization of high-speed I/O systems," *in IEEE Symposium on Electromagnetic Compatibility*, pp. 173-178, Aug. 2012.

[6] T. Rahal-Arabi, G. Taylor, M. Ma, and C. Webb, "Design and validation of the Pentium® III and Pentium® 4 processors power delivery," *in Symposium on VLSI Circuits Digest of Technical Papers*, June 2002, pp. 220-223.

[7] K. L. Wong, T. Rahal-Arabi, M. Ma, and G. Taylor, "Enhancing microprocessor immunity to power supply noise with clock-data compensation," *IEEE Journal of Solid-State Circuits*, pp. 749-758, Apr. 2006.

[8] I. Kantorovich and C. Houghton, "Effectiveness of on-die decoupling capacitance in improving chip performance," *in Proceedings of IEEE Electrical Performance of Electronic Packaging and Systems Conference*, Oct 2008, pp. 165-168.

[9] D. Jiao, J. Gu, and C. H. Kim, "Circuit techniques for enhancing the clock data compensation effect under resonant supply noise," *Proceedings of IEEE Custom Integrated Circuits Conference*, Sep. 2009, pp. 29-32.

[10] S. Sun, L. Smith, and P. Boyle, "On-chip PDN noise characterization and modeling," presented at the *IEC DesignCon*, Santa Clara, CA. 2010.

Power Integrity of a 4.8Gbps-per-link Low-swing Single-ended-I/O Server Memory Interface

Chris Madden, Hai Lan, Zhuo Yan*, Ravi Kollipara, and Jihong Ren

Rambus Inc.
1050 Enterprise Way, Suite 700, Sunnyvale, CA, USA
chris_madden@rambus.com, (408)462-8454

Abstract— **The power integrity characterization of a high-capacity, compute-server memory system operating at 4.8Gbps-per-link is presented. The design robustness of the low-swing, single-ended signaling is verified as the system has excellent immunity to the noise from simultaneously switching outputs (SSO) and a low power-supply-induced jitter (PSIJ) at the primary chip-package resonance frequency.**

Keywords— power integrity, noise, jitter.

I. INTRODUCTION

To meet the projected needs of future high-performance multi-core-processor-based servers, a prototype memory system was constructed of a socketed controller physical-layer interface (CPHY), a flip-chip-packaged memory PHY (MPHY) shown to be suitable for assembly on a dual inline memory module (DIMM), and a 12-layer FR-4 motherboard fitted with standard 240-pin DDR3 DIMM connectors. The two PHYs were fabricated in 28nm CMOS and shown to perform well at 6.4Gb/s-per-link[1], a data rate that is well beyond what is currently expected of future DDR4 memory systems. Using high-speed single-ended signaling for high pin efficiency and aggregate interface bandwidth, good timing and voltage margins were also shown at 6.4Gb/s for a system with two dual-rank 16-device (16-D or single-row) DIMMs[2,3]. For the increased memory–system capacity required in the most demanding applications, higher-capacity, 36-memory-device, dual-rank DIMMs were also fabricated but demonstrated at a lower data rate (4.8Gb/s) due to the added loss from extending the high-speed clock lines to drive a second row of MPHYs. Future systems could avoid this limitation with per-rank clocking. A high-level view of this last system configuration is shown in Fig. 1.

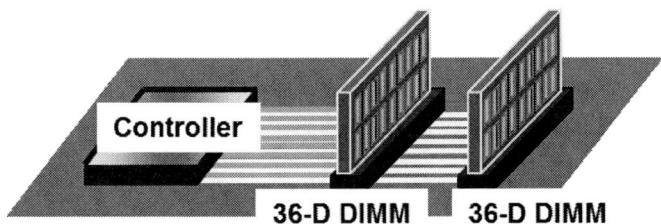

Fig. 1. High-capacity system with a pair of 36-D DIMMs.

To provide for low signaling power dissipation while still optimizing the signal integrity for high-data-rate bi-directional operation in a difficult passive channel having a pin-through-hole DIMM connector, a unique low-swing, near-ground signaling is implemented using an N-over-N output driver with per-pin Tx regulator and impedance calibration in both output states. The simplified signaling model for WRITE operations is shown in Fig. 2, including the calibrated Rx ODT to ground. Advanced features needed to support the signaling up to a data rate of 6.4Gb/s, include transmitter and receiver (decision-feedback) equalization on the CPHY and receiver offset voltage calibration (for better sensitivity) implemented, on both the CPHY and MPHY, after the Rx linear equalizer/pre-amp.

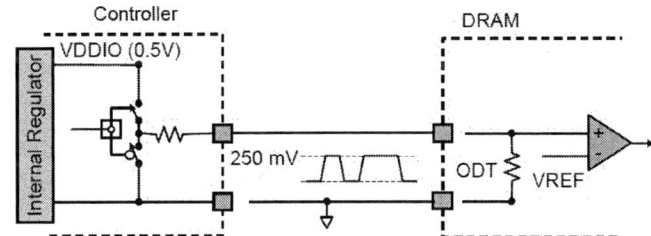

Fig. 2. Simplified signaling model for the low-swing memory interface.

II. CHALLENGES IN POWER INTEGRITY

A. Supply Partitioning

In the PHYs, it is desirable to use power supply partitioning to separate circuits that are sensitive to supply noise from circuits that can generate significant noise. The main partition strategy was to place all the digital logic, high-speed datapath, and I/O circuits on one, potentially noisy, supply (VDDR) and keep it separate from the supply (VDDA) for the more sensitive clock distribution circuits. The clock distribution circuits are a mix of CML- and CMOS-type, and running at 4.8GHz, so PSIJ simulations indicated they would be up to 3x more sensitive than the circuits on VDDR[4]. On the CPHY, the PLL circuitry was expected to be of high sensitivity but quite localized and low power, so a dedicated supply (VDDP) and a single BGA ball was easily separated off and further filtered to reduce its sensitivity. Isolation for both VDDA and VDDP from the noise on VDDR was found to be better than 23dB in the frequency range where that noise is largest.

On the MPHY, the high-speed clock, which was forwarded from the CPHY, is a 4-phase(quadrature) clock running at ¼ the data rate. The lower frequency of this clock and the smaller physical size of the MPHY both helped with lowering its VDDA PSIJ w.r.t. the CPHY but, because of the phase

---

*Now with North Carolina State University, Raleigh, N.C., USA.

978-1-4799-0708-3/13 $31.00 © 2013 IEEE

correction circuits required as a result of the fly-by topology of the clock routing on the DIMMs, the net sensitivity was actually slightly worse.

B. PDN Optimization

Each of the power distribution networks (PDNs) were modeled for the contribution of on-die decoupling capacitors and power grid as well as package inductances and parasitic capacitances. The resulting impedance to ground seen by the active circuitry is what converts that activity to self-generated voltage power supply noise (PSN). It is particularly important to model the package inductances accurately as the effective package inductance and on-die capacitance largely determine the "chip-package" resonant frequency at which the PDN impedance is highest and where the worst-case noise will occur if the circuit activity is also concentrated there. Optimization of the PDN usually involves minimizing the ground and parasitic inductances in the package and maximizing the placement and effectiveness of the bypass capacitors. This often requires rationing package and on-die resources among the supplies but allows for flexibility in managing the PDN resonance frequency and maximum impedance for reduced worst-case noise. These optimized impedance profiles for the three CPHY supplies are shown in Fig. 3 after on-die decaps were allocated 61% to VDDR, 37% to VDDA, and just 2% to VDDP.

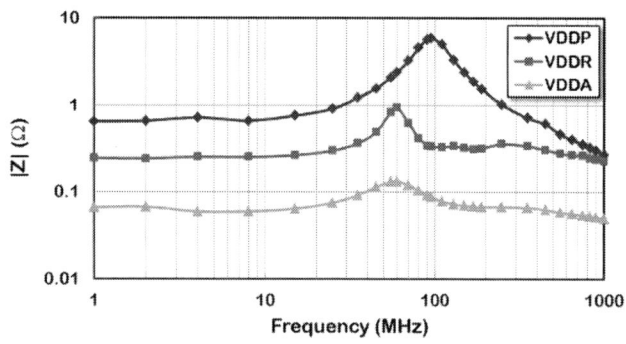

Fig. 3.   Measured impedance of the CPHY PDNs.

C. Power Supply Noise

Each power supply has a unique voltage noise spectrum that depends on the activity present in its circuits. The activity on CPHY VDDA and VDDP is the clocking so the noise spectrum is almost entirely dependent on the data rate, not the data activity. The PDN impedances of these supplies, at such frequencies well above resonance, are low enough to effectively suppress the PSN to less than 10mV. For VDDR, the noise is very strongly dependent on the data activity, as intended from the partitioning. The on-die-measured power supply noise spectra for two very different activities are shown in Figs. 4(a) and 4(b), respectively. To facilitate measurement of this noise, an independently-clocked, on-die noise generation circuit was designed for each power supply to give sufficient signal over the relevant frequency range – typically up to 500MHz or more. A separate on-die measurement circuit[5] was used to find the actual signal amplitude generated. The peak-to-peak noise is somewhat less for the random data case (81 vs. 116mV) as only a fraction of the activity occurs near the 60MHz PDN resonance peak.

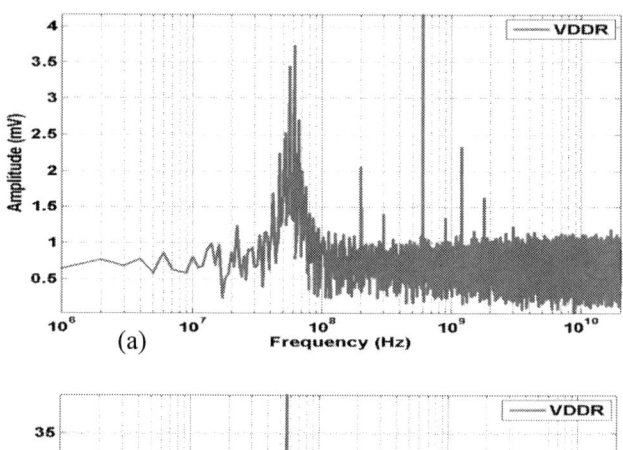

(a)

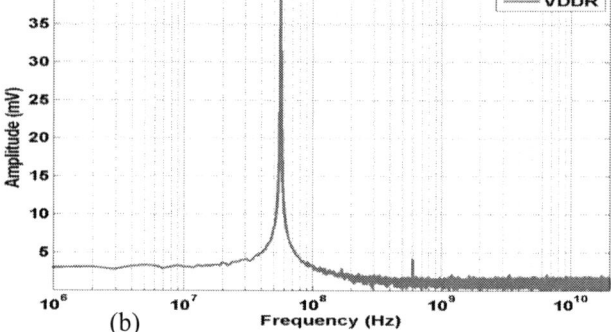

(b)

Fig. 4.   Self-generated CPHY VDDR supply noise spectrum under (a) the same random data on all pins (b) the same clock-like data tuned to PDN resonance.

III. TX REGULATOR

The Tx switching regulator, shown in Fig. 2, maintains 500mV at the node above the pull-up device. The resistance of the pull-up and pull-down devices are also calibrated, using replica circuits, to provide good matching to the PCB channel impedance and ODT resistance (40ohms). The power-supply-rejection-ratio (PSRR) of the transmitter output w.r.t. VDDR is shown in Fig. 5. It was measured using the calibrated on-die noise generator and an oscilloscope while the output was continuously high. There is a bit of peaking in the curve at about 150MHz due to some splitting of the regulator's primary

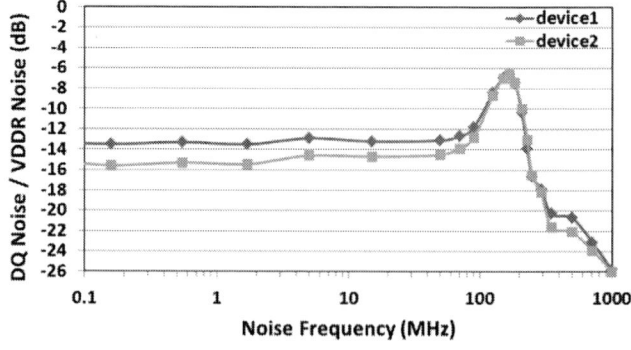

Fig. 5.   Measured PSRR of the Tx (regulator alone would be ~6dB higher).

pole and zero, but the key design goal was to have a good PSRR (< -12dB) at the PDN resonance frequency of 60MHz.

For the Tx eye diagram of Fig. 6(a), the CPHY has just the one pin active so there is minimal PSN. The ~25mV(p-p) higher noise on the "1" bits is the result of the switching regulator's own noise. In Fig. 6(b), the degradation from the worst case SSO is just 15mV additional noise. There is also a clear crosstalk signature on the bottom rail below the eye crossing, due almost entirely to the on-die nearest neighbor, but it is not having any appreciable effect on the transmit jitter.

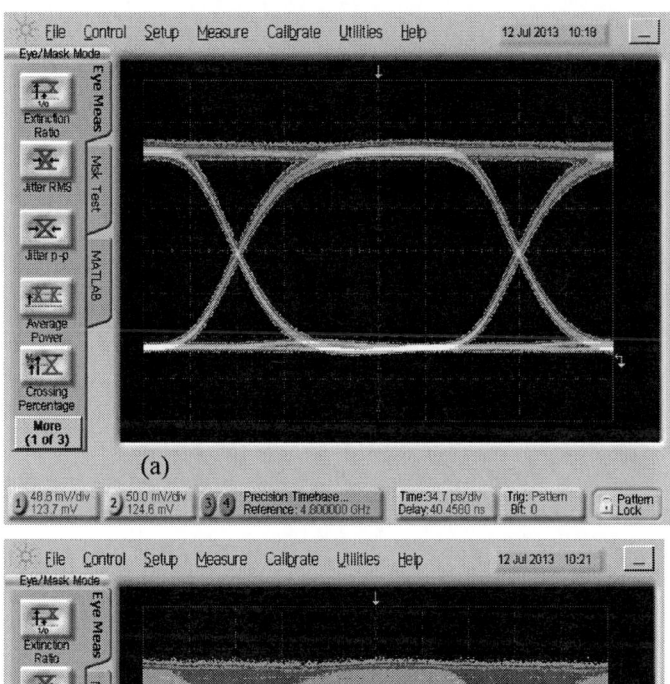

(a)

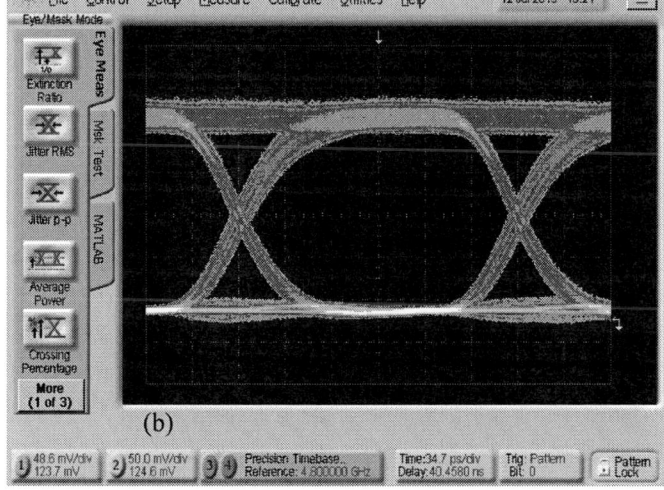

(b)

Fig. 6.   Tx eye diagram (PRBS2^11-1 pattern) for CPHY, (a) with minimal activity (PSN) and (b) with worst-case SSO noise at 60MHz.

IV.   TX PSIJ AND LINK PSIJ

The PSN impact on transmitter jitter is the product of noise spectral content (in mV) and the Tx PSIJ sensitivity as a function of frequency (in ps/mV)[4,6]. The Tx PSIJ sensitivity is simply the ratio of the Tx jitter, measured on a sampling oscilloscope with a frequency-domain jitter analysis feature, and the on-die power supply "noise" amplitude added by the noise generator.

The measured Tx VDDR PSIJ sensitivity for the CPHY is shown in Fig. 7. Overall, the sensitivity is very low except for a peak (due to the Tx local clocking) at about 8 MHz and a smaller peak (likely related to the peaking in the PSRR of the Tx regulator) at 150MHz. The key to the PDN design optimization is to recognize that the VDDR noise will always be concentrated around the chip-package impedance resonance peak at 60MHz, and at this frequency, the PSIJ is a very low 0.1ps/mV. The impact of the worst-case (100mVp-p) SSO noise, due to data activity tuned to the impedance resonance, is then just 10ps(the dual rising edges on Fig. 6(b)).

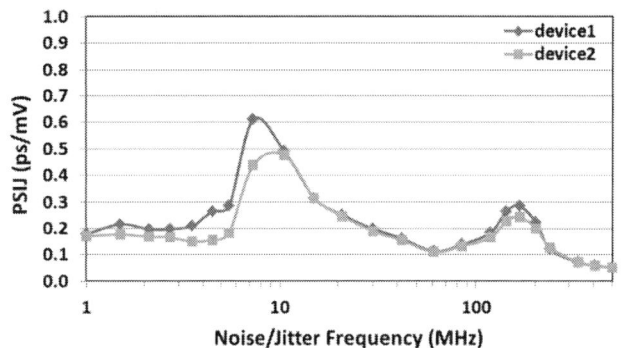

Fig. 7.   Measured CPHY Tx PSIJ sensitivity to VDDR.

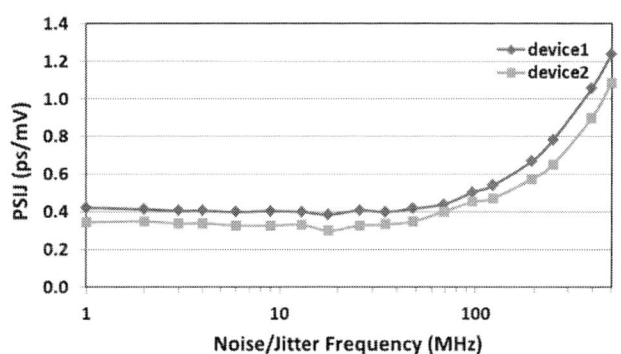

Fig. 8.   Measured MPHY Tx PSIJ sensitivity to VDDA.

The MPHY Tx VDDA sensitivity is shown in Fig. 8. The curves match adequately to the simulated results in [4] and demonstrate that, if there is a small amount of VDDR noise coupled over to VDDA because both supplies have resonance around 85MHz, the sensitivity there is less than 0.5ps/mV.

At the memory-system level, both DQ data and the corresponding clock will have PSIJ but it is the net impact on link timing margin that is critical. If, at the link receiver, a sinusoidal jitter on data and clock are of the same amplitude and phase, then the data jitter is considered "tracked" by the clock as there is no net effect on the system[6,7]. Jitter tracking can occur at very low jitter frequency and each time the difference in path delays of data and clock cause a rotation through 360° of relative phase. In addition, the data jitter is said to be "anti-tracked" at those frequencies where the data and clock jitters are 180° out of phase and the net effect on the link margin is double the amplitude. Finding this (net, system-level) link PSIJ requires the ability to measure the link timing margin at a target bit error ratio, through active adjust of the clock to data timing across a unit interval, and to measure the peak-to-peak, on-die PSN. The ratio of these gives the link PSIJ sensitivity as a function of the noise generator's clock frequency, recognizing it does have a second-order dependence on the shape of the voltage noise waveform, if it is not exactly sinusoidal.

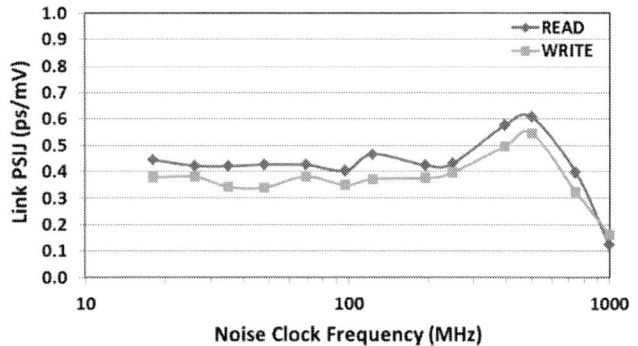

Fig. 9.   Link PSIJ sensitivity to MPHY VDDA.

Figure 9 shows the link PSIJ sensitivities to MPHY VDDA noise. Since the added noise in this case is only on the MPHY clock circuits, the resulting jitter not subject to tracking, so the link sensitivities (for both WRITE to, and READ from, the MPHY device) correlate pretty well with the Tx sensitivity of Fig. 8, as expected.

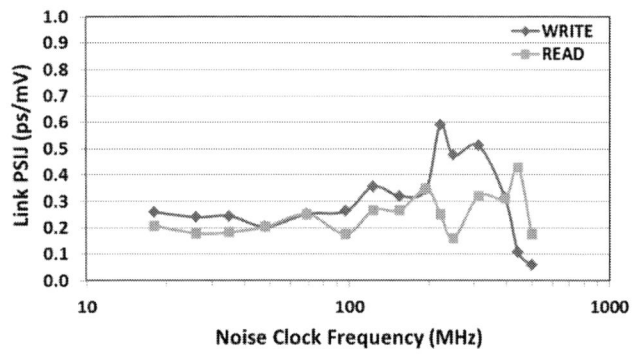

Fig. 10.   Link PSIJ sensitivity to CPHY VDDR.

Figure 10 shows the Link PSIJ sensitivity to CPHY VDDR noise. Since the added noise is now on the PDN for both the DQ data circuits and the DCK clock path as it is forwarded to the MPHY, the resulting data jitter is subject to potential tracking. The main causes for the difference in the delay for clock and data in the WRITE direction are the longer routing trace for DCK on the DIMM (last clocked device in the fly-by topology) and the received clock distribution delay (DCK to

DQ) on the MPHY. Based on the system physical design, the estimated DQ-DCK skew for WRITE is 2.65ns, which results in jitter tracking at 380MHz and anti-tracking at 190- and 570MHz, and correlates well with the peak and minimum in the WRITE sensitivity. In the READ direction, the clock jitter is just the local VDDR sensitivity in the CPHY DQ receiver while the data jitter originates in the DCK transmitter and is delayed by the forwarding to the MPHY plus the returning on the DQ data path. Again, estimating based on the design, a round-trip delay of 4.65ns results in tracking at 220MHz and 440MHz which correlates well with the curve. In both data directions, the link PSIJ sensitivity is less than 0.3ps/mV at the VDDR PDN impedance resonance of 60MHz.

V.   SUMMARY

A high-capacity memory system architecture and physical design, suitable for server applications to 4.8Gbps/link, relies on single-ended, low-swing, near-ground signaling for good signal integrity in a very DDR3-like DIMM physical channel. Control of both voltage and timing degradation of the high-speed signaling by power supply noise is achieved through careful PDN design and low-PSIJ (sub-1ps/mV) system architecture and circuit design.

ACKNOWLEDGMENTS

The authors would like to thank Wendem Beyene and Michael Bucher for helpful discussions.

REFERENCES

[1]   K. Kaviani et al., " A 6.4-Gb/s near-ground single-ended transceiver for dual-rank DIMM memory interface systems," in *IEEE International Solid-State Circuits Conference* (ISSCC), San Francisco, CA, February 2013.

[2]   Y. Lu, R. Kollipara, and A. Vaidyanath, "System design of a bufferless 6.4Gb/s server multi-DIMM memory interface system," presented at IPC Electronic System Technologies Conference, Las Vegas, NV, May 2013.

[3]   R. Kollipara et al., "Characterization of a low-power 6.4Gbps DDR DIMM memory interface system", in Proc. of 63rd IEEE Electronic Components and Technology Conference, Las Vegas, NV, May 2013.

[4]   H. Lan et al., "Power supply noise induced jitter in a 6.4Gbps/Link memory interface system," presented at the IEC DesignCon, Santa Clara, CA, February 2012.

[5]   E. Alon, V. Stojanovic, and M. Horowitz, "Circuits and techniques for high-resolution measurement of on-chip power supply noise," IEEE J. Solid-State Circuits, vol. 40, no. 4, pp. 820-828, April 2005.

[6]   H. Lan, X. Jiang, and J. Ren, "Analysis of power integrity and its jitter impact in a 4.3Gbps low-power memory interface", in Proc. of 63rd IEEE Electronic Components and Technology Conference, Las Vegas, NV, May 2013.

[7]   Y. Shim et al., "System-level clock jitter modeling for DDR systems", in Proc. of 63rd IEEE Electronic Components and Technology Conference, Las Vegas, NV, May 2013.

Power Distribution Network Design Optimization with On-Die Voltage-Dependent Leakage Path

Xiang Zhang[*], Yang Liu[+], Ryan Coutts[*], and Chung-Kuan Cheng[**]

[*]ECE Dept., University of California, San Diego, CA, USA, Email: xiz110@ucsd.edu, rcoutts@gmail.com
[+]Institute of Electronic CAD, Xidian University, Xi'an, China, Email:liuyang@mail.xidian.edu.cn
[**]CSE Dept., University of California, San Diego, CA, USA, Email: ckcheng@ucsd.edu

Abstract—**Leakage current has become a significant source of power consumptions of CMOS circuit, as the technology node continues to shrink. Our study shows that the equivalent on-die leakage resistance monotonically decreases as the supply voltage increases and exceeds MOSFET threshold voltage. We propose a system-level power distribution network (PDN) design optimization with voltage-dependent leakage resistance considered in a standard RLC tank model. Our results show that the voltage-dependent leakage resistance can impact on the PDN noise and affect the optimal value of the circuit parameters to minimize the noise. An equivalent constant leakage resistor is proposed to replace the voltage-dependent model for quick noise prediction.**

I. INTRODUCTION

Power distribution network (PDN) has become one of the most critical topics in nano-scale VLSI design. According to [1], the current density of a single chip keeps increasing while the operating voltage of high performance processors is gradually dropping. This results in the target impedance of a PDN in 2026 to drop more than five-fold from that value in 2011 (Figure 1), which brings us an even tighter noise margin requirement. Meanwhile, the full-chip leakage power in 2016 is predicted almost three times as that in 2011 as shown in Table I [1], [2], indicating that on-die leakage is no longer negligible for PDN analysis. Therefore, minimizing IR drop and simultaneous switching noise (SSN) of a PDN caused by leakage and parasitic resistance, loop inductance and transient currents have become extremely important.

A power distribution network may consist of a voltage regulator module (VRM), broad/package parasitics and on-die power grid with decoupling capacitors. Lumped model is widely used in system-level PDN analysis [3]. The passive components are modelled as cascaded RLC tanks.

TABLE I.   FULL-CHIP LEAKAGE POWER (NORMALIZED TO FULL-CHIP LEAKAGE POWER DISSIPATION IN 2011).

Yr. of Production	2011	2012	2013	2014	2015	2016
Leakage Power	1.00	1.00	1.27	1.45	2.18	2.91

Several works have been performed for PDN modelling and noise minimization. Zhang *et al.* [4] utilized power transmission line based design to minimize power supply noise for 3D IC. Charles *et al.* [5] studied the PDN effect of four chip-stacking topologies. Tanaka *et al.* [6] measured the PDN impedance profile and SSN of 3D system in a single package. Smith [7] proposed a method to characterize on-die PDN noise and generate the worst-case current stimuli. Kim [8] estimated

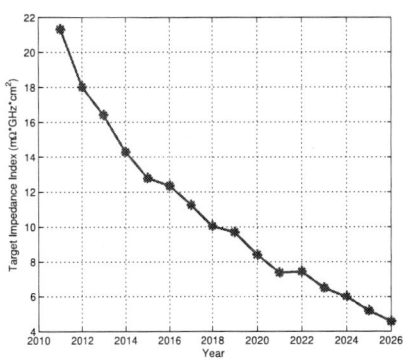

Fig. 1.   Target impedance prediction from [1]. Assume $Z_{target} = \frac{V_{dd} \times 5\%}{I_{load}}$

the supply noise from the frequency-domain PDN impedance profile. Kim *et al.* [9] gave the closed-form expressions for the power supply noise caused by IC switching current for a complete PDN structure.

In previous works, the system-level PDN design takes no consideration of the on-die leakage resistance effect [8], [9] or only considers the leakage resistance as a constant [7], [10]. This assumption is good as along as leakage resistance is always much larger than the target impedance. However, the ITRS roadmap shows that leakage current keeps increasing in the future (Table I) , resulting in a smaller leakage resistance which is no longer negligible in the PDN characterization and affecting the accuracy of the PDN noise prediction.

To our knowledge, none of the previous work has considered the on-die leakage resistance as a function of supply voltage in the PDN design. In this paper, we propose a method to calculate the PDN noise of a RLC tank model with the voltage-dependent leakage resistance $R_{leak}(v(t))$ considered. We demonstrate the relation of the optimal resistor value of RLC tank and the leakage resistance R_{leak}. We propose an equivalent constant leakage resistance model for the quick prediction of the PDN noise while it maintains the good accuracy compared to the voltage-dependent leakage model.

The remainder of the paper is organized as follows. We discuss the definition of the voltage-dependent leakage resistance model in Section II. We analyse RLC tank with leakage resistance models in Section III. A complete PDN path is studied in Section IV. Finally, we conclude this paper in Section V.

978-1-4799-0708-3/13 $31.00 © 2013 IEEE

II. VOLTAGE DEPENDENT LEAKAGE RESISTANCE MODEL

On-die leakage current comes from three main contributors: subthreshold leakage, gate leakage and band-to-band leakage (BTBT) [11]. Gate leakage has been substantially reduced as the high-k dielectrics in massive CMOS production and band-to-band leakage is relatively small compared to the other two. Therefore, we focus on subthreshold leakage in this paper.

Subthreshold leakage is a weak inversion current between source and drain in a MOS transistor when the gate voltage is below the threshold voltage V_t. In digital design, we can analyse the subthreshold leakage by setting the gate voltage $V_g = Gnd$ for NMOS and $V_g = V_{dd}$ for PMOS. The weak inversion current I_{ds} is a function of the threshold voltage V_t. V_t is mainly determined by two factors.

- Body effect: $V_t = V_{t0} + \gamma(\sqrt{\phi_s + V_{sb}} - \sqrt{\phi_s}) \approx V_{t0} + k_\gamma V_{sb}$, where $\phi_s = 2v_T ln\frac{N_A}{n_i}$, $\gamma = \frac{t_{ox}}{\varepsilon_{ox}}\sqrt{2q\varepsilon_{si}N_A} = \frac{\sqrt{2q\varepsilon_{si}N_A}}{C_{ox}}$ and $k_\gamma = \frac{\gamma}{2\sqrt{\phi_s}}$.

- Drain-induced barrier lowering (DIBL): $V_t = V_{t0} - \eta V_{ds}$, where η is on the order of 0.1.

Therefore, the subthreshold leakage can be expressed as,

$$I_{ds} = I_{ds0}e^{\frac{V_{gs}-V_t}{nv_T}}(1 - e^{-\frac{V_{ds}}{v_T}}), \qquad (1)$$

where $I_{ds0} = \beta v_T^2 e^{1.8}$, $n = 1.3 \sim 1.7$, $v_T = \frac{kq}{T}$, $V_t = V_{t0} + k_\gamma V_{sb} - \eta V_{ds}$ and $\beta = \mu_0 \frac{\varepsilon_{ox}}{T_{ox}}\frac{W}{L}$. (All the parameters are explained in [12].) By setting $V_{ds} = V_{dd}$, it can be inferred that I_{ds} is superlinear proportional to the supply voltage.

The leakage resistance R_{leak} becomes a function of V_{dd},

$$R_{leak} = \frac{V_{dd}}{I_{ds}}. \qquad (2)$$

We compare the theoretical model from Eq. 1 with an industrial 28nm HPm NMOS Spice model. We set the voltage of each port of NMOS: $V_d = V_{dd}$, $V_s = Gnd$, $V_g = Gnd$ and $V_b = Gnd$. The nominal V_{dd} is 0.9V and the operating temperature is set to 25 deg C. The results are shown in Figure 2. As the supply voltage is swept from 0.1V to 1.3V, we observe that R_{leak} slightly increases when $V_{dd} < V_t$, reaches a peak value when $V_{dd} \approx V_t$ and then decreases when $V_{dd} > V_t$. Results show that the theoretical model from Eq. 1 can accurately match the industrial model when $V_{dd} > 0.5V$.

Figure 2(b) shows that the leakage resistance of a single transistor is on the order of $10^7\Omega$. Meanwhile, the transistor count of a single high performance CPU had topped 5 billion in 2012 [13]. Suppose 10% of transistors contribute to the on-die leakage, the equivalent full-chip leakage resistance can be found on the order of $100m\Omega$. Since $18m\Omega$ target impedance for a 1GHz chip with $1cm^2$ die area in 2012 (Figure 1), the ratio of leakage resistance over target impedance can be approximate to five. To cover all the possible leakage resistance over target impedance cases in various IC designs, we analyse such resistance ratio in a wide range (from 1 to 100) in this paper.

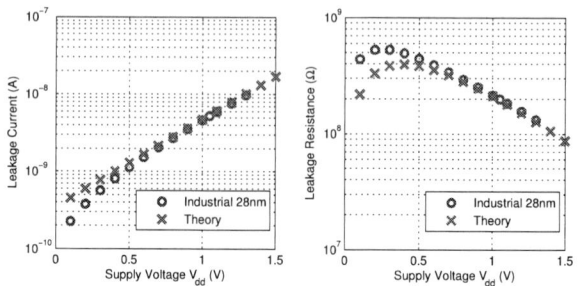

Fig. 2. (a) Leakage current vs supply voltage (b) Equivalent leakage resistance vs supply voltage

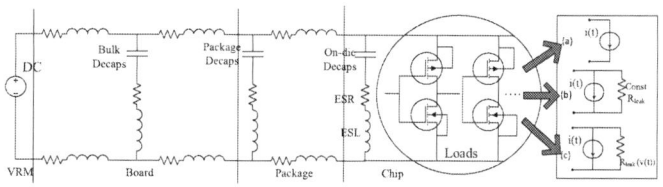

Fig. 3. A circuit diagram characterizes the impedance of PDN. On-chip load can be modelled as (a) a single current source, (b) a current source with constant leakage resistor, (c) a current source with voltage-dependent leakage resistor.

III. RLC TANK MODEL WITH LEAKAGE RESISTANCE

We discuss the impact of leakage resistance on PDN noise of the RLC tank model in this section. Figure 3 shows a complete PDN path for system-level analysis. Previous studies show that RLC tank model is a basic element of PDN and the worst-case noise is a summation of the worst-case noise from each individual tank [14]. Traditionally, the on-chip load is modelled as a current source (Figure 3(a)). Here we model the on-chip load as a current source in parallel with a constant leakage resistor R_{leak} (Figure 3(b)) or a current source in parallel with a voltage-dependent leakage resistor $R_{leak}(v(t))$ (Figure 3(c)).

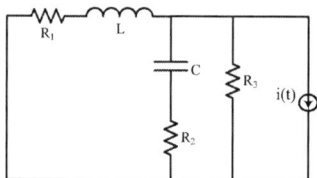

Fig. 4. A RLC tank model with leakage resistance

Figure 4 shows a RLC tank model with leakage resistance R_3. Its impedance profile $Z(s)$ in Laplace domain can be expressed as,

$$\begin{aligned} Z(s) &= \frac{s^2LCR_2 + s(R_1R_2C + L) + R_1}{s^2LC + s(R_1 + R_2)C + 1}//R_3 \\ &= \frac{(s^2LCR_2 + s(R_1R_2C + L) + R_1)R_3}{s^2LC + s(R_1 + R_2)C + 1 + R_3} \end{aligned} \qquad (3)$$

The PDN noise $v(t)$ is calculated from the convolution of the load current $i(t)$ and the impulse responses $h(t)$ when R_3 is a constant model,

$$v(t) = \int_0^\infty h(\tau)i(t-\tau)d\tau \qquad (4)$$
$$\forall t : 0 \le i(t) \le a$$

where $h(t)$ is from the inverse Laplace transform of $Z(s)$. Numerically, the worst-case noise (voltage drop) V_{max} can be obtained by setting $i(t-\tau) = a$ when $h(\tau) > 0$ and $i(t-\tau) = 0$ when $h(\tau) \le 0$. We analyse the problem by setting R_3 as a constant value or a voltage-dependent variable. The design objective is to minimize the worst-case noise. We also define the overshoot of a PDN to be $min(v(t))$.

A. Constant Leakage Resistance

If R_3 is set to a constant, Eq. 3 is simplified to a second-order system. When the leakage resistance is much greater than the impedance of the rest circuit (e.g. two order of magnitude difference), the leakage path can be ignored and the worst-case noise can be predicted from [14]. Otherwise, the leakage path needs to be included in the worst-case noise calculations.

For example, we extract a RLC tank with ($C = 0.1\mu F$, $L = 0.1nH$) from a PDN. The upper bound of $i(t)$ is set to 1. From various combinations of R_1 and R_2, we search for the minimum worst-case noise from Eq. 4 in Matlab. Simulation results show that the minimum value of the worst-case noise is 0.0282V, where the $R_1 = 0.018\Omega$ and $R_2 = 0.022\Omega$. The peak impedance is $35.1m\Omega$ without leakage resistance R_3. Based on the peak impedance value, we sweep R_3 from 30Ω to $30m\Omega$. As R_3 decreases, the worst-case noise monotonically drops as well. When R_3 falls in the same magnitude of the original target impedance without R_3, R_2 gradually decreases and R_1 drops dramatically for the minimal worst-case noise. Our observation of the minimum value of the worst-case noise and its corresponding R_1 and R_2 are shown in Figure 5.

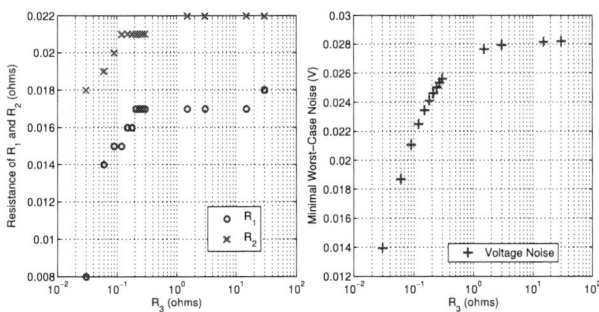

Fig. 5.   (a)The optimal value R_1 and R_2 (with minimum worst-case noise) as leakage R_3 decreases. (b)The minimum worst-case noise of a RLC tank as leakage R_3 decreases.

B. Voltage-Dependent Leakage Resistance

R_3 is modelled as a function of the voltage at the load($V_{dd}-v(t)$) (Eq. 2) in this subsection. The nominal voltage V_{dd} is set to 0.9V and the tolerance of supply noise is set to $\pm 10\%$ of V_{dd}. We keep the same parameters as the previous case: $C = 0.1\mu F$, $L = 0.1nH$, $R_1 = 0.018\Omega$, $R_2 = 0.022\Omega$ and increase the upper bound of $i(t)$ to 3.17A to scale up the noise to $V_{noise} = 0.09V$.

Eq. 4 cannot be directly applied to calculate $v(t)$ in this case as the impulse response of the system $h(t)$ changes dynamically due to the variations from the leakage resistance. Instead, we use the Backward Euler method to analyse this model. We set inductor current $i_L(t)$ and capacitor voltage $v_C(t)$ as two variables and derive two equations from Figure 4. R_3 is updated in each time step according the current supply voltage level.

$$\begin{cases} L\dfrac{di_L}{dt} + i_L R_1 = v_C + C\dfrac{dv_C}{dt}R_2 \\ i_L + C\dfrac{dv_C}{dt} + \dfrac{1}{R_3(t)}(L\dfrac{di_L}{dt} + i_L R_1) = i(t) \end{cases} \qquad (5)$$

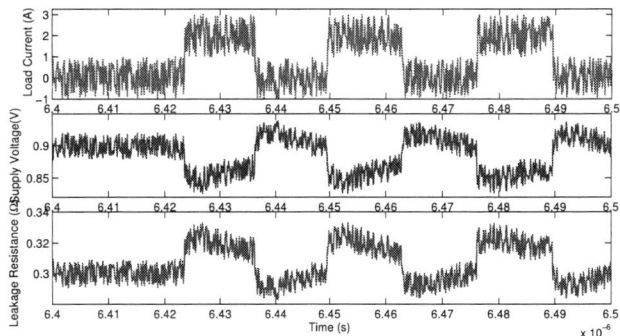

Fig. 6.   Leakage resistance R_3 as the load current $i(t)$ changes.

Figure 6 shows how leakage resistance R_3 changes in real-time as the load current $i(t)$ changes. Assume $R_3 = 300m\Omega$ at nominal voltage $V_{dd} = 0.9V$.

We compare the PDN noise with same load current pattern for both constant leakage resistance model and voltage-dependent leakage resistance model. We set R_3 at the nominal voltage equal across all the models.

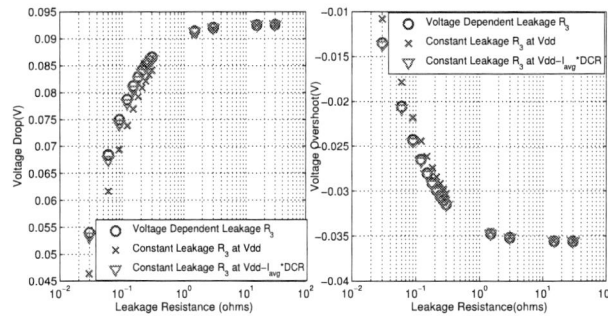

Fig. 7.   Voltage noise of a RLC tank with different leakage resistance models

The simulation results are shown in Figure 7. Voltage noise is divided into two categories: overshoot and drop. Results show that the constant R_3 model underestimates voltage drop/overshoot for more than 16% compared to voltage-dependent model when R_3 approaches the impedance of the rest circuit without R_3.

We also observe that when R_3 is set to the value at $V_{dd}-I_{avg}*DCR$ in the constant leakage model, where I_{avg} is the average load current and DCR is DC resistance of PDN, it provides similar noise value as the voltage dependent model. It slightly underestimates the drop and overestimates the overshoot (both differences are less than 2%). This approximation method can greatly reduce the simulation time since there is no need to update R_3 in Eq. 5 for each time step.

IV. CASE STUDY: A COMPLETE PDN PATH

A complete PDN path with on-die leakage is set up from Figure 3(c). Its impedance profile is shown in Figure 8 with different leakage resistors. As the leakage resistance R_3 drops, the magnitude of all the impedance peaks is reduced.

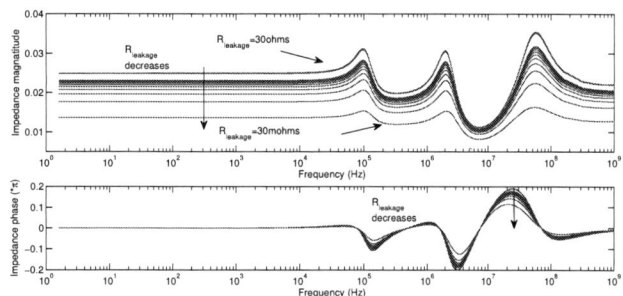

Fig. 8. Impedance profile of a complete PDN path with various leakage resistance values

Suppose that R_3 is $300m\Omega$, we compare the results of the voltage noise between the constant and voltage-dependent leakage models in time-domain in Figure 9. The constant leakage model at V_{dd} underestimates the peak voltage noise 5% compared to the voltage-dependent leakage resistance model. Figure 10 shows the voltage noise (drop and overshoot) with different leakage resistance models from Figure 8. The constant leakage at V_{dd} model underestimates the maximum voltage drop(overshoot) for up to 16% (25%) compared to voltage-dependent model, while the constant leakage at $V_{dd} - I_{avg} * DCR$ model underestimates the drop for only 2% and overestimates the overshoot up to 3%.

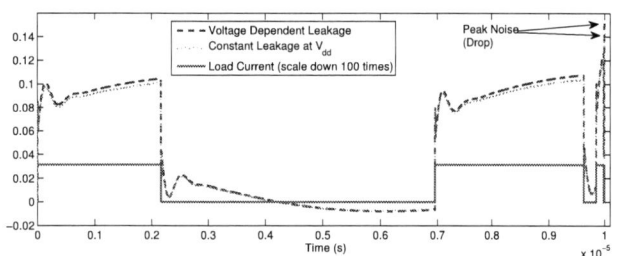

Fig. 9. The peak voltage noise (drop) of a complete PDN path in time-domain

V. CONCLUSION

Future power distribution network requires additional attention to leakage resistance as the on-die leakage current keeps increasing. In this paper, we propose to design and optimize the power distribution network with the consideration of constant or voltage-dependent leakage resistance path. We demonstrate that the leakage resistance can effectively affect the optimal resistor values in RLC tank model, when it is close to the same scale of the target impedance. Our results show that the

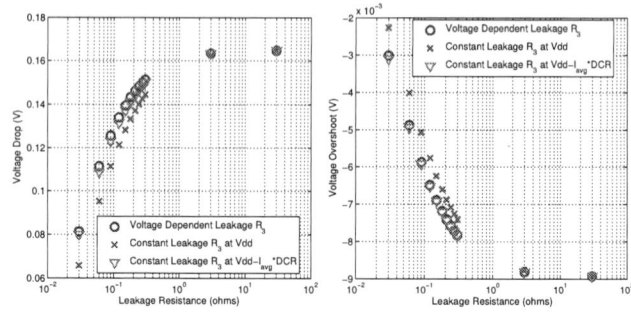

Fig. 10. Voltage noise of a complete PDN path with different leakage resistance models

constant leakage at supply voltage minus average IR drop can provide a quick way to estimate the voltage noise caused by voltage dependent leakage resistance with less than 3% error. Our future work includes temperature variations to the leakage resistance model for system-level PDN analysis.

ACKNOWLEDGEMENT

The authors would like to acknowledge the support of NSF CCF-1017864.

REFERENCES

[1] http://www.itrs.net.

[2] N. Kim, T. Austin, D. Baauw, T. Mudge, K. Flautner, J. Hu, M. Irwin, M. Kandemir, and V. Narayanan, "Leakage current: Moore's law meets static power," *Computer*, vol. 36, no. 12, pp. 68–75, 2003.

[3] M. Popovich, A. Mezhiba, and E. Friedman, *Power distribution networks with on-chip decoupling capacitors.* New York: Springer, 2008.

[4] D. Zhang, M. Swaminathan, and S. Huh, "New power delivery scheme for 3d ics to minimize simultaneous switching noise for high speed i/os," *EPEPS*, pp. 87–90, 2012.

[5] G. Charles and P. Franzon, "Comparison of tsv-based pdn-design effects using various stacking topology methods," *EPEPS*, pp. 83–86, 2012.

[6] Y. Tanaka, H. Takatani, H. Fujita, Y. Oizono, Y. Nabeshima, T. Sudo, A. Sakai, S. Uchiyama, and H. Ikeda, "Measurement of sso noise and pdn impedance of 3d sip with 4k-io widebus structure," *EPEPS*, pp. 91–94, 2012.

[7] L. Smith, S. Sun, P. Boyle, and B. Krsnik, "System power distribution network theory and performance with various noise current stimuli including impacts on chip level timing," *CICC*, pp. 621–628, 2009.

[8] W. Kim, "Estimation of simultaneous switching noise from frequency-domain impedance response of resonant power distribution networks," *IEEE Trans. Compon. Packag. Manuf. Technol.*, vol. 1, no. 9, pp. 1359–1367, sept. 2011.

[9] J. Kim, L. Li, S. Wu, H. Wang, Y. Takita, H. Takeuchi, K. Araki, J. Fan, and J. Drewniak, "Closed-form expressions for the maximum transient noise voltage caused by an ic switching current on a power distribution network," *IEEE Trans. Electromagn. Compat.*, vol. 54, no. 5, pp. 1112–1124, oct. 2012.

[10] I. Novák, *Power Distribution Network Design Methodologies.* International Engineering Consortium, 2008.

[11] S. Mukhopadhyay and K. Roy, "Modeling and estimation of total leakage current in nano-scaled-cmos devices considering the effect of parameter variation," in *ISLPED*, 2003, pp. 172–175.

[12] N. Weste and D. Harris, *CMOS VLSI Design: A Circuits and Systems Perspective.* ADDISON WESLEY Publishing Company Inc., 2011.

[13] http://en.wikipedia.org/.

[14] X. Zhang, Y. Liu, and C.-K. Cheng, "Worst-case noise prediction using power network impedance profile," *SLIP*, 2013.

Design, Implementation and Measurement of Board-to-Board Wireless Power Transfer (WPT) for Low Voltage Applications

Sukjin Kim, Bumhee Bae, Sunkyu Kong, Daniel H. Jung, Jonghoon J. Kim and Joungho Kim

Terahertz Interconnection and Package Laboratory, EE, Korea Advanced Institute of Science and Technology (KAIST),
373-1 Guseong-dong, Yuseong-gu, Daejeon, KOREA[1]sukjinkim@kaist.ac.kr; teralab@kaist.ac.kr

Abstract— **In this paper, we present measurement results of board-to-board wireless power transfer (WPT) for low voltage applications using resonant coupling. WPT system consists of source coil, receiver coil, rectifier and load on printed circuit board (PCB). Among them, spiral coil is expressed as a simple equivalent circuit, of which the components are calculated using its physical configurations. The measurement result in frequency-domain shows a good correlation with simulation result using the simple equivalent circuit. In addition, CMOS full bridge rectifier with low turn-on voltage is designed and fabricated to be mounted on PCB. In this system, it is observed that turn-on voltage of the rectifier is very small and DC level at the load is sufficient to be adapted to low voltage applications. Consequently, coil-to-coil voltage transfer ratio (VTR) and DC level at the load of board-to-board WPT are achieved to be 0.50 and 1.32V, respectively. Power transfer efficiency of 30% is calculated using circuit simulation.**

Keywords—wireless power transfer (WPT); board-to-board level; spiral coil; CMOS full bridge rectifier

I. INTRODUCTION

Recently, the wireless power transfer (WPT) technology is applied to multiple applications such as wireless charging of mobile phones, note PCs and other handheld devices [1]. This technology is able to be adapted to multi-board system as well. Conventionally, power is linked from a board to another board by connector; however, as more and more components are mounted on the board, the number of power lines continuously increases, while the number of signal lines even exceeds that of power lines. Increasing number of connector in a limited board area leads to more complicated system design. Therefore, as a solution to replace the use of connector as power lines, the required power can be transferred wirelessly.

A simplified diagram of board-to-board WPT system using magnetic resonance is shown in Fig. 1. The power source board consists of inductive coils and DC-AC inverter. The power receiver board consists of inductive coils, full bridge rectifier and chips, such as an application processor (AP), which is supplied with DC. In board-to-board WPT system, the magnetic resonant coupling between the inductive coils is a key factor because it dominantly determines the overall performance of the system. The magnetic resonance in an inductively coupled system efficiently increases the amount of magnetic flux linked between coils, which results in

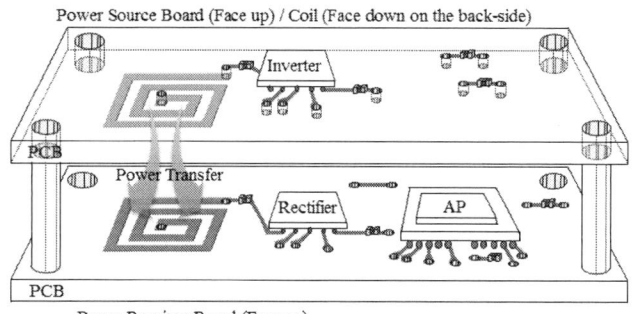

Fig. 1. Simplified diagram of board-to-board wireless power transfer system using magnetic resonance.

significant improvement of coil-to-coil voltage transfer ratio (VTR) [2]-[3]. In addition, low voltage application is appropriate because total size of board-to-board WPT system needs to be small in order to be adapted to a multi-board system. In low voltage applications, a turn-on voltage of a full bridge rectifier is the key value in board-to-board WPT system. As the turn-on voltage of the full bridge rectifier decreases, the DC level at the load increases. Hence, CMOS full bridge rectifier is necessary for current to flow through the channel of MOSFET instead of the PN junction of diode [4].

In this paper, a board-to-board WPT system using magnetic resonance with the spiral coils and the CMOS full bridge rectifier is introduced. The spiral coil is modeled with self-inductance and parasitic resistance for equivalent circuit simulation [5]-[6]. Further, a board-to-board WPT system is manufactured on board-level for experimental verification by comparison with the simulation results in frequency-domain. In this system, coil-to-coil VTR and DC level at the load are presented through time-domain measurement. In addition, power transfer efficiency is calculated from current waveforms using circuit simulation.

II. SPIRAL COIL ON BOARD-TO-BOARD WIRELESS POWER TRANSFER SYSTEM

In this section, the spiral coil on the manufactured printed circuit board (PCB) for board-to-board WPT system is presented. The most important component in this system is the coil, which determines the coil-to-coil VTR of the entire WPT system. Therefore, analyzing the equivalent circuit model of the spiral coil is very important, of which the parameters can

978-1-4799-0708-3/13 $31.00 © 2013 IEEE

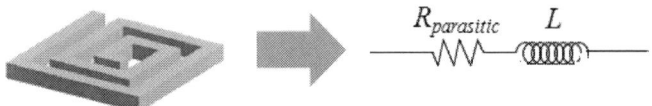

Fig. 2. Conventional equivalent circuit model of spiral coil.

TABLE I
PHYSICAL CONFIGURATION OF SPIRAL COIL

Physical Configuration	Value
Exterior Diameter, d_e	10 mm
Interior Diameter, d_i	3.4 mm
Turns, N	5 turns
Metal Width, w	0.5 mm
Metal Space, s	0.2 mm
Metal Thickness, t	0.018mm

be obtained by calculations using the physical configuration. The explanation of the equivalent circuit model is provided using the equations below.

As shown in Fig. 2, in the conventional equivalent circuit model, the spiral coil is modeled as a self-inductance, L, and a parasitic resistance, $R_{Parasitic}$. In order to calculate the self-inductance, the physical configuration of the spiral coil should first be known, as listed in Table I. From these configurations, the self-inductance, L, is given by

$$L = \frac{\mu N^2 d_m 1.27}{2}\left[\ln\left(\frac{2.07}{\rho}\right)+0.18\rho+0.13^2\rho\right] \quad (1)$$

$$d_m = (d_e+d_i)/2 \ , \ \rho=(d_e-d_i)/(d_e+d_i) \quad (2)$$

where d_m is the average diameter and ρ is the fill ratio [5].

From (1) and (2) above, it can be calculated that the self-inductance, L, is 208nH. Moreover, the parasitic resistance, $R_{Parasitic}$, of the metal, copper is given by

$$R_{Parasitic} = \frac{l}{\sigma w \cdot t_{eff}} \quad (3)$$

$$t_{eff} = \delta \cdot (1-e^{-t/\delta}) \ , \quad \delta=\sqrt{\frac{1}{\pi f \mu \sigma}} \quad (4)$$

where t_{eff} is the effective metal thickness and δ is the skin depth of copper at 110MHz [6].

From (3) and (4), it can be calculated that the parasitic resistance, $R_{Parasitic}$ is 0.65Ω. Therefore, the equivalent circuit model is represented only by the self-inductance, L, and the parasitic resistance, $R_{Parasitic}$.

III. COMPARISON OF COIL TO COIL VOLTAGE TRANSFER RATIO BETWEEN SIMULATION AND MEASUREMENT

In this section, comparison of coil-to-coil VTR between the simulation and measurement results in board-to-board WPT system, as well as the analysis, is presented.

The coil-to-coil VTR in frequency-domain can be obtained from the equivalent circuit and the measurement. It is given by

$$Coil-to-Coil \ VTR = \frac{V_R}{V_S}=\frac{V_R/I_S}{V_S/I_S}=\frac{Z_{21}}{Z_{11}} \quad (5)$$

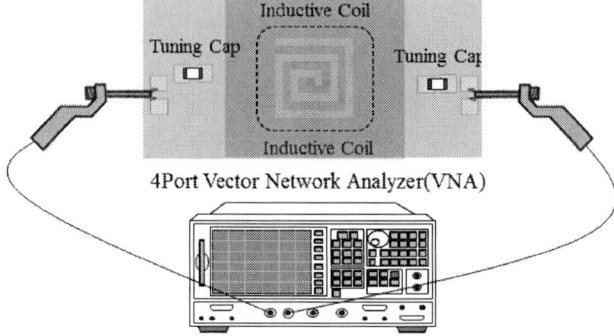

Fig. 3. Frequency-domain measurement setup for the WPT system.

$$k = M/\sqrt{(L_S L_R)}$$

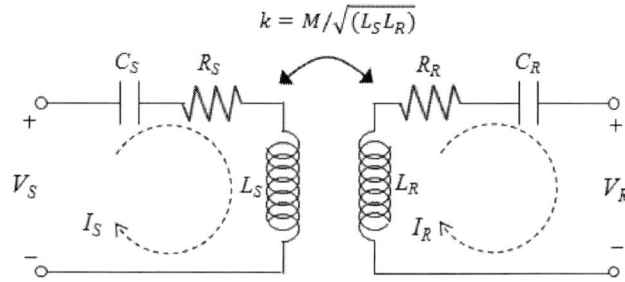

Fig. 4. Simplified equivalent circuit model of board-to-board WPT system

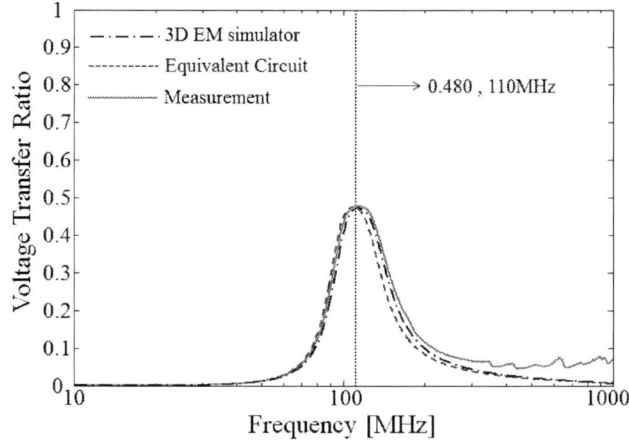

Fig.5. Coil-to-coil voltage transfer ratio at the board-to-board distance of 3mm.

where Z_{21} is a transfer impedance and Z_{11} is an input impedance.

The transfer impedance and the input impedance are extracted from the frequency-domain measurement, which is performed as shown in Fig. 3. Fig. 4 shows the simplified equivalent circuit model for frequency-domain simulation. The source board is modeled as a voltage source and the load of receiver is modeled as a resistor, R_L. The tuning capacitors connected in series to the source and receiver coils are C_S and C_R, respectively. R_S and R_R are the parasitic resistances of the source and receiver coils, respectively. M is the mutual inductance between the self-inductance of the coils, L_S and L_R, and k is the coupling coefficient of the coils.

For comparison, simulations are performed using both the equivalent circuit and 3D EM simulator, ANSYS HFSS. The

978-1-4799-0708-3/13 $31.00 © 2013 IEEE

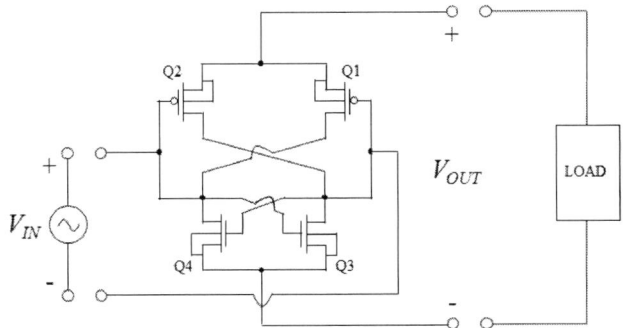

Fig. 6. Simplified equivalent circuit model of board-to-board WPT system

Fig. 7. Layout and photomicrograph of CMOS full bridge rectifier

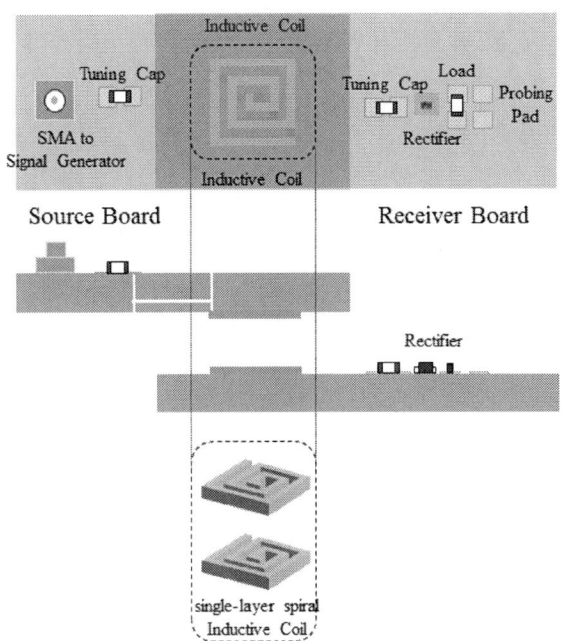

Fig. 8. Board-to-board wireless power transfer system on printed circuit board.

7. Its area is 127um by 192um. The transistors are 5V MV devices and total widths of N/PMOS are 1200/600um. The operating frequency is under 2GHz. Pads for wire-bonding to be mounted on PCB are added.

results are shown in Fig. 5, and coil-to-coil VTR results obtained from the simulations show a good correlation with the measurement results. The maximum VTR value is 0.48 at the frequency of 110MHz.

From the coil-to-coil VTR curve of the WPT system shown in Fig. 5, the peak is found to be at 110 MHz, the frequency where the input impedance is minimized using a tuning capacitor. The maximum value of the coil-to-coil VTR is expected at the frequency which has the minimum input impedance [7]-[8].

IV. DESIGN OF CMOS FULL BRIDGE RECTIFIER

As previously mentioned, the turn-on voltage of a full bridge rectifier is important in low voltage applications. Therefore, full bridge rectifier with low turn-on voltage is needed. The circuit diagram of CMOS full bridge rectifier is depicted in Fig. 6. The upper side consists of PMOS transistors and the lower side consists of NMOS transistors. A high voltage potential at $V_{IN}(+)$ and a low voltage potential at $V_{IN}(-)$ lead to the conducting stage of Q1 and Q3. Thus current can flow from $V_{IN}(+)$ over Q1 to the load and then back to $V_{IN}(-)$ over Q3. In the opposite voltage case Q2 and Q4 are conductive [4]. Turn-on voltage of CMOS full bridge rectifier is much smaller than turn-on voltage, which is about 0.7V, of PN junction using a diode because on resistance of the transistor is small.

CMOS full bridge rectifier is fabricated using 0.18um CMOS process. Layout and photomicrograph is shown in Fig.

V. IMPLEMENTATION AND MEASUREMENT OF BOARD-TO-BOARD WIRELESS POWER TRANSFER SYSTEM

For implementation and measurement of board-to-board wireless power transfer system, the spiral coil and CMOS full bridge rectifier are combined on PCB as shown in Fig. 8. The spiral coil is printed and CMOS full bridge rectifier is mounted by wire-bonding on PCB. The distance of board-to-board is 3mm, which is an applicable distance considering a height of the mounted chips in multi-board system. The source board is composed of an SMA connector to the signal generator that is used as the voltage source, a tuning capacitor for matching the resonant frequency and a source spiral coil. The receiver board is composed of a tuning capacitor, a receiver spiral coil, CMOS rectifier and a load. Two cases of the load are applied: 50Ω and 50 Ω with 1nF shunt capacitor.

In this system, the time-domain measurement is conducted. A sinusoidal voltage is supplied as the source at the frequency of 110MHz, where the coil-to-coil VTR is maximized, and the voltage waveform at the load is detected. High-z probe is used for measuring the waveform at the load, while 4V peak voltage source is supplied from the signal generator with 50 Ω source resistance. The voltage waveforms are shown in Fig. 9(a). In case of peak voltage of 2V before the rectifier, coil-to-coil VTR in time-domain is 0.5 at the frequency of 110MHz. The time-domain measurement value is similar to 0.48 at the frequency of 110MHz in frequency-domain. After the rectifier, the voltage drop due to turn-on voltage of the rectifier is 0.2V,

978-1-4799-0708-3/13 $31.00 © 2013 IEEE   93

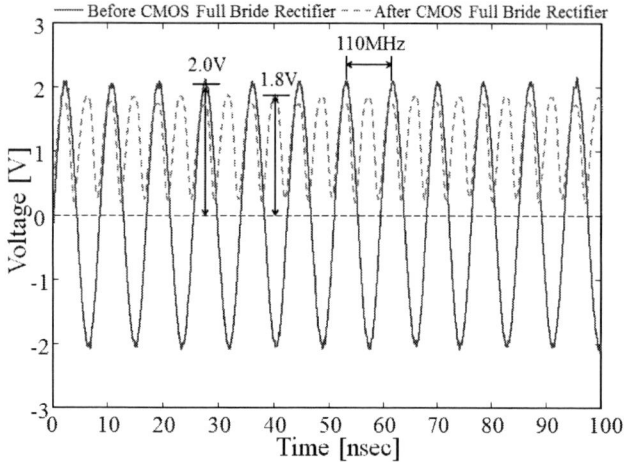

(a)

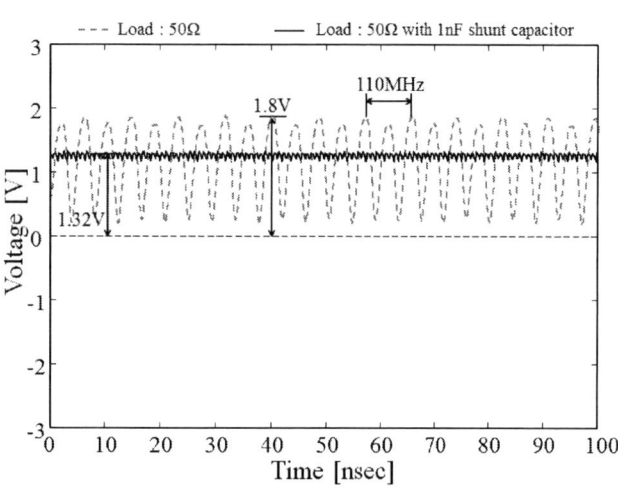

(b)

Fig. 9. Voltage waveform in time-domain measurement

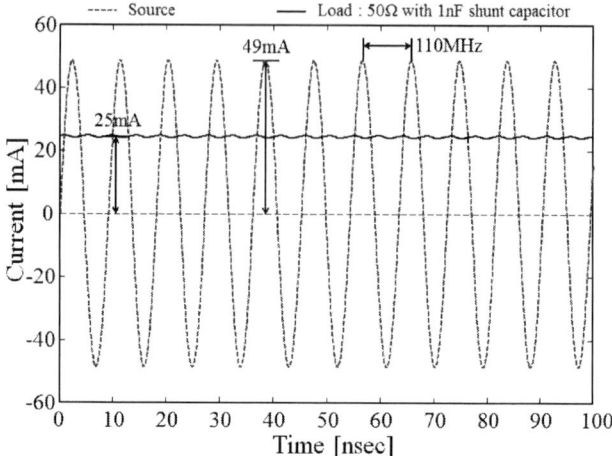

Fig. 10. Current waveform extracted from circuit simulation

which is significantly small compared to 0.7V of PN junction diode.

In order to obtain the DC level at the load, 50 Ω with 1nF shunt capacitor is used. The waveforms are plotted in Fig. 9(b) for comparison of the results at the load of 50 Ω with and without 1nF shunt capacitor. DC level of 1.32V is observed applying the load 50 Ω with 1nF shunt capacitor. The resulting DC level is sufficient to operate the low voltage application, which consists of 1.2V devices.

VI. SIMULATION AND ANALYSIS OF POWER TRANSFER EFFICIENCY

Fig. 10 shows the current waveforms extracted from SPICE circuit simulation of board-to-board WPT system. To experimentally obtain current waveforms from time-domain measurement, a monitoring pattern, which can cause parasitic inductance and capacitance, is required if we were to use a current probe. In addition, the performance of WPT system, such as coil-to-coil VTR and power transfer efficiency (PTE) can be altered due to the inclusion of extra parasitic components. Therefore, in this paper, PTE of WPT system is

calculated with the current waveforms extracted from SPICE circuit simulation and voltage waveforms from time-domain measurement. Power at the source is the product of the root-mean-square voltage and current at the source, whereas the power at the load is the product of the mean value voltage and current at the load. Since the waveforms we obtain at the source and load are AC and DC, respectively, we need to take the root-mean-square at the source, and mean value at the load. Power at the source and load are calculated to be 100mW and 30mW, respectively, and from these values, power transfer efficiency is found to be 30%. Power transfer efficiency in this system cannot exceed 50%, due to the 50 Ω source resistance of the signal generator and 50 Ω load resistance. Hence, power transfer efficiency can be improved using low source resistance of the signal generator or designed circuit acting like the source.

VII. CONCLUSION

Board-to-board WPT system for low voltage application was proposed to measure the load voltage. We have modeled the equivalent circuit of the spiral coil using R and L, and coil-to-coil VTR has presented the relationship between the simulation and the measurement results in frequency-domain. We presented the design of CMOS full bridge rectifier to reduce turn-on voltage for low voltage applications. Fabricated CMOS full bridge rectifier was mounted on PCB with the spiral coil and time-domain measurement was performed. The maximum value of coil-to-coil VTR was predictable using the analysis in frequency-domain. The maximum coil-to-coil VTR of board-to-board WPT was achieved to be 0.5 when the distance between the boards was 3 mm. Also, it was observed that turn-on voltage of the rectifier was very small and DC level at the load of 50 Ω with 1nF shunt capacitor was sufficient to be adapted to low voltage application. Power transfer efficiency of 30% was calculated from current waveforms using circuit simulation. Therefore, with further research to increase coil to coil VTR and DC level at the load, such WPT system can be widely adapted to low voltage applications with multi-board system.

978-1-4799-0708-3/13 $31.00 © 2013 IEEE

REFERENCES

[1] H. J. Brockmann and H. Turtiainen, "Charger with inductive power transmission for batteries in a mobile electrical device", US Patent 6 118 249, 1999.

[2] A. Kurs, A. Karalis, R. Moffatt, J. D. Joannopoulos, P. Fisher, and M. Soljačić, "Wireless Power Transfer via Strongly Coupled Magnetic Resonances," *Science*, vol. 317, no. 5834, pp. 83–86, 2007.

[3] A. Karalis, J. Joannopoulos, and M. Soljačić, "Efficient wireless nonradiative mid-range energy transfer," *Annals of Physics*, vol. 323, no. 1, pp. 34–48, 2008.

[4] Peters, C.; Kessling, O.; Henrici, F.; Ortmanns, M.; Manoli, Y., "CMOS Integrated Highly Efficient Full Wave Rectifier," *Circuits and Systems, 2007. ISCAS 2007. IEEE International Symposium on* , vol., no., pp.2415,2418, 27-30 May 2007

[5] Pacurar, C.; Topa, V.; Racasan, A.; Munteanu, C.; , "Inductance calculation and layout optimization for planar spiral inductors," *Optimization of Electrical and Electronic Equipment (OPTIM), 2012 13th International Conference on* , vol., no., pp.225-232, 24-26 May 2012

[6] Yue, C.P.; Wong, S.S.; , "Physical modeling of spiral inductors on silicon," *Electron Devices, IEEE Transactions on* , vol.47, no.3, pp.560-568, Mar 2000

[7] S. Kong; M. Kim; K. Koo; S. Ahn; B. Bae; and J. Kim; , "Analytical expressions for maximum transferred power in wireless power transfer systems," *Electromagnetic Compatibility (EMC), 2011 IEEE International Symposium on* , vol., no., pp.379-383, 14-19 Aug. 2011

[8] S. Kim; M. Kim; S. Kong; J.J. Kim; and J. Kim; , "On-chip magnetic resonant coupling with multi-stacked inductive coils for chip-to-chip wireless power transfer (WPT)," Electromagnetic Compatibility (EMC), 2012 IEEE International Symposium on , vol., no., pp.34,38, 6-10 Aug. 2012.

978-1-4799-0708-3/13 $31.00 © 2013 IEEE

Advanced CAD Techniques

978-1-4799-0708-3/13 $31.00 © 2013 IEEE

SPICE-Based Statistical Assessment of Interconnects Terminated by Nonlinear Loads with Polynomial Characteristics

Paolo Manfredi*, Alessandro Biondi†, Dries Vande Ginste†, Daniël De Zutter† and Flavio G. Canavero*

*EMC Group, Department of Electronics and Telecommunications, Politecnico di Torino
Corso Duca degli Abruzzi 24, 10129 Torino, Italy
E-mail: paolo.manfredi@polito.it

†Electromagnetics Group, Department of Information Technology, Ghent University
Sint-Pietersnieuwstraat 41, 9000 Gent, Belgium
E-mail: alessandro.biondi@ugent.be

Abstract—This paper proposes an exact formalism for the inclusion of nonlinear elements with polynomial I-V characteristic into the polynomial chaos framework for statistical circuit simulation, which was so far limited to linear circuits. The formulation is SPICE-compatible, thus allowing the convenient integration of such nonlinear elements into standard circuit solvers. This contribution represents a considerable step forward towards the inclusion of nonlinear terminations into the SPICE- and polynomial chaos-based statistical analysis of interconnects with stochastic parameters. The theory is illustrated and validated by means of an application example.

Index Terms—Circuit modeling, circuit simulation, nonlinear, polynomial chaos, stochastic analysis, transmission lines, uncertainty.

I. INTRODUCTION

With the increasing shrinking of device dimensions, the impact of process variability on circuit performance is becoming more and more important. Therefore, statistical approaches are usually preferred in circuit simulation in order to provide variation-aware results and thus more robust designs [1], [2]. In the packaging and manufacturing community, much attention has been devoted so far to polynomial chaos (PC)-based techniques [3]–[8], according to which statistical information is obtained via the projection of stochastic variables onto orthogonal polynomials [9].

Specifically, the authors of this contribution developed a PC-based framework for the statistical simulation of distributed networks that include lossy and dispersive lines with variability in their cross-sectional parameters [10]. The formulation is compatible with standard SPICE-type simulators, thus easily enabling the simulation of arbitrary network topologies. Nevertheless, a considerable limitation of the approach is that it hitherto applies exclusively to linear circuits.

As far as the extension towards nonlinear networks is concerned, a novel formalism has been proposed allowing for the inclusion of general nonlinear terminations into the PC framework [11]. This new formulation has been implemented into MATLAB in conjunction with a finite-difference time-domain (FDTD) scheme and applies to arbitrary I-V character-

istics. However, although very good accuracy was established, the approach is approximate. In this paper, we show that an alternative and *exact* formulation can be derived when the nonlinear terminations have a polynomial I-V characteristic. We also demonstrate that its integration in SPICE solvers is possible, thus allowing the analysis of arbitrary network topologies as well as the inclusion of lossy and dispersive interconnects, which is rather cumbersome via the FDTD technique.

The paper is organized as follows: Section II summarizes the rationale of the PC-based circuit simulation; Section III presents the formalism for nonlinear elements with polynomial I-V characteristic and discusses the integration into SPICE solvers; numerical results and validations are provided in Section IV; finally, conclusions are drawn in Section V.

II. POLYNOMIAL CHAOS-BASED SIMULATION OF STOCHASTIC INTERCONNECTS

For the sake of simplicity and ease of notation, the discussion is based on a single transmission line characterized by stochastic per-unit-length (p.u.l.) parameters $R(\boldsymbol{\xi})$, $L(\boldsymbol{\xi})$, $G(\boldsymbol{\xi})$, $C(\boldsymbol{\xi})$, and loaded with a given termination, as illustrated in Fig. 1. Here, $\boldsymbol{\xi} \in \mathbb{R}^d$ is a d-variate vector collecting all the independent random variables (RVs) affecting the line properties and is used to highlight the parameters that exhibit variability. Generalization to multiconductor interconnects, as well as to larger network topologies, is straightforward.

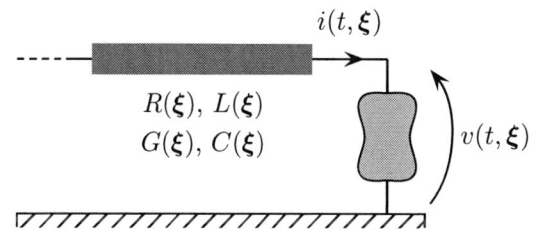

Fig. 1. Stochastic transmission line and its termination.

The underlying idea of the PC-based simulation of stochastic interconnects is to express terminal voltages and currents as PC expansions:

$$v(t, \boldsymbol{\xi}) \approx \sum_{k=0}^{P} v_k(t)\phi_k(\boldsymbol{\xi}), \quad i(t, \boldsymbol{\xi}) \approx \sum_{k=0}^{P} i_k(t)\phi_k(\boldsymbol{\xi}), \quad (1)$$

where $\{\phi_k\}_{k=0}^{P}$ is an *orthonormal* basis of polynomial functions constructed based on the inner product

$$\langle \phi_k, \phi_m \rangle = \int_{\mathbb{R}^d} \phi_k(\boldsymbol{\xi})\phi_m(\boldsymbol{\xi})w(\boldsymbol{\xi})d\boldsymbol{\xi} = \delta_{km}, \quad (2)$$

with $w(\boldsymbol{\xi})$ the joint probability density function (PDF) of $\boldsymbol{\xi}$ and δ_{km} the Kronecker delta. For standard distributions, the classes of polynomials are well-known and correspond, e.g., to Hermite polynomials (for Gaussian RVs), Legendre polynomials (for uniform RVs), and so on.

The advantage of having a representation like (1) is that, thanks to the orthogonality properties, the first two statistical moments are readily given as

$$\mathrm{E}\{v(t, \boldsymbol{\xi})\} \approx v_0(t), \qquad \mathrm{Var}\{v(t, \boldsymbol{\xi})\} \approx \sum_{k=1}^{P} v_k^2(t), \quad (3)$$

and this of course also holds for the current $i(t, \boldsymbol{\xi})$. Moreover, higher order moments as well as distribution functions can be obtained by randomly sampling (1), this step being extremely fast because (1) are merely polynomials.

The problem therefore reduces to the determination of the unknown coefficients $v_k(t)$ and $i_k(t)$. It can be proven that such coefficients are the line voltages and currents of a *deterministic* multiconductor transmission line described by the following telegrapher's equations (see e.g., [10])

$$\frac{\partial}{\partial z}\tilde{\mathbf{v}}(z, t) = -\tilde{\mathbf{R}}\tilde{\mathbf{i}}(z, t) - \tilde{\mathbf{L}}\frac{\partial}{\partial t}\tilde{\mathbf{i}}(z, t)$$
$$\frac{\partial}{\partial z}\tilde{\mathbf{i}}(z, t) = -\tilde{\mathbf{G}}\tilde{\mathbf{v}}(z, t) - \tilde{\mathbf{C}}\frac{\partial}{\partial t}\tilde{\mathbf{v}}(z, t), \quad (4)$$

where $\tilde{\mathbf{v}} = [v_0, \ldots, v_P]^T$ and $\tilde{\mathbf{i}} = [i_0, \ldots, i_P]^T$ and with the entries of the pertinent p.u.l. matrices given as e.g.

$$\tilde{R}_{jm} = \sum_{k=0}^{P} R_k \langle \phi_k \phi_m, \phi_j \rangle, \quad (5)$$

$j, m = 0, \ldots, P$. In (5), $\langle \phi_k \phi_m, \phi_j \rangle$ is merely a real number and the R_k are the PC-expansion coefficients of the random p.u.l. parameter, which can be computed — based on the random geometric and material properties of the line — via numerical integration techniques [10]. The remaining p.u.l. matrices $\tilde{\mathbf{L}}$, $\tilde{\mathbf{G}}$ and $\tilde{\mathbf{C}}$ are constructed analogously.

Once the expansion coefficients of the p.u.l. parameters are known, the modified matrices can be constructed and the corresponding multiconductor transmission line can be simulated e.g. in a SPICE-type circuit analysis tool to retrieve the sought-for coefficients for the voltage and current variables [10], provided that proper boundary conditions are enforced and the line terminations, as illustrated in Fig. 2 for the case

$P = 2$. This step is discussed in the next section. As a result, a *single*, deterministic simulation of the modified circuit allows to collect statistical information substantially faster compared to, e.g., performing a Monte Carlo analysis, i.e. simulating a large number of samples (realizations) of the original network.

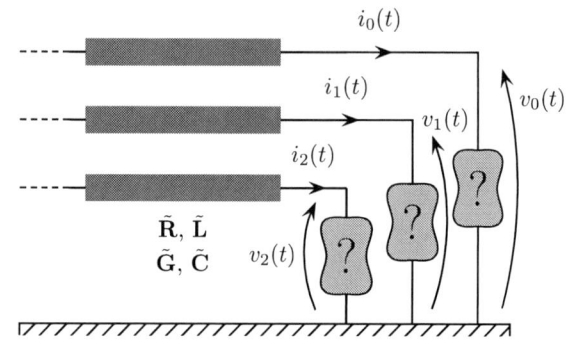

Fig. 2. Modified, deterministic transmission line and its new terminations to be determined.

III. Boundary Conditions at Line Terminations

In order to simulate the transmission line characterized by (4), suitable boundary conditions, expressed in terms of current-voltage relationships, must be derived for the line terminations. These stem from the I-V characteristic of the original termination. The result turns out to be trivial when the load is linear, because in that case such a load is simply replicated on all the terminations of the multiconductor line. Unfortunately, this is not the case when the termination is nonlinear.

Now, we relax the assumption of linearity and assume a nonlinear current-voltage relationship at the line termination:

$$i(t, \boldsymbol{\xi}) = F(v(t, \boldsymbol{\xi})). \quad (6)$$

Replacing the voltage and current with their PC expansions (1) and weighting the resulting equation using the basis functions $\{\phi_m\}$ yields

$$\sum_{k=0}^{P} i_k(t)\phi_k(\boldsymbol{\xi})\phi_m(\boldsymbol{\xi}) = F\left(\sum_{k=0}^{P} v_k(t)\phi_k(\boldsymbol{\xi})\right)\phi_m(\boldsymbol{\xi}) \quad (7)$$

($m = 0, \ldots, P$). Then, integrating with the inner product (2) produces

$$i_m(t) = \int_{\mathbb{R}^d} F\left(\sum_{k=0}^{P} v_k(t)\phi_k(\boldsymbol{\xi})\right)\phi_m(\boldsymbol{\xi})w(\boldsymbol{\xi})d\boldsymbol{\xi} \quad (8)$$

where, in general, no exact closed-form expression exist for the right-hand side. In [11], a closed-form, but approximate, expression is obtained by discretizing the integral using a numerical quadrature with a given number of nodes. The method provides very good accuracy and excellent efficiency. However, in this paper, we provide an alternative and exact solution that applies when the nonlinear characteristic $F(v)$ can be expressed as a polynomial function, i.e.

$$i(t, \boldsymbol{\xi}) = F(v(t, \boldsymbol{\xi})) = \sum_{n=0}^{N} G_n (v(t, \boldsymbol{\xi}))^n. \qquad (9)$$

Substituting (9) into (8) yields

$$i_m(t) = \int_{\mathbb{R}^d} \sum_{n=0}^{N} G_n \left(\sum_{k=0}^{P} v_k(t) \phi_k(\boldsymbol{\xi}) \right)^n \phi_m(\boldsymbol{\xi}) w(\boldsymbol{\xi}) d\boldsymbol{\xi}. \qquad (10)$$

The multinomial theorem allows to write

$$\left(\sum_{k=0}^{P} v_k(t) \phi_k(\boldsymbol{\xi}) \right)^n = \\ \sum_{k_0 + \ldots + k_P = n} \binom{n}{k_0, \ldots, k_P} \prod_{0 \le r \le P} (v_r(t) \phi_r(\boldsymbol{\xi}))^{k_r}, \qquad (11)$$

with the multinomial coefficient defined as

$$\binom{n}{k_0, \ldots, k_P} = \frac{n!}{k_0! \cdots k_P!}. \qquad (12)$$

Substituting (11) into (10) and rearranging leads to

$$i_m(t) = \sum_{n=0}^{N} \sum_{k_0 + \ldots + k_P = n} G_n \binom{n}{k_0, \ldots, k_P} \\ \times \prod_{0 \le r \le P} (v_r(t))^{k_r} \int_{\mathbb{R}^d} \prod_{0 \le r \le P} (\phi_r(\boldsymbol{\xi}))^{k_r} \phi_m(\boldsymbol{\xi}) w(\boldsymbol{\xi}) d\boldsymbol{\xi}. \qquad (13)$$

Despite the somewhat bulky equation, it is now possible to note that:

1) the term $\binom{n}{k_0, \ldots, k_P} G_n$ is merely a constant number;
2) the integral $\int_{\mathbb{R}^d} \prod_{0 \le r \le P} (\phi_r(\boldsymbol{\xi}))^{k_r} \phi_m(\boldsymbol{\xi}) w(\boldsymbol{\xi}) d\boldsymbol{\xi}$ also yields a constant number that can be calculated analytically, at least for standard classes of orthogonal polynomials ϕ_k;
3) the argument $\prod_{0 \le r \le P} (v_r(t))^{k_r}$ is still a $(P+1)$-variate polynomial, depending on all the PC coefficients of the controlling voltage, and of total degree at most N.

As an example, the three nonlinear terminal conditions for the case of a Gaussian RV with $d = 1$, $P = 2$, and $N = 2$, reduce to

$$i_0(t) = G_0 + G_1 v_0(t) + G_2[v_0^2(t) + v_1^2(t) + v_2^2(t)] \\ i_1(t) = G_1 v_1(t) + G_2[2v_0(t)v_1(t) + 2\sqrt{2}v_1(t)v_2(t)] \\ i_2(t) = G_1 v_2(t) + G_2[\sqrt{2}v_1^2(t) + 2v_0(t)v_2(t) + 2\sqrt{2}v_2^2(t)] \qquad (14)$$

As already pointed out, the new terminal conditions (13) preserve a polynomial characteristic. This allows to take advantage of the capability, available in circuit solvers like HSPICE or PSPICE, of handling multivariate polynomial functions of a given set of controlling node voltages (using, e.g., the POLY keyword, cfr. [12]). The new line terminations can then be straightforwardly implemented in SPICE-type simulators as voltage-dependent current sources ("G-elements") that are a polynomial function of all the line terminal voltages.

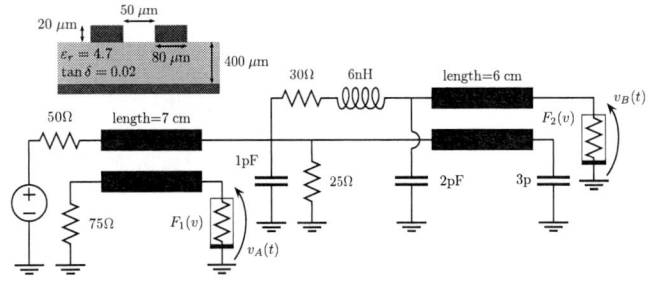

Fig. 3. Transmission-line network considered for the application example.

IV. VALIDATION AND NUMERICAL RESULTS

This section proposes an application example in order to illustrate and validate the proposed theoretical formulation. The network in Fig. 3 includes two coupled copper microstrip lines with two nonlinear terminations. The microstrip cross-section is also shown on the top. The I-V characteristics of the nonlinear terminations are $i = F_1(v) = 0.02v - 0.1v^2 + 0.6v^3$ and $i = F_2(v) = 0.01v - 0.15v^2 + 0.8v^3$. The variability is provided by the substrate thickness and by the trace-to-trace separation, which are considered as two independent Gaussian random variables with relative standard deviations of 10% and 5% w.r.t. their nominal values as indicated on the figure, respectively. The voltage source produces a pulse with an amplitude of 5 V, a duration of 4 ns and rise/fall times of 100 ps. All the simulations are carried out using HSPICE on an ASUS U30S laptop with an Intel(R) Core(TM) i3-2330M, CPU running at 2.20 GHz and 4 GB of RAM. For additional details on how to setup a PC-based simulation of such a transmission-line network, the reader is referred to [10].

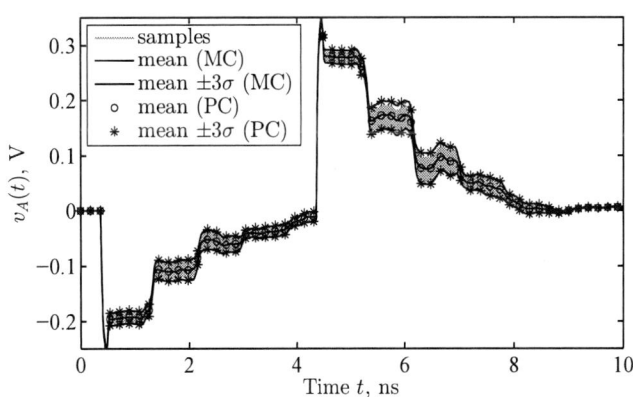

Fig. 4. Statistical transient analysis of $v_A(t)$. Gray lines: samples of the random response; black lines: mean response and $\pm 3\sigma$ limits obtained with Monte Carlo analysis; markers: mean response (circles) and $\pm 3\sigma$ limits (asterisks) estimated with PC.

Fig. 4 shows the transient simulation of the voltage $v_A(t)$ across the first nonlinear termination, $F_1(v)$. A 1000-sample Monte Carlo analysis is performed first, using the available feature in HSPICE, and the black lines display the corresponding estimated mean response as well as the $\pm 3\sigma$ limits (σ

978-1-4799-0708-3/13 $31.00 © 2013 IEEE

denoting the standard deviation). The microstrip lines are characterized with the internal field solver, which allows to take losses and dispersion into account. A reduced set of response samples is also plotted in gray to visualize the fluctuation due to the variability of the line parameters. Then, a PC-based simulation is run to compute the PC-expansion coefficients. The markers compare the same statistical information obtained from the voltage coefficients according to (3). A remarkable accuracy can be appreciated. Furthermore, Fig. 5 provides similar results, this time for the voltage across the second nonlinear termination, $F_2(v)$. The meaning of the curves is the same as in Fig. 4, and again excellent accuracy is established.

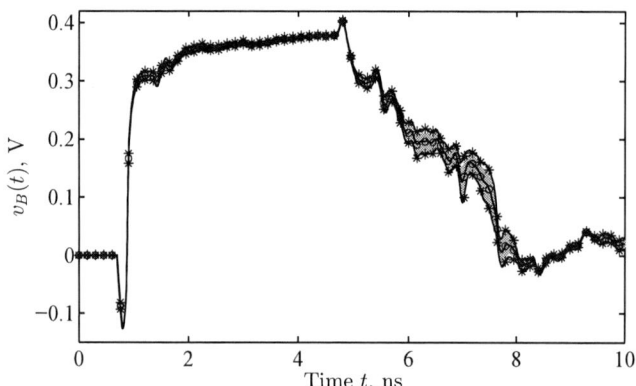

Fig. 5. Statistical transient analysis of $v_B(t)$. Curve identification is similar as Fig. 4.

Finally, Fig. 6 shows the PDF of $v_A(t)$ at 6 ns, when the fluctuation of the voltage is quite large. The gray bars are the histogram constructed based on the 1000 Monte Carlo samples, whilst the black line is obtained by randomly sampling the PC expansion of $v_A(t)$. It is worthwhile to point out that this post-processing step takes less than 1 s for 10^6 PC-samples, as such allowing a much smoother reproduction of the PDF.

As far as the simulation times are concerned, the Monte Carlo analysis required 26 min 5 s, whereas the PC simulation took 11.4 s. An impressive speed-up of $140\times$ is thence achieved.

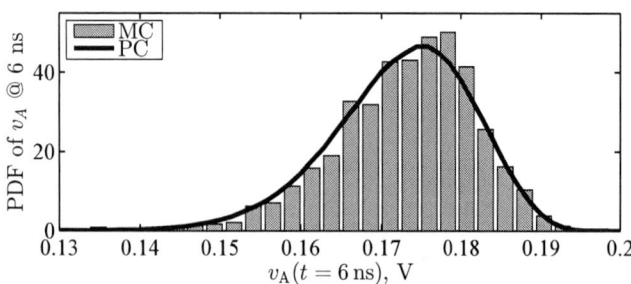

Fig. 6. Probability density function of v_A at $t = 6$ ns obtained from both the Monte Carlo samples (gray bars) and the PC expansion (black line).

V. CONCLUSIONS

This paper presents an exact and closed-form formulation to include nonlinear terminations with polynomial I-V characteristics into the PC framework of statistical interconnect simulation. Contrary to the Monte Carlo approach, a single simulation of a modified network allows to extract statistical information much faster than analyzing a large number of random circuit configurations. The formulation is SPICE compatible, thus enabling the designer to perform stochastic analyses using standard circuit solvers. As such, it represents a first step towards the inclusion of nonlinear elements in the SPICE- and PC-based circuit analysis. The theory is illustrated by means of an application example involving the simulation of a network with two lossy and dispersive coupled microstrip sections having random variations in the substrate thickness and trace separation, and terminated by nonlinear, polynomial loads.

REFERENCES

[1] T. Mikazuki and N. Matsui, "Statistical design techniques for high-speed circuit boards with correlated structure distributions," *IEEE Trans. Compon. Packag. Manuf. Technol. A*, vol. 17, no. 1, pp. 159–165, Mar. 1994.

[2] A. H. Zaabab, Qi-Jun Zhang, and M. Nakhla, "A neural network modeling approach to circuit optimization and statistical design," *IEEE Trans. Microw. Theory Tech.*, vol. 43, no. 6, pp. 1349–1358, Jun. 1995.

[3] I. S. Stievano, P. Manfredi, and F. G. Canavero, "Parameters variability effects on multiconductor interconnects via Hermite polynomial chaos," *IEEE Trans. Compon. Packag. Manuf. Techol.*, vol. 1, no. 8, pp. 1234–1239, Aug. 2011.

[4] P. Manfredi, I. S. Stievano, and F. G. Canavero, "Alternative SPICE implementation of circuit uncertainties based on orthogonal polynomials," in *Proc. IEEE 20th Conf. Elect. Perform. Electron. Packag. Syst.*, San Jose, CA, Oct. 2011, pp. 41–44.

[5] A. Rong and A. C. Cangellaris, "Interconnect transient simulation in the presence of layout and routing uncertainty," in *Proc. IEEE 20th Conf. Elect. Perform. Electron. Packag. Syst.*, San Jose, CA, Oct. 2011, pp. 157–160.

[6] D. Vande Ginste, D. De Zutter, D. Deschrijver, T. Dhaene, P. Manfredi, and F. Canavero, "Stochastic modeling-based variability analysis of on-chip interconnects," *IEEE Trans. Compon. Packag. Manuf. Technol.*, vol. 2, no. 7, pp. 1182–1192, Jul. 2012.

[7] J. S. Ochoa and A. C. Cangellaris, "Fast analysis of the impact of interconnect routing variability on signal degradation," in *Proc. IEEE 21th Conf. Elect. Perform. Electron. Packag. Syst.*, Tempe, Az, Oct. 2012, pp. 315–318.

[8] T.-A. Pham, E. Gad, M. Nakhla, and R. Achar, "Efficient Hermite-based variability analysis using approximate decoupling technique," in *Proc. IEEE 17th Workshop on Signal and Power Integrity*, Paris, France, May 2013, pp. 111–114.

[9] D. Xiu, "Fast numerical methods for stochastic computations: a review," *Commun .Comput. Physics,* vol. 5, no. 2–4, pp. 242–272, Feb. 2009.

[10] P. Manfredi, D. Vande Ginste, D. De Zutter, and F. G. Canavero, "Uncertainty assessment of lossy and dispersive lines in SPICE-type environments," *IEEE Trans. Compon. Packag. Manuf. Techol. (in press)*, DOI: 10.1109/TCPMT.2013.2259295.

[11] A. Biondi, D. Vande Ginste, D. De Zutter, P. Manfredi, and F. G. Canavero, "Variability analysis of interconnects terminated by general nonlinear loads," *IEEE Trans. Compon. Packag. Manuf. Techol. (in press)*, DOI: 10.1109/TCPMT.2013.2259896.

[12] *HSPICE User Guide: Signal Integrity, Version B-2008.09*, Synopsys, Inc., Mountain View, CA, USA, Sep. 2008.

Cost Effective Modeling Methodologies and Evaluating Electrical Interaction in FCBGA Packages

Hyunho Baek*, Julius Delino** and William R. Eisenstadt*
*Department of Electrical and Computer Engineering, University of Florida
1064 Center Dr. 505 NEB, Gainesville, FL 32611, USA
**Intel Corporation, Folsom, CA 95630, USA
hhbaek@ufl.edu

Abstract —This paper demonstrates a modeling methodology for the die-to-package connectivity in a Flip-chip Ball Grid Array (FCBGA) so that the die and package are simulated separately instead of co-simulation of the FCBGA structure for saving a computational cost. In addition, a cost saving polynomial regression modeling method is proposed to find electrical behavior for huge and complex package structures. In addition, various types of patterns were simulated to investigate and evaluate electrical interactions between the die and package through C4 bumps.

Index Terms — Modeling methodology, FCBGA, Electrical behavior, S-parameter, Polynomial regression.

I. INTRODUCTION

As the integration technology advances, both package modeling and simulation are more critical in FCBGA package design since physical structures are more complex. Also, not only modeling die and package accurately but also analyzing the electrical behavior of the package structures is important in order to anticipate higher performance and functionality for power integrity and signal integrity in FCBGA packages. However, system-level considerations have become increasingly important as ITRS modeling and simulation roadmaps identify in crosscut issues [1] that 3D modeling and simulating the entire IC-package structure is a huge challenge [2]; the large system model size requires expertise across design disciplines; as well as having unacceptably high computational cost due to in the hundreds and thousands of physical links with connectivity [3].

In this paper, a modeling methodology is proposed for die-to-package connectivity that shows good agreement with the power, ground grids and C4 bump performance. Both a single 3D and a combined model will be demonstrated and the S-parameters extracted from both models are compared to verify that the methodology can be used for die-to-package connectivity. In addition, polynomial modeling is proposed to represent an electrical behavior in terms of S-parameters for a huge structure. The evaluation of the electrical behavior of the package physical structure between the first metal layer on the die and package is demonstrated.

II. MODELING METHODOLOGY FOR DIE-PKG CONNECTIVITY

Although co-simulation of the entire structure gives accurate modeling simulation results [4], the high computational cost is a key challenge. In this work, the authors

show a modeling methodology for the FCBGA structures that can be a cost saving as seen in Fig. 1.

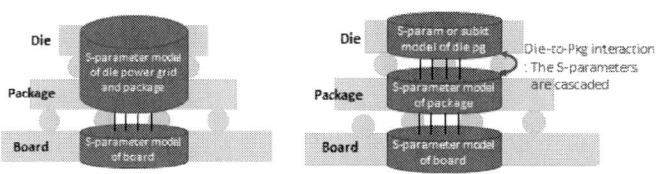

(a) Single 3D model – Die + Pkg   (b) Combined model - Separated Die & Pkg

Fig 1. Modeling Stratigies

Fig. 1(a) shows that the S-parameter is derived from a single 3D model connected between die and package model. And, Fig. 1(b) shows a system build from both a die and a package model separately so that the S-parameters are cascaded through C4 bumps. For verification of the die to package connectivity, S-parameters are extracted from both a single 3D model in Fig. 1(a) and separated planar models in Fig. 1(b). The connectivity can be demonstrated if the S-parameter of a single 3D model is matched with the combined S-parameters for planar substrates which are connected through C4 bumps for the separated die and package.

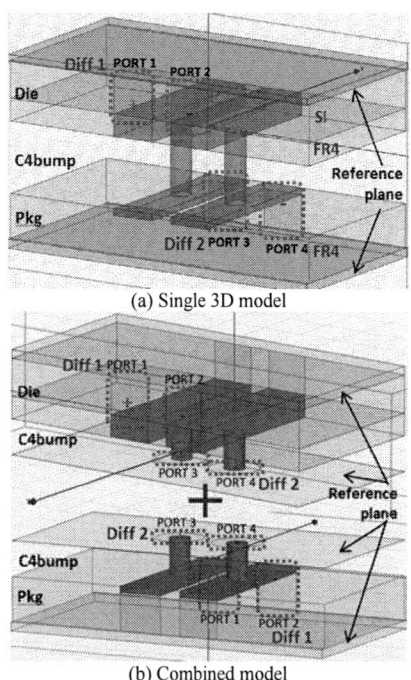

(a) Single 3D model

(b) Combined model

Fig. 2. Port definition in HFSS for modeling strategies

978-1-4799-0708-3/13 $31.00 © 2013 IEEE   103

Fig. 2 illustrates the simplified FCBGA structure and port definitions for the die and package for Fig. 1. The authors designed the structure in HFSS [5] and especially the C4 bump in Fig. 2(b) is cut in half to simulate the die and package separately and then to combine each S-parameter of die and package through the C4 bumps. The solution type of this modeling is set by the choice of the driven terminal and lumped ports are used for all ports in Fig. 2. Especially, the impedance of the each terminal has to be same between the port 3 and the port 4 on the C4 bumps for die and pkg so that the S-parameters are cascaded. Also, the ports are defined as differential pairs for a symmetrical differential port between a die and a package as a microwave interconnection. In order to approach the simulation of this system as a two-port network, the 4-port network that is defined above is treated as two-port network using post processing in ADS [6].

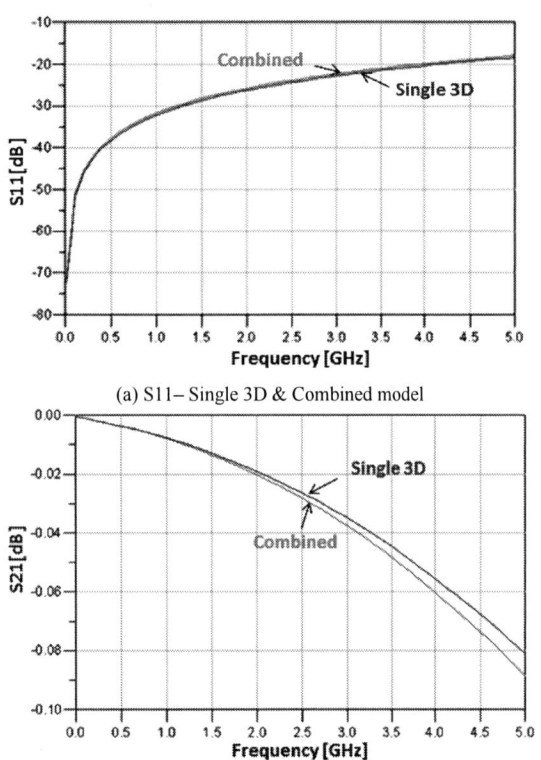

(a) S11– Single 3D & Combined model

(b) S21 - Single 3D & Combined model

Fig 3. S-parameters between single 3D model and combined planar models

Using these modeling strategies, the die and package connectivity can be verified in terms of S_{21} and S_{11} parameters and the magnitudes are almost matched as shown in Fig. 3. In addition, Ansys Q3D [7] which is specialized for low frequency electromagnetic modeling is used to simulate between DC and 10MHz and the S_{11} curve is matched to the S_{11} of HFSS at 10MHz for wide-band modeling. Hence, the methodology we propose allows simulating the die and packaging separately to save computational cost for a complex FCBGA model.

III. POLYNOMIALS USING POLYNOMIAL REGRESSION

Polynomial regression is a form of linear regression in which data is used to find polynomials representing the transfer function of each model. In this work, the polynomials are derived from the transfer function which is the gain from the incident voltage, V_a, to the output voltage, V_l of the 2-port network in the preceding figure. As mentioned previously, a 4-port network is treated as a 2-port network. Fig. 4 shows how to compute the V_a from the source voltage V_s [8] and the magnitude of transfer function of each model can be equal to S_{21} parameter to determine the polynomials using the equations (1)-(3).

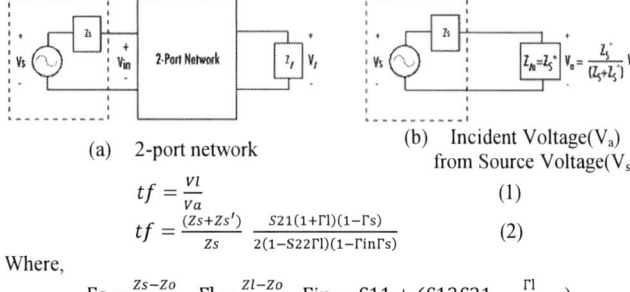

(a) 2-port network

(b) Incident Voltage(V_a) from Source Voltage(V_s)

$$tf = \frac{Vl}{Va} \quad (1)$$

$$tf = \frac{(Zs+Zs')}{Zs} \frac{S21(1+\Gamma l)(1-\Gamma s)}{2(1-S22\Gamma l)(1-\Gamma in\Gamma s)} \quad (2)$$

Where,

$$\Gamma s = \frac{Zs-Zo}{Zs+Zo}, \quad \Gamma l = \frac{Zl-Zo}{Zl+Zo}, \quad \Gamma in = S11 + (S12S21\frac{\Gamma l}{(1-S22\Gamma l)})$$

Hence,

$$tf = \frac{Vl}{Va} = S21 \quad where, Zs = Zs' = Zl = Zo \quad (3)$$

Fig 4. Proof the transfer function is equal to S_{21} of 2-port network

As shown in Fig. 5, three-different models with identical port setup in Fig. 2(a) are designed. Each model has different number of N, that is a number of pairs of C4 bumps within the ports, and the length of model is proportional to the N since N=1 has a unit length of the model.

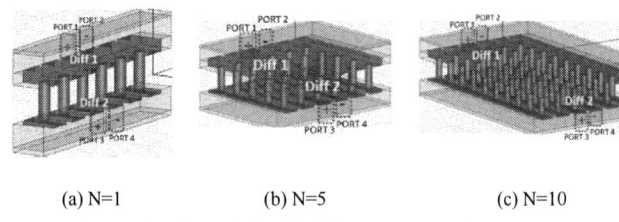

(a) N=1   (b) N=5   (c) N=10

Fig 5. Die-Pkg models with different number of N (N=1,5,10)

To find a polynomial coefficients of each model, the transfer function needs to be derived from the S_{21} parameters of each model and the curve of transfer function can be defined by 4th-degree polynomials using polynomial regression as shown in Fig. 6.

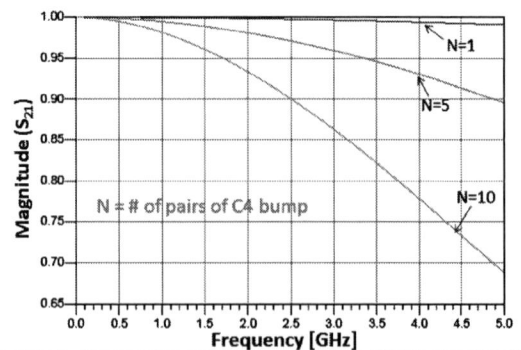

N	4th degree polynomials of each curve
1	f(x)=-7.74e-07x⁴ + 5.21e-06x³ -0.0003x² -0.0002x + 0.9999
5	f(x)=3.997e-06x⁴ + 7.668e-05x³ -0.004x² -0.0007x + 0.9999
10	f(x)=2.368e-05x⁴ + 0.0015x³ -0.017x² -0.0002x + 0.9996

Fig 6. Transfer function curve of each model (N=1,5,10)

Once the coefficients of each degree are derived from the polynomials of each curve, the coefficients can be set as P1, P2, P3, P4 and P5 using polynomial regression as shown in Fig. 7(a). Hence, the electrical behavior of complex model can be captured using the polynomials in Fig. 7(b) instead of 3D co-simulation with highly computational cost.

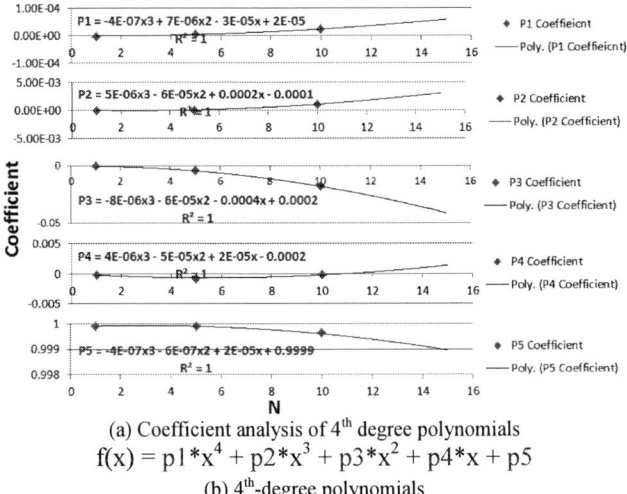

(a) Coefficient analysis of 4th degree polynomials

$$f(x) = p1*x^4 + p2*x^3 + p3*x^2 + p4*x + p5$$

(b) 4th-degree polynomials

Fig 7. Coefficient analysis of 4th degree polynomials

In Fig. 8, complex models are simulated using the polynomial coefficients and the S_{21} is matched between the curve of polynomial and the simulation result in HFSS. And, the comparison of the computational cost between the HFSS and the polynomial for N=15 and 30 are shown under the Intel Core i7 Quad-core processor desktop in Table. I.

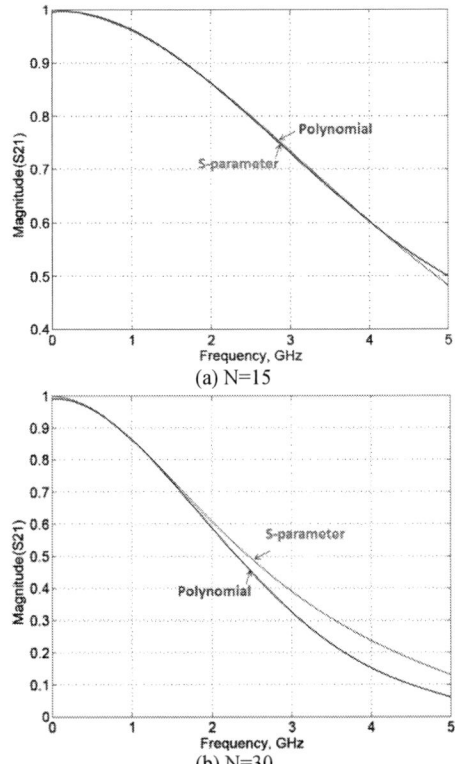

Fig 8. Comparison S-parameters and calculated polynomials

TABLE I.   COMPARISON THE COMPUTATIONAL COST BETWEEN HFSS AND POLYNOMIAL

	N=15		N=30	
Type	HFSS	Polynomial	HFSS	Polynomial
Time	14hrs	less than a min.	30hrs	less than a min.

IV. EVALUATING ELECTRICAL INTERACTION BETWEEN DIE AND PACKAGE IN FCBGA

Conventional modeling methods separate the IC and the package during characterization since they inherently assume no significant electrical interaction between the first layer IC metal and flip-chip package systems. However, the worst case IC upper level metal to package interconnect signal coupling interactions are anticipated through these simulations.

A. Different pattern between Die and Package planes

In Fig. 9,10 the structures modeled in HFSS are simulated to analyze the factors such as coupling, insertion and return loss that can be accounting for during early phase design decisions about the first level metal signal lines of die and package. The C4 bumps patterns are identical for all models and the pattern between the die and package metal is different.

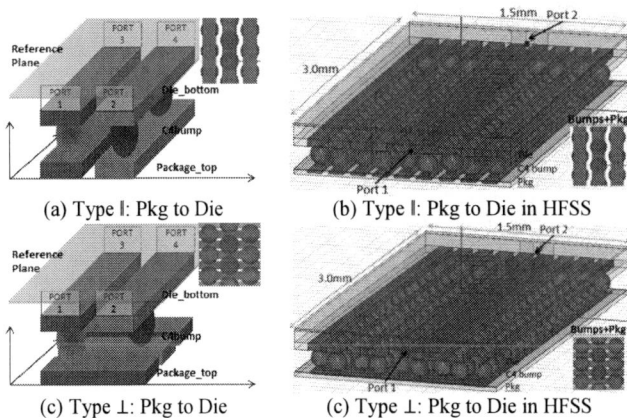

(a) Type ‖: Pkg to Die   (b) Type ‖: Pkg to Die in HFSS

(c) Type ⊥: Pkg to Die   (c) Type ⊥: Pkg to Die in HFSS

Fig 9. Different patterns between the first layer of die and package

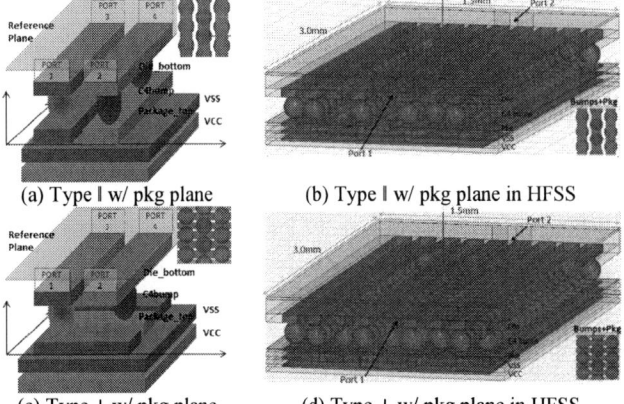

(a) Type ‖ w/ pkg plane   (b) Type ‖ w/ pkg plane in HFSS

(c) Type ⊥ w/ pkg plane   (d) Type ⊥ w/ pkg plane in HFSS

Fig 10. Different patterns between th first layer of die and package planes

In Fig. 11(a), a parallel type has less insertion loss than a perpendicular type since C4 bumps are connected by other grid in the first layer of package. Through EM simulation, we found that Fig. 9(a) is more inductive than Fig. 9(b). In case of Fig.

978-1-4799-0708-3/13 $31.00 © 2013 IEEE   105

11(b), both insertion losses are similar to each other because all grids are connected by package planes.

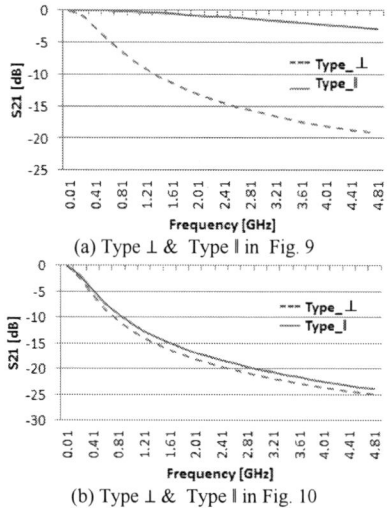

(a) Type ⊥ & Type ‖ in Fig. 9

(b) Type ⊥ & Type ‖ in Fig. 10

Fig 11. Simulations of electrical interations of Fig. 9 and Fig. 10

Hence, the coupling between Die bottom layer and Pkg top layer shows different values depending on the pattern and the metal layers in the packages. For reference, the package modeling properties are shown in Table II.

TABLE II.   PROPERTIES OF THE MATERIAL IN THE PACKAGE

Layer	Thickness (um)	Permittivity	Loss Tangent	Conductivity (s/m)
Solder Mask	5	4	0.015	
Top(Pwr/Gnd)	15			5.91E+07
Build-up 1	30	3.5	0.026	
Gnd_L01	15			5.91E+07
Build-up 2	30	3.5	0.018	
Pwr_L02	15			5.91E+07
Build-up 3	30	3.5	0.018	

B. Electrical behavior between C4 bump pairs

Using the port setup in Fig. 2, the electrical interaction among C4 bumps pair can be determined how the return losses are varied depending on the length of the sequence of each C4 bumps pairs. This allows evaluating the performance and finding which route is efficient among the bumps. In Fig. 12 and 13, Diff2 can be switching to other pairs (1st, 2nd, 3rd and 4th) in order to measure the S-parameter between Diff2 and Diff1 sequentially. The return loss of S_{11_4th} is the lowest and S_{11_1st} is the highest in all frequencies range. In addition to, the die to package connectivity can be demonstrated by the separated die and package model through the ports on the C4 bumps.

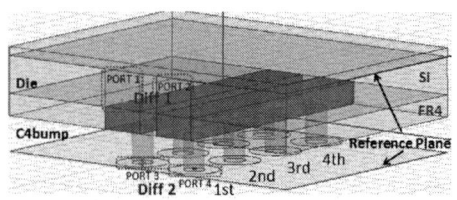

(a) Die metal layer w/ 4 C4 bump pairs

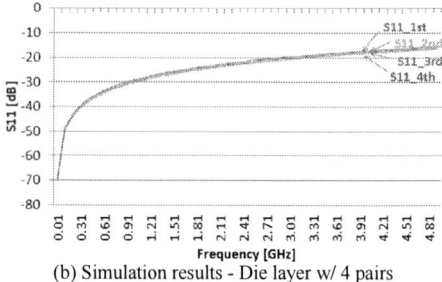

(b) Simulation results - Die layer w/ 4 pairs

Fig 12. S11 among C4 bumps pair on a die

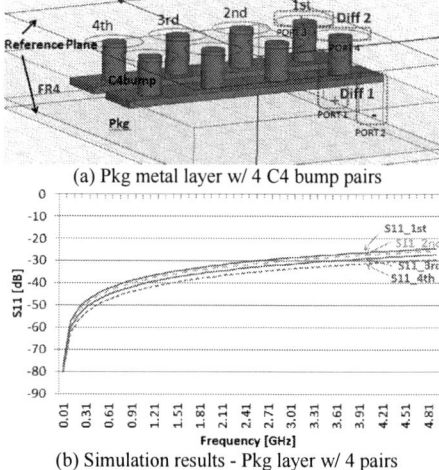

(a) Pkg metal layer w/ 4 C4 bump pairs

(b) Simulation results - Pkg layer w/ 4 pairs

Fig 13. S11 among C4 bumps pair on a pkg

V. CONCLUSION

In this paper, a proposed modeling methodology is demonstrated for the die-to-package connectivity in FCBGA so that the die and package can be simulated separately, not as a co-simulation. Building a FCBGA model using polynomials is proposed and we demonstrated the matched S-parameters between the curve of polynomial of the model and the S-parameter of the model which is extracted from HFSS. In addition, various patterns between the top of metal layer in a die and package layers are modeled in HFSS to evaluate the electrical interaction so that die–level and package-level designer can make early decisions. All of above techniques can be used for a saving computational cost in case of huge package and complex C4 bumps modeling in FCBGAs and other applications.

References

[1] International Technology Roadmap for Modeling and Simulation (ITRS) reports, http://www.itrs.net/reports.html.

[2] S. Wane and A. Kao, "Chip-Package Co-Design Methodology for Global Co-Simulation of Re-Distribution Layers (RDL)," 2008 IEEE-EPEP. 3, pp. 59-62, Oct. 2008.

[3] M. Kowalsky and P. Codd, "Co-Simulation of IC, Package and PCB Power Delivery Networks in Ultra-Low Voltage Power Rail Designs," 2007 IEEE-ECTC, pp. 798-803, May 2007.

[4] M. Ha, et al., "Chip-Package Co-Simulation with Multiscale Structure," 2007 IEEE-EPEP, pp. 27-29, Oct. 2008.

[5] High Frequency Structure Simulator(HFSS) v13.0, ANSYS, Inc.

[6] Advanced Design Systems(ADS) 2008, Agilent Technologies.

[7] Q3D Extractor v13.0, ANSYS, Inc.

[8] Matlab R2009a, The MathWorks

978-1-4799-0708-3/13 $31.00 © 2013 IEEE

Electrical-Thermal Co-Simulation for DC IR-Drop Analysis of Large-Scale Integrated Circuits

Tianjian Lu and Jian-Ming Jin

Department of Electrical and Computer Engineering
University of Illinois at Urbana-Champaign
Urbana, Illinois 61801
Email: {tlu7, j-jin1}@illinois.edu

Abstract—**For accurate DC IR-drop analysis of integrated circuits, electrical-thermal co-simulation is necessary in order to take into account the effect of Joule heating. A suitable numerical method for this co-simulation is the finite element method due to its capabilities in modeling complex geometries and materials. In order to deal with large-scale problems, a domain decomposition scheme is applied to the finite element method to enable simulation with multiple processors in parallel and achieve significant reduction in computation time.**

I. INTRODUCTION

Through silicon via (TSV) technology, which enables multiple dies in a single stack, is believed to smooth the path for continuing the interconnect scaling. On one hand, the continuous integration leads to high packing density and in general high performance; on the other hand, it brings reliability concerns for the designers: the supply voltage continues to drop with the increased scaling and the devices are more vulnerable to power noise; additionally, Joule heating is likely to cause significant rise in temperature if the heat generated from a large amount of current flowing through interconnects is not removed efficiently [1]. Because of the temperature-dependent resistivity, the aforementioned two concerns should be addressed simultaneously, bringing out the need for electrical-thermal co-simulation in the DC IR-drop analysis. This has been implemented with the finite volume method [2] and the equivalent circuit model [3]. In this work, the finite element method (FEM) is adopted owing to its capability of dealing with complex geometries [4].

This work is not limited to the electrical-thermal co-simulation, but also aims at developing a more efficient approach for large-scale problems. One way of improving the efficiency while dealing with large-scale problems is to use domain decomposition; in this work, the finite element tearing and interconnecting (FETI) method [5]–[7] is employed. By FETI, the entire computation domain is decomposed into non-overlapping subdomains of smaller size which allows the possibility of parallel computing. The continuity enforcement at subdomain interfaces through Dirichlet and Neumann boundary conditions via Lagrange multipliers yields a reduced-order interface problem. The interface problem can be solved by a Krylov subspace method, and the obtained solutions can serve as the boundary condition to solve for the unknowns within individual subdomains. FETI is by nature highly parallelizable; through parallel computing with multiple processors significant reduction in computational time can be achieved.

II. ITERATIVE SCHEME OF ELECTRICAL-THERMAL CO-SIMULATION

The boundary-value problem of the DC IR-drop analysis consists of two parts: the governing equation

$$\nabla \cdot \sigma \nabla \phi = 0, \tag{1}$$

where σ is the temperature-dependant electrical conductivity and ϕ is the electrical potential and the boundary conditions

$$\phi = \phi_c \qquad \text{on } \Gamma_{\text{vc}}, \tag{2}$$
$$\sigma \frac{\partial \phi}{\partial n} = \frac{\phi}{RS} \qquad \text{on } \Gamma_{\text{load}}, \tag{3}$$

where Γ_{vc} is the Dirichlet boundary, Γ_{load} is the impedance boundary of external load, and R and S are the resistance and the cross-sectional area of the impedance boundary, respectively.

The boundary-value problem of the steady-state thermal analysis consists of two parts: the governing equation

$$\nabla \cdot k \nabla T = -P, \tag{4}$$

where k is the thermal conductivity, T is the temperature distribution, and P is the heat source; and the boundary conditions

$$T = T_c \qquad \text{on } \Gamma_{\text{tc}}, \tag{5}$$
$$k \frac{\partial T}{\partial n} = -h(T - T_a) \qquad \text{on } \Gamma_{\text{conv}}, \tag{6}$$

where Γ_{tc} is the isothermal boundary, Γ_{conv} is the convection boundary, h and T_a are convective heat transfer coefficient and ambient temperature, respectively.

The heat source P in Equation (4) includes both the external heat source and that from Joule heating P_{joule} as

$$P_{\text{joule}} = \boldsymbol{J} \cdot \boldsymbol{E} = \sigma \boldsymbol{E}^2. \tag{7}$$

The temperature-dependant resistivity can be written as

$$\rho = \rho_0 \left[1 + \alpha(T - T_0)\right], \tag{8}$$

where ρ_0 is the resistivity at temperature T_0 and α is the temperature coefficient of the material. The electrical analysis and thermal analysis are coupled through Equations (7) and (8). The iterative scheme of solving Equations (1), (4), (7), and (8) concurrently is illustrated in Figure 1.

978-1-4799-0708-3/13 $31.00 © 2013 IEEE

INPUT   initial resistivity ρ_0, tolerance *tol*, number of iterations N
OUTPUT   voltage distribution and temperature profile
Step 1   Set $i = 1$ and $\tilde{\rho} = \rho_0$
Step 2   While $i \leq$ number of iterations do Steps 3-8
Step 3   Compute voltage via electrical analysis
Step 4   Calculate power
Step 5   Compute temperature via thermal analysis
Step 6   Update resistivity ρ
Step 7   If $|\rho - \tilde{\rho}| < tol$
                Output and Exit
Step 8   Set $i = i + 1$ and $\tilde{\rho} = \rho$
Step 9   Output ('Not converged after N iterations')

Fig. 1: The iterative scheme of electrical-thermal co-simulation.

III. INCORPORATION OF FETI INTO THE FINITE ELEMENT FORMULATION

In this section, we incorporate FETI into the finite element formulation. Since the boundary-value problems of the DC IR-drop analysis and the steady-state thermal analysis are similar to each other, we take that of the DC IR-drop analysis as example. By applying FETI, the entire computation domain is torn into N_s non-overlapping subdomains. The neighboring subdomains Ω^i and Ω^j share the interface denoted as Γ_{ij}. To enforce the continuity of the voltage and the current at the interfaces, the Dirichlet-type transmission condition

$$\phi_i = \phi_j \tag{9}$$

and the Neumann-type transmission condition

$$\hat{n}^i \cdot \sigma_i \nabla \phi_i = -\hat{n}^j \cdot \sigma_j \nabla \phi_j = \Lambda \tag{10}$$

are applied accordingly, where $\hat{n}^i$ is the unit normal vector at Γ_{ij} pointing to the exterior region, and Λ is the unknown Neumann boundary data.

The finite element discretization is applied to individual subdomains, and one obtains the linear system of the sth subdomain

$$[K^s]\{\phi^s\} + [B^s]^T\{\lambda\} = \{f^s\}, \tag{11}$$

where $[K^s]$ includes the volumetric discretization information, $[B^s]^T\{\lambda\}$ includes the information of the subdomain interfaces, $\{\lambda\}$ is the dual variable, called the Lagrange multiplier, and $\{f^s\}$ contains the information of the excitation. Assembling the system equations from all the subdomains yields a global system as follows

$$\begin{bmatrix} [K^1] & \cdots & 0 & [B^1]^T \\ \vdots & \ddots & \vdots & \vdots \\ 0 & \cdots & [K^{N_s}] & [B^{N_s}]^T \\ [B^1] & \cdots & [B^{N_s}] & 0 \end{bmatrix} \begin{bmatrix} \{\phi^1\} \\ \vdots \\ \{\phi^{N_s}\} \\ \{\lambda\} \end{bmatrix} = \begin{bmatrix} \{f^1\} \\ \vdots \\ \{f^{N_s}\} \\ 0 \end{bmatrix}. \tag{12}$$

It is worth mentioning that the last equation in Equation (12) is the result of enforcing the Dirichlet-type boundary condition at subdomain interfaces. By eliminating $\{\phi^s\}$, $s = 1, 2, ..., N_s$, one arrives at an interface system for $\{\lambda\}$ as

$$[F]\{\lambda\} = \{d\}, \tag{13}$$

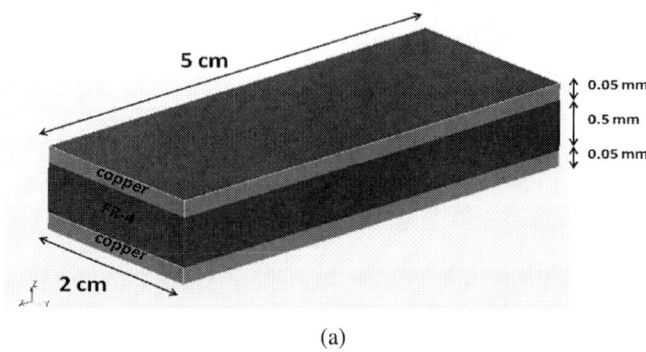

(a)

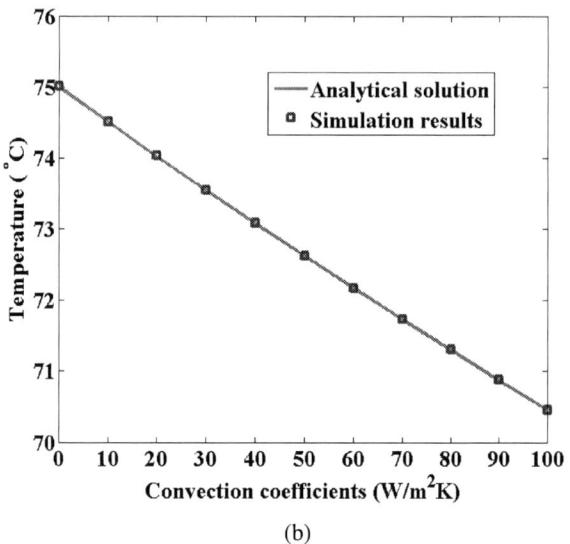

(b)

Fig. 2: Two-layered PCB: (a) geometry, (b) temperature varying with convection coefficients.

where

$$[F] = \sum_{s=1}^{N_s} [B^s][K^s]^{-1}[B^s]^T \tag{14}$$

$$\{d\} = \sum_{s=1}^{N_s} [B^s][K^s]^{-1}\{f^s\}. \tag{15}$$

The interface system in Equation (13) can be solved using a Krylov subspace method such as the biconjugate gradient stabilized (BiCGSTAB) method together with the Dirichlet preconditioner [8]. After iteratively solving the interface problem for the dual variables, the voltage unknowns of individual subdomains can be computed with the obtained Neumann-type boundary condition.

IV. NUMERICAL EXAMPLE AND MODEL VERIFICATION

A two-layered PCB board [2] in Figure 2(a) is taken as a benchmark example to verify the implementation of heat conduction and convection, as well as the iterative scheme of electrical-thermal co-simulation. The detailed information of the structure can be found in [2]. The variation of the temperature on the top layer with different convection coefficients from both the simulation and analytical solution is plotted in Figure 2(b). Voltage distributions and temperature profiles of

978-1-4799-0708-3/13 $31.00 © 2013 IEEE   108

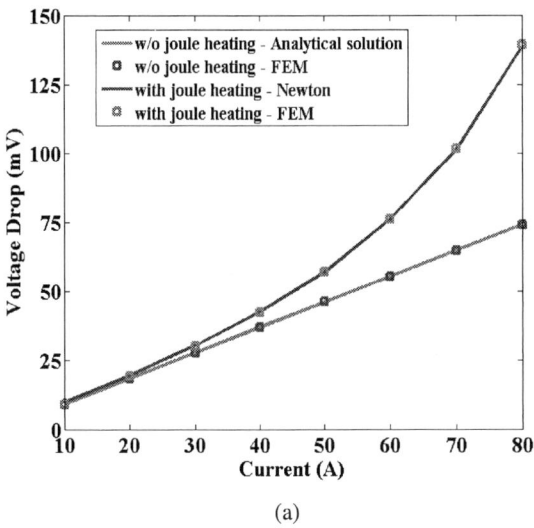

(a)

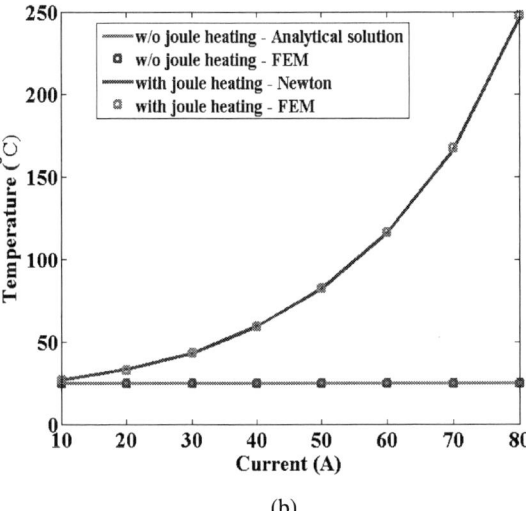

(b)

Fig. 3: Electrical-thermal co-simulation of two-layered PCB: (a) voltage distribution; (b) temperature profile.

the top layer with and without the Joule heating effect are compared with those in [2] and shown in Figure 3.

V. PARALLEL IMPLEMENTATION OF FETI IN THE ELECTRICAL-THERMAL CO-SIMULATION

In this section, the parallel implementation of FETI in the electrical-thermal co-simulation using Message Passing Interface (MPI) is investigated. All computations are performed with Intel Xeon 2.67 GHz hex-core processors. Intel MKL pardiso is used whenever it is necessary to perform matrix factorization and solve linear systems.

Consider the two-layered TSV array in Figure 4: the copper-filled TSV has radius 8 μm and height 30 μm; the thickness of the copper plane is 5 μm; the separation between two neighboring TSVs is 30 μm along the x-axis and 34 μm along the y-axis, both of which are measured between the centers of the vias. Assume the bottom surface of the copper plane has constant temperature 313 K and is exposed to

Fig. 4: Two-layered TSV array.

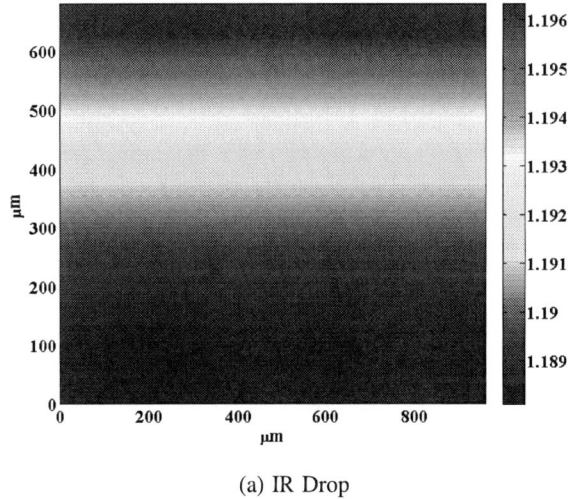

(a) IR Drop

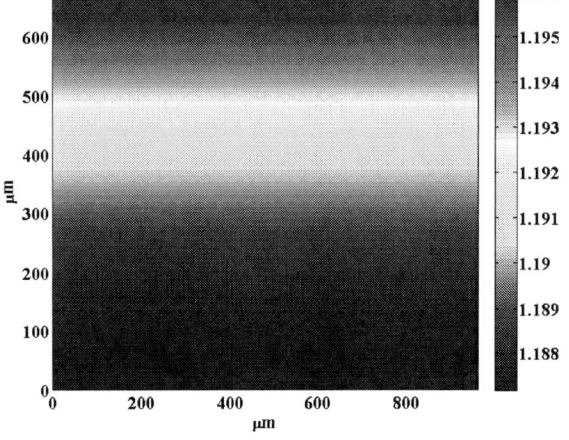

(b) IR Drop with Joule heating

Fig. 5: Voltage distribution at $z = 30$ μm of the two-layered of TSV array in Figure 4.

convection cooling with ambient temperature of 293 K and the convective heat transfer coefficient is 10 W/m²K. Voltage of 1.2 V is applied to the upper edge of the lower plane. Loads of 40 Ω are connected through the vias on the top layer. This

TABLE I: Speed-up of FETI with MPI

Total Number of DOFs: 2.03×10^6				
Number of processors	4	8	16	32
Number of subdomains	4	8	16	32
Number of DOFs per subdomain	5.1×10^5	2.6×10^5	1.4×10^5	6.4×10^4
Total computation time (seconds)	166.74	64.04	26.76	14.70

TABLE II: Unit-load-per-processor parallel effieciency of FETI with MPI

Total number of DOFs	2.6×10^5	5.1×10^5	1.0×10^6	2.0×10^6
Number of processors	4	8	16	32
Number of subdomains	4	8	16	32
Total computation time (seconds)	13.30	14.05	13.85	14.70

example has a total number of 2.03 million DOFs.

First, we explore the trend of the reduction in computation time with the increase in the number of processors. Four cases are generated for comparison where the entire structure is cut into 4, 8, 16, and 32 subdomains, respectively. Each subdomain is assigned to one processor. The computation time of all these four cases is recorded in Table I. The speed-up is defined as the ratio of the computation time of two different cases. It is observed that when the number of processors is doubled from 4 to 8, a speed-up of 2.6 is achieved. The reason that the speed-up is larger than 2 lies in the factorization of subdomain matrices; in the case of 4 subdomains, the time taken to factorize the matrix is far more than twice of that for the case of 8 subdomains. It is also observed that when the number of processors is doubled from 16 to 32, the speed-up is 1.82, which is less than 2. This is because with the increased number of subdomains, the size of the interface problem and the time consumed in communication over different processors increase accordingly. In this investigation, by continuously doubling the number of processors from 4 to 32 while keeping the size of the problem, speed-up around 2 is achieved. This is owing to the high scalability of FETI.

The second investigation of parallel efficiency is carried out based on the idea of unit load per processor where the number of DOFs assigned to individual processors is kept the same. There are 32 columns of vias along the x-axis. Each column of vias forms one subdomain and is assigned to one processor. Four cases are generated for a comparison; the sizes of the problems and the computation time are recorded in Table II. It is observed that the four cases of different numbers of DOFs take relatively the same amount of computation time. From the case of 4 subdomains to the case of 32 subdomains, the total number of DOFs is increased by 8 times whereas the computational time has only an increment of 1.4 seconds. It can be drawn from Table II that FETI-enabled parallel implementation with MPI is capable of solving a large-scale problem with a similar amount of time to that of a much smaller problem by simply employing more processors.

The voltage distribution at $z = 30$ μm with and without the Joule heating effect are depicted in Figure 5: without considering Joule heating, the voltage drop to the plane of $z = 30$ μm is 7 mV; with Joule heating, the voltage drop

becomes 9 mV; the thermal effect on the voltage drop is 28%. It can be seen from the color bars of Figure 5 that Joule heating causes lower voltage level on the plane of $z = 30$ μm , which indicates a higher voltage drop from the source.

VI. CONCLUSION

In this paper, the electrical-thermal co-simulation is implemented with the finite element method. The Joule heating effect is taken into account for an accurate prediction of voltage distribution and temperature rise in the integrated circuits. This paper also incorporates FETI, a domain decomposition method, into the electrical-thermal co-simulation in order to handle large-scale problems. Numerical examples have demonstrated high parallel efficiency of FETI with MPI as well as significant reduction in computation time by using FETI with multiple processors in parallel.

REFERENCES

[1] K. Banerjee and A. Mehrotra, "Global (interconnect) warming," *Circuits and Devices Magazine, IEEE*, vol. 17, no. 5, pp. 16–32, 2001.

[2] J. Xie and M. Swaminathan, "Electrical-thermal co-simulation of 3D integrated systems with micro-fluidic cooling and Joule heating effects," *Components, Packaging and Manufacturing Technology, IEEE Transactions on*, vol. 1, no. 2, pp. 234 –246, Feb 2011.

[3] Y. Zhong and M. D. F. Wong, "Thermal-aware IR drop analysis in large power grid," in *Quality Electronic Design*, San Jose, CA, 2008, pp. 194–199.

[4] J. M. Jin, *The finite element method in electromagnetics*. Hoboken, NJ: Wiley, 2002.

[5] C. Farhat and F.-X. Roux, "A method of finite element tearing and interconnecting and its parallel solution algorithm," *International Journal for Numerical Methods in Engineering*, vol. 32, no. 6, 1991.

[6] C. Farhat and J. Mandel, "The two-level FETI method for static and dynamic plate problems part i: An optimal iterative solver for biharmonic systems," *Computer Methods in Applied Mechanics and Engineering*, vol. 155, no. 1, pp. 129–151, 1998.

[7] Y. Li and J.-M. Jin, "A vector dual-primal finite element tearing and interconnecting method for solving 3-D large-scale electromagnetic problems," *Antennas and Propagation, IEEE Transactions on*, vol. 54, no. 10, pp. 3000–3009, 2006.

[8] D. J. Rixen and C. Farhat, "A simple and efficient extension of a class of substructure based preconditioners to heterogeneous structural mechanics problems," *International Journal for Numerical Methods in Engineering*, vol. 44, no. 4, pp. 489–516, 1999.

978-1-4799-0708-3/13 $31.00 © 2013 IEEE

Reliable Detection of Causality Violations in Tabulated Scattering Parameters through Filtered Dispersion Relations

Piero Triverio

Edward S. Rogers Sr. Department of Electrical and Computer Engineering, University of Toronto
Sandford Fleming Bldg, 10 King's College Rd, Toronto, ON M5S 3G4
Email: piero.triverio@utoronto.ca   Phone: +1 (416) 978 0562

Abstract—We introduce a new form of dispersion relations useful to detect causality violations in sampled scattering parameters. Compared to existing solutions, the new approach is simpler to implement, more intuitive, and provides a better insight on the detected violation.

Keywords—*Macromodeling, Signal Integrity, CAD techniques.*

I. INTRODUCTION

The quest for faster and smaller electronic products has dramatically increased the impact on electronic design of signal integrity and electromagnetic compatibility issues, such as crosstalk, signal dispersion and interference [1]. Typically, these issues arise in interconnect structures like board traces, connectors, cables, and power/ground structures. Therefore, designers need accurate and efficient interconnect models to be able to predict and minimize them. Interconnect modeling involves two steps. First, the scattering parameters of the structure are measured with a vector network analyzer or computed with a full-wave simulator. Then, they are included into transient circuit simulators like SPICE using macromodeling, fast convolution, or model order reduction techniques [1]. The accuracy and stability of the final transient simulation is significantly influenced by the *physical consistency* of the initial samples [2]. Consistency holds if samples respect the fundamental properties of causality, stability, and passivity [2]. While real interconnects always satisfy these properties because of physical reasons, their measured or computed scattering parameters may violated them. Violations can be induced, for example, by an improper calibration of the vector network analyzer, or by the use of non-physical models for the permittivity and permeability of materials.

In this paper, we focus on causality violations. Causal systems cannot produce an output (effect) before an input has been applied (cause) [2]. Clearly, all systems in nature are causal. A linear system with impulse response $h(t)$ is causal if [3]

$$h(t) = 0 \text{ for } t < 0. \qquad (1)$$

Equivalently, its frequency response[1] $H(j\omega)$ must satisfy Kramers-Krönig dispersion relations [4], a pair of integral

transforms that link the real part and imaginary part of any causal frequency response. By verifying if Kramers-Krönig relations are respected, one can test the causality of a given set of frequency samples [4]–[7]. This practice is useful since non-causal frequency samples can impede their conversion into an accurate model for transient simulations [2], even when this step is performed with robust macromodeling techniques like Vector Fitting [8]. In other cases, non-causal samples can lead to divergent transient simulations [2].

In practice, Kramers-Krönig relations cannot be computed exactly, but must be truncated to the bandwidth spanned by the available samples. The resulting *truncation error* can severely bias the verification of causality, unless rigorously estimated as in [4], [7], [9]. These works, however, rely upon a generalized form of Kramers-Krönig relations which, although powerful, is not straightforward to implement. In this work, we introduce a new form of dispersion relations *with filtering*, extending to the frequency domain the time-domain approach of [10]. Thanks to the presence of a filter, the new relations allow for a precise control of truncation error, and lead to a robust method to test causality. Compared to state-of-the-art solutions [4], [7], the new method is easier to implement, more intuitive, and provides a better estimate of the detected violation. This information is useful to quantify the impact of the violation on subsequent modeling steps, as numerical examples will demonstrate.

II. CAUSALITY AND DISPERSION RELATIONS

We consider a linear device with N ports, and we denote one of the entries of its scattering matrix with $H(j\omega)$. The following Theorem [3] shows that $H(j\omega)$ is causal if it satisfies Kramers-Krönig relations.

Theorem 1. *A square-integrable function $H(j\omega)$ is causal if and only if it satisfies*

$$H(j\omega) = \underbrace{\frac{1}{j\pi} \int_{-\infty}^{+\infty} \frac{H(j\alpha)}{\omega - \alpha} d\alpha}_{\mathcal{R}\{H(j\omega)\}}, \qquad (2)$$

where the integral is defined using Cauchy's principal value.

The right hand side of (2) can be seen as a reconstruction operator $\mathcal{R}\{H(j\omega)\}$ that, when applied to a causal function

---

This research was undertaken, in part, thanks to funding from the Canada Research Chairs program.

[1] The frequency response $H(j\omega)$ is the Fourier transform of $h(t)$.

$H(j\omega)$, will return the same function. Conversely, when applied to a non-causal $H(j\omega)$, the reconstruction operator will return a different function. Therefore, a frequency response $H(j\omega)$ is causal if the *causality error*

$$\Delta(j\omega) = \frac{1}{2}\left[\mathcal{R}\{H(j\omega)\} - H(j\omega)\right] \qquad (3)$$

is vanishing at all frequencies [4]

$$\Delta(j\omega) = 0 \quad \forall \omega. \qquad (4)$$

When one tries to verify (4) numerically, a major difficulty arises. Since measurements and simulations return scattering parameters only up to a maximum frequency $\omega = \omega_{\mathrm{m}}$, the infinite integral in (2) must be truncated to the available bandwidth. The induced *truncation error* can severely distort the verification of (4), causing "false positives" or hiding actual causality violations [4]. Since scattering parameters do not in general decay to zero as $\omega \to \infty$, truncation error is typically quite large and may actually diverge! Therefore, strictly speaking, the causality of scattering parameters known on a finite bandwidth cannot be understood with (4). In the next section, we address this issue in a rigorous way. In order to make the paper more readable, we always represent the response under test as a continuous function of ω, rather than referring to its individual samples returned by measurement or simulation. The method that will be proposed, however, is meant to be applied to sampled scattering parameters.

III. CAUSALITY VERIFICATION VIA FILTERED DISPERSION RELATIONS

In order to minimize truncation error, one must reduce the sensitivity of (2) to the behaviour of $H(j\omega)$ beyond ω_{m}. This can be achieved by applying a low pass filter $F(j\omega)$ to the given response [10]

$$H_F(j\omega) = F(j\omega)H(j\omega). \qquad (5)$$

The following Theorem shows that, if the filter meets certain requirements, the causality of $H(j\omega)$ can be fully verified through the filtered response $H_F(j\omega)$.

Theorem 2. *Let $F(j\omega)$ be minimum phase[2] and such that $H_F(j\omega)$ is square integrable. The function $H(j\omega)$ is causal if and only if $H_F(j\omega)$ satisfies (2), ie.*

$$H_F(j\omega) = \frac{1}{j\pi}\int_{-\infty}^{+\infty}\frac{H_F(j\alpha)}{\omega - \alpha}d\alpha. \qquad (6)$$

Proof: The Theorem can be proved with an argument similar to the one given in [10]. One must exploit the fact that Theorem 1 can be applied to $H_F(j\omega)$ since the function is square integrable. ∎

If we multiply (6) by $F^{-1}(j\omega)$, we obtain a *dispersion relation with filtering*

$$H(j\omega) = \underbrace{\frac{F^{-1}(j\omega)}{j\pi}\int_{-\infty}^{+\infty}\frac{F(j\alpha)H(j\alpha)}{\omega - \alpha}d\alpha}_{\mathcal{R}_F\{H(j\omega)\}}, \qquad (7)$$

---

[2]$F(j\omega)$ is minimum phase if both the function itself and its inverse $F^{-1}(j\omega) = 1/F(j\omega)$ are causal [11].

which can be interpreted as follows. The original samples $H(j\alpha)$ are first filtered with $F(j\alpha)$ and then processed by the reconstruction operator $\mathcal{R}\{H(j\omega)\}$. Finally, the inverse filter $F^{-1}(j\omega)$ is applied. The right hand side of (7) leads to the definition of the *reconstruction operator with filtering* $\mathcal{R}_F\{H(j\omega)\}$ and the corresponding reconstruction error

$$\Delta_F(j\omega) = \frac{1}{2}\left[\mathcal{R}_F\{H(j\omega)\} - H(j\omega)\right]. \qquad (8)$$

Owing to Theorem 2, $H(j\omega)$ is causal if and only if (8) vanishes at all frequencies. This powerful result holds for any filter which meets the hypotheses of Theorem 2. All Butterworth and Chebyshev (type I) low-pass filters satisfy the minimum phase requirement [11]. Also, when applied to scattering parameters, these filters lead to a square integrable $H_F(j\omega)$. This holds since scattering parameters are bounded functions, ie. we can always find $M > 0$ such that

$$|H(j\omega)| \leq M \quad \forall \omega. \qquad (9)$$

In comparison with (3), the filter in (8) enables for a rigorous control of truncation error, as discussed in the next section.

A. Truncation Error

When $H(j\omega)$ is available only up to $\omega = \omega_{\mathrm{m}}$, one can only compute a truncated reconstruction operator

$$\widehat{\mathcal{R}}_F\{H(j\omega)\} = \frac{F^{-1}(j\omega)}{j\pi}\int_{-\omega_{\mathrm{m}}}^{+\omega_{\mathrm{m}}}\frac{F(j\alpha)H(j\alpha)}{\omega - \alpha}d\alpha \qquad (10)$$

which differs from the exact operator in (7) because of the truncation error

$$E(j\omega) = \widehat{\mathcal{R}}_F\{H(j\omega)\} - \mathcal{R}_F\{H(j\omega)\} = \\ -\frac{F^{-1}(j\omega)}{j\pi}\int_{|\alpha|>\omega_{\mathrm{m}}}\frac{F(j\alpha)H(j\alpha)}{\omega - \alpha}d\alpha. \qquad (11)$$

In order to unbias the causality check from this truncation error, we find its worst-case value $T(\omega)$. By applying the triangular inequality to (11) and accounting for (9), we obtain

$$|E(j\omega)| \leq \frac{M\left|F^{-1}(j\omega)\right|}{\pi}\int_{|\alpha|>\omega_{\mathrm{m}}}\frac{|F(j\alpha)|}{|\omega - \alpha|}d\alpha = T(\omega). \qquad (12)$$

The bound $T(\omega)$ is always finite and can be computed numerically with the help of the change of variable $\beta = 1/\alpha$ which maps the infinite integration interval $|\alpha| > \omega_{\mathrm{m}}$ onto the finite range $|\beta| < 1/\omega_{\mathrm{m}}$.

B. Robust Causality Check for Finite-Bandwidth Responses

The proposed causality check proceeds as follows. From the samples of $H(j\omega)$, we compute the truncated reconstruction operator $\widehat{\mathcal{R}}_F\{H(j\omega)\}$ using numerical integration to evaluate (10). Then, we form the causality error

$$\widehat{\Delta}_F(j\omega) = \frac{1}{2}\left[\widehat{\mathcal{R}}_F\{H(j\omega)\} - H(j\omega)\right], \qquad (13)$$

which is the "ideal" causality error (8) plus a truncation error $E(j\omega)/2$. In the worst case, this error reaches $T(\omega)/2$. Therefore, if the computed causality error (13) exceeds this threshold for some $\omega = \omega'$

$$\left|\widehat{\Delta}_F(j\omega')\right| > \frac{T(\omega')}{2}, \qquad (14)$$

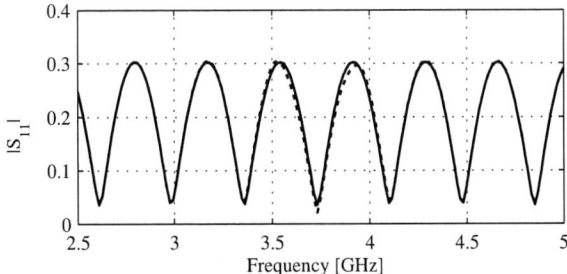

Fig. 1. Original (solid line) and corrupted (dotted line) S_{11} samples for the line of Sec. IV-A. For readability, only frequency points above 2.5 GHz are shown.

we can state with full confidence that $\Delta_F(j\omega') \neq 0$ and, by Theorem 2, that the given samples are not causal. Conversely, if causality error (13) is below the truncation error bound at all frequencies

$$\left|\widehat{\Delta}_F(j\omega)\right| \leq \frac{T(\omega)}{2} \quad \forall |\omega| < \omega_{\mathrm{m}}, \qquad (15)$$

the given samples are considered causal. If any causality violation is present, it is not detectable since it falls below the uncertainty caused by the missing out-of-band samples. The only way to detect such violation would be to acquire further measurements beyond ω_{m}. In comparison with (4), the proposed condition (15) does not require any user interaction to establish whether the detected causality error is small enough to be considered zero. Since a rigorous threshold for truncation error is provided, the proposed test is robust, reliable, and can be fully automated.

IV. NUMERICAL RESULTS

A. Analytic Example

We first validate the proposed technique on an analytic example where we can precisely control the presence of causality violations. We consider the S_{11} parameter of a lossy transmission line (length 10 cm, per-unit-length resistance $0.8\,\Omega/\mathrm{cm}$, inductance $4.73\,\mathrm{nH/cm}$, capacitance $3.8\,\mathrm{pF/cm}$) which was computed with Matlab at 300 frequency points from DC up to 5 GHz. The initial samples were corrupted with a causality violation mostly concentrated in the band 3.5 - 4 GHz. The applied violation mimics a systematic error that affected the measurement of the line. We obtained the corrupted samples as

$$S_{11}(j\omega) = S_{11,c}(j\omega) + A \cdot P^*(j\omega), \qquad (16)$$

where $S_{11,c}(j\omega)$ is the true line response, $A = 2 \cdot 10^{-2}$ is the violation amplitude and $*$ denotes complex conjugation. The violation was created from the response $P(j\omega)$ of a bandpass Chebyshev filter of order 4 and maximum gain of 1 within the passband 3.5-4 GHz. While the filter response $P(j\omega)$ is perfectly causal, its conjugate $P^*(j\omega)$ is a pure causality violation, since conjugation corresponds to the inversion of the time axis. Fig. 1, which depicts both original and corrupted samples, shows that the applied violation is very small and barely noticeable.

We applied the proposed causality test to the corrupted samples using a Chebyshev low-pass filter $F(j\omega)$ of order 6,

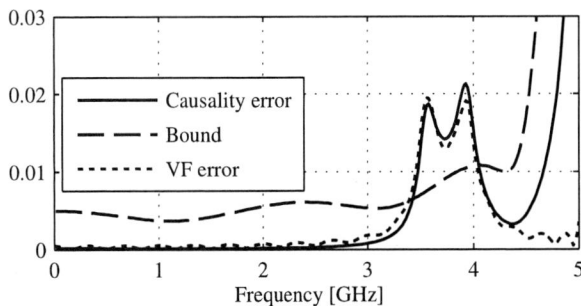

Fig. 2. Application of the proposed causality test to the structure of Sec. IV-A: causality error (solid line) and truncation error bound (dashed line). The macromodeling error obtained when Vector Fitting (VF) is applied to the corrupted samples is also shown (dotted line).

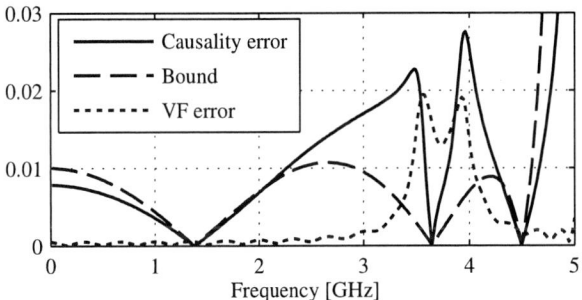

Fig. 3. As in Fig. 3, but for the method of [4].

cut-off frequency 4.5 GHz, and maximum passband ripple of 3 dB. The test took 0.1 s on a 2.7 GHz CPU, and led to the results shown in Fig. 2. The corrupted samples fail the proposed test (15) since causality error (13) exceeds the truncation error bound. These results confirm that the proposed check is sensitive to weak violations which would not be detectable through a naïve test of (4). For comparison, the causality error detected by the state-of-the-art technique [4] is provided in Fig. 3, together with the corresponding truncation error bound. The figure confirms the presence of a causality violation since causality error is above the truncation error bound.

If Vector Fitting (VF) is applied to convert the raw samples into a macromodel compatible with circuit simulators, the presence of causality violations limits the achievable fitting error to $1.9 \cdot 10^{-2}$. The error between the best macromodel we could extract and the S_{11} samples is depicted in both Fig. 2 and Fig. 3. We notice that the proposed causality error (13) is closer to the VF error than the causality error provided by previous techniques [4], [7], [9]. This happens because these techniques use a generalized form of dispersion relations with anchor points. At anchor points, causality error vanishes by construction, as noticeable in Fig. 3 at $f \simeq 1.4, 3.8, 4.5$ GHz. The presence of an anchor point in the band of the violation is responsible for the difference between the causality errors shown in Figs. 2 and 3. By avoiding anchor points, the proposed causality provides a causality error which is a better predictor of the accuracy degradation that the violation may cause if one tries to macromodel the flawed samples.

978-1-4799-0708-3/13 $31.00 © 2013 IEEE

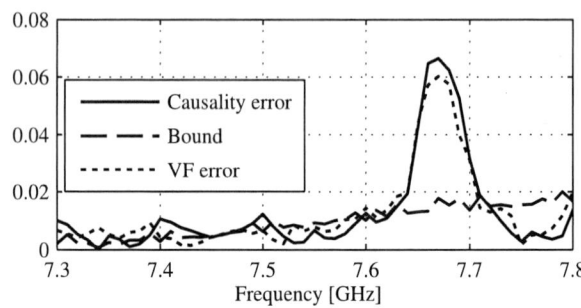

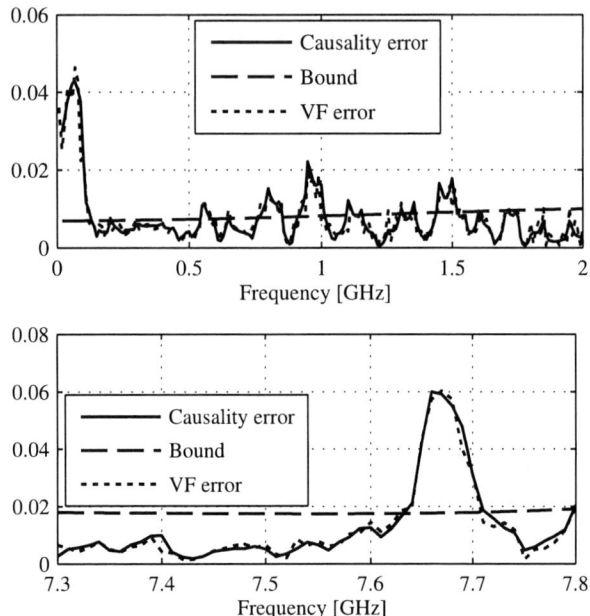

Fig. 4. Application of the proposed causality test to the measured scattering parameters of a long link (Sec. IV-B): causality error (solid line) and truncation error bound (dashed line). The macromodeling error obtained with Vector Fitting (VF) is also shown (dotted line).

Fig. 5. As in the bottom panel of Fig. 4, but using the method of [4].

B. Qualification of Measured Scattering Parameters

In this example, we used the proposed method to qualify the measured scattering parameters of a long interconnect link (courtesy of IBM). Samples were obtained with a network analyzer from 10 MHz up to 10 GHz. We applied the proposed technique to asses the physical consistency of one return loss coefficient, here called S_{11}. In (10), we used a Chebyshev low-pass filter of the 5th order, with cut-off frequency of 9 GHz and maximum ripple in the passband of 3 dB. Two violations were detected near DC and near $f = 7.65$ GHz, as shown in the panels of Fig. 4. The test took 0.29 s on a 2.7 GHz processor. The detected violations are in accordance with those found with dispersion relations with anchor points [4], as illustrated in Fig. 5 for the violation near $f = 7.65$ GHz. In both Fig. 4 and Fig. 5 we reported the lowest fitting error we could achieve when applying VF to the samples. As in previous example, the causality error from the proposed method is closer to the VF error than the error returned by previous techniques [4], [7], [9].

V. CONCLUSION

We proposed a new technique for the robust detection of causality violations in sampled scattering parameters. Leveraging a novel form of dispersion relations with filtering, the new technique rigorously accounts for the truncation error induced by the finite bandwidth of the samples. The new method is simpler to implement than existing techniques, more intuitive, and facilitates the interpretation of the causality check outcome. Numerical results indeed show that the proposed causality error is a good predictor of the macromodeling error that may arise if one tries to extract a macromodel from the flawed samples. Once the method will become available

in Electronic Design Automation tools, it will allow for a quick and reliable assessment of the quality of a measurement or full-wave simulation. By discarding inconsistent datasets, and improving the acquisition process, designers will avoiding wasting precious design time in trying to obtain good models from flawed data.

REFERENCES

[1] R. Achar and M. Nakhla, "Simulation of high-speed interconnects," *Proc. IEEE*, vol. 89, no. 5, pp. 694–728, 2001.

[2] P. Triverio, S. Grivet-Talocia, M. S. Nakhla, F. Canavero, R. Achar, "Stability, causality, and passivity in electrical interconnect models," *IEEE Trans. Adv. Packag.*, vol. 30, no. 4, pp. 795–808, 2007.

[3] E. C. Titchmarsh, *Introduction to the theory of Fourier integrals*, 2nd ed. Oxford university press, 1948.

[4] P. Triverio and S. Grivet-Talocia, "Robust Causality Characterization via Generalized Dispersion Relations," *IEEE Transactions on Advanced Packaging*, vol. 31, no. 3, pp. 579–593, 2008.

[5] K. R. Waters, J. Mobley, J. G. Miller, "Causality-Imposed (Kramers-Krönig) Relationships Between Attenuation and Dispersion," *IEEE Trans. Ultrason., Ferroelectr., Freq. Control*, vol. 52, pp. 822–833, 2005.

[6] F. M. Tesche, "On the Use of the Hilbert Transform for Processing Measured CW Data," *IEEE Trans. Electromagn. Compat.*, vol. 34, pp. 259–266, 1992.

[7] S. Lalgudi, "On checking causality of tabulated s-parameters," *IEEE Transactions on Components, Packaging and Manufacturing Technology*, vol. 3, no. 7, pp. 1204–1217, 2013.

[8] B. Gustavsen and A. Semlyen, "Rational approximation of frequency domain responses by Vector Fitting," *IEEE Trans. Power Del.*, vol. 14, no. 3, pp. 1052–1061, July 1999.

[9] P. Triverio and S. Grivet-Talocia, "A robust causality verification tool for tabulated frequency data," in *10th IEEE Workshop on Signal Propagation on Interconnects, Berlin, Germany*, May 9–12, 2006.

[10] P. Triverio, "Robust Causality Check for Sampled Scattering Parameters via a Filtered Fourier Transform," *IEEE Microwave and Wireless Components Letters*, 2013, (submitted).

[11] J. D. Rhodes, *Theory of electrical filters*. Wiley London, 1976.

A Parallel, Adaptive Multi-Point Model Order Reduction Algorithm

G. De Luca [1], G. Antonini [2], P. Benner [3]

[1,2] *Dipartimento di Ingegneria Industriale e dell'Informazione e di Economia*
Universita' degli Studi dell'Aquila, Via G. Gronchi 18, L'Aquila, Italy,

[1] *giulio.antonini@univaq.it,* [2] *giovanni.deluca84@gmail.com*

[3] *Computational Methods in Systems and Control Theory*
Max Planck Institute for Dynamics of Complex Technical Systems, Sandtorstr. 1, Magdeburg, Germany,

[3] *benner@mpi-magdeburg.mpg.de*

Abstract—This paper describes a model order reduction technique for circuit simulation, based on the parallelization of the well-known multi-point PRIMA algorithm. In order to obtain an optimal accuracy of the reduced-order model in the entire frequency range of interest, the reduced models are computed on different expansion points in correspondence of which the errors, between the transfer functions of the original model and of the actual reduced one, exhibit the largest value, in a recursive way. Moreover, since the computation of the error is a computationally expensive routine, this task is parallelized, assuming that each error value is independent of the others and to work with modern multi-core computers or a cluster of workstations. The numerical results show that the parallelized model order reduction algorithm is able to provide accuracy and speed up with respect to the sequential one, for both dense and sparse data sets.

I. INTRODUCTION

The rapid growth in circuit complexity and miniaturization of electronic components, along with the increase of the working frequency, make the analysis of the behavior of electrical circuits a critical and relevant aspect in order to ensure the reliability, stability and speed of operations of the systems under consideration. In the design of integrated circuits at high speed, therefore, the electromagnetic analysis is essential.

The starting point is an electromagnetic system which has to be simulated. Maxwell's equations describe the behavior of that system continuously in time and space. For complex 3D geometries, they are discretized in space, by means of, e.g., finite difference (FD) methods, the finite element method (FEM) or the partial element equivalent circuit (PEEC) method. Independently of the approach, the direct simulation of the resulting model is computationally expensive in terms of time and memory storage. The generation of a compact model preserving the properties of the physical system is an important step because it allows to use the model in a virtual prototyping, avoiding the creation of a physical one. Thus, it is possible to verify the correct functioning of the designed system before it goes into production, testing different operating conditions and interconnection effects when linked to other devices. Moreover, the creation of a virtual prototype is definitely cheaper and faster than creating a physical prototype.

In order to speed up the simulations and save memory space, model order reduction (MOR) techniques have bee developed and are efficiently used.

MOR has been particularly successful in reducing the complexity of large linear sub-circuits modeling parasitic effects of interconnect, and it is becoming an increasingly useful tool also in other areas of circuit design [1] (e.g., for a list of methodologies and benchmarks used in this field, the interested reader may refer to the website [2]).

In particular, MOR techniques aim to generate a dynamical system of reduced size which accurately reproduces the original one in a prescribed frequency range of interest. In this paper, we focus on the use a multi-point version of the well-known PRIMA algorithm [3], which is made adaptive and parallel [4]), by exploiting the power of modern multi-core processors.

Section II provides a brief review of the PRIMA algorithm. Section III describes the parallelization of the multi-point PRIMA algorithm; Section IV presents the numerical results obtained from simulations using different data sets. Finally, in Section V we draw the conclusions and point out future perspectives.

II. THE PRIMA ALGORITHM

A. Basic formulation

Let us assume the electromagnetic system be represented in descriptor form as

$$\Psi : \begin{cases} C\dot{x}(t) & = -Gx(t) + Bu(t) \\ y(t) & = Lx(t) \end{cases} \tag{1}$$

with zero initial condition $x(0) = 0$. Matrices $C, G \in \mathbb{R}^{n \times n}$ contain memory and memoryless elements, respectively, $x(t) \in \mathbb{R}^n$ denotes the vector of MNA [5] variables (e.g. nodal potentials and currents flowing in inductances and voltage sources of the equivalent circuit). Also, $B \in \mathbb{R}^{n \times m}$ and $L \in \mathbb{R}^{m \times n}$ are the input and output matrices, respectively.

978-1-4799-0708-3/13 $31.00 © 2013 IEEE

The corresponding complex-valued, matrix transfer function in the Laplace domain reads

$$H(s) = L(G + sC)^{-1}B \qquad (2)$$

The PRIMA [3] algorithm is a projection-based MOR technique which searches for an approximation $\tilde{x}(t)$ of $x(t)$, providing an orthogonal projection-matrix V (where $V^H V = I$, with I the identity matrix of appropriate dimension), and then the approximation $\tilde{x}(t)$ can be represented by $\hat{x}(t) = Vz$. Therefore, $x(t)$ can be approximated by $x(t) \approx Vz(t)$. Here $z(t)$ is a vector of length $k \ll n$. Once $z(t)$ is computed, the approximate solution $\tilde{x}(t) = Vz(t)$ for $x(t)$ can be obtained. The vector $z(t)$ can be computed from the reduced model

$$\tilde{\Psi} : \begin{cases} \tilde{C}\dot{z}(t) & = -\tilde{G}z(t) + \tilde{B}u(t), \\ \tilde{y}(t) & = \tilde{L}z(t). \end{cases} \qquad (3)$$

Besides, the reduced model preserves the main properties of the original system (stability, passivity) under some assumptions on the structure of the matrices C and G [3]. The reduced MNA matrices are

$$\begin{aligned} \tilde{C} &= V^H C V, & \tilde{G} &= V^H G V, \\ \tilde{B} &= V^H B, & \tilde{L} &= LV. \end{aligned} \qquad (4)$$

The unitary projection matrix V of dimension $n \times k$ is obtained using the block Arnoldi process, such that its k column-vectors span the Krylov subspace $\mathcal{K}_l(A, R)$ induced by the l-block moments, as the sequel $colspan\{V\} = \mathcal{K}_l(A, R) = span\{R, AR, \dots, A^{l-1}R\}$, where $R \equiv (sC + G)^{-1}B$, with $s \in \mathbb{C}$ a selected expansion point, and $A \equiv -(sC + G)^{-1}C$.

III. PARALLEL MULTI-POINT PRIMA

The PRIMA algorithm approximates the original system just locally, around a selected expansion point. Wideband modeling would require a large number of moments to be matched to obtain a prescribed accuracy. Also, in some applications too many derivatives need to be matched, and then higher orders, to obtain a sufficient accuracy. In order to reach a better accuracy in all the frequency range of interest, without resorting to high order moments, multiple expansion points have been considered in [6], so that a limited number of moments need to be matched in each expansion point.

There have been many efforts and ideas in this direction: Benner et al. developed in [7] an efficient MOR scheme which chooses the expansion points and the number of moments in an fully adaptive way; an adaptive algorithm to automatically identify the expansion points and set the order of the model has been proposed in [4]; among the others, Lee et al. proposed an adaptive-order rational Arnoldi (AORA) method [8] to be applied to large-scale linear systems.

Our algorithm is based on the recursive exploration of the frequency range of interest and additional expansion points are identified through the exploration step by comparing the frequency response of the original system and the reduced one. When the size of the modeled system exceeds few thousands, the repeated computation of the frequency response of the original system is typically time consuming.

The most obvious way to parallelize the multi-point expansion reduction, exploiting the power of modern multi-core processors, is to assign an expansion point to each process. The efficiency of such an approach strongly depends on the data transfer which can be quite time consuming and will be investigated in forthcoming reports.

In this work, we propose an algorithm that adaptively chooses the expansion points where the PRIMA model is computed. As a preliminary step, the frequency response of the original system is pre-computed and saved in a vector (called $vec_{linDistr}$) of n_{ep} expansion points linearly distributed in the frequency range. Such a task is embarrassingly parallel since each evaluation is totally independent of the others. In correspondence of each of these points the algorithm computes the weighted RMS-error, defined in [4] as

$$err = \sqrt{\frac{\sum_{i=1}^{n_p^2} \mid \omega_{Z_i}(s)(Z_{k,i}(s) - Z_i(s)) \mid^2}{(n_p)^2}} \qquad (5)$$

with

$$\omega_{Z_i}(s) = \mid (Z_i(s))^{-1} \mid$$

(where, $Z_{k,i}$ and Z_i are the impedances of the current ROM and of the original model, respectively, for the i-th output-input pair, n_p is the number of ports of the system, and s is the processed expansion point) in a parallel fashion, selecting as additional expansion point that one affected by the maximum error. Then, a reduced model is generated for the new expansion point, the projection matrix is block-column concatenated with the previous one and the point is removed from the list $vec_{linDistr}$. The parallel computation of the error is repeated for the remaining points, and, as before, other points are extracted from the vector $vec_{linDistr}$ and used as expansion points until, for all the frequency points in $vec_{linDistr}$, the error is smaller than the a-priori fixed tolerance. The pseudo-code of the parallel and adaptive multi-point PRIMA is shown below.

The algorithm receives as inputs the matrices of the original system in descriptor form (G,C,B,L), the start and end points of the frequency range of interest ($sStart$ and $sStop$, respectively), the number of moments to be matched ($nMom$) and the desired tolerance (tol) for the accuracy of the ROM. The projection-matrix V_{fin} is produced as output.

Profiling the code, we observed that the most time consuming part of the algorithm is represented by the computation of the error (the time computation of PRIMA is negligible respect to it, when choosing few moments to be matched for each expansion point). Since it is inherently time consuming, in this work we chose to parallelize this task, replacing the for-loop in Algorithm 1 with its parallel counterpart. In the MATLAB® [9] framework, the *par-for*-loop can be efficiently used to this aim.

As a measure of the parallelization, we refer to the *speed up* factor

$$SP(n) = \frac{t_{seq}(n)^*}{t_{par}(n)} \qquad (6)$$

978-1-4799-0708-3/13 $31.00 © 2013 IEEE

Algorithm 1 Compute the projection-matrix V_{fin}

$sMid = (sStart + sStop)/2;$
$V = PRIMA(G, C, B, sMid, nMom);$
$Vfin = V;$
$vecLinDistr = linspace(sStart, sStop, n_{ep});$
$cycle = 1;$
while $cycle$ **do**
for $i = 1 : size(vecLinDistr)$ **do**
compute the error for each expansion points into $vecLinDistr$, in parallel;
end for
select that expansion point exhibiting the maximum value ($maxVal$) of the error;
if $maxVal > tol$ **then**
remove from $vecLinDistr$ the frequency point with the maximum error;
use PRIMA on the selected point to
block-column concatenate the computed projection-matrix V with V_{fin};
else
$cycle = 0;$
end if
end while
return V_{fin}

---

where $t_{seq}(n)^*$ is the execution time of the fastest sequential program solving the same problem (the code of the sequential and parallel versions are the same, but while the latter runs in single-thread mode, the first one runs in multi-thread mode). Besides, as a measure of utility of the selected number of processors that work in parallel, we define the *efficiency* of the parallelization :

$$E_p = \frac{t_{seq}(n)^*}{p \cdot t_{par}(n)} \qquad (7)$$

with p the number of processors [10].

IV. NUMERICAL EXPERIMENTS

The proposed algorithm is implemented in MATLAB r2012b, running on a desktop equipped with Intel® Xeon Quad-Core Processor, CPU operating at 3.47 GHz, with 48 GB of RAM available. Four data sets have been used: three of them are characterized by dense matrices obtained by the PEEC method [11]; the fourth data set is sparse (with dimension 6134) and has been gently provided by the Max Plank Institute of Magdeburg. We set a tolerance $tol = 10^{-2}$ for all the data sets, which is considered reasonable for engineering applications; the number of moments matched for each expansion point equal to 1 .

Fig. 1 shows magnitude, phase and error spectra of the input impedance Z_{11} of an interconnect modeled with $n = 1600$ internal states and $m = 14$ inputs, of both the original system and the reduced model of order $k = 126$, obtained with the parallel version of the multi-point PRIMA algorithm. For this data set, the algorithm retained 9 expansion points that are plotted in the same Fig. 1. Fig. 2 plots magnitude, phase and error spectra of the transfer impedance $Z_{14,15}$ of an interconnect model of dimension $n = 2624$. We can appreciate a very good approximation, since the method just retains the expansion points with the larger error values respect to the desired tolerance.

Finally, Table I reports for each data set the size n of

the original model and the number of ports m, the simulation time results, speed-up and efficiency obtained running the sequential and parallel codes. The last two columns report the number of expansion points and the order of the reduced model for each data set.

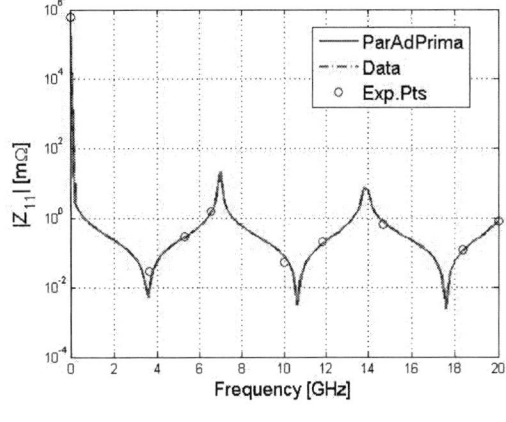

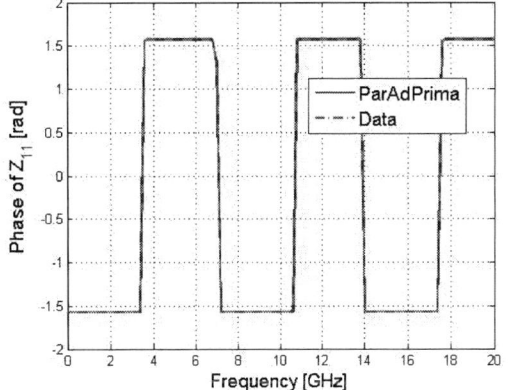

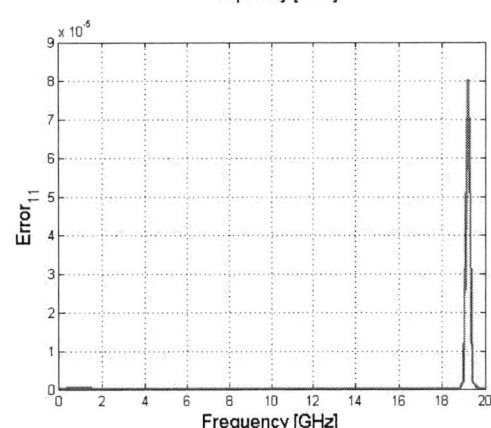

Fig. 1
MAGNITUDE (TOP), PHASE (MIDDLE) AND ERROR (BOTTOM) SPECTRA OF THE INPUT IMPEDANCE Z_{11} OF AN INTERCONNECT MODEL WITH DIMENSION $n = 1600$.

Furthermore, it has to be pointed out that a better speed-up can be obtained if a larger number of tasks is run in parallel so to optimize the parallelization of the computation of the error function which is quite intensive and to min-

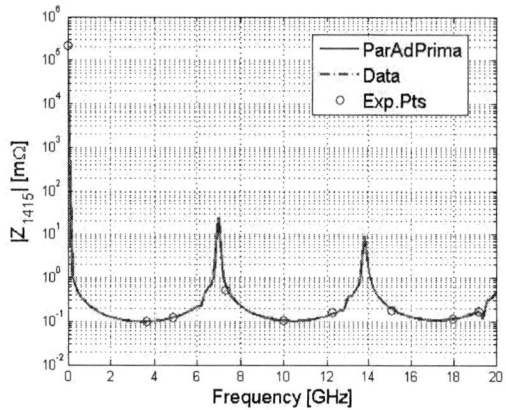

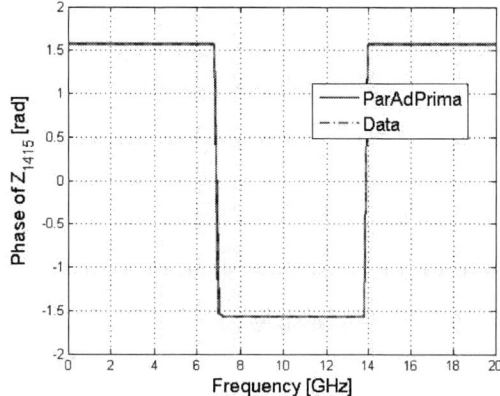

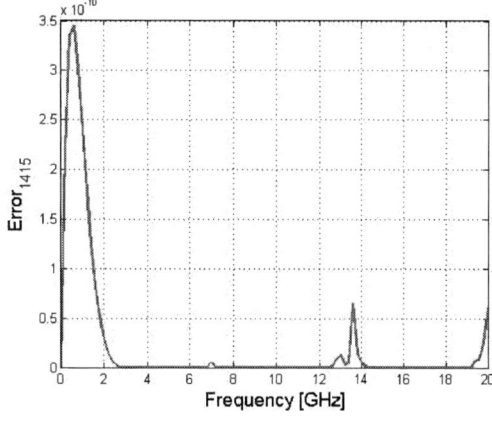

Fig. 2

MAGNITUDE (TOP), PHASE (MIDDLE) AND ERROR (BOTTOM)
SPECTRA OF THE TRANSFER IMPEDANCE $Z_{14,15}$ OF AN
INTERCONNECT MODEL WITH DIMENSION $n = 2624$.

imize the impact of the data transfer. In our experiments, the number of error evaluations is set to 50 and we work mostly with dense data sets, so that the data transfer from the client to the workers is time consuming. This issue clearly limits the speed-ups. Gathering data from all the workers to the client is less expensive, since just one scalar, the error, needs to be transferred. Indeed, a good speed up of almost a factor of 3.82 is obtained for the sparse data set, with size $n = 6134$, where a smaller quantity of data is in-

TABLE I

TABLE OF RESULTS (TOLERANCE OF 0.01, NUMBER OF MOMENTS
MATCHED FOR EACH EXPANSION POINTS EQUALS TO 1, WORKING
WITH 8 PROCESSORS IN PARALLEL).

n/m	$t_{seq}(sec)$	$t_{par}(sec)$	SP	E_p	ep_{ret}	k
1600/14	199.39	84.06	2.37	0.29	9	126
2624/30	625.34	348.97	1.79	0.22	9	270
6134/10	112.59	29.47	3.82	0.47	15	150
5536/8	1445.80	1084.34	1.33	0.17	4	32

volved in the transfer with respect to the dense counterparts (anyway, a linear speed up with the number of processors can practically never be reached, according to Amdahl's law [12], due to the necessity to transfer large data and the different speed between CPUs and memory access [13]).

V. CONCLUSIONS

In this paper, a parallel adaptive multi-point model order reduction is proposed, improving the searching method of [4], and obtaining a minimum order of reduction in a parallel fashion, exploiting the modern architectures of multi-core processors. The numerical results confirm the efficiency and accuracy of the proposed approach. Larger examples will be presented in the final paper. The numerical experiments have also demonstrated that the speed-up can be limited by the time spent to transfer data from the client to all the workers involved in the parallelization, that slowdowns the operation. In the future, the usage of a cluster of workstations will be investigated, which, when powered by fast communication links, most likely will allow to achieve higher speed up factors.

REFERENCES

[1] M. H. P. Benner and E. J. W. ter Maten, Eds., *Model Reduction for Circuit Simulation*. Springer, 2011.
[2] "The Model Order Reduction Wiki," http://morwiki.mpi-magdeburg.mpg.de/morwiki/index.php/Main_Page, 2013.
[3] A. Odabasioglu, M. Celik, and L. T. Pileggi, "PRIMA: passive reduced-order interconnect macromodeling algorithm," *IEEE Transactions on Computer-Aided Design*, vol. 17, no. 8, pp. 645–654, Aug. 1998.
[4] F. Ferranti, M. Nakhla, G. Antonini, T. Dhaene, L. Knockaert, and A. Ruehli, "Multipoint full-wave model order reduction for delayed PEEC models with large delays," *IEEE Transactions on Electromagnetic Compatibility*, vol. 53, no. 4, pp. 959–967, Nov. 2011.
[5] C. Ho, A. Ruehli, P. Brennan, "The modified nodal approach to network analysis," *IEEE Transactions on Circuits and Systems*, vol. 22, no. 6, pp. 504–509, Jun. 1975.
[6] J. R. Phillips, E. Chiprout, and D. D. Ling, "Efficient full-wave electromagnetic analysis via model-order reduction of fast integral transforms," in *DAC '96: Proceedings of the 33rd annual Design Automation Conference*, New York, NY, USA, 1996, pp. 377–382.
[7] L. Feng, J. G. Korvink, and P. Benner, "A fully adaptive scheme for model order reduction based on moment-matching," Max Planck Institute Magdeburg Preprint MPIMD/12-14, September 2012, available from http://www.mpi-magdeburg.mpg.de/preprints/.
[8] H.-J. Lee, C.-C. Chu, and W.-S. Feng, "An adaptive-order rational arnoldi method for model-order reductions of linear time-invariant systems," *Linear Algebra and its Applications*, vol. 415, no. 2-3, pp. 235 – 261, 2006.
[9] "*Matlab User's Guide,*," The Mathworks, Inc., Natick, 2001.
[10] D. P. Bertsekas and J. N. Tsitsiklis, *Parallel and Distributed Computation: Numerical Methods.* Upper Saddle River, NJ, USA: Prentice-Hall, Inc., 1989.
[11] A. E. Ruehli, "Equivalent circuit models for three dimensional multiconductor systems," *IEEE Transactions on Microwave Theory and Techniques*, vol. MTT-22, no. 3, pp. 216–221, Mar. 1974.
[12] M. Creeland William L. Goffe, "Multi-core cpus, clusters, and grid computing: A tutorial," *Computational Economics*, vol. 32, no. 4, pp. 353–382, 2008.
[13] S. A. McKee, "Reflections on the memory wall," in *Proceedings of the 1st conference on Computing frontiers.* New York, NY, USA: ACM, 2004, pp. 162–170.

Multiple and Non-Existent Barnes-Hut Center-of-Charge (CoC) Solution in Lossy Layered Substrates: Half Space Analytic and Numerical Study

Khalid Butt, Jonatan Aronsson, Vladimir Okhmatovski

University of Manitoba, Dept. of Electrical and Computer Engineering, 75A Chancellor Circle, Winnipeg, MB R3T 5V6, Canada
email: umbuttk@cc.umanitoba.ca

Abstract

For rapid extraction of capacitance in lossless substrates Barnes-Hut algorithm is an effective technique for evaluating N charges in $O(N \log N)$ operations. When substrates are lossy the hierarchical application of the CoC approximation, apart from resulting in unique CoC, can result in cases like there does not exist CoC or there exist multiple CoC solutions. In this paper we analytically and numerically study CoC approximation in lossy half space media.

Introduction

In an effort to construct an efficient $O(N \log N)$ algorithm for N-body problem in layered media which does not introduce extra degrees of freedom due to the presence of dielectric layers and which is amenable to iterative mesh refinements, we generalized the key ideas originated in works of Appel [1] as well as Barnes Hut [2] to directly account for in homogeneity of layered media [3], [4]. In [4] we showed that the CoC approximation in lossless multilayered media preserves the same properties in stratified media which it has in free space. Namely, for any box of size b crossing multiple dielectric interfaces and containing arbitrary charged sources there exists a unique CoC approximating the field of original sources with $O(1/R^3)$ error at distances $R \gg b$. Also, the layered media CoCs was shown to allow for hierarchical grouping [4]. The above properties allow to directly apply the BH algorithm to acceleration of electrostatic interactions in *lossless* layered substrates provided the effect of stratification on the location of CoC is properly accounted for. As the permittivities of layers become complex both the charge values and their respective locations become complex as well. Though the mathematical formalism for determining the complex CoC locations remains the same as in lossless case [4], the situations become possible when for a box containing charges the CoC does not exist or when there exist multiple CoCs.

We can treat the phenomenon for the case of a lossy half-space where an analytic expression for the CoC location is available [3]. Namely, we can show that when grouping of two complex charges q_1 and q_2 at complex elevations z_1' and z_2' occurs across dielectric interface separating complex permittivities ϵ_1 and ϵ_2, the complex plane q_1/q_2 gets partitioned into four distinct regions by two partially overlapping circles: region where the CoC does not exist, region where two CoCs co-exist, and two regions corresponding to the unique CoCs cases. As the dielectric losses become negligible the off-centering of circles partitioning q_1/q_2 complex plane disappears and the four regions collapse to two - the region inside the circle corresponding to the case of unique CoCs situated below the interface and the region outside the circle corresponding to the unique CoCs situated above the interface or vice versa.

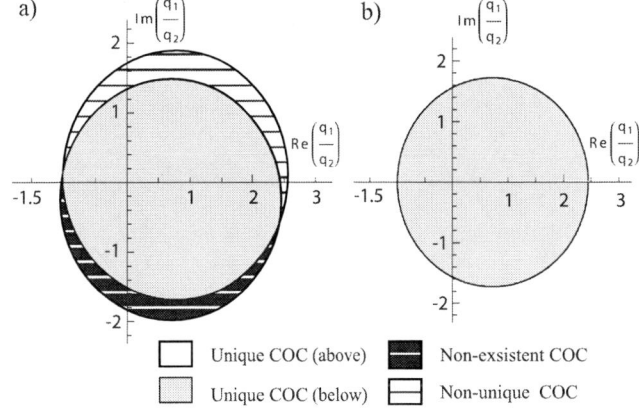

Figure 1: An example of the four regions of solution cases to (3) in the complex plane of $Q = q_1/q_2$. Here, we have set $\epsilon_1 = 4 - i0.88$, $\epsilon_2 = 2.2 - i0.0088$, $z_1' = 0.52$, and $z_2' = -0.7$.

Analytic Study of CoC Approximation in Lossy Half-Space Media

In [3] it is shown that for two real charges q_1 and q_2 situated at real coordinate $z_1' > 0$ and $z_2' < 0$ above and below the dielectric interface at $z = 0$ separating lossless materials with permittivities ϵ_1 and ϵ_2 the CoC location z_0' is either below the interface, when

$$z_0' = \frac{q_1 z_1' + q_2 z_2' + q_1 z_1'(\epsilon_2/\epsilon_1 - 1)}{q_1 + q_2}, \quad z_0' < 0, \quad (1)$$

or above the interface when

$$z_0' = \frac{q_1 z_1' + q_2 z_2' + q_2 z_2'(\epsilon_1/\epsilon_2 - 1)}{q_1 + q_2}, \quad z_0' > 0. \quad (2)$$

It can be proven that in lossless case: a) either (1) or (2) is always satisfied (*existence* of CoC) and b) (1) and (2) cannot be satisfied simultaneously (*uniqueness* of CoC).

When dielectrics exhibit loss both the permittivities ϵ_1, ϵ_2, charge values q_1, q_2 and charge elevations z_1', z_2' become complex for the purposes of BH algorithm. Since the formalism of spectral domain solution for Poisson equation [3] remains the same for the complex-valued permittivities, charges and coordinates, the complex CoC location z_0' for the half-space case is governed by similar expressions

$$z_0' = \begin{cases} \frac{q_1 z_1' + q_2 z_2' + q_1 z_1'(\epsilon_2/\epsilon_1 - 1)}{q_1 + q_2}, & \text{if } \Re(z_0') < 0, \\ \frac{q_1 z_1' + q_2 z_2' + q_2 z_2'(\epsilon_1/\epsilon_2 - 1)}{q_1 + q_2}, & \text{if } \Re(z_0') > 0. \end{cases} \quad (3)$$

Solution of (3), however, may now yield one of four possible cases:

- Unique CoC exists and is located below the interface (top equation in (3) is satisfied);

- Unique CoC exists and is located above the interface (bottom equation in (3) is satisfied;

- Two CoCs exit - one below and one above the interface (both equations in (3) are satisfied);

- No CoC exists either below or above the interface (neither of equation in (3) is satisfied).

The boundaries separating these four cases are governed by

$$\Re\left[\frac{\frac{\epsilon_2}{\epsilon_1}q_1 z_1' + q_2 z_2'}{q_1 + q_2}\right] = 0,$$
$$\Re\left[\frac{q_1 z_1' + \frac{\epsilon_1}{\epsilon_2}q_2 z_2'}{q_1 + q_2}\right] = 0. \quad (4)$$

Let us introduce notations $\gamma = \frac{\epsilon_1}{\epsilon_2}$, $Q = q_1/q_2$ and rewrite boundary equations (4) as

$$\Re\left[\frac{1}{\gamma}\frac{z_1'}{1 + 1/Q} + \frac{z_2'}{Q+1}\right] = 0,$$
$$\Re\left[\frac{z_1'}{1 + 1/Q} + \gamma\frac{z_2'}{Q+1}\right] = 0. \quad (5)$$

With subscripts r and i denoting the real and imaginary parts, equations (5) governing the boundary between the four CoC regions become

$$\Re\left[\frac{1}{\gamma_r + i\gamma_i}\frac{z_{1,r}' + iz_{1,i}'}{1 + 1/(Q_r + iQ_i)} + \frac{z_{2,r}' + iz_{2,i}'}{Q_r + iQ_i + 1}\right] = 0,$$
$$\Re\left[\frac{z_{1,r}' + iz_{1,i}'}{1 + 1/(Q_r + iQ_i)} + (\gamma_r + i\gamma_i)\frac{z_{2,r}' + iz_{2,i}'}{Q_r + iQ_i + 1}\right] = 0. \quad (6)$$

Expanding the terms in (6) and extracting the real parts, we obtain equations separating the four CoC solution regions

$$(z_{1,i}'\gamma_i + z_{1,r}'\gamma_r)Q_i^2 + (z_{2,i}'|\gamma|^2 + z_{1,r}'\gamma_i - z_{1,i}'\gamma_r)Q_i$$
$$+ (Q_r + 1)(z_{2,r}'|\gamma|^2 + Q_r z_{1,i}'\gamma_i + Q_r z_{1,r}'\gamma_r) = 0,$$
$$z_{1,r}'Q_i^2 + (-z_{1,i}' + z_{2,r}'\gamma_i + z_{2,i}'\gamma_r)Q_i$$
$$+ (Q_r + 1)(Q_r z_{1,r}' - z_{2,i}'\gamma_i + z_{2,r}'\gamma_r) = 0. \quad (7)$$

These conditions show that for the fixed complex elevations of charges z_1' and z_2' and permittivities ratio $\gamma = \epsilon_1/\epsilon_2$ the four regions corresponding to the unique, non-unique, and non-existent CoCs are separated by the two off-centered circles in the complex plane of parameter $Q = q_1/q_2$. Fig. 1a shows these boundaries for a particular case of $\epsilon_1 = 4 - i0.88$, $\epsilon_2 = 2.2 - i0.0088$, $z_1' = 0.52$, and $z_2' = -0.7$. For the case of vanishing imaginary part in both permittivities ϵ_1, ϵ_2 and charge locations z_1', z_2' both equations in (7) become the same

$$z_{1,r}'Q_i^2 + (Q_r + 1)(Q_r z_{1,r}' + z_{2,r}'\gamma_r) = 0, \quad (8)$$

and the off-centering in the circular boundaries separating the four regions disappears (Fig. 1b). Hence, there exist only two possible CoC scenarios:

Table 1: Charges and their coordinates in the case of multiple or non-unique CoC solutions for a box in lossy substrate

$q_i^{(l)}$, [nC]	$z_i'^{(l)}$, [mm]
$q_0^{(1)} = 0.57899 + i1.15625$	$z_0'^{(1)} = 0.5547$
$q_1^{(1)} = 0.00726 + i1.55252$	$z_1'^{(1)} = 0.5495$
$q_0^{(2)} = 0.03311 + i0.84841$	$z_0'^{(2)} = 0.5691$
$q_1^{(2)} = 0.02863 + i0.86009$	$z_1'^{(2)} = 0.5689$

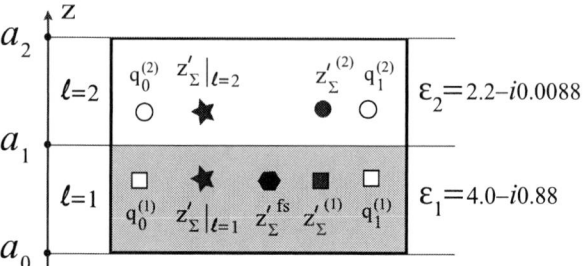

Figure 2: Distribution of four point charges per Table 1 which produce two non-unique CoC solutions $z_\Sigma'|_{\ell=1} = 0.561033 - i0.00363169[mm]$ and $z_\Sigma'|_{\ell=2} = 0.562678 - i0.00139544[mm]$. The elevations $z_\Sigma'^{(1)}$ and $z_\Sigma'^{(2)}$ of partial per-layer CoCs are also depicted. Dielectric interface locations a_0 through a_2 are equal to (in mm) 0.162, 0.562 and 0.616 respectively.

- Unique CoC exists and is located below the interface (top equation in (3) is satisfied);

- Unique CoC exists and is located above the interface (bottom equation in (3) is satisfied;

just as required by the conditions of existence and uniqueness of CoC solution in lossless layered media. The above analytical study shows that when applying the CoC approximation in lossy layered media one may encounter situations in which complex values of layer permittivities, charges, and their locations lead non-existing CoC or availability of multiple CoCs which can be associated with a particular group of charges in the box crossing interface.

Numerical Study of CoC Approximation in Lossy Half-Space Media

A detailed derivation of the CoC approximation in a general multilayered media has been discussed in [4] for the case of lossless dielectrics. Here we do not repeat the derivation for the CoC shift in case of lossy layered media since the formalism for the Poisson equation solution in case of complex charges, charge locations, and permittivity values presented in [4] remains the same. It leads to the same expressions determining CoC elevation z_Σ' (formula (57) in [4]), its lateral coordinates ρ_Σ', and net charge Q_Σ (see (18) in [4]).

When the dielectrics feature loss, however, expression determining the complex CoC elevation z_Σ' (formula (57) in [4]) may produce either a unique solution as in the case of lossless dielectrics or an anomalous situation in which the solution for

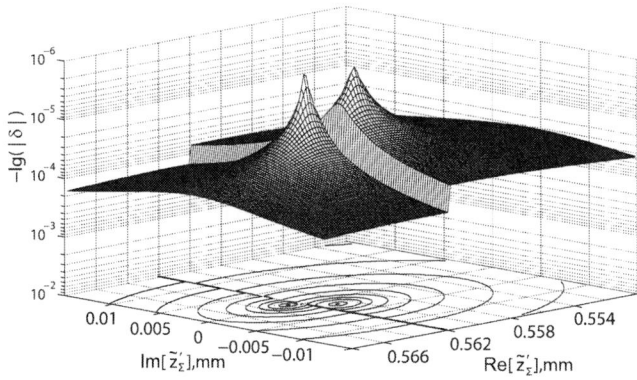

Figure 3: Plot of the relative error $\delta(R, \tilde{r}_\Sigma)$ on complex plane $\tilde{z}_\Sigma$ depicting two error minima coinciding with non-unique CoC solutions $z'_\Sigma|_{\ell=1}$ and $z'_\Sigma|_{\ell=2}$. Observation radius is $R = 3.90625[mm]$ from the box center. The dot markers on the contour sub-plot correspond to the theoretical locations of the CoC elevations $z'_\Sigma|_{\ell=1}$ and $z'_\Sigma|_{\ell=2}$ obtained from (57) in [4].

z'_Σ does not exist or there exist multiple solutions for z'_Σ. To study these possibilities we hierarchically partitioned the cube situated in the lossy substrate depicted in Fig. 2 and applied formula (57) in [4] to determine CoC elevations. To conserve space, the numerically detected scenarios of the multiple and non-existent solutions have been studied and their typical representative are discussed below.

1 The case of multiple CoC solutions in lossy substrate

In order to study the anomalous situation in which multiple CoC solutions satisfying (57) in [4] exist. We look at one such box in the oct-tree, the size of the box $b = 0.0390625[mm]$, which contains four point charges described in Table 1 and according to formula (57) in [4] this distribution of charges results in 2 non-unique CoC solutions as depicted in Fig. 2. To prove that all two complex CoC locations $z'_\Sigma|_{\ell=1}$ and $z'_\Sigma|_{\ell=2}$ are indeed valid solutions we numerically computed the error $\delta(R, \tilde{r}_\Sigma)$ for all possible complex elevations $\tilde{z}'_\Sigma$ of the CoC with the real and imaginary parts of complex CoC location $\tilde{z}'_\Sigma$ varying over the intervals $\Re[\tilde{z}'_\Sigma] \in [0.55, 0.568][mm]$ and $\Im[\tilde{z}'_\Sigma] \in [-0.015, 0.015][mm]$ which contain two CoC solutions. A plot of the error $|\delta(R, \tilde{r}_\Sigma)|$ on the complex plane $\tilde{z}'_\Sigma$ is shown in Fig. 3 for the radial distance $R = 3.90625[mm]$. One can observe two distinct minima in the error behaviour which exactly coincide with the above complex CoC solutions $z'_\Sigma|_{\ell=1}$ and $z'_\Sigma|_{\ell=2}$ obtained from formula (57) in [4]. Per-layer CoCs which do not involve grouping of charges across the dielectric interfaces their locations $z'^{(1)}_\Sigma$ and $z'^{(2)}_\Sigma$ are determined using conventional formula for CoC location in free-space (formula (36) in [4]). Fig. 4 shows the relative error

$$\delta_\Sigma(R) =$$

$$\int_0^{2\pi} \int_0^\pi \frac{|\Phi(r(R,\theta,\phi)) - \sum_{l=0}^L Q_\Sigma^{(l)} G_m^{(l)}(r(R,\theta,\phi), r'^{(l)}_\Sigma)|}{4\pi|\Phi(r(R,\theta,\phi))|}$$

$$\sin\theta d\theta d\phi,$$

(9)

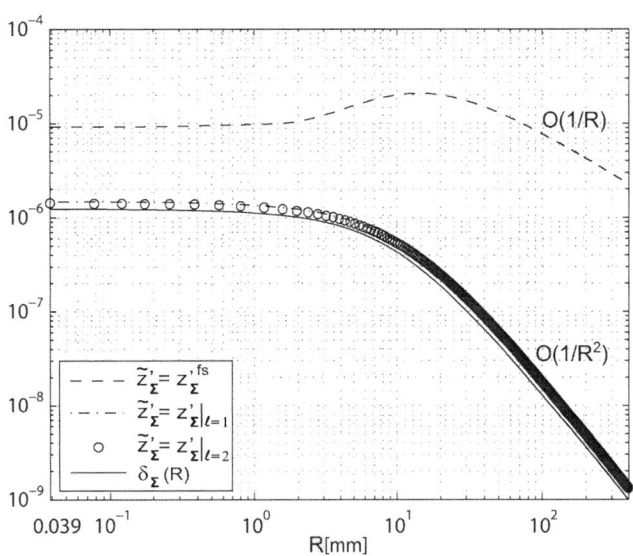

Figure 4: Plot of the average relative error $\delta(R, r'_\Sigma|_{\ell=1})$ and $\delta(R, r'_\Sigma|_{\ell=2})$ versus R for non-unique CoC elevation $z'_\Sigma|_{\ell=1}$ and $z'_\Sigma|_{\ell=2}$ satisfying (57) in [4]. For comparison, the error $\delta(R, r'^{fs}_\Sigma)$ corresponding the free-space CoC elevation $z'^{fs}_\Sigma = 0.558353 + i0.000344[mm]$ and the relative error $\delta_\Sigma(R)$ of field approximation with all per-layer CoC solutions (9) are included.

Table 2: Charges and their coordinates in the case of non-existent CoC solution for a box in lossy substrate

$q_i^{(l)}$, [nC]	$z_i'^{(l)}$, [mm]
$q_0^{(1)} = 0.094927 - i0.381086$	$z_0'^{(1)} = 0.09933$
$q_1^{(1)} = 0.103089 - i0.367126$	$z_1'^{(1)} = 0.09933$
$q_0^{(2)} = 0.094927 - i0.381086$	$z_0'^{(2)} = 0.11475$
$q_1^{(2)} = 0.103089 - i0.367126$	$z_1'^{(2)} = 0.11475$

as a function of radial distance R for the situation when all non-zero per-layer CoCs are utilized in the approximation of the field. In (9), m is the index of the layer containing the observation point r, and $G_m^{(l)}(r, r')$ is the layered media Green's function (for exact definition see (1) and (2) in [4]). One can see from Fig. 4 that not only the relative error $\delta_\Sigma(R)$ exhibits the desired $O(1/R^2)$ attenuation it also lower than either of the error levels $\delta(R, r'_\Sigma|_\ell)$, $\ell = 1, 2$, due to two valid CoC solutions $z'_\Sigma|_\ell$, $\ell = 1, 2$.

2 The case of non-existent CoC solution in lossy substrate

As was previously mentioned, when complex charges in lossy layered media are grouped it is possible that the CoC solution does not exist. This is in contrast to the case of lossless substrate where the CoC solution always exist. The implication of this is that under certain conditions there cannot be found a single point source which approximates the far field of a given set of charges with $O(1/R^3)$ absolute error. To study the non-existent CoC scenario numerically, we look at one such box in the oct-tree, the size of the box $b = 0.01953125[mm]$,

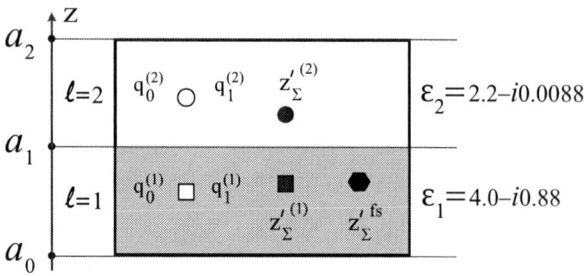

Figure 5: Distribution of four point charges per Table 2 which produce non-existent CoC solution. The elevations $z_\Sigma'^{(1)}$ and $z_\Sigma'^{(2)}$ of partial per-layer CoCs are also depicted. Dielectric interface locations a_0 through a_2 are equal to (in mm) 0.062, 0.108 and 0.162 respectively.

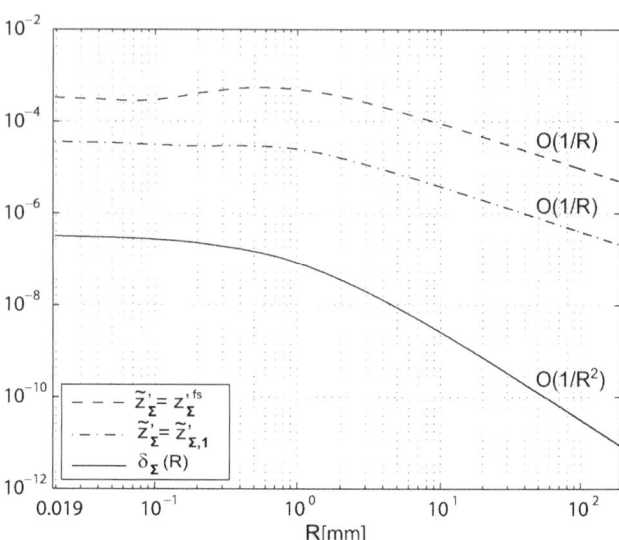

Figure 7: Plot of the average relative error $\delta(R, \tilde{r}_{\Sigma,1}')$ versus radius R corresponding to the locations of minima 'P_1' at complex elevations $\tilde{z}_{\Sigma,1}'$ depicted in Fig. 6. For comparison, the error $\delta(R, r_\Sigma'^{fs})$ corresponding the free-space CoC elevation $z_\Sigma'^{fs} = 0.104498 + i0.003555[mm]$ and the relative error $\delta_\Sigma(R)$ of field approximation with all per-layer CoC solutions (9) are included.

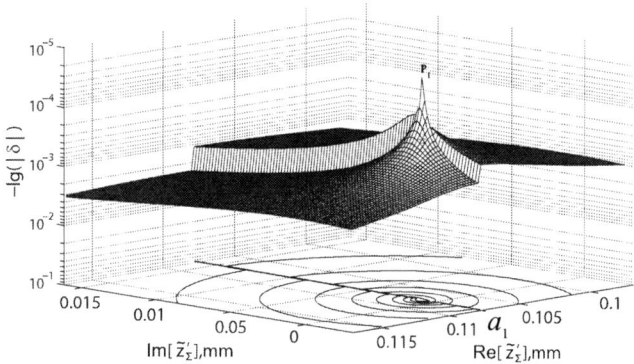

Figure 6: Plot of the relative error $\delta(R, \tilde{r}_\Sigma')$ on complex plane $\tilde{z}_\Sigma'$ at observation radius $R = 1.95[mm]$. The error minima peaks 'P_1' is at the point $\tilde{z}_{\Sigma,1}' = 0.108 - i0.00061945[mm]$.

which contains four point charges described in Table 2 which exhibit such behaviour. We take a single point charge at location $\tilde{r}_\Sigma' = x_\Sigma'\hat{x} + y_\Sigma'\hat{y} + \tilde{z}_\Sigma'\hat{z}$ and the complex elevation $\tilde{z}_\Sigma'$ is varied over the area $0.0976[mm] < \Re[\tilde{z}_\Sigma'] < 0.117[mm]$, $-0.00328[mm] < \Im[\tilde{z}_\Sigma'] < 0.0162[mm]$ as shown in Fig. 6. The same Fig. 6 shows the behaviour of the relative error $\delta(R, \tilde{r}_\Sigma')$ averaged over the angles at radial distance from the box centre $R = 1.95$ [mm]. Though the error behaviour exhibits a minima at complex elevations $\tilde{z}_{\Sigma,1}'$ (corresponding to peak P_1 in Fig. 6) the relative error $\delta(R, \tilde{r}_{\Sigma,1}')$, where $\tilde{r}_{\Sigma,i}' = x_\Sigma'\hat{x} + y_\Sigma'\hat{y} + \tilde{z}_{\Sigma,i}'\hat{z}$, $i = 1$, does not exhibit the desirable $O(1/R^2)$ attenuation of the field approximation error (Fig. 7). If, however, we use both per-layer CoCs with net charges $Q_\Sigma^{(1)}$, $Q_\Sigma^{(2)}$ elevated at $z_\Sigma'^{(1)}$, $z_\Sigma'^{(2)}$ for the approximation of the far field the averaged relative error $\delta_\Sigma(R)$ given by (9) exhibits the desired $O(1/R^2)$ decay as shown in Fig. 7.

Conclusion

The paper discusses modification of Barnes-Hut algorithm in lossy layered media for the acceleration of electrostatic interactions. It is shown that for a set of complex charges located in lossy substrates anomalous situations are possible in which the center-of-charge solution may not exist or there may

exist multiple such solutions. This is opposite to the case of grouping charges in lossless layered media in which the center-of-charge solution always exists and it is unique. Rigorous treatment of such center-of-charge solutions is given for lossy case. The modification of the Barnes-Hut algorithm utilizing per-layer centers-of-charge for each anomalous box exhibiting non-existent or multiple center-of-charge solutions is proposed.

References

[1] A. W. Appel, "An efficient program for many-body simulation," *SIAM Journal on Scientific and Statistical Computing*, vol. 6, no. 1, pp. 85–103, Jan. 1985.

[2] J. Barnes and P. Hut, "A hierarchical o(nlogn) force-calculation algorithm," *Nature*, vol. 324, pp. 446–449, Dec. 1986.

[3] I. Jeffrey and V. Okhmatovski, "Effect of multilayered substrate on the BH CoC clustering approximation: Half-space case study," *Signal Propagation on Interconnects, 2008. SPI 2008. 12th IEEE Workshop on*, pp. 1–4, May 2008.

[4] J. Aronsson, K. Butt, I. Jeffrey, and V. Okhmatovski, "The Barnes-Hut hierarchical CoC approximation for fast capacitance extraction in multilayered media," *Microwave Theory and Techniques, IEEE Transactions on*, vol. 58, no. 5, pp. 1175-1188, May 2010.

Macromodeling

978-1-4799-0708-3/13 $31.00 © 2013 IEEE

An iterative reweighting process for macromodel extraction of power distribution networks

S. Grivet-Talocia, A. Ubolli
Politecnico di Torino
C. Duca degli Abruzzi 24, 10129 Torino, Italy
e-mail {grivet,andrea.ubolli}@polito.it

M. Bandinu, A. Chinea
IdemWorks s.r.l.
C. Trento 13, 10129 Torino, Italy
e-mail {m.bandinu,a.chinea}@idemworks.com

Abstract—**This paper introduces a new algorithm for the generation of optimal time-domain macromodels of power distribution networks, starting from a set of tabulated scattering responses and given a nominal termination scheme for active blocks, decoupling capacitors, and voltage regulator module. The new concept being introduced is a modified metric to characterize and optimize the accuracy of the macromodel, which takes into account the operation conditions that will be applied to run transient simulations for power integrity assessment. This metric is applied through an iterative frequency-dependent reweighting scheme in a fully automated flow. Two examples illustrate the performance of the proposed algorithm.**

I. INTRODUCTION

The Power Distribution Network (PDN) is a fundamental block of any electronic system. The main purpose of the PDN is to connect a stablized voltage source, the Voltage Regulator Module (VRM), to all devices in the system, in order to supply power and allow their operation [1]. A stable voltage is guaranteed only at the VRM location, but any device will operate properly if its own supply voltage is within a prescribed small range. Therefore, the parasitics of the entire PDN at board, package and chip levels must be carefully controlled via proper design rules, in order to minimize their impact on the supply voltage variations at the device locations [2], [3].

The most common approach for designing PDN's relies on a frequency-domain characterization and optimization of a target impedance, which relates the supply voltage variations to the current loading from active devices [1], [2]. The smaller is this impedance, the less sensitive is the device voltage with respect to device supply currents. When this impedance is detected to be too large within some frequency range, sets of decoupling capacitors are connected in parallel to the PDN, in order to locally decrease the PDN impedance. The precise location and the actual values of the capacitors are often determined by complex optimization processes [4], [5].

Although design and the optimization of PDN's are performed in the frequency domain, the actual verification that the device supply voltages are within allowed ranges must be performed in the time domain, by running a transient simulation on a system-level circuit that includes suitable models for PDN, decoupling capacitors, VRM, and active device blocks. Macromodeling schemes are very useful for

this task, since they allow to convert frequency-domain descriptions (typically in form of tabulated scattering responses computed by full-wave solvers) into a stable and passive time-domain macromodel in state-space form, ready for transient simulation [6]– [17].

Assuming a scattering form of the initial PDN frequency-domain responses, the state-space macromodel will represent with good accuracy the dynamics of the PDN in terms of the transient scattering outgoing waves at its ports, when excited by transient incident scattering waves. Equivalently, the "nominal" termination scheme for which the macromodel accuracy is optimized during the fitting is a set of resistances R_0 equal to the reference impedance used to define the scattering parameters. This termination scheme is very different from the actual terminations that will be applied to the macromodel during transient simulation, mainly consisting of decoupling capacitors, localized or distributed current sources, and at least one VRM model. It was verified for several application cases that even when the macromodel is very accurate in the scattering representation, the corresponding derived PDN impedance may differ significantly from the nomimal impedance computed from the raw scattering data. The main reason for this difference is due to the sensitivity of the transformation that converts the scattering PDN responses into the PDN impedance under realistic loading conditions. The main objective of this work is in fact to elminate this problem by optimizing the accuracy of the macromodel in terms of the PDN impedance, by explicitly including this sensitivity into account during model extraction. This is achieved by iteratively optimizing a set of frequency-dependent weights based on this sensitivity during the rational fitting stage.

This work is organized as follows. In Section II we provide evidence of PDN impedance sensitivity on a simple test case, and we formally state the problem under investigation. In Section III we describe our proposed iteratively reweighted rational fitting scheme. Finally, in Section IV we demonstrate the performance of the proposed algorithm on some application examples.

II. PROBLEM STATEMENT

Our starting point is a set of frequency samples of the PDN scattering matrix

$$\hat{\mathbf{S}}_k = \hat{\mathbf{S}}(j\omega_k), \quad k = 1, \dots, K, \tag{1}$$

978-1-4799-0708-3/13 $31.00 © 2013 IEEE

available from a field solver. Each of the samples $\hat{\mathbf{S}}_k$ is a complex-valued $P \times P$ matrix, where P denotes the number of ports of the PDN description. These ports are grouped into three different subsets $P = P_a + P_c + P_v$, where

- the first P_a ports are connected to suitable models of the active device blocks; the simplest model for these blocks are current sources drawing a fixed time-dependent signal representing the cumulative transient supply current of the active devices switching within the block;
- the second P_c ports are connected to suitable models of the decoupling capacitors, including their series resistance and inductance parasitics;
- the last P_v ports are connected to suitable VRM models, often simply represented by a constant voltage source with a small series impedance. We will replace this voltage source with a short circuit to characterize only the voltage fluctuations around the nominal value.

The above described network loading the PDN can be represented by its frequency-dependent non-homogeneous admittance representation

$$-\boldsymbol{I}(s) = \mathbf{Y}_L(s)\boldsymbol{V}(s) - \boldsymbol{J}(s), \tag{2}$$

where $\boldsymbol{I}(s)$ collects the currents entering each PDN port, $\boldsymbol{V}(s)$ collects PDN port voltages, and where $\mathbf{Y}_L(s)$ collects in its diagonal entries the admittances connected to each port. The source vector $\boldsymbol{J}(s)$ collects the active device currents $\boldsymbol{J}_a(s)$ in its first P_a entries and is vanishing otherwise. Combining (1) with (2) and solving for the port voltages at each frequency point ω_k leads to

$$\boldsymbol{V}_a(\mathrm{j}\omega_k) = \hat{\mathbf{Z}}_{a,k}\boldsymbol{J}_a(\mathrm{j}\omega_k), \tag{3}$$

where $\hat{\mathbf{Z}}_{a,k}$ denotes the upper-left $P_a \times P_a$ block of matrix $[\hat{\mathbf{Y}}_k + \mathbf{Y}_L(\mathrm{j}\omega_k)]^{-1}$, with

$$\hat{\mathbf{Y}}_k = R_0^{-1}[\mathbf{I} - \hat{\mathbf{S}}_k][\mathbf{I} + \hat{\mathbf{S}}_k]^{-1} \tag{4}$$

Each element (i, j) of matrix $\hat{\mathbf{Z}}_{a,k}$ represents the voltage response at the active block location i, excited by the switching current of the active block j, with all other ports terminated by decoupling capacitors and VRM models.

Suppose now that we compute a rational macromodel in scattering pole-residue form

$$\mathbf{S}(s) = \sum_{n=1}^{N} \frac{\mathbf{R}_n}{s - p_n} + \mathbf{R}_0 \tag{5}$$

by minimizing the cumulative least squares fitting error with respect to the original data (1)

$$E^2 = \sum_{k=1}^{K} E_k^2 = \sum_{k=1}^{K} \|\mathbf{S}(\mathrm{j}\omega_k) - \hat{\mathbf{S}}_k\|^2. \tag{6}$$

This process is standard in Vector Fitting (VF) applications; the global fitting error E can be reduced below a prescribed tolerance δ with a suitable model order N. Once this accurate scattering model is computed, we combine (5) with the terminations (2) to obtain the model of the PDN impedance $\mathbf{Z}_a(s)$,

following the same procedure of (2)-(5). We are interested in the characterization of the frequency-dependent error of this impedance model

$$\Delta_k = \|\mathbf{Z}_a(\mathrm{j}\omega_k) - \hat{\mathbf{Z}}_{a,k}\|. \tag{7}$$

We remark that enforcing E_k to be small during the fitting process does not guarantee that Δ_k will be small. In fact, error amplification may occur due to the transformation expressed by (2)-(4), which is termination-dependent. Using a first-order approximation, one may write

$$\Delta_k \approx \mathcal{S}_k E_k, \tag{8}$$

where $\mathcal{S}_k$ can be interpreted as a sensitivity of the PDN impedance with respect to perturbations in the scattering PDN responses under nominal termination conditions (2).

This sensitivity is illustrated on a simple canonical structure. Consider a template PDN formed by a 10×10 mesh of impedances $sL + Z_i(s)$, where $L = 0.25$ nH and $Z_i(s)$ is an equivalent lumped network for the internal conductor impedance and representing both DC and skin effect losses, with identical capacitances $C = 4.43$ pF connecting each node to a reference ground. At one corner (port 3) we connect a 1 mΩ resistor to represent a VRM, in parallel with a 470 μF large bulk capacitor. A single decoupling capacitor ($C = 1$ μF, $R_s = 25$ Ω, $L_s = 0.23$ nH) is connected at the center node (port 2) of the mesh. We are interested in the input impedance $Z_a(s)$ at the opposite corner (port 1) with respect to the VRM. A reduced-order macromodel is computed via standard VF from the 3×3 scattering matrix of the PDN up to 5 GHz. The top panel of Fig. 1 demonstrates the very good accuracy of this macromodel. However, due to the large sensitivity of $\hat{Z}_a(s)$ at low frequency to the model fitting error in the scattering domain (middle panel), the impedance $Z_a(s)$ computed from the macromodel (bottom panel of Fig. 1, thick solid black line) is very different from the nominal $\hat{Z}_a(s)$. Even enforcing an exact DC value in the fit results in a loss of accuracy around 10 MHz (dashed red line). This canonical example shows that even for simple structures the sensitivity $\mathcal{S}_k$ may be large, leading to a complete loss of accuracy when simulating an accurate scattering-based macromodel under typical PDN termination conditions.

III. FORMULATION

The error amplification issues that were observed in Sec. II can be significantly alleviated by a reweighting process during the macromodel generation. Consider the frequency-dependent sensitivity $\mathcal{S}_k$ depicted in Fig. 1. This plot tells us that the difference between the fitted macromodel $\mathbf{S}(\mathrm{j}\omega_k)$ and the original data $\hat{\mathbf{S}}_k$ is amplified at the frequencies where $\mathcal{S}_k$ is large. A compensation for this effect is achieved by making the macromodel more accurate at these frequencies, i.e., by emphasizing their contribution in the overall fitting error (6). More precisely, we define a modified cost function

$$E_w^2 = \sum_{k=1}^{K} E_{w,k}^2 = \sum_{k=1}^{K} w_k^2 \|\mathbf{S}(\mathrm{j}\omega_k) - \hat{\mathbf{S}}_k\|^2 \tag{9}$$

978-1-4799-0708-3/13 $31.00 © 2013 IEEE

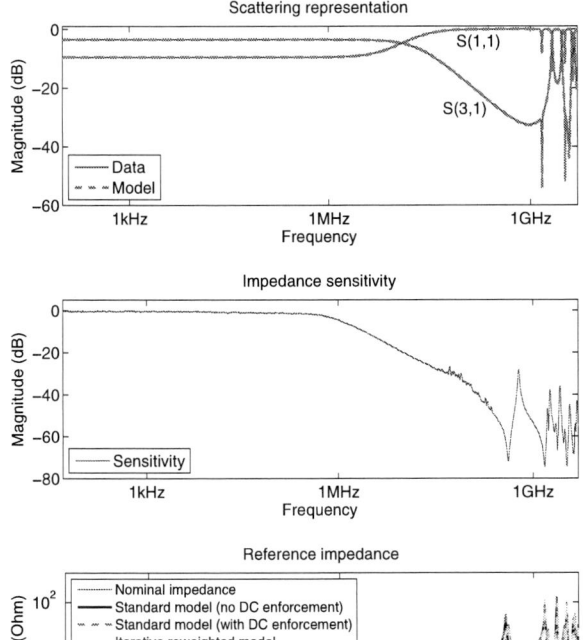

Fig. 1. Illustration of macromodel sensitivity on a canonical PDN structure. Top panel: comparison between macromodel and original responses in the scattering representation. Middle panel: frequency-dependent impedance sensitivity. Bottom panel: comparison between nominal impedance and the corresponding macromodels (standard and iteratively reweighted).

by multiplying the error contribution of each individual frequency by a frequency-dependent weight w_k. This modified cost function is then used as a target accuracy within standard VF iterations (both pole relocation and residue identification stages). In practice, any row k in the linear least squares systems to be solved during VF iterations is multiplied by the corresponding weight w_k, see [6] for more details on practical implementations.

A reasonable choice for the weight w_k is the frequency-dependent sensitivity $\mathcal{S}_k$. However, it may be the case that even this weighted fitting process is not able to attain the desired accuracy. Therefore, we setup the following iterative scheme for $\mu = 0, 1, \ldots$ until the desired accuracy is met.

1) As initialization for $\mu = 0$ we define an initial weight $w_k^0 = \mathcal{S}_k$. The sensitivity can be computed at each frequency ω_k in a closed form by a first-order expansion of Δ_k in (7). Alternatively, before starting the iterations, the raw data $\hat{\mathbf{S}}_k$ are perturbed through the multiplication of all matrix elements by $(1+\varepsilon_k)$, where ε_k is a zero-mean Gaussian variable with prescribed standard deviation σ, and the resulting perturbation $\partial Z_{a,k}$ induced on the target impedance $\hat{\mathbf{Z}}_{a,k}$ is computed. The ratio between the standard deviation of $\partial Z_{a,k}$ and σ in a Monte-Carlo run provides a good estimate of the sensitivity $\mathcal{S}_k$.

2) At each iteration $\mu = 0, 1, \ldots$, a macromodel $\mathbf{S}^\mu(s)$ is computed by weighted VF using the cost function (9) with weight w_k^μ.

3) When macromodel $\mathbf{S}^\mu(s)$ is available, the corresponding PDN impedance $\mathbf{Z}_a^\mu(s)$ is derived and its frequency-dependent deviation Δ_k^μ from the nominal impedance is computed as in (7). If $\Delta_k^\mu < \delta$ at all frequencies, where δ is the desired target accuracy, the iteration is stopped.

4) Otherwise, a new frequency-dependent weight for next iteration is defined as

$$w_k^{\mu+1} = w_k^\mu \cdot \mathcal{F}(\Delta_k^\mu), \qquad (10)$$

where $\mathcal{F} : \mathbb{R}^+ \mapsto \mathbb{R}^+$ denotes a non-decreasing function such that $\mathcal{F}(\delta) = 1$. This choice guarantees that the weight of next iteration will further emphasize those frequencies for which the impedance error is significant, and will not modify the weight for those frequencies that are already accurate. The iteration count is then increased $\mu \leftarrow \mu + 1$ and the scheme is restarted from step 2).

This simple weight updating process is able to converge to a good solution even if the initialization step is skipped and the initial weight is set to $w_k^0 = 1$, due to the automatic detection of the most sensitive frequency points in step 4).

IV. NUMERICAL EXAMPLES

We first illustrate the performance of the proposed technique on the canonical PDN example of Fig. 1. The bottom panel reports with a green dash-dot line the PDN impedance $Z_a(s)$ computed from the iteratively reweighted macromodel, showing that the loss of accuracy that was observed for the standard macromodels has been successfully compensated.

The second example is part of a real product (courtesy of Yan Shen Fen, Intel). The structure of interest is a cut out of a PDN with a total of $P = 13$ ports. The first $P_a = 12$ ports correspond to C4 bumps and the last port represents the reference board. The nominal termination network includes a lumped 1 mΩ resistor connected to the board port and suitable models for the on-die capacitance connected to the C4 ports. The actual C4 terminations are all series RC branches with individual values of resistance and capacitance in the range 1.5–3 mΩ and 10–15 nF, respectively. Excitation of the PDN is provided by a total normalized current of $J_{\text{tot}} = 1$ A, uniformly spread through all C4 ports, i.e., $J_{a,k} = J_{\text{tot}}/P_a$ for $k = 1, \ldots, P_a$. The voltage at the C4 ports is considered as the output to define the target impedances $Z_{a,k} = V_k/J_{\text{tot}}$.

The macromodeling results depicted in Fig. 2 confirm that, although the scattering-based macromodel matches closely the scattering responses of the PDN (top panel), the corresponding target impedance is very inaccurate (bottom panel) due to the large sensitivity, especially at low and medium frequencies (middle panel). Conversely, the iteratively reweighted macromodel provides a very accurate impedance response (bottom panel). The middle panel in Fig. 2 also reports the final weight w_k resulting from the proposed iteration. This weight was

978-1-4799-0708-3/13 $31.00 © 2013 IEEE

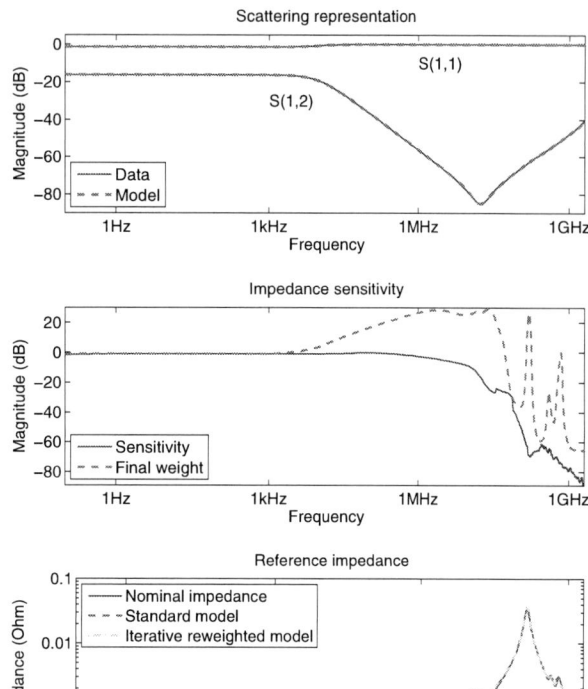

Fig. 2. As in Fig. 1, but for a real PDN structure.

at the most sensitive frequency points. This application will be documented in a forthcoming report.

REFERENCES

[1] M. Swaminathan, A.E. Engin, *Power Integrity Modeling and Design for Semiconductors and Systems*, Prentice Hall, 2007.

[2] Swaminathan, M.; Joungho Kim; Novak, I.; Libous, J.P., "Power distribution networks for system-on-package: status and challenges," *IEEE Trans. on Advanced Packaging*, vol. 27, no. 2, pp. 286–300, May 2004.

[3] Swaminathan, M.; Daehyun Chung; Grivet-Talocia, S.; Bharath, K.; Laddha, V.; Jianyong Xie, "Designing and Modeling for Power Integrity," *IEEE Trans. on Electromagnetic Compatibility*, vol. 52, no. 2, pp. 288–310, May 2010.

[4] Haihua Su; Sapatnekar, S.S.; Nassif, S.R., "Optimal decoupling capacitor sizing and placement for standard-cell layout designs," *IEEE Trans. on Computer-Aided Design of Integrated Circuits and Systems*, vol. 22, no. 4, pp. 428–436, Apr 2003.

[5] Kose, S.; Friedman, E.G., "Distributed On-Chip Power Delivery," *IEEE Journal on Emerging and Selected Topics in Circuits and Systems*, vol. 2, no. 4, pp. 704–713, Dec. 2012

[6] B. Gustavsen, A. Semlyen, "Rational approximation of frequency responses by vector fitting", *IEEE Trans. Power Delivery*, Vol. 14, N. 3, pp. 1052–1061, July 1999.

[7] D. Deschrijver, B. Haegeman, T. Dhaene, "Orthonormal Vector Fitting: A Robust Macromodeling Tool for Rational Approximation of Frequency Domain Responses", *IEEE Transactions on Advanced Packaging*, Vol. 30, No. 2, pp. 216–225, May 2007.

[8] D. Deschrijver, M. Mrozowski, T. Dhaene, D. De Zutter, "Macromodeling of Multiport Systems Using a Fast Implementation of the Vector Fitting Method," *IEEE Microwave and Wireless Components Letters*, Vol. 18, N. 6, June 2008, pp.383–385.

[9] A. Chinea, S. Grivet-Talocia, "On the Parallelization of Vector Fitting Algorithms," *IEEE Trans. on Components, Packaging, and Manufacturing Technology*, Vol. 1, n. 11, Nov. 2011, pp. 1761–1773.

[10] S. Grivet-Talocia, S.B. Olivadese, P. Triverio, "A compression strategy for rational macromodeling of large interconnect structures," EPEPS 2011, San Jose, CA (USA), October 23-26, 2011, pp. 53–56.

[11] C. P. Coelho, J. Phillips, L. M. Silveira, "A Convex Programming Approach for Generating Guaranteed Passive Approximations to Tabulated Frequency-Data", *IEEE Trans. CAD*, Vol. 23, No. 2, pp. 293–301, Feb. 2004.

[12] S. Grivet-Talocia, "Passivity enforcement via perturbation of Hamiltonian matrices", *IEEE Trans. CAS-I*, Vol. 51, No. 9, pp. 1755-1769, Sept. 2004.

[13] D. Saraswat, R. Achar and M. Nakhla, "Global Passivity Enforcement Algorithm for Macromodels of Interconnect Subnetworks Characterized by Tabulated Data", *IEEE Trans. VLSI Systems*, Vol. 13, No. 7, pp. 819–832, July 2005.

[14] S. Grivet-Talocia, A. Ubolli "On the Generation of Large Passive Macromodels for Complex Interconnect Structures", *IEEE Trans. Adv. Packaging*, vol. 29, No. 1, pp. 39–54, Feb. 2006.

[15] B. Gustavsen, A. Semlyen, "Enforcing passivity for admittance matrices approximated by rational functions", *IEEE Trans. Power Systems*, Vol. 16, N. 1, pp. 97–104, Feb. 2001.

[16] L. De Tommasi, M. de Magistris, D. Deschrijver, T. Dhaene, "An algorithm for direct identification of passive transfer matrices with positive real fractions via convex programming,", *International Journal of Numerical Modelling: Electronic Networks, Devices and Fields*, Vol. 24, N. 4, pp. 375–386, 2011.

[17] D. Deschrijver, T. Dhaene, "Fast Passivity Enforcement of S-Parameter Macromodels by Pole Perturbation," *IEEE Trans. MTT*, Vol. 57, no. 3, pp. 620–626, 2009.

obtained by starting the iterations with $w_k^0 = 1$, i.e., without any prior knowledge of the actual sensitivity response. The figure shows that the algorithm can optimize the weight even with a poor initialization, thus proving quite robust.

V. CONCLUSIONS

This paper presented a very simple yet effective iterative process to optimize the accuracy of macromodels when a nominal termination scheme is known. Although special emphasis was put here on PDN structures, the approach is general and in principle applicable to more general structures. The main idea is to exploit knowledge of the nominal terminations by enhancing the fitting accuracy at those frequencies that happen to amplify the inevitable approximation errors that occur during macromodel extraction. The numerical results on both canonical and real structures demonstrate the validity of the approach.

As a final remark, we note that the proposed reweighting scheme was applied here only for the macromodel generation by rational VF, and no discussion on passivity verification and enforcement was carried out. In fact, passivity is a fundamental requirement to guarantee stable and reliable time-domain verifications, which should never be omitted. It turns out that, once the final weighting factor w_k is obtained at the end of the reweighting iterations, the same weighted cost function (9) can be used in any perturbation-based passivity enforcement schemes [11]– [17], resulting in implicit accuracy preservation

Noise compliant macromodel synthesis for RF and Mixed-Signal applications

Salvatore Bernardo Olivadese[*†], Stefano Grivet-Talocia[†], Pietro Brenner[*]

[†] Dept. Electronics and Telecommunications, Politecnico di Torino
[*] Intel - Mobile and Communications Group
e-mail `salvatore.b.olivadese@intel.com`

Abstract—**This paper proposes a compact synthesis approach for reduced-order behavioral macromodels of linear circuit blocks for RF and Mixed-Signal design. The proposed approach revitalizes the classical synthesis of lumped linear and time-invariant multiport networks by reactance extraction, which is here exploited to obtain reduced-order equivalent SPICE netlists that can be used in any type of system-level simulations, including transient and noise analysis. The effectiveness of proposed approach is demonstrated on a real design application.**

I. INTRODUCTION

Design and Analysis of System on Chip (SoC) and System in Package (SiP) components for mobile applications becomes more challenging from day to day. Parasitic effects from on-chip, package and Printed Circuit Board (PCB) multi-layer interconnect stacks are increasingly relevant for performance assessment and verification and need to be carefully evaluated in pre-tapeout (Chip, SoC) and module (Package, PCB) design phases to avoid complete system fails or even yield losses in the production ramp-up phase. In such complex systems, many lossy passive devices are already integrated in the metal stacks, like shielded transmission lines, integrated coils, transformers, inductors and capacitors. All these passive devices and interconnect structures are lossy, that means they have beneath distributed capacitance C/m and distributed inductance L/m also a distributed resistivity R/m per unit length. This internal device resistivity per unit length significantly contributes to the overall noise behaviour of the communication system and must be thoroughly taken in account.

Nowadays, it is a common practice to assess system performances by means of Computer Aided Design (CAD) simulation methodologies. To cope with the complexity of system level simulations, divide and conquer approaches prove quite effective. The whole system is first decomposed into macro-blocks. By clever partitioning, the designer can split the system into mainly nonlinear (active) circuit blocks and linear (passive) circuit blocks interacting through well defined ports. For the linear building blocks, Linear Time Invariant (LTI) models with low complexity can be extracted by means of standard techniques [1] and synthesized into a reduced-order, linear network using resistors (R), capacitors (C) and controlled sources (CS) [2]. This technique leads to tremendous complexity reduction and can be readily extended to model soft non-linear devices under suitable biasing [3], [4]. The main drawback of this methodology lies in the noise analysis

simulation.

The RCCS synthesis from [2] is quite popular in Model Order Reduction applications because it mimics accurately the input-output response of the LTI model using a small number of network elements. Unfortunately, simulated noise spectra are often orders of magnitude away from the real physical behaviour of system under analysis, mainly because the resistors extracted by the RCCS synthesis of black-box or behavioral macromodels are not "physical", but only dummy elements used to cast the state-space equations of the macromodel in a SPICE-compatible form. Therefore, any physics-based noise model applied to such components is intrinsically ill-defined. Moreover, dependent sources do not have an associated noise model and inevitably lead to incorrect results in circuit-based noise characterization.

In order to extract proper resistors preserving the noise behaviour of the system, the RCCS synthesis from [2] is replaced in this work by a classical RLCT (Resistors, Inductors, Capacitors and ideal Transformers) synthesis from [5]. This work demonstrates in practice that this synthesis method not only reproduces accurately the input-output response of the LTI system but also preserves the noise response of the original system extracting the resistors associated to the unique passive invariants of the reciprocal lumped linear network [11].

This noise-preserving synthesis is very important for a widespread application of LTI models in RF and Mixed-Signal (MS) applications. The relevance of the RLCT synthesis was underestimated for several years due to the need of ideal transformers. As it was demonstrated by McMillan [6] the usage of transformers can not be avoided in the synthesis of LTI networks. Thus, the RLCT synthesis is used here to exploit the benefits of this classic method to automatically construct reduced-order LTI macromodels for RF applications, which can be applied in all SPICE analysis types (DC, AC TRAN, NOISE, HBB, PSS) preserving the correct physical noise behaviour. The availability of a noise compliant synthesis opens the way for notable simplifications and improvements in model-based analysis and design verification methodology.

II. PRELIMINARIES

Consider a given linear and passive circuit block. It is common practice to model this network via frequency-dependent scattering parameters, which are easily extracted by common EDA tools from layout data. A system level circuit analysis

978-1-4799-0708-3/13 $31.00 © 2013 IEEE

however requires the conversion of such representation into an equivalent circuit which can be directly employed in transient analysis using solvers of the SPICE class. This conversion, usually denoted as *macromodeling*, amounts to fitting the scattering parameter data with a rational function of frequency specified in terms of poles and residue matrices. This task is easily achieved by widespread tools such as Vector Fitting (VF) [1]. Once available, the rational macromodel can be converted to state-space form

$$\dot{x}(t) = \mathbf{A}x(t) + \mathbf{B}u(t), \tag{1}$$

$$y(t) = \mathbf{C}x(t) + \mathbf{D}u(t), \tag{2}$$

with $\mathbf{A} \in \mathbb{R}^{n \times n}$, $\mathbf{B} \in \mathbb{R}^{n \times p}$, $\mathbf{C} \in \mathbb{R}^{p \times n}$ and $\mathbf{D} \in \mathbb{R}^{p \times p}$, which in turn is readily synthesized into an equivalent netlist made of resistors, capacitors, and controlled sources [2].

The above process can be directly applied to admittance or impedance data. In the following, the models are assumed to be in the impedance input-output representation, so that

$$\mathbf{Z}(s) = \mathbf{C}(s\mathbf{I} - \mathbf{A})^{-1}\mathbf{B} + \mathbf{D} \leftrightarrow \left(\begin{array}{c|c} \mathbf{A} & \mathbf{B} \\ \hline \mathbf{C} & \mathbf{D} \end{array} \right), \tag{3}$$

where s is the complex frequency (Laplace) variable.

The McMillan degree [12] of $\mathbf{Z}(s)$ is equal to n (the size of $\mathbf{A}$) when the state-space realization (3) is minimal, i.e., when the system is both controllable and observable. Although state-space realizations are not unique, two minimal state-space realizations of the same system

$$\left(\begin{array}{c|c} \mathbf{A} & \mathbf{B} \\ \hline \mathbf{C} & \mathbf{D} \end{array} \right) \leftrightarrow \left(\begin{array}{c|c} \tilde{\mathbf{A}} & \tilde{\mathbf{B}} \\ \hline \tilde{\mathbf{C}} & \tilde{\mathbf{D}} \end{array} \right) \tag{4}$$

are equivalent to each other through a change of basis in the state space [12], applied as a similarity transformation as

$$\tilde{\mathbf{A}} = \mathbf{T}^{-1}\mathbf{A}\mathbf{T}, \qquad \tilde{\mathbf{B}} = \mathbf{T}^{-1}\mathbf{B}, \tag{5}$$

$$\tilde{\mathbf{C}} = \mathbf{C}\mathbf{T}, \qquad \tilde{\mathbf{D}} = \mathbf{D}, \tag{6}$$

with $\mathbf{T} \in \mathbb{R}^{n \times n}$ invertible.

A minimal system is passive [8] if and only if the Kalman-Yakubovich-Popov (KYP), also known as Positive-Real Lemma (PRL) [9] holds,

$$\exists \ \mathbf{P} = \mathbf{P}^T \succ 0 \ \text{s.t.} \begin{bmatrix} \mathbf{A}^T\mathbf{P} + \mathbf{P}\mathbf{A} & \mathbf{P}\mathbf{B} - \mathbf{C}^T \\ \mathbf{B}^T\mathbf{P} - \mathbf{C} & -\mathbf{D} - \mathbf{D}^\mathbf{T} \end{bmatrix} \preceq 0 \tag{7}$$

If (7) is not fulfilled after the rational fitting process, a passivity enforcement algorithm can be applied, see [10], [16] and references therein.

III. SYNTHESIS BY REACTANCE EXTRACTION

The RLCT macromodel synthesis used in this work is based on the well-known procedure of *reactance extraction* [5]. Suppose to start with a LTI multiport structure $\mathcal{N}$ with p ports and n internal dynamic elements. This structure is interpreted as the connection of a purely adynamic part $\mathcal{N}_t$ with $p + n$ ports, connected to a purely dynamic part $\mathcal{N}_d$ including all capacitors, inductors and mutual couplings, see Fig. 1.

Each port of the dynamic part $\mathcal{N}_d$ is formed by a single capacitor or inductor. This implies that the hybrid matrix of $\mathcal{N}_d$

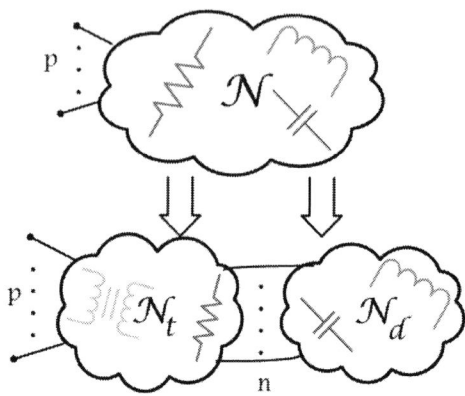

Fig. 1. Decomposition of a general p-port LTI network $\mathcal{N}$ into an adynamic $(p + n)$-port subnetwork $\mathcal{N}_t$ connected to a dynamic n-port (lossless) part $\mathcal{N}_d$ through n internal ports.

is simply $s\mathbf{\Lambda}$, where all ports connected to a capacitor (resp. inductor) are defined as voltage-controlled (resp. current-controlled), and where $\mathbf{\Lambda}$ is a block-diagonal matrix, which reduces to a diagonal matrix in case of no inductive mutual coupling elements.

The constant non-dynamic network $\mathcal{N}_t$ can be described by its hybrid matrix

$$\mathbf{M} = \begin{bmatrix} \mathbf{M}_{1,1} & \mathbf{M}_{1,2} \\ \mathbf{M}_{2,1} & \mathbf{M}_{2,2} \end{bmatrix} \tag{8}$$

with $\mathbf{M} \in \mathbb{R}^{(p+n) \times (p+n)}$, $\mathbf{M}_{1,1} \in \mathbb{R}^{p \times p}$ and $\mathbf{M}_{2,2} \in \mathbb{R}^{n \times n}$. The first block of p ports of $\mathbf{M}$ are defined as current-controlled, whereas the second block of n ports connected to $\mathcal{N}_d$ are defined as voltage- or current-controlled, according to their corresponding definition in $\mathbf{\Lambda}$.

These premises imply that the impedance matrix observed at the input p ports of $\mathcal{N}_t$ when loaded by $\mathcal{N}_d$ reads

$$\mathbf{Z}(s) = \mathbf{M}_{1,1} - \mathbf{M}_{1,2}(s\mathbf{\Lambda} + \mathbf{M}_{2,2})^{-1}\mathbf{M}_{2,1}. \tag{9}$$

Assume now that the circuit block is known via a state-space realization of its impedance matrix (3). A direct comparison between (9) and (3) suggests that the circuit synthesis can be performed by defining

$$\mathbf{M} = \begin{bmatrix} \mathbf{D} & -\mathbf{C} \\ \mathbf{B} & -\mathbf{A} \end{bmatrix}, \quad \mathbf{\Lambda} = \mathbf{I}_n \tag{10}$$

and by interpreting $\mathbf{M}$ as the constant hybrid matrix of a $p+n$ adynamic multiport element, and $s\mathbf{I}_n$ as the hybrid matrix of a dynamic (lossless) multiport element constructed by unitary capacitances and inductances directly connected at its n ports. The synthesis of $\mathbf{M}$ using a network of controlled sources is straightforward and well known from network theory [11].

The main objective is here to synthesize the state-space system (3) into an equivalent circuit which can be used in noise analysis. This requires that, due to the limitations imposed by SPICE-like solvers, the only elements that should be responsible for power dissipation, hence noise generation, should be resistors. In fact, only resistors are equipped by a suitable noise model within standard SPICE solvers, whereas other components such as controlled sources, although they may dissipate power, unfortunately do not have a corresponding noise model. For this reason controlled sources are not allowed in our synthesis, and the attention is restricted to a RLCT synthesis, where the T stands for ideal (multiport) *transformers*. Since the latter are neutral elements for what concerns power [11], their contribution to a noise analysis is null. Since inductors and capacitors do not contribute as noise sources, the overall RLCT netlist is thus guaranteed to provide consistent noise results.

Any RLCT network is passive and reciprocal. Therefore, the proposed approach will be applicable only to reciprocal systems, for which

$$\mathbf{Z}(s) = \mathbf{Z}(s)^T . \tag{11}$$

Since $\mathbf{\Lambda}$ in (10) represents a passive (lossless) and reciprocal network by construction, passivity and reciprocity of $\mathbf{Z}(s)$ obtained by reactance extraction is guaranteed by the passivity and reciprocity of its adynamic part $\mathcal{N}_t$. These conditions are guaranteed when

$$\mathbf{M} + \mathbf{M}^T \succeq 0 \tag{12}$$

$$\widehat{\mathbf{\Sigma}}\mathbf{M} = \mathbf{M}^T\widehat{\mathbf{\Sigma}}, \tag{13}$$

where $\widehat{\mathbf{\Sigma}} = \text{blkdiag}\{\mathbf{I}_p, \mathbf{\Sigma}\}$, and where $\mathbf{\Sigma} \in \mathbb{R}^{n \times n}$ is the so-called *reactance signature* matrix [5], basically a diagonal matrix of 1 and -1 placed in correspondence of the current-controlled and voltage-controlled ports, respectively. Since a general state-space realization (3) does not satisfy those constraints, the main enabling factor for this synthesis process is to find a transformation matrix $\mathbf{T}$ that converts via (4) a given state-space system into an equivalent realization for which (12)-(13) hold. This transformation is detailed next, following [12].

Before discussing further details, we should emphasize that the main drawback of the RLCT synthesis, despite being canonical [12], lies in the complexity of the resulting netlist, intended as number of primitive network elements, which scales as $\mathcal{O}(n^2p^2)$. For the RCCS synthesis complexity scales as $\mathcal{O}(np)$. As a consequence, this method is of practical relevance only for models with small or medium number of ports and dynamical order.

IV. RECIPROCAL POSITIVE REAL REALIZATIONS

We start from the generic (passive and reciprocal) state-space realization $\{\mathbf{A}, \mathbf{B}, \mathbf{C}, \mathbf{D}\}$ in (3), and we explicitly solve the Positive Real Lemma (7) for $\mathbf{P}$. Also, we form the dual system $\{\mathbf{A}^T, \mathbf{C}^T, \mathbf{B}^T, \mathbf{D}^T\}$ and we solve the PRL for the corresponding matrix $\mathbf{Q}$. Restricting now the analysis to the case $\mathbf{R} = \mathbf{D} + \mathbf{D}^T \succ 0$ (corresponding to asymptotic strict dissipativity), it follows that the matrices $\mathbf{P}$ and $\mathbf{Q}$ can be found by solving the Continuous Algebraic Riccati Equations (CARE) [13]

$$\mathbf{A}^T\mathbf{P} + \mathbf{P}\mathbf{A} + (\mathbf{P}\mathbf{B} - \mathbf{C}^T)\mathbf{R}^{-1}(\mathbf{B}^T\mathbf{P} - \mathbf{C}) = 0, \tag{14}$$

$$\mathbf{A}\mathbf{Q} + \mathbf{Q}\mathbf{A}^T + (\mathbf{Q}\mathbf{C}^T - \mathbf{B})\mathbf{R}^{-1}(\mathbf{C}\mathbf{Q} - \mathbf{B}^T) = 0, \tag{15}$$

with $\mathbf{P} = \mathbf{P}^T \succ 0$ and $\mathbf{Q} = \mathbf{Q}^T \succ 0$. This calculation can be performed through the Laub's method [14], based on the evaluation of the invariant subspaces of the Hamiltonian matrices associated to (14) and (15).

The next step is to compute the Cholesky factorization [15] of $\mathbf{P}$ and $\mathbf{Q}$

$$\mathbf{P} = \mathbf{F}^T\mathbf{F}, \tag{16}$$

$$\mathbf{Q} = \mathbf{G}^T\mathbf{G}, \tag{17}$$

with $\mathbf{F}, \mathbf{G} \in \mathbb{R}^{n \times n}$ triangular matrices. The matrix product $\mathbf{F}\mathbf{G}^T$ is then subject to a Singular Value Decomposition [15]

$$\mathbf{F}\mathbf{G}^T = \mathbf{U}\mathbf{S}\mathbf{V}^T, \tag{18}$$

with $\mathbf{U}, \mathbf{V} \in \mathbb{R}^{n \times n}$ orthogonal, where the diagonal matrix $\mathbf{S} \in \mathbb{R}^{n \times n}$ stores the singular values in decreasing order on its main diagonal. An invertible similarity transformation matrix $\mathbf{T}$ is now defined as

$$\mathbf{T} = \mathbf{G}^T\mathbf{V}\mathbf{S}^{-1/2} \tag{19}$$

and applied as in (4). A complete proof that the resulting state-space realization verifies (12)-(13) can be found in [7].

At this point the synthesis of the constant hybrid matrix $\mathbf{M}$ (10) associated to the network $\mathcal{N}_t$ can be performed according to [11]. The outer diagonal blocks in (10), i.e. $\mathbf{M}_{1,2}$ and $\mathbf{M}_{2,1}$, are directly synthesized via multiport ideal transformers. The main diagonal blocks, i.e. $\mathbf{M}_{1,1}$ and $\mathbf{M}_{2,2}$, are first diagonalized through their orthogonal eigenvector matrices [15], and then synthesized via multiport ideal transformers and unitary resistors. The connection of n internal ports of $\mathcal{N}$ on unitary inductors and capacitors, as depicted in Fig. 1, concludes the synthesis process.

V. EXAMPLE

Several tests were performed to assess the reliability of the proposed RLCT synthesis for RF and MS applications. A representative case is illustrated below, namely the centrally involved LC-tank coil of a single-coil Digitally Controlled Oscillator (DCO). DCOs can be tuned very accurately: their noise behaviour is a key figure of merit and requires therefore accurate noise modeling of all involved design parts. Thus modeling of the centrally involved LC-tank coil is a good benchmark for the noise compliant synthesis.

The Scattering parameters obtained from full-wave simulations of the centrally involved LC-tank coil of real DCO for RF applications have been used as starting point for derivation and synthesis of a macromodel. An initial macromodel was obtained by the Vector Fitting scheme [1]. This macromodel was already passive, as confirmed by a check via [16]. The port count for this example is 25 and the McMillan degree of the resulting model is 350. The state-space transformation (5) discussed in this work was then performed, leading to a synthesized RLCT netlist in SPICE form. A frequency (AC analysis) sweep of the macromodel in SPICE led to the frequency responses depicted and compared to the original data in Fig. 2. The frequency dependent noise analysis results

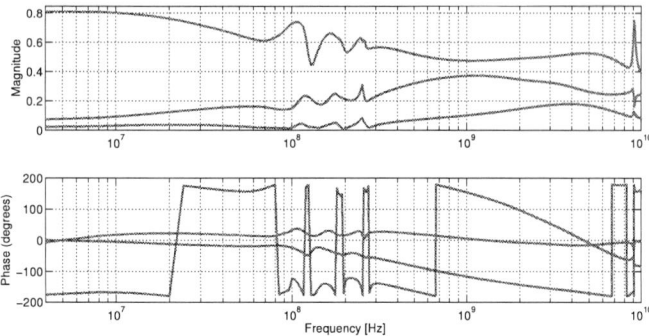

Fig. 2. Selected S-parameters from the centrally involved LC-tank coil of a real RF Digitally Controlled Oscillator. The blue continuous lines are the S-parameters obtained from full-wave simulation; the red dashed lines are obtained by SPICE simulation of the RLCT synthesized macromodel.

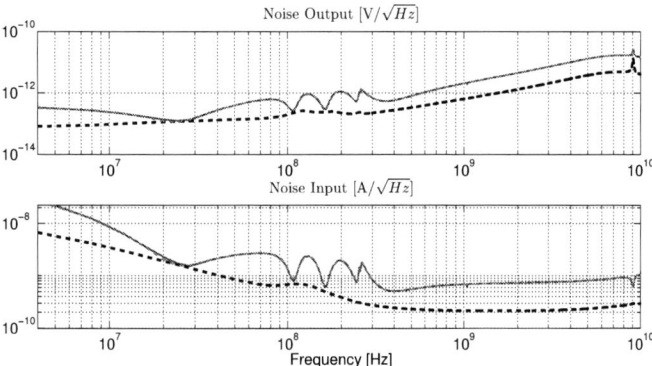

Fig. 3. Port 1 frequency dependent input-output noise spectral density. The input noise is depicted in the bottom plot while the output noise is in the top plot. The blue continuous lines are the noise results obtained using the S-parameter raw data [17]; the red dashed lines are obtained by SPICE simulation of the RLCT synthesized macromodel, while the black dashed lines come from the SPICE simulation of the RCCS synthesized macromodel.

are depicted in Fig. 3. This figures confirms the excellent accuracy of the macromodel.

The RLCT macromodel was used to perform a noise analysis of the structure, whose results are reported in Table I. The Table also reports the results from the same noise simulation using directly the S-parameters [17], as well as the results obtained using the macromodel synthesized as a RCCS network. As expected, the RLCT synthesis produces the same results that can be obtained directly from the S-parameters analysis, while the RCCS provides wrong answers.

TABLE I
DC INPUT NOISE SPECTRAL DENSITY $[A/\sqrt{Hz}]$ AND OUTPUT NOISE SPECTRAL DENSITY $[V/\sqrt{Hz}]$ FOR AN RF SINGLE COIL DIGITALLY CONTROLLED TRANSFORMER.

Port		SP data		RLCT synth		RCCS synth	
In	Out	Input	Output	Input	Output	Input	Output
1	1	4.7e-8	3.6e-13	4.7e-8	3.6e-13	1.1e-8	7.7e-14
1	2	4.8e-8	3.4e-13	4.8e-8	3.4e-13	1.1e-8	7.7e-14
2	1	4.8e-8	3.4e-13	4.8e-8	3.4e-13	1.1e-8	7.7e-14
2	2	1.1e-7	1.4e-13	1.1e-7	1.4e-13	1.1e-8	7.7e-14

VI. CONCLUSION

A noise compliant RLCT synthesis method for linear behavioral macromodels based on the classical reactance extraction technique is presented. Relying solely on the use of network elements possessing a proper noise model in SPICE based solvers, the proposed strategy is able to reproduce properly the noise behaviour of the system. The accuracy of the results obtained from the noise analysis is assessed by comparing the proposed synthesis with standard methods [17], [2].

The availability of a noise compliant network synthesis can be of paramount importance in analog behavioural modeling for devices and complete building blocks. Noise-preserving modeling is a must for simulation-based design and design verification purposes of complex analog systems. Unfortunately the RLCT synthesis requires a number of networks elements which scales as $\mathcal{O}(n^2 p^2)$. Further investigations are needed in order to reduce the complexity of the synthesized network while preserving the noise behaviour.

REFERENCES

[1] B. Gustavsen, A. Semlyen, "Rational approximation of frequency domain responses by vector fitting", *IEEE Trans. Power Del.*, vol. 14, no. 3, pp. 1052-1061, July, 1999.

[2] M. Celik, L. Pileggi, A. Odabasioglu, "Tutorial lecture on Characterization and Macromodeling of 3D Interconnects", 8th IEEE Workshop on Signal Propagation on Interconnects (SPI), Heidelberg (Germany), May 9-12, 2004.

[3] S. Grivet-Talocia, P. Brenner, F.G. Canavero, "Fast Macromodel-based Signal Integrity Assessment for RF and Mixed-Signal Modules", IEEE International Symposium on Electromagnetic Compatibility 2007, pp.1-6, 9-13 July, 2007.

[4] S.B. Olivadese, P. Brenner, S. Grivet-Talocia, "DC-Compliant Small-Signal Macromodels of Non-Linear Circuit Blocks", 17th IEEE Workshop on Signal and Power Integrity (SPI), Paris (France), May 12-15, 2013.

[5] D.C. Youla and P. Tissi, "n-Port Synthesis via Reactance Extraction - Part I", IEEE International Convention, New York, N.Y., March 21-25, 1966, Part 7, pp. 183-207.

[6] B. McMillan, "Passive Multiterminal Networks Without Transformers", IEEE Trans. on Circuits and Systems, Vol. 54, No. 10, October 2007.

[7] T. Reis, J.C. Willems, "A balancing approach to the realization of systems with internal passivity and reciprocity", System e Control Letters, 60, pp. 69-74, Elsevier, 2011.

[8] P. Triverio, S. Grivet-Talocia, M.S. Nakhla, F. Canavero, R. Achar, "Stability, causality, and passivity in electrical interconnect models", IEEE Trans. on Advanced Packaging, vol. 30, no. 4, pp. 795-808, 2007.

[9] B.D.O. Anderson, "A system theory criterion for positive real matrices", SIAM J. Control 5 (2), 1967, pp. 171-182.

[10] S. Grivet-Talocia, "Passivity Enforcement via Perturbation of Hamiltonian Matrices" , IEEE Trans. Circuits and Systems I: Fundamental Theory and Applications, pp. 1755-1769, vol. 51, n. 9, September, 2004.

[11] V. Belevitch, "Classical network theory", Holden-Day, 1968.

[12] B.D.O. Anderson, S. Vongpanitlerd, "Network Analysis and Synthesis", Prentice Hall, Englewood Cliff, 1973.

[13] P. Lancester, L. Rodman, "Algebraic Riccati Equations", Oxford University Press, Sept. 1995.

[14] A.J. Laub, "A Schur method for solving Algebraic Riccati Equations", IEEE Trans. Automat. Control., vol. AC-24, pp. 913-921, Dec. 1979.

[15] G. H. Golub, C.F. Van Loan, "Matrix computations", 3rd ed., Baltimore: John Hopkins University Press, 1996.

[16] S. Grivet-Talocia, "An Adaptive Sampling Technique for Passivity Characterization and Enforcement of Large Interconnect Macromodels", IEEE Trans. on Advanced Packaging, vol. 30, n. 2, pp. 226-237, May, 2007.

[17] J.F. Harris, "On the Use of Windows for Harmonic Analysis with Discrete Fourier Transform", Proceedings of the IEEE, vol. 66, no. 1, Jan. 1978.

978-1-4799-0708-3/13 $31.00 © 2013 IEEE

Connecting vector fitting to barycentric interpolation and the Loewner matrix

Sanda Lefteriu
INRIA Sophia Antipolis Mditerranée
2004 Route des Lucioles, 06902 Valbonne, France
Email: s.lefteriu@gmail.com

Athanasios C. Antoulas
Rice University
6100 Main St, Houston, TX 77251, US
and Jacobs University Bremen
Campus Ring 12, 28759 Bremen, Germany
Email: aca@rice.edu

Abstract—**Vector fitting (VF) [1] is a popular algorithm among electronic engineers for building rational models from measured or full-wave electromagnetic simulated frequency domain responses (S-, Y-, or Z- parameters). This paper exhibits a connection between the vector fitting formulation, on one hand, and the barycentric form of a rational interpolant and the Loewner matrix, on the other hand. This connection helps address the issues of choosing the starting poles and determining the true model order.**

I. INTRODUCTION

The vector fitting algorithm [1], [2] relies on a particular formulation, with the numerator and denominator of the rational model expressed in pole-residue form, sharing the same poles. Their initial location is either specified according to the heuristic suggested in [1] or chosen by the user based on insights into the actual location of the system poles. Next, the iterative process called *pole relocation* moves them around in the complex plane until their location no longer changes. In the case of noise-free data and if the desired order is greater than the order of the underlying system, the algorithm converges to the true poles of the system [3]. This pole identification stage is followed by a residue identification step.

This paper shows that the vector fitting formulation is related to the barycentric form of a rational interpolant, which can be obtained by considering the Lagrange basis for expressing the numerator and denominator polynomials. In this formulation, the denominator is no longer a proper rational function [2], but rather a strictly proper one. Moreover, we exhibit a relationship between vector fitting and the Loewner framework [4]–[6] which allows us to address issues related to the choice of starting poles and determining the true model order. The initial poles can be chosen as some of the measurements due to the fact that interpolation is exact at those points. Currently, the user guesses the order of the model, but due to the relationship with the Loewner matrix, the underlying model order can be deduced from the rank of the matrix used in setting up the linear system in the pole identification stage. Barycentric interpolation was also considered by the authors in [7] who proposed the Barycentric Vector Fitting algorithm to improves the run time over existing Vector Fitting algorithms.

II. REVIEW OF RELAXED VECTOR FITTING

Assuming measurements generated by a SISO linear time-invariant system with transfer function $H(s)$, VF aims at finding an approximant $f(s)$ expressed in pole-residue form:

$$H(s) \approx f(s), \ f(s) = \sum_{i=1}^{n} \frac{r_i}{s - p_i} + d.$$

The residues r_i and poles p_i are either real quantities or come in complex conjugate pairs, while d is real. VF assumes a pole-residue representation for the numerator $\mathbf{n}(s)$ and denominator $\mathbf{d}(s)$ of the unknown rational model $f(s)$:

$$f(s) = \frac{\mathbf{n}(s)}{\mathbf{d}(s)} \Rightarrow \begin{cases} \mathbf{n}(s) = \sum_{i=1}^{n} \frac{c_i}{s - a_i} + d \\ \mathbf{d}(s) = \sum_{i=1}^{n} \frac{\tilde{c}_i}{s - a_i} + \tilde{d} \end{cases}. \quad (1)$$

Note that $\mathbf{n}(s)$ and $\mathbf{d}(s)$ share the same poles a_i. Multiplying by the denominator yields the usual linearization

$$f(s) = \frac{\mathbf{n}(s)}{\mathbf{d}(s)} \Rightarrow \underbrace{\sum_{i=1}^{n} \frac{c_i}{s - a_i} + d}_{\mathbf{n}(s)} = \underbrace{\left(\sum_{i=1}^{n} \frac{\tilde{c}_i}{s - a_i} + \tilde{d} \right)}_{\mathbf{d}(s)} f(s). \quad (2)$$

Equation (2) is linear in c_i, d, $\tilde{c}_i$, $\tilde{d}$, but nonlinear in the quantities a_i, so by specifying a_i, it becomes linear. A least squares problem can be set up by writing (2) at the sample points where measurements are provided:

$$\underbrace{\mathbf{A}}_{\in \mathbb{C}^{N \times 2(n+1)}} \underbrace{\mathbf{x}}_{\in \mathbb{C}^{2(n+1) \times 1}} = \underbrace{\mathbf{0}}_{\in \mathbb{C}^{N \times 1}}, \quad (3)$$

$$\mathbf{A} = [\ \Phi \quad \mathbf{1} \quad -diag(\mathbf{b})\Phi \quad \mathbf{b} \ ],$$

$$\Phi = \begin{bmatrix} \frac{1}{s_1 - a_1} & \cdots & \frac{1}{s_1 - a_n} \\ \vdots & \ddots & \vdots \\ \frac{1}{s_N - a_1} & \cdots & \frac{1}{s_N - a_n} \end{bmatrix}, \ \mathbf{b} = \begin{bmatrix} H_1 \\ \vdots \\ H_N \end{bmatrix}$$

and $\mathbf{1}$ is a column vector containing 1's, while

$$\mathbf{x} = \begin{bmatrix} \mathbf{c} \\ d \\ \tilde{\mathbf{c}} \\ \tilde{d} \end{bmatrix}, \ \text{where } \mathbf{c} = \begin{bmatrix} c_1 \\ \vdots \\ c_n \end{bmatrix}, \ \tilde{\mathbf{c}} = \begin{bmatrix} \tilde{c}_1 \\ \vdots \\ \tilde{c}_n \end{bmatrix},$$

with $s_i \in \mathbb{C}$ and $H_i = H(s_i) \in \mathbb{C}$, for $i = 1, \ldots, N$. The relaxed non-triviality constraint consists of adding the equation $\Re \left\{ \sum_{i=1}^{N} \left(\sum_{j=1}^{n} \frac{\tilde{c}_j}{s_i - a_j} + \tilde{d} \right) \right\} = N$ as the last row in (3). At the

978-1-4799-0708-3/13 $31.00 © 2013 IEEE

next iteration, a_i are assigned as the zeros of the denominator $\mathbf{d}(s)$, which are found by solving an eigenvalue problem set up using the already determined quantities $\tilde{c}_i$ and $\tilde{d}$. This iteration constitutes the pole relocation iteration, at the end of which the model's poles are found. The residues are determined by setting up another linear system similar to that in (3).

III. LAGRANGE BASIS AND BARYCENTRIC INTERPOLATION

Like any rational expression, the model $f(s)$ can be written as a ratio of two polynomials:

$$f(s) = \frac{N(s)}{D(s)} = \frac{\alpha_0 + \alpha_1 s + \alpha_2 s^2 + \ldots + \alpha_n s^n}{\beta_0 + \beta_1 s + \beta_2 s^2 + \ldots + \beta_n s^n}. \quad (4)$$

One can express polynomials in various bases. Above, $N(s)$ and $D(s)$ are expressed in the monomial basis, known to be ill-conditioned for a high degree n. An alternative basis for polynomials of degree up to n is that of Lagrange polynomials:

$$q_i(s) = \prod_{0 \le j \le n, j \ne i} \frac{s - \alpha_j}{\alpha_i - \alpha_j}, \; i = 0, \ldots, n,$$

for α_j chosen a-priori. A polynomial can be expressed in the Lagrange basis as $N(s) = \sum_{i=0}^{n} \beta_i q_i(s)$ and it satisfies the following interpolation conditions $N(\alpha_i) = \beta_i, \, i = 0, \ldots, n$. The rational function in (4) can also be expressed in the Lagrange basis as:

$$f(s) = \frac{N(s)}{D(s)} = \frac{\beta_0 q_0(s) + \beta_1 q_1(s) + \ldots + \beta_n q_n(s)}{\tilde{\beta}_0 q_0(s) + \tilde{\beta}_1 q_1(s) + \ldots + \tilde{\beta}_n q_n(s)}. \quad (5)$$

Changing notation, we can express the transfer function as

$$f(s) = \frac{N(s)}{D(s)} = \frac{c_0 \hat{q}_0(s) + c_1 \hat{q}_1(s) + \ldots + c_n \hat{q}_n(s)}{\tilde{c}_0 \hat{q}_0(s) + \tilde{c}_1 \hat{q}_1(s) + \ldots + \tilde{c}_n \hat{q}_n(s)},$$

with the numerator and denominator still in the Lagrange basis. The quantities $\hat{q}_i(s)$ are in fact $\hat{q}_i(s) = \prod_{0 \le j \le n, j \ne i} (s - a_j)$, while the constant factor $\prod_{0 \le j \le n, j \ne i} \frac{1}{(a_i - a_j)}$ has been incorporated in the coefficients c_i and $\tilde{c}_i$, respectively. Dividing both the numerator and denominator by the common factor $\prod_{0 \le j \le n} (s - a_j)$, we obtain:

$$f(s) = \frac{N(s)}{D(s)} = \frac{\frac{c_0}{s - a_0} + \ldots + \frac{c_n}{s - a_n}}{\frac{\tilde{c}_0}{s - a_0} + \ldots + \frac{\tilde{c}_n}{s - a_n}} = \frac{\sum_{i=0}^{n} \frac{c_i}{s - a_i}}{\sum_{i=0}^{n} \frac{\tilde{c}_i}{s - a_i}}. \quad (6)$$

The expression above is the barycentric form of a rational interpolant and it is similar to the VF formulation in (1). The main difference is that, in VF, the denominator is a proper rational function of order n, while in the barycentric one, the denominator is a strictly proper function of order $n + 1$.

Remark III.1. *Interpolation at $s = a_i$ is exact for both VF and the regular Lagrange expansion: when evaluating $f(s)$ at $s = a_i$, the result is $f(a_i) = \frac{c_i}{\tilde{c}_i}$, as also pointed out in [8].*

Next, we focus on choosing a_i at the first iteration. To enforce interpolation at a desired point, one can select a_0 freely. For the other starting poles, VF uses a heuristic [1].

Remark III.2. *We propose to choose a_i at the first step as complex conjugate imaginary quantities, namely $\pm j\omega_i = \pm 2\pi f_i \jmath$, where f_i are some of the frequencies where measurements are provided, equally spaced in the frequency band.*

Multiplying (6) by the denominator, a similar system as (3): $\hat{\mathbf{A}} \hat{\mathbf{x}} = \mathbf{0}$, is obtained, with the unknowns c_i and $\tilde{c}_i$ and a null right hand side. This can solved by computing the nullspace of $\hat{\mathbf{A}}$ (or by selecting the right singular vector corresponding to the smallest singular value of $\hat{\mathbf{A}}$ in the case of noisy measurements [9]).

Remark III.3. *At this point, in case of noise-free measurements, the original system is recovered, provided that n, the approximation order, is higher than or equal to the order of the underlying system and the number of data points $N \ge 2(n+1)$. Lemma III.1 provides a realization.*

Lemma III.1. *[10]* $f(s) = \tilde{\mathbf{C}}(s\tilde{\mathbf{E}} - \tilde{\mathbf{A}})^{-1}\tilde{\mathbf{B}}$, where
$$\tilde{\mathbf{C}} = [\, c_0 \quad c_1 \quad \ldots \quad c_n \,], \quad \tilde{\mathbf{B}}^T = [\, 0 \quad \ldots \quad 0 \quad -1 \,],$$
$$(s\tilde{\mathbf{E}} - \tilde{\mathbf{A}}) = \begin{bmatrix} (s-a_0) & (a_1-s) & & & \\ (s-a_0) & 0 & (a_2-s) & & \\ \vdots & & \ddots & \ddots & \\ (s-a_0) & & & 0 & (a_n-s) \\ \tilde{c}_0 & \tilde{c}_1 & \tilde{c}_2 & \ldots & \tilde{c}_n \end{bmatrix}.$$

For noisy measurements, an approximate fit is obtained. To improve it, one can employ the pole relocation iteration: the n finite poles of the present model, which are found by solving a generalized eigenvalue problem for the pencil $(\tilde{\mathbf{A}}, \tilde{\mathbf{E}})$, together with the additional starting pole a_0, are used as the new a_i.

IV. CONNECTION TO THE LOEWNER MATRIX

Select a_i, $i = 0, \ldots, n$ as some of the measurements, denoted as λ_i. Thus, $H(\lambda_i)$, which we denote as w_i, can be expressed as $\frac{c_i}{\tilde{c}_i}$. Rewriting (6), we have that $f(s)$ satisfies

$$\sum_{i=0}^{n} \tilde{c}_i \frac{f(s) - w_i}{s - \lambda_i} = 0, \quad \tilde{c}_i \ne 0. \quad (7)$$

To determine $\tilde{c}_i$ (and consequently, c_i, from $c_i = w_i \tilde{c}_i$), write (7) for the remaining $N - (n+1)$ measurements:

$$H(\mu_k) = v_k, \; k = 0, \ldots, N - (n+1) - 1, \quad (8)$$

where μ_k satisfy $\mu_i \ne \mu_k$, $i \ne k$, $\lambda_i \ne \mu_k$, $\forall\, i, k$. Substituting (8) in (7), we obtain the following condition for $\tilde{c}_i$: $\mathbb{L}\tilde{\mathbf{c}} = \mathbf{0}$,

$$\mathbb{L} = \begin{bmatrix} \frac{v_0 - w_0}{\mu_0 - \lambda_0} & \ldots & \frac{v_0 - w_n}{\mu_0 - \lambda_n} \\ \vdots & \ddots & \vdots \\ \frac{v_m - w_0}{\mu_m - \lambda_0} & \ldots & \frac{v_m - w_n}{\mu_m - \lambda_n} \end{bmatrix}, \quad \tilde{\mathbf{c}} = \begin{bmatrix} \tilde{c}_0 \\ \vdots \\ \tilde{c}_n \end{bmatrix}, \quad (9)$$

with $m = N - (n+1) - 1$. The matrix $\mathbb{L}$ is the Loewner matrix [4], [5], [11], [12] associated with the *row array* $(\mu_j, \mathbf{v}_j)$, and the *column array* $(\lambda_i, \mathbf{w}_i)$. Thus $f(s)$ is determined if $\tilde{\mathbf{c}}$ is in the right kernel of $\mathbb{L}$ [11], [13]. A key fact is that the *degree* of the underlying rational function, defined as the maximum between the degrees of the polynomials in the numerator and denominator, is equal to the rank of $\mathbb{L}$.

Outline of the proposed approach:

1) Determine the numerical rank of a square Loewner matrix (constructed using half of the measurements as λ_i and the rest as μ_k). The rank indicates the order of the system, so the user does not need to guess an approximation order. For high noise levels, the numerical rank is difficult to determine, so use an educated guess.

2) Compute the Loewner matrix with the column array $(\lambda_i, \mathbf{w}_i)$, chosen as $n+1$ samples and the corresponding measurements, and the row array $(\mu_k, \mathbf{v}_k)$ as the remaining samples and measurements. This step eliminates the need for the heuristic [1] to choose the starting poles.

3) For all remaining iterations, a linear system $\hat{\mathbf{A}}\hat{\mathbf{x}} = \mathbf{0}$ is set up with a_0 chosen freely and $a_1, \ldots, a_n$, the generalized eigenvalues of $(\tilde{\mathbf{A}}, \tilde{\mathbf{E}})$ in Lemma III.1.

Besides the advantages of determining the approximation order and not having to specify starting poles, the proposed approach is also more computationally efficient. One needs to compute the nullspace of the Loewner matrix of dimension $(N - (n + 1)) \times (n + 1)$, in the first step, and of $\hat{\mathbf{A}}$ of dimension $N \times 2(n + 1)$, in the subsequent steps, while in the relaxed VF formulation, one had to solve a least squares problem involving a matrix of size $(N + 1) \times 2(n + 1)$. The nullspace of a matrix can be computed via a QR or SVD decomposition, while the least squares problem in (3) can be solved by first computing the QR or SVD decomposition of $\mathbf{A}$, followed by other operations (for QR, solving a linear system of size $2(n+1)$ by back substitution, and for SVD, performing matrix-vector multiplications).

In the SIMO (single input, multiple output) case, namely the case in which the transfer function is a column vector with p outputs, the residues in the numerator, the c_i quantities, become column vectors with p components, while the residues in the denominator, namely $\tilde{c}_i$, are still scalars. Likewise, the v_k and w_i quantities are column vectors, so to obtain a square matrix in step 1 of the proposed approach, one needs to use $m = \left\lceil \frac{N}{p+1} \right\rceil - 1$ measurements in the row array and the remaining $n = \left\lceil \frac{Np}{p+1} \right\rceil - 1$ in the column array. In the general MIMO case, fitting can be done element-, vector- or matrix-wise or the concept of tangential interpolation [5] can be used.

V. NUMERICAL EXAMPLE

Consider the transfer function (adapted from [1]) with 17 poles and residues given in Table I. We sample it at 100 frequencies linearly distributed between 1 and 10^5Hz. We also append the complex conjugates of these measurements, so $N = 200$. Next, the measurements are corrupted with additive noise which is 10^{-6} relative to the magnitude of each measurement.

Figure 1 shows the singular value drop (4 orders of magnitude) of the square Loewner matrix built using the odd measurements and their complex conjugates as the column array (λ_i, w_i) and the even ones as the row array (μ_k, v_k).

Poles	residues
-41000	-83000
-100±5000$\jmath$	-5±7000$\jmath$
-120±15000$\jmath$	-20±18000$\jmath$
-3000±35000$\jmath$	6000±45000$\jmath$
-200±45000$\jmath$	40±60000$\jmath$
-15e2±45e3$\jmath$	90±10000$\jmath$
-5e2±7e4$\jmath$	50000±80000$\jmath$
-1e3±73e3$\jmath$	1000±45000$\jmath$
-2e3±90e3$\jmath$	-5000±92000$\jmath$

TABLE I.   POLES AND RESIDUES OF THE SYSTEM

The drop indicates an underlying system of size 17, which is indeed the case.

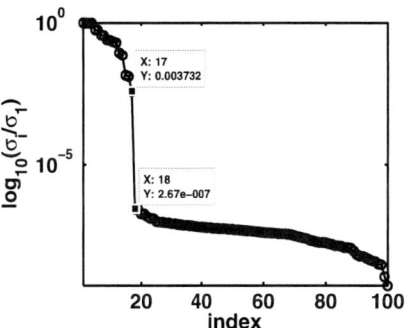

Fig. 1.   Drop of the singular values of the square Loewner matrix

We compare RVF to the proposed reformulation in terms of the normalized errors in the poles at each iteration. For RVF, the starting poles were chosen according to the heuristic [1], together with a real pole located at -50000. Figure 2(a) shows the normalized error in the poles, namely $\frac{|p_i - \tilde{p}_i|}{|p_i|}$, where p_i and $\tilde{p}_i$ are the true and approximated pole, respectively, for $i = 1, \ldots, 17$. The first pole is the real one, while the rest appear in complex conjugate pairs. A few remarks are in place. Two different values for the extra pole a_0, namely 10^4 and 10^6, were selected. The first step in the proposed formulation consists of computing the tall Loewner matrix according to (9), therefore the resulting poles are independent of a_0, which is only needed starting from the second iteration. Last, the pole estimates after the first step obtained with RVF using the heuristic are about two orders of magnitude worse than with the proposed approach.

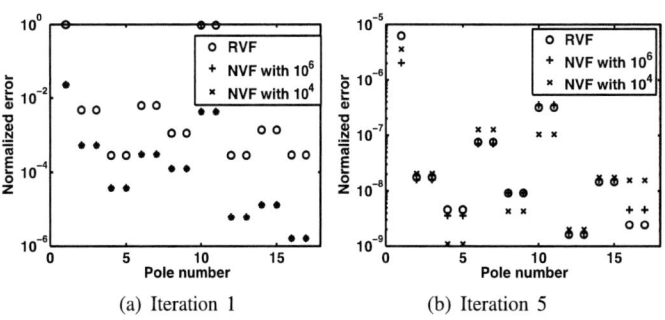

(a) Iteration 1   (b) Iteration 5

Fig. 2.   Error in the poles at different iterations

At the end of 5 iterations, the normalized errors in the poles were plotted in Fig. 2(b). At this stage, all methods are able to provide estimates which are between 10^{-5} and 10^{-9} away from the true poles.

The accuracy of the models was assessed using two error measures: $\mathcal{H}_\infty$ error $= \dfrac{\max\limits_{i=1...k} |f(\jmath\omega_i) - H_i|}{\max\limits_{i=1...k} |H_i|}$, $\mathcal{H}_2$ error $= \sqrt{\dfrac{\sum_{i=1}^{k} |f(\jmath\omega_i) - H_i|^2}{\sum_{i=1}^{k} |H_i|^2}}$. Table II presents the $\mathcal{H}_\infty$ and $\mathcal{H}_2$ errors. While all approaches yielded errors smaller than the noise values, the proposed approach with the extra pole located at 10^6 yielded errors smaller than RVF. When choosing the extra pole at 10^4, the resulting $\mathcal{H}_\infty$ error is the smallest.

	Noise	Relaxed VF	New VF with a_0 at 10^6	New VF with a_0 at 10^4
$\mathcal{H}_\infty$	6.9123e-7	2.6012e-7	2.5596e-7	2.3343e-7
$\mathcal{H}_2$	7.7772e-7	2.7741e-7	2.7535e-7	4.3319e-7

TABLE II.   ERRORS WITH THE VARIOUS APPROACHES

VI. Conclusion

This paper presents a connection between the vector fitting formulation on one hand, and barycentric interpolation and the Loewner matrix on the other hand. As a consequence: (1) the asymptotic condition may be replaced by the introduction of an additional pole; (2) the order of the approximation can be determined from the rank of a square Loewner matrix; (3) an educated guess for the choice of the starting poles in the pole relocation iteration is provided.

References

[1] B. Gustavsen and A. Semlyen, "Rational approximation of frequency domain responses by vector fitting," *IEEE Trans. Power Del.*, vol. 14, pp. 1052–1061, Jul. 1999.

[2] B. Gustavsen, "Improving the pole relocation properties of vector fitting," *IEEE Trans. Power Del.*, vol. 21, no. 3, pp. 1587–1592, Jul. 2006.

[3] S. Lefteriu and A. Antoulas, "On the convergence of the vector-fitting algorithm," *IEEE Trans. Microw. Theory Tech.*, vol. 61, no. 4, pp. 1435–1443, 2013.

[4] A. C. Antoulas, *Approximation of Large-Scale Dynamical Systems*. Philadelphia: SIAM, 2005.

[5] A. J. Mayo and A. C. Antoulas, "A framework for the solution of the generalized realization problem," *Linear Algebra and Its Applications*, vol. 405, no. 2-3, pp. 634–662, 2007.

[6] S. Lefteriu and A. C. Antoulas, "A new approach to modeling multiport systems from frequency domain data," *IEEE Trans. Comput.-Aided Design Integr. Circuits Syst.*, vol. 29, no. 1, pp. 14 –27, Jan. 2010.

[7] D. Deschrijver, L. Knockaert, and T. Dhaene, "A barycentric vector fitting algorithm for efficient macromodeling of linear multiport systems," *IEEE Microw. Wireless Compon. Lett.*, vol. 23, no. 2, pp. 60–62, 2013.

[8] L. Knockaert, F. Ferranti, and T. Dhaene, "Vector fitting vs. Levenberg-Marquardt : Some experiments," in *IEEE Workshop on Signal Propagation on Interconnects*, may 2009, pp. 1 –4.

[9] G. H. Golub and C. F. Van Loan, *Matrix computations (3rd ed.)*. Baltimore, MD, USA: Johns Hopkins University Press, 1996.

[10] A. Antoulas, A. Ionita, and S. Lefteriu, "On two-variable rational interpolation," *Linear Algebra and its Applications*, vol. 436, no. 8, pp. 2889 – 2915, 2012, issue dedicated to Danny Sorensen's 65th birthday.

[11] A. C. Antoulas and B. D. O. Anderson, "On the scalar rational interpolation problem," *IMA J. of Mathematical Control and Information*, vol. 3, no. 2-3, pp. 61–88, 1986, special issue on Parametrization problems.

[12] B. D. O. Anderson and A. C. Antoulas, "Rational interpolation and state-variable realizations," *Linear Algebra and Its Applications*, vol. 137/138, pp. 479–509, 1990, special issue on Matrix problems.

[13] L. Knockaert, "A simple and accurate algorithm for barycentric rational interpolation," *IEEE Signal Processing Letters*, vol. 15, pp. 154 –157, 2008.

A Novel Algorithm for Optimum Order Estimation of Nonlinear Reduced Macromodels

Behzad Nouri, Michel S. Nakhla, and Ramachandra Achar

Department of Electronics, Carleton University, Ottawa, Canada K1S 5B6

Email: {sbnouri, msn, achar}@doe.carleton.ca

Abstract—**Estimation of an optimal order for reduced models is a challenging task and is often based on heuristics. In this paper, a new systematic algorithm is presented for estimating the minimum acceptable order for reduced models of nonlinear systems to ensure accurate and efficient transient behavior. The methodology incorporates the techniques developed in nonlinear time-series analysis, nonlinear model order reduction and computational geometry for a precise determination of the optimum order for a reduced nonlinear system.**

I. INTRODUCTION

Signal and power integrity analysis of high-speed interconnects and packages are becoming increasingly important; however, extremely challenging due to the large circuit sizes and mixed frequency-time domain analysis issues. Model order reduction (MOR) has proven successful to tackle these issues and hence, has been an active research topic in the CAD area, in the recent years. The goal of model order reduction is to extract a smaller but accurate model for a given system, to enable fast simulations of large complex designs. So far, MOR techniques for linear time invariant systems have been well-developed and widely used [1]. On the other hand, nonlinear systems present numerous challenges for MOR, and the area still suffers from the lack of robust and efficient tools and methodologies.

An important and practical common problem in prominently used nonlinear order-reduction techniques is that of "selection of order" for the reduced model. The selection of an optimum order is important to achieve a pre-defined accuracy while not over-estimating the order, which otherwise can lead to inefficient transient simulations. The reduced-order estimation issue for linear circuits has been recently addressed in [2], [3].

In this paper, we propose a novel algorithm to obtain an optimally minimum order for a nonlinear circuit reduction. The proposed methodology is based on the idea of monitoring the behavior of the projected nonlinear trajectory in the reduced space. To serve this purpose, a mathematical algorithm is devised to observe the behavior of near neighboring points, lying on the low-dimensional nonlinear trajectory, when increasing the dimension of a reduced-space. The order is determined such that the projected trajectory is unfolded properly in the reduced space, while monitoring the count of the "False Nearest Neighbor (FNN)" points on the projected trajectory. The reduced model in this optimally reduced subspace captures the major dynamical properties of the original system.

The rest of the paper is organized as follows. Section II provides the background and preliminary formulations and Section III describes the development of the proposed algorithm. Examples are presented in Section IV that demonstrate the validity and performance of the proposed algorithm. Section V presents the conclusion.

II. BACKGROUND

A. Formulation of Circuit Equations

A general nonlinear circuit can be described in the time-domain using the Modified Nodal Analysis (MNA) as follows [4]

$$\mathbf{C}\frac{d}{dt}\mathbf{x}(t) + \mathbf{G}\mathbf{x}(t) + \mathbf{f}\left(\mathbf{x}(t)\right) = \mathbf{B}\mathbf{u}(t) \qquad (1)$$

$$\mathbf{i}(t) = \mathbf{L}\mathbf{x}(t) \qquad (2)$$

where $\mathbf{C}$ and $\mathbf{G} \in \mathbb{R}^{n \times n}$ are susceptance and conductance matrices including the contribution of linear elements, respectively, $\mathbf{x}(t) \in \mathbb{R}^n$ denotes the vector of MNA variables (the nodal voltages, some branch currents and electrical charges) of the circuit. $\mathbf{f}(\mathbf{x}) \in \mathbb{R}^n$ is a vector of real-valued functions including the stamps of all nonlinear elements in the circuit. $\mathbf{B}$ and $\mathbf{L}$ are the input and output matrices, respectively.

B. Model order reduction of nonlinear systems

The basic idea of model order reduction of a circuit is to replace the original system by an approximating system with a state-space dimension of order m, significantly smaller than the original order n. Model reduction algorithms seek a proper order m for which the outputs from the reduced system $\hat{\mathbf{i}}(t)$ and the original responses $\mathbf{i}(t)$ from (1) and (2) are approximately equal for inputs of interest $\mathbf{u}(t)$.

Some of the well-known methods for the reduction of nonlinear systems are trajectory piecewise-linear techniques (TPWL) [5], Proper Orthogonal Decomposition (POD) [6], and extensions of the truncated balanced realization (TBR) [7]. The last two methods are projection based approaches.

The key idea in any projection based reduction process is to project the original n-dimensional state space to a m-th order subspace ($m \ll n$). This requires creation of some projection operators $\mathbf{W}$ and $\mathbf{Q} \in \mathbb{R}^{n \times m}$, where $\mathbf{W}^{\mathrm{T}}\mathbf{Q} = \mathbf{I}_{m \times m}$. Assume that, there exists $\mathbf{z}(t)$ in a reduced subspace such that $\mathbf{x}(t) = \mathbf{Q}\mathbf{z}(t)$. Due to the orthogonality of the projection matrices ($\mathbf{W}$ and $\mathbf{Q}$), $\mathbf{z}(t) = \mathbf{W}^{\mathrm{T}}\mathbf{x}(t)$. By a variable change

978-1-4799-0708-3/13 $31.00 © 2013 IEEE

from (1) and (2) we have

$$\mathbf{W}^{\mathrm{T}} \frac{d}{dt} \mathbf{C} \left(\mathbf{Q} \mathbf{z}(t) \right) = \mathbf{W}^{\mathrm{T}} \mathbf{f} \left(\mathbf{Q} \, \mathbf{z}(t) \right) + \left(\mathbf{W}^{\mathrm{T}} \mathbf{B} \right) u \left(t \right) \quad (3)$$

$$\mathbf{i} \left(t \right) = \left(\mathbf{L} \mathbf{Q} \right) \mathbf{z} \left(t \right) \quad (4)$$

The approximate response $\tilde{\mathbf{x}}(t)$ is obtained by solving the reduced-order dynamical model in (3) as $\tilde{\mathbf{x}}(t) = \mathbf{Q} \, \mathbf{z}(t)$. The error between the original state variables and its approximation is $\zeta = \mathbf{x} - \mathbf{Q} \mathbf{W}^{\mathrm{T}} \mathbf{x}$. For a Galerkin projection, $\mathbf{W} = \mathbf{Q}$.

In the rest of this paper, we will use the classical POD [6], [8] to describe the proposed algorithm for order estimation. *However, it should be emphasized that, the proposed algorithm is not limited to a specific nonlinear model order reduction method and can be used in conjunction with any of the above mentioned methods.*

Using the POD method, for a given representative input $\mathbf{u}(t)$, the "time-snapshots" of the response are collected in a data matrix as $\mathbf{X} = [\mathbf{x}(t_0), \mathbf{x}(t_1), \dots, \mathbf{x}(t_N)] \in \mathbb{R}^{n \times N}$. The POD method seeks to find a projection basis $\mathbf{Q}$ to accurately approximate the original response with an approximate representation of data points by minimizing the overall projection error $||\zeta||_2 = ||\mathbf{X} - \mathbf{Q} \mathbf{Q}^{\mathrm{T}} \mathbf{X}||_2$. The solution to this optimization problem is obtained by performing singular value decomposition (SVD) on the data matrix as $\mathbf{X} = \mathbf{V} \mathbf{\Sigma} \mathbf{U}^{\mathrm{T}}$ [8]. The POD basis for a Galerkin projection is given by the first m columns in $\mathbf{V}$ ($\in \mathbb{R}^{n \times n}$) as $\mathbf{Q} = [\mathbf{v}_1, \dots, \mathbf{v}_m] \in \mathbb{R}^{n \times m}$. POD is known as a promising method to provide efficient and accurate transient simulation (e.g. more accurate than the TPWL [8]). This is for the output responses corresponding to a family of excitation signals close to the one used to form the POD basis.

III. THE PROPOSED ALGORITHM FOR ORDER ESTIMATION FOR NONLINEAR CIRCUIT REDUCTION

A. Preliminaries

Let $\mathbf{x}(t)$ be an analytical solution to (1) for a given input $\mathbf{u}(t)$ and an initial condition $\mathbf{x}_0$. This solution would be a set of functions $\{\mathbf{x}_i(t) : \text{for } i = 1, 2, \dots, n\}$, which can be used as canonical-coordinates system in state-space ($\subseteq \mathbb{R}^n$) embedding the response of the system. The variables x_i ($1 \leq i \leq n$) are obtained as a set of time-sequenced (time-series) data. The response at each time instant t_i, represented by $\mathbf{x}(t_i) = [x_1(t_i), x_2(t_i), \dots, x_n(t_i)]^{\mathrm{T}}$ defines a point in the multidimensional response space. The locus in this space for all ($t_i \in \mathcal{D} \subset \mathbb{R}_+$), henceforth referred to as the "trajectory" of the system, is a time-parametrized directional path $\mathbf{\Phi}_t : \mathcal{D} \mapsto \mathbb{R}^n$ that starts at the point $\mathbf{x}(t_0) = \mathbf{x}_0$. It is to be noted that, response (solution) of a dynamic circuit is a real-valued continuously-differentiable function [9] from an open interval $\mathcal{D} \subset \mathbb{R}_+$ into the response space ($\subseteq \mathbb{R}^n$). Under practical assumptions [10], (1) is guaranteed to have a unique analytical solution over any finite time interval, that passes through the initial state at $t = t_0$. This establishes that the trajectories (a) do not intersect each other, (b) do not have self-crossing points, and (c) do not have over-folding sections [11]. This implies that, the trajectory curves possess certain geometric properties and structure. In this paper we consider the trajectory as a geometrical model to study the dynamic behavior of the nonlinear circuits and to develop the proposed order estimation algorithm for nonlinear circuit reduction.

These concepts are illustrated in Fig. 1 for a Chua's circuit [12].

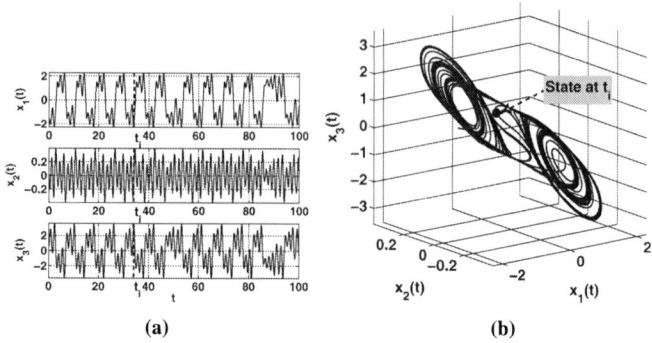

Fig. 1. (a) The time-series plot of the system variables ($x_i(t)$) as coordinates of state-space; (b) trajectory of the Chua's circuit in the state-space (scaled time: $0 \leq t \leq 100$) for a given initial condition.

B. Geometrical Framework for the Projection

Using a projection operator $\mathbf{Q}_{n \times m}$, an image of the trajectory is obtained through a point-wise projection of the original trajectory into a low-dimensional subspace as $\mathbf{z}(\cdot) = \mathbf{Q}^{\mathrm{T}} \mathbf{x}(\cdot)$. The coordinate system defining the reduced subspace are the functions $z_i(t)$ for $i = 1, \dots, m$ that are linear combinations of the original state functions; i.e. $z_i(t) = \sum_{j=1}^{n} q_{ji} \, x_j(t)$, for $i = 1, 2 \dots m$, where $m << n$.

For the sake of simplicity in the notation, hereafter, we drop "t" in the equations (e.g. $\mathbf{x}(t_i)$ is referred to as $\mathbf{x}(i)$). The Euclidean distance between any two points on the trajectory $\mathbf{x}(i)$ and $\mathbf{x}(j)$ can be expressed using the Euclidean norm as

$$d_n \left(i, j \right) = ||\mathbf{x}(i) - \mathbf{x}(j)|| = \left[\sum_{r=1}^{n} \left(x_r(i) - x_r(j) \right)^2 \right]^{0.5} \quad (5)$$

and similarly, $d_m \left(i, j \right) = ||\mathbf{z}(i) - \mathbf{z}(j)||$ for the reduced subspace.

C. Nearest Neighborhood

In the proposed approach, we consider the pairwise closeness of the states on the trajectories (in Euclidean sense) as a measure to characterize the local geometrical structure of the trajectories. For this purpose, we define the ε_n-neighborhood of $\mathbf{x}(i)$ as $\mathbf{U}(\mathbf{x}(i), \varepsilon_n) = \{\mathbf{x}(t) \in \mathbb{R}^n \mid d_n \left(\mathbf{x}(i), \mathbf{x}(t) \right) < \varepsilon_n\}$. It is geometrically visualized as an n-dimensional open ball centered at $\mathbf{x}(i)$ with a radius of ε_n. To study the local geometry, ε_n needs to be small in a certain sense, hence, $\mathbf{U}(\mathbf{x}(i), \varepsilon_n)$ is referred to as "(local) nearest neighborhood" of $\mathbf{x}(i)$ and neighbors are defined as "nearest neighboring points". These concepts are illustrated in Fig. 2. By mapping the trajectory curve from the original space to an m-dimensional subspace ($\mathbb{R}^n \mapsto \mathbb{R}^m$), the trajectory curve is contracted to reside in the reduced subspace [3]. When

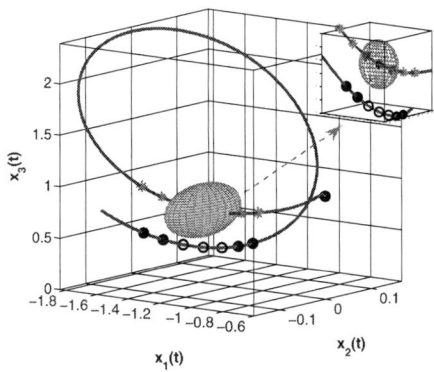

Fig. 2. Illustration of a multidimensional adjacency ball centered at $\mathbf{x}(t_i)$ (✳), accommodating its two nearest neighboring points (▼) on the trajectory of the Chua's circuit (for $0 \leq t \leq 2$).

m is too small, an over-contraction of the projected object in the target subspace is inevitable, which means that, the projected curve passes a particular point more than once (self-intersections). Fig. 3 illustrates this fact, from a geometrical perspective. It depicts a self-intersection point (✳, ◯) in the projected curve, while the corresponding original states (✳, ◯) were not neighbors. Also, the points (e.g. ◯) on the projected trajectory are close to a candidate point (✳), solely because we are viewing the projected path in a dimension that is too small.

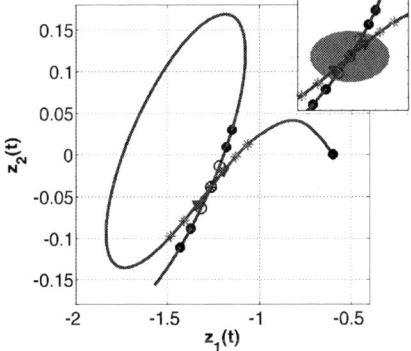

Fig. 3. Illustration of false nearest neighbor (FNN), where the 3-dimensional trajectory of the Chua circuit in Fig.2 is projected.

D. Order Estimation for Nonlinear Reduced Models

Definition 1. *The points which are neighbors in the reduced space are defined as "false neighbors" when the corresponding states are not neighbors in the original state space, and are "true neighbors" when the corresponding original states are also neighbors in the original state space.*

The underlying idea in the proposed algorithm is to geometrically observe the behavior of near neighboring points that are lying on the projected nonlinear trajectory. Starting from a low-dimensional subspace, the order of the reduced space is consecutively increased. In each step, the projected trajectory is expanded into higher dimensions. Consequently, some neighboring points move far apart and reveal themselves as false nearest neighbors. This is illustrated in Fig. 4. It depicts how neighborhood relations may change by going from

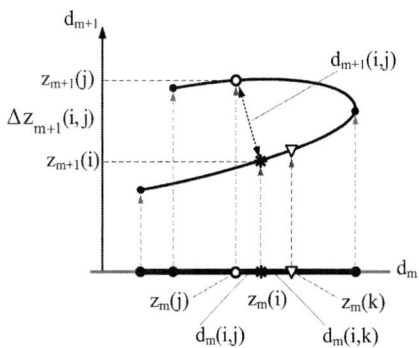

Fig. 4. Displacement between two false nearest neighbors in the unfolding process.

m to $m+1$ (note that in Fig. 4 $m = 1$). The neighboring point (◯) that is closely located to the reference point (✳) in $\mathbb{R}^m$ is noticeably displaced by the transition to $\mathbb{R}^{m+1}$ and hence is revealed as false neighbor. As a first step towards a quantitative measure of this effect, we form the ratio

$$R_{ij} = \frac{\Delta z_{m+1}(i,j)}{d_m(i,j)} . \qquad (6)$$

This process of expansion, ultimately leads to an order for which and also for other higher orders, only neighbors on the projected trajectory in the reduced space are true neighbors. When m is the adequate order, the count of false nearest neighbors drops to zero. After constructing a subspace of sufficient order in which an unfolded projected trajectory can be embedded, further increasing the order does not lead to revealing any new false nearest neighbors.

Corollary 1. *For nonlinear systems, the order m_0 is an optimally minimum reduction order if increasing the order of the reduced subspace to m_1, where $m_1 > m_0$ does not reveal any false nearest neighboring (FNN) points on the nonlinear reduced trajectory.*

The proof is omitted here due to the lack of space.

The objective of the proposed method is to determine this optimum dimension m_0 for the reduced subspace while preserving desired accuracy. The steps of the proposed algorithm are summarized in Algorithm-1.

Algorithm 1: Proposed Order Estimation Algorithm

Input: $\mathbf{X} \in \mathbb{R}^{n \times N}$ (data matrix from original trajectory)
output: Optimal minimum reduction order (m_0)

1 Using the POD algorithm, the projection matrix $\mathbf{Q}$ is formed with a small initial number (m) of orthonormal basis;
2 The time-series data from the projected trajectory is stored in the form $\mathbf{Z} = [\mathbf{z}(t_0), \dots, \mathbf{z}(t_N)] \in \mathbb{R}^{m \times N}$;
3 A set of close points $\mathbf{\Pi}_i$ to each point on the projected trajectory is found based on the following criteria $d_m(i,j) < R$, for $j = 1, \cdots, N$ and $j \neq i$, where R is a search radius;

4 The number of orthogonal basis is increased from m to $m + 1$ and the new dimension of the subspace is computed as $\mathbf{z}_{(m+1)}(\cdot) = \mathbf{q}_{(m+1)}^T \mathbf{x}(\cdot)$;

5 All the close points $\mathbf{z}(j) \in \Pi_i$ to $z(i)$ $i = 1, \ldots, N$, that satisfies the following ratio test are marked as false neighbors $R_{ij} = \dfrac{\Delta z_{m+1}(i, j)}{d_m(i, j)} > \rho_t$, where ρ_t is a pre-specified threshold value;

6 By repeating the steps (4)-(5), the projected trajectory is expanded into higher dimensions. Ultimately, at a particular order m_0, the count of false nearest neighbors in step (5) drops to zero, such that, further increasing the order does not lead to revealing any new false nearest neighbors and m_0 is selected as the minimum acceptable order for the reduced model.

IV. NUMERICAL EXAMPLES

In this example, the diode chain network shown in Fig. 5 is considered [8]. The circuit consists of 302 sections. A sample of representative input excitations $u(t)$ is shown in Fig. 5. Applying the proposed method, Fig. 6 shows the percentage

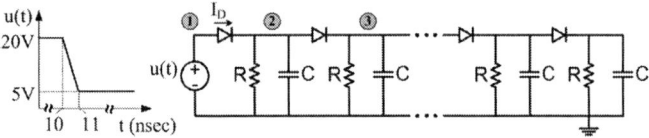

Fig. 5. Diode chain circuit.

of the false nearest neighbors on the projected trajectory, while the dimension of the model is changed from m to $m + 1$. As seen from Fig. 6, $m \geq 13$ completely unfolds the

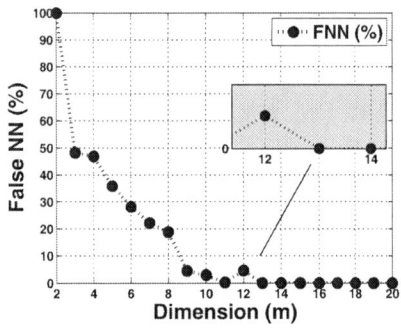

Fig. 6. The percentage of the false nearest neighbors on the projected trajectory

projected trajectory with no false neighbors. Hence, according to Corollary-1, $m = 13$ is selected as the optimum order.

To verify that m=13 provides an accurate reduced model, the original circuit is reduced using the POD algorithm with order 13. Comparison of the simulation results obtained from the original circuit and from the reduced circuit are shown in Fig. 7-(a), which shows an excellent agreement. Also, for the purpose of demonstrating the selected order is the optimum order, Fig. 7-(b) depicts that $m_0 = 13$ is the minimum order that ensures the accuracy of the reduced model, beyond which

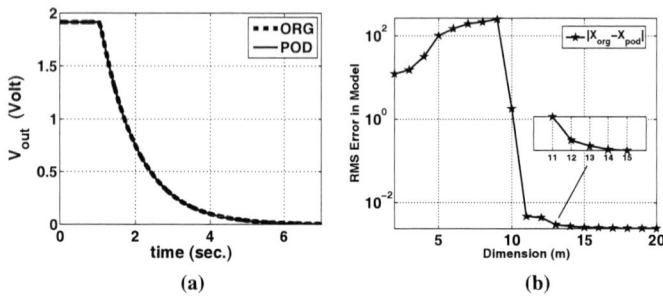

Fig. 7. (a) Comparison of the responses at node-27, (b) RMS error comparison between the response of different orders (m) of reduced models and the original circuit response.

the error does not change significantly. Also, for the purpose of validity demonstration in this paper, Fig. 7-(b) depicts that $m_0 = 13$ is the the minimum order to ensure the accuracy of the reduced model.

V. CONCLUSION

Estimating an optimal order for the reduced model is of crucial importance to ensure accurate and efficient transient behavior. In this paper, guided by geometrical considerations, a novel and efficient algorithm is presented to obtain the minimum sufficient order that ensures the accuracy and efficiency of the reduced nonlinear model. The proposed algorithm is based on evaluation of the number of false nearest neighbors (FNN) in an iterative process of increasing the order of a projection subspace. The method is not dependent on any specific nonlinear order reduction algorithm and can work in conjunction with any intended reduced modeling scheme such as: TPWL with a global reduced subspace, TBR, or POD, etc. The proposed algorithm is suitable for parallel implementation leading to even further reduction in its execution time.

REFERENCES

[1] S. X.-D. Tan and L. He, *Advanced Model Order Reduction Techniques in VLSI Design.* Cambridge, MA: Cambridge University Press, 2007.

[2] B. Nouri, M. Nakhla, and R. Achar, "A novel algorithm for optimum order estimation of reduced order macromodels," in *Proc. 15th IEEE Workshop on Signal and Power Integrity*, Naples, Italy, May 2011, pp. 33–36.

[3] B. Nouri, M. S. Nakhla, and R. Achar, "Optimum order estimation of reduced macromodels based on a geometric approach for projection-based mor methods," *IEEE T-CPMT*, pp. 1218–1227, May 2013.

[4] C.-W. Ho, A. Ruehli, and P. Brennan, "The modified nodal approach to network analysis," *IEEE T-CAS*, pp. 504–509, Jun. 1975.

[5] M. J. Rewieński and J. White, "A trajectory piecewise-linear approach to model order reduction and fast simulation of nonlinear circuits and micromachined devices," *IEEE T-CAD*, pp. 155–170, Feb. 2003.

[6] P. Astrid, "Reduction of process simulation models: a proper orthogonal decomposition approach," Ph.D. dissertation, Eindhoven University of Technology, Eindhoven, Netherlands, Nov. 2004.

[7] J. M. Aleida, "Balancing for nonlinear systems," Ph.D. dissertation, University of Twente, Enschede, the Netherlands, 1994.

[8] M. Rathinam and L. Petzold, "A new look at proper orthogonal decomposition," *SIAM J. Numer. Anal.*, vol. 41, no. 5, pp. 1893–1925, 2003.

[9] S. Smale, "On the mathematical foundation of electrical circuit theory," *Journal of Differential Geometry*, vol. 7, no. 1-2, pp. 193–210, 1972.

[10] L. Chua, "Nonlinear circuits," *IEEE T-CAS*, vol. 31, no. 1, pp. 69–87, 1984.

[11] H. K. Khalil, *Nonlinear Systems*, 3rd ed. New Jersey, NJ: Prentice Hall, 2002.

[12] R. N. Madan, Ed., *Chuas's Circuit: A Paradigm for Chaos.* Singapore: World Scientific, 1993.

Fixed-Order Parametric Macromodeling of Interconnects from S-parameter Data using Loewner Matrix based Method

Muhammad Kabir [1] and Roni Khazaka [2]

[1] [2] Department of Electrical and Computer Engineering, McGill University, Montréal, Québec, Canada

Email: [1] muhammad.kabir@mail.mcgill.ca and [2] roni.khazaka@mcgill.ca

Abstract—A Parametric macromodeling algorithm based on standard Loewner Matrix (LM) method was introduced recently for generating parametric time-domain macromodels based on S-parameter data. The method was shown to be efficient and accurate for systems with large number of ports and poles. However, the method is not suitable for the parametric problems where the parameter has a direct impact on the order *i.e.* the order changes considerably with the parameter. In this paper, we proposed an extension of this parametric algorithm in order to handle this kind of problems.

Index Terms—Parametric macromodel, Loewner Matrix, Matrix Format Tangential Interpolation, Orthonormal bases, Principal Component Analysis, Least square fitting.

I. INTRODUCTION

In microwave and packaging applications a closed-form time-domain macromodel of a structure is often difficult to obtain analytically. A common solution is to extract a macromodel from the measured/simulated S/Y-parameter data. A number of such algorithms have been proposed in the literature [1–4] to achieve this goal. Furthermore, these techniques have been extended to generate parametric macromodels, suitable for optimization and design space exploration [5–9]. One of the key challenges of these approaches is that a constant order is required for the macromodel. For example, in the parameterized Loewner Matrix method in [9] a constant order is maintained by truncating to the minimum required order over the parameter range. This results in an acceptable loss of accuracy in many cases, however for cases where the macromodel order is a strong function of the parameter, the loss of accuracy can be unacceptable. An example of such parameter is the length of a transmission line or the relative permittivity of the dielectric substrate. In this paper the algorithm in [9] is extended to handle such parameters for the case of interconnect structures by exploiting the relationship between the frequency bandwidth and macromodel order of a transmission line. The bandwidth is varied along with the parameter (e.g. transmission line length) in order to ensure that the macromodel order remains relatively stable over the range of interest while maintaining accuracy over the required bandwidth. As can be seen from the example, this approach provide a much better accuracy than simple order truncation for such parameters.

The authors would like to thank Natural Sciences and Engineering Research Council of Canada (NSERC) and the Regroupement Stratégique en Microsystéme du Québec (ReSMiQ) for supporting this project.

II. PROBLEM FORMULATION

Consider a p-port interconnect module $\mathcal{N}$. The objective of the algorithm described in this paper is to construct a parametric macromodel for the module from the measured/simulated parametric S-parameter data.

A. Parametric S-parameter Data

A parametric S-parameter data as a function of a parameter, α can be represented as follows:

$$\{(s_k, \alpha_l), \mathbf{S}(s_k, \alpha_l)\}; \quad k = 1, \ldots, n, \quad l = 1, \ldots, q \quad (1)$$

where n is the number of frequency samples at each value of the parameter and q is the number of parameter values considered within the prescribed range. $\mathbf{S}(s_k, \alpha_l) \in \mathbb{C}^{p \times p}$ is the S-parameter matrix at point (s_k, α_l).

B. Parametric Macromodel

The objective is to obtain a SPICE-compatible time-domain parametric macromodel of the module $\mathcal{N}$ that matches the frequency-domain data in (1):

$$s\mathbf{E}(\alpha)\dot{\mathbf{x}} = \mathbf{A}(\alpha)\mathbf{x} + \mathbf{B}(\alpha)\mathbf{u},$$
$$\mathbf{y} = \mathbf{C}(\alpha)\mathbf{x}. \quad (2)$$

where, $\mathbf{u} \in \mathbb{R}^p$ and $\mathbf{y} \in \mathbb{R}^p$ contain the vectors of input and output power waves respectively, $\mathbf{E}(\alpha), \mathbf{A}(\alpha) \in \mathbb{R}^{m \times m}$, $\mathbf{B}(\alpha) \in \mathbb{R}^{m \times p}$ and $\mathbf{C}(\alpha) \in \mathbb{R}^{p \times m}$ and m is the order of the system. In other words, the goal is to obtain the parametric model, $\{\mathbf{E}(\alpha), \mathbf{A}(\alpha), \mathbf{B}(\alpha), \mathbf{C}(\alpha)\}$.

III. LOEWNER MATRIX (LM) BASED PARAMETRIC MACROMODELING ALGORITHM

In this section, we will review the parametric macromodeling algorithm based on standard LM algorithm, proposed in [9]. In order to obtain a parametric macromodel, first a macromodel in the form shown in (2) for every $\alpha = \alpha_l$ is generated. This process must be repeated for every value of the parameter α in the data set, and as such the parameter α is omitted from some of the equations below for simplicity.

A. Splitting the Data

First the S-parameter data is appended with the complex conjugates at the negative frequencies, resulting in $2n$ data points, and then split into two sets as follows.

$$\{s_k, \mathbf{S}(s_k)\} \rightarrow \{\gamma_i, \mathbf{S}(\gamma_i)\}, \{\mu_j, \mathbf{S}(\mu_j)\} \quad (3)$$

There are a number of strategies for splitting the data and these are outlined in [2–4,10]. The technique, used in this work, is Matrix Format Tangential Interpolation (MFTI) [3,4].

978-1-4799-0708-3/13 $31.00 © 2013 IEEE

B. Loewner Matrices (LMs)

The next step is to construct the matrices $\mathbb{L}$, $\sigma\mathbb{L}$, $\mathbb{F}$ and $\mathbb{W}$ as follows [2–4,10]:

$$[\mathbb{L}_{j,i}] = \frac{\mathbf{S}(\mu_j) - \mathbf{S}(\gamma_i)}{\mu_j - \gamma_i}, \quad [\sigma\mathbb{L}_{j,i}] = \frac{\mu_j\mathbf{S}(\mu_j) - \gamma_i\mathbf{S}(\gamma_i)}{\mu_j - \gamma_i}. \quad (4)$$

where $j = 1, \ldots, n$, $i = 1, \ldots, n$ and $[\mathbb{L}_{j,i}]$, $[\sigma\mathbb{L}_{j,i}]$ represent the $(j,i)^{th}$ block entry of $\mathbb{L}$ and $\sigma\mathbb{L}$ respectively. Similarly the matrices $\mathbb{F}$ and $\mathbb{W}$ are constructed as follows:

$$\mathbb{F} = \left[\mathbf{S}(\mu_1)^T \ldots \mathbf{S}(\mu_n)^T\right]^T, \quad \mathbb{W} = \left[\mathbf{S}(\gamma_1) \ldots \mathbf{S}(\gamma_n)\right]. \quad (5)$$

Note that $[\mathbb{L}_{j,i}]$ and $[\sigma\mathbb{L}_{j,i}]$ are block matrices of size $p \times p$. The matrices as constructed in (4) and (5) are complex. The equivalent real Matrices $\{\mathbb{L}_r, \sigma\mathbb{L}_r, \mathbb{F}_r, \mathbb{W}_r\}$ can be computed using a similarity transformation [2,4].

The first two steps are repeated for every value of $\alpha = \alpha_l$. Note that these matrices $\mathbb{L}_r$, $\sigma\mathbb{L}_r$, $\mathbb{F}_r$ and $\mathbb{W}_r$ are continuous functions of the S-parameter matrices and in turn, the S-parameter matrices are continuous functions of α. As such, the LMs are continuous functions of α.

C. Extract Regular Part

By extracting the regular part of the LMs, the macromodel for each $\alpha = \alpha_l$, $\{\mathbf{E}, \mathbf{A}, \mathbf{B}, \mathbf{C}\}^l$ can be obtained. In order to extract the regular part a singular value decomposition (SVD) is performed on the LM pencil $(x\mathbb{L}_r^l - \sigma\mathbb{L}_r^l)$ [2,4,10]:

$$x\mathbb{L}_r^l - \sigma\mathbb{L}_r^l = \mathbf{\Lambda}^l \mathbf{\Sigma}^l (\mathbf{\Psi}^l)^* \quad (6)$$

where, $()^l$ denotes the value for $\alpha = \alpha_l$.

1) Determine Order of the Model for $\alpha = \alpha_l$: The first step of extracting the regular part is to determine the order. The order of each individual model ($\alpha = \alpha_l$), m_l is determined by examining the normalized singular values of the LM pencil $(x\mathbb{L}_r^l - \sigma\mathbb{L}_r^l)$ [2,10]. Note that these orders are not guaranteed to be the same.

2) Compute Orthonormal Bases of the Model for $\alpha = \alpha_l$: Once the order is determined, the next step is to compute the orthonormal bases $\mathbf{\Lambda}_R^l$ and $\mathbf{\Psi}_R^l$ which are the first m_l columns of the orthonormal matrices $\mathbf{\Lambda}^l$ and $\mathbf{\Psi}^l$ respectively.

3) Compute Common Orthonormal Bases: A common set of orthonormal bases is computed in this step that can be used for each $\alpha = \alpha_l$. This can be obtained by performing a Principal Component Analysis (PCA) [11] on the set of orthonormal bases. In order to do that the correlation matrices $\mathbf{G}$ and $\mathbf{H}$ are formed as follows:

$$\mathbf{G} = \left[\mathbf{\Lambda}_R^1 \ \mathbf{\Lambda}_R^2 \ \ldots \ \mathbf{\Lambda}_R^q\right], \quad \mathbf{H} = \left[\mathbf{\Psi}_R^1 \ \mathbf{\Psi}_R^2 \ \ldots \ \mathbf{\Psi}_R^q\right]. \quad (7)$$

where, $\mathbf{\Lambda}_R^i \in \mathbb{R}^{n \times m_i}$ and $\mathbf{\Psi}_R^i \in \mathbb{R}^{n \times m_i}$ represent the orthonormal bases for $\alpha = \alpha_i$. In order to obtain the orthonormal matrices spanning the dominant subspace (*i.e.* the principal components) an SVD is then performed on the matrices $\mathbf{G}$ and $\mathbf{H}$:

$$\mathbf{G} = \mathbf{X}\mathbf{\Sigma}_x \mathbf{V}_x^*, \quad \mathbf{H} = \mathbf{Y}\mathbf{\Sigma}_y \mathbf{V}_y^*. \quad (8)$$

where $\mathbf{X}$ and $\mathbf{Y}$ are orthonormal matrices spanning the dominant subspace of $\mathbf{G}$ and $\mathbf{H}$ respectively.

In order to make sure of a constant order the smallest of all the orders is chosen as the overall order m_a of the parametric macromodel.

$$m_a = min(m_1, m_2, \ldots, m_q) \quad (9)$$

The common orthonormal bases $\mathbf{X}_R$ and $\mathbf{Y}_R$ are then formed by taking the first m_a columns of $\mathbf{X}$ and $\mathbf{Y}$ respectively.

4) Extract Macromodels for each α: The macromodels are then extracted for each $\alpha = \alpha_l$ as follows.

$$\begin{aligned}\mathbf{E}^l &= -\mathbf{X}_R^* \mathbb{L}_r^l \mathbf{Y}_R, \quad \mathbf{A}^l = -\mathbf{X}_R^* \sigma\mathbb{L}_r^l \mathbf{Y}_R, \\ \mathbf{B}^l &= \mathbf{X}_R^* \mathbb{F}_r^l, \quad \mathbf{C}^l = \mathbb{W}_r^l \mathbf{Y}_R.\end{aligned} \quad (10)$$

The resulting macromodel would be continuous with respect to parameter α since the LMs are continuous and the same common bases are used to extract the macromodel.

D. Continuous Macromodel

Finally the discrete macromodels $\{\mathbf{E}^l, \mathbf{A}^l, \mathbf{B}^l, \mathbf{C}^l\}$ for $l = 1, \ldots, q$ are fitted into a continuous function of parameter α by using the least square approximation.

E. Inaccuracy Caused by Order Truncation

The discrete orders are not guaranteed to be the same in the stated LM based parametric macromodeling algorithm. A constant order is maintained by truncating to the minimum required order over the parameter range as shown in (9). For cases where the the parameter has a strong impact on the macromodel order, the loss of accuracy can be unacceptable. An example of such parameter is the length of a transmission line or the relative permittivity of the dielectric substrate.

IV. PROPOSED FIXED-ORDER PARAMETRIC MACROMODELING ALGORITHM

In this section, we propose a bandwidth adjustment scheme of the input parametric S-parameter data for which LM based parametric macromodeling algorithm generates discrete models with a stable order which are likely to parameterize.

A. Change of Variable in Parametric Data

Parameter α in the parametric data in (1) is redefined by another variable in this step.

$$\alpha_1 = \mathbf{r}_1\beta_0, \ldots, \alpha_i = \mathbf{r}_i\beta_0, \ldots, \alpha_q = \mathbf{r}_q\beta_0. \quad (11)$$

where, $\mathbf{r}$ is the new variable, $\mathbf{r}_1 = 1$ and $\mathbf{r}_i \geq 1$, $\beta_0 = \alpha_1$ is a constant. The parametric data can be redefined as follows:

$$\{(s_k, \mathbf{r}_l), \mathbf{S}(s_k, \mathbf{r}_l)\}; \quad l = 1, \ldots, q \quad (12)$$

where, q is the number of parameter values considered within the prescribed range.

978-1-4799-0708-3/13 $31.00 © 2013 IEEE

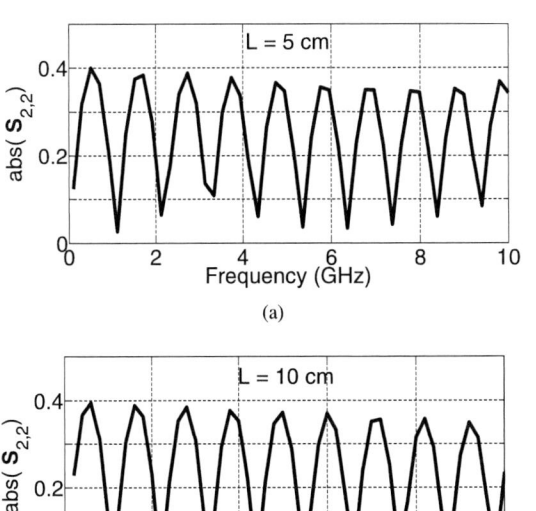

Fig. 1. (a) S-parameter data with L=5 cm, (b) S-parameter data with L=10 cm

B. Variable Frequency Bandwidth

In a typical transmission line structure modeled over a fixed frequency bandwidth (e.g. 0 to 5 GHz) the order of the required macromodel increases as the length of the line increases. For example a 10 cm line would require a macro-model approximately twice the size of that of a 5 cm line when both are matched over the same 0 to 5 GHz bandwidth. However, as can be seen from Fig. 1 a 10 cm transmission line modeled from 0 to 5 GHz would have approximately the same order macromodel as a 5 cm line modeled from 0 to 10 GHz. The strategy proposed in this paper is to increase the bandwidth as the line length is decreased such that the order of the macromodel remains relatively stable. In this case, any required order truncation will not result in large loss of accuracy. The frequency bandwidth is selected such that the minimum bandwidth corresponds to the required bandwidth of the macdromodel. A possible mapping for the bandwidth with the parameter, length of the line, is

$$f_i^m = \frac{\mathbf{r}_q \times f_0}{\mathbf{r}_i}; \quad i = 1, \ldots, q \tag{13}$$

where, 0 to f_0 GHz is the minimum bandwidth, required for the macdromodel, $\mathbf{r}_i$ is the new parameter defined in (12), $\mathbf{r}_q$ is the largest value of the parameter and f_i^m is the ith frequency bandwidth.

C. New Parametric S-parameter Data

With the variable frequency bandwidth the input parametric S-parameter data can be restated as follows:

$$\{(s_l^k, \mathbf{r}_l), \mathbf{S}(s_l^k, \mathbf{r}_l)\}; \quad k = 1, \ldots, n, \ l = 1, \ldots, q \tag{14}$$

TABLE I
SIMULATION RESULTS

	EX1	EX2
Data Points, q	11	11
Overall order, m_a	26	54
Polynomial order, h	6	6
CPU time (macromodeling) sec	0.83	1.3
CPU time (curve fitting) sec	0.2	0.7

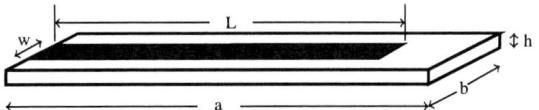

Fig. 2. Sample Micrstrip line structure with 1 line

where, s_l^k is the set of frequencies for $\mathbf{r} = \mathbf{r}_l$ and n is the number of frequency samples. $\mathbf{S}(s_l^k, \alpha_l) \in \mathbb{C}^{p \times p}$ is the S-parameter matrix at point $(s_l^k, \mathbf{r}_l)$.

Using the S-Parameter data defined above, the order of the macromodel remains approximately constant with respect to the parameter value. As such, any order truncation required in the LM based parametric macromodeling algorithm described in Sec. III will extract a parametric macromodel with acceptable accuracy.

V. SIMULATION RESULTS

In this section we present some interconnect examples to demonstrate the accuracy of our proposed approach. For all examples the minimum frequency bandwidth is 0 to 5 GHz. The frequency bandwidth is varied to the corresponding value and the S-parameter data spanning the varied bandwidth were obtained using HFSS, a commercial full-wave simulation tool.

A. Example Structures

Example 1 is a 2 port microstrip line structure with one line of width 0.2 cm fabricated on a substrate with thickness $h = 0.1$ cm, length, $a = 10$ cm, width, $b = 2$ cm and $\epsilon_r = 10.2$ shown in Fig. 2. Example 2 is a 4 port microstrip line with 2 lines of width 0.2 cm and 0.1 cm separation. The width of the substrate is $b = 2.3$ cm and all other dimensions/specifications are same as Example 1. The length of the lines, L for both examples are varied from 5 cm to 10 cm in order to obtain the parametric data with $\mathbf{r} = 1, \ldots, 2$. The S-parameter data was generated for 11 values of $\mathbf{r}$ between 1 and 2 spanning frequency bandwidth ranging from 5 to 10 GHz.

In order to illustrate the change in order with the length of a transmission line, a sample parametric S-parameter data from 0 to 5 GHz is shown in Fig. 3. As can be seen the response squeezes (i.e. the order increases) with the length of the transmission line.

B. Accuracy and Efficiency Check

As the length of the line increases from 5 cm to 10 cm, the order changes from 16 to 26 and 34 to 54 for Example 1 and 2 respectively. The usual order truncation will result in

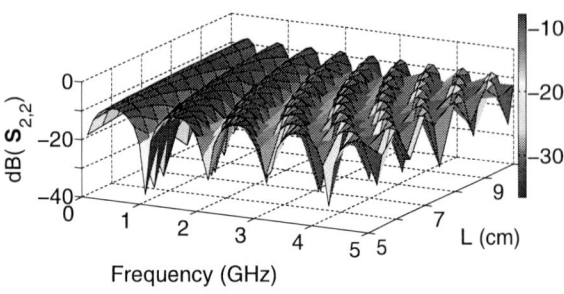

Fig. 3. Parametric S-parameter data for Ex 1

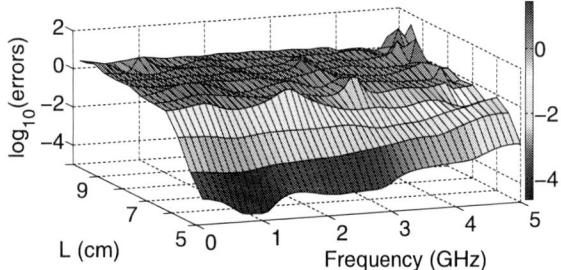

Fig. 4. Relative errors for Ex 1 using order truncation

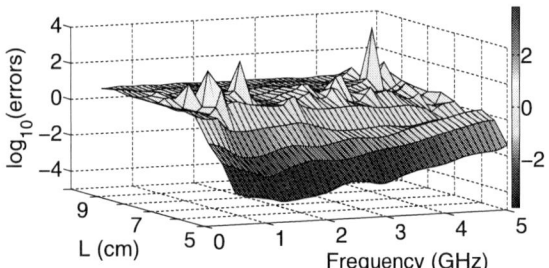

Fig. 5. Relative errors for Ex 2 using order truncation

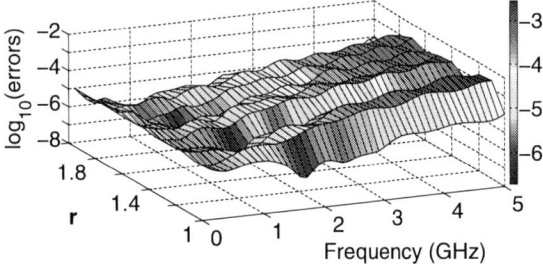

Fig. 6. Relative errors for Ex 1

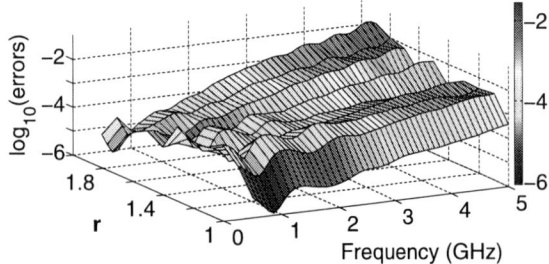

Fig. 7. Relative errors for Ex 2

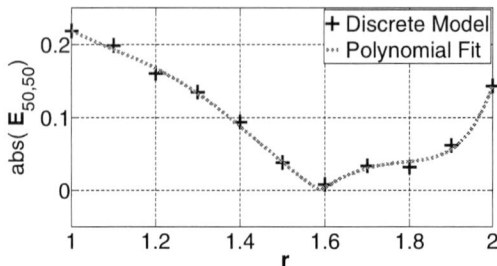

Fig. 8. Comparison between the discrete and continuous model for Ex 2

a large inaccuracy as shown in Fig. 4 and 5. On the other hand, the order remains relatively constant (Example 1: 26 to 28, Example 2: 54) for the proposed parametric data which results in parametric macromodel with an acceptable accuracy (10^{-2} to 10^{-6}). The relative error plots are show in Fig. 6 and 7. The simulation results are shown in Table I. As we can see from the simulation results, the proposed approach can achieve the parametric macromodels efficiently. A sample comparison between the discrete and the fitted model is shown in 8. As can be seen, the discrete models are very suitable for generating a parametric macromodel using a curve-fitting algorithm.

VI. CONCLUSION

In this paper, a fixed-order parametric macromodeling algorithm based on standard Loewner Matrix method was proposed. The method is suitable for the interconnects with varying order and was shown to be accurate and efficient.

REFERENCES

[1] B. Gustavsen and A. Semlyen, "Rational approximation of frequency domain responses by vector fitting," *IEEE Trans. Power Del.*, vol. 14, no. 3, pp. 1052–1061, Jul. 1999.
[2] S. Lefteriu and A. C. Antoulas, "A new approach to modeling multiport systems from frequency-domain data," *IEEE Trans. Comput.-Aided Design Integr. Circuits Syst.*, vol. 29, no. 1, pp. 14–27, Jan. 2010.
[3] Y. Wang, C. Lei, G. Pang, , and N. Wong, "MFTI: Matrix-format tangential interpolation for modeling multi-port systems," in *Proceedings IEEE/ACM Design Automation Conference (DAC'10)*, Anaheim, CA, 2010, pp. 683–686.
[4] M. Kabir and R. Khazaka, "Macromodeling of distributed networks from frequency-domain data using the loewner matrix approach," *IEEE Trans. Microw. Theory Tech.*, vol. 60, no. 12, pp. 3927–3938, 2012.
[5] F. Ferranti, L. Knockaert, and T. Dhaene, "Guaranteed passive parameterized admittance-based macromodeling," *IEEE Trans. Adv. Packag.*, vol. 33, no. 3, pp. 623–629, 2010.
[6] P. Triverio, S. Grivet-Talocia, and M. Nakhla, "A parameterized macromodeling strategy with uniform stability test," *IEEE Trans. Adv. Packag.*, vol. 32, no. 1, pp. 205–215, 2009.
[7] D. Deschrijver, T. Dhaene, and D. De Zutter, "Robust parametric macromodeling using multivariate orthonormal vector fitting," *IEEE Trans. Microw. Theory Tech.*, vol. 56, no. 7, pp. 1661–1667, 2008.
[8] S. Lefteriu, A. Antoulas, and A. C. Ionita, "Parametric model reduction in the loewner framework," in *International Federation of Automatic Control (IFAC) 18th World Congress*, 2011, pp. 12 751–12 756.
[9] M. Kabir and R. Khazaka, "Parametric macromodeling of high-speed modules from frequency-domain data using loewner matrix based method," in *Microwave Symposium Digest, 2012 IEEE MTT-S International*, 2013.
[10] A. J. Mayo and A. C. Antoulas, "A framework for the solution of the generalized realization problem," *Linear Algebra and its Application*, vol. 425, no. 2-3, pp. 634–662, Sep. 2007.
[11] I. T. Jolliffel, *Principal Component Analysis*. New York: Springer-Verlag, 2002.

High Speed Links I

978-1-4799-0708-3/13 $31.00 © 2013 IEEE

978-1-4799-0708-3/13 $31.00 © 2013 IEEE

On-die Supply-inducecd Jitter Behavioral Modeling

Xiaoqing Wang

Intel Corporation
Santa Clara, CA
Xiaoqing.j.wang@intel.com

Aaron Martin

Intel Corporation
Folsom, CA
Aaron.martin@intel.com

Abstract— **This paper describes a simple, yet efficient supply-induced jitter modeling methodology for high-speed I/O circuits. The proposed model uses the average supply noise and a linear factor derived from Spice simulations to estimate the jitter for a circuit block. For circuits with bias voltages, the transfer function of the biasing network is included. The model is implemented in Simulink and closely correlated with Spice simulations. The modeling accuracy is further validated to be within ±15% of the silicon measurement for the period jitter of a ring oscillator. The jitter modeling technique is applied to multiple memory I/O designs and can be extended to other high-speed interface designs and their timing budgeting.**

Keywords— *timing jitter; power supply noise; high-speed integrated circuits;*

I. INTRODUCTION

The deviations of the zero crossing of the waveform from their ideal points in time domain, often characterized as timing jitter, cause the uncertainties of data recovery in high-performance link designs. With the increasing data rate and design complexity, the jitter of I/O circuit dramatically degrades link quality and ultimately limits bandwidth in many systems. Therefore, the accurate estimation of timing jitter at the interface level becomes critical to a quality link design. While the timing uncertainty in I/O circuit can come from various other sources, such as device noise, duty cycle distortion, supply variation-induced jitter usually dominates in modern single-ended I/O design. That is why this work mainly focuses on power supply noise induced jitter for I/O circuits.

Most of existing publications on jitter analysis focus on phase noise and jitter in ring oscillators and PLLs [1]-[9]. Common approaches include the impulse sensitivity function-based modeling technique [2]-[7], closed-form equations by the autocorrelation function [8], [9], and transistor-level circuit analysis [10], [11]. These reported modeling techniques are either very specific to a certain type of circuit, or relying heavily on multiple technology-dependent device-level parameters, making the modeling for large circuits inefficient. In [12], [13], it reports a frequency-domain approach that relies on supply noise spectrum and frequency-dependent supply noise jitter sensitivity, of which characterization requires extensive spice-level simulations including all I/O circuits.

Our proposed modeling method uses an alpha factor that captures the linear transfer function from the supply noise to delay variation. This alpha factor is derived from transistor-level Spice simulation by applying a DC offset or a low-frequency AC noise to the nominal supply. Afterwards, both

the alpha factor and the nominal delay of the circuit block are used to build the jitter model. For AC supply noise, the average supply variation is calculated and used. For circuits that require bias voltages, the AC transfer function of supply noise to the bias voltage needs to be characterized and included in the model to capture its impact on the circuit delay. Once the individual blocks are accurately modeled, they can be cascaded together in Simulink to construct a full-interface model. The direct simulation result is the time-variant jitter for each signal. Based on that output, other types of jitter can be easily derived, such as N-cycle jitter, as well as the relative jitter between data and clock. Here the supply noise refers to both power supply and ground variations, although the following analysis mainly focuses on power supply noise for the purpose of illustration.

The rest of the paper is organized as follows: it starts with the introduction of the alpha model, followed by the detailed models of an inverter chain and a ring oscillator as well as their correlations with Spice simulations. Finally, the silicon correlation is presented for the ring oscillator.

II. ALPHA MODEL

A. Model Definition

To enable jitter behavioral simulation for large circuits and interface-level applications, a generic circuit model suitable for various circuit topologies is desired. It is also necessary to use a simplified model with reasonable approximation to achieve a good balance between the speed and the accuracy. Inverters are the most common building blocks for any IO circuits. The detailed analysis of the propagation delay of inverters can be found in [14], [15] for deep sub-micrometer process. Based on these studies, to the first order, the delay change of an inverter chain can be approximated as:

$$\Delta t_d = K_d \cdot \Delta V_{dd} \qquad (1)$$

In (1), Δt_d is the change of delay resulting from a small perturbation to the supply voltage, ΔV_{dd}. K_d is the static coefficient that is related to device size and technology. This equation signifies the delay change is linearly dependent on the small supply variation for typical inverter designs with fast signal transitions. Such approximation is valid as long as the supply noise is small and the nonlinear effects of the transistors do not dominate.

The alpha factor is defined as the ratio of the normalized delay and V_{dd} variation, as shown in (2). It is convenient since in most systems, the supply noise is usually characterized with

978-1-4799-0708-3/13 $31.00 © 2013 IEEE

respect to the nominal supply voltage. For any circuit block, its alpha factor can be derived by measuring its delay separately for supply voltages with and without a small DC variation in the spice-level simulation. Once it is calculated, given any supply variations, the jitter can be computed by (2).

$$\alpha = \frac{\Delta t_d / t_d}{\Delta V_{dd} / V_{dd}} ; \Delta t_d = \alpha \cdot \frac{\Delta V_{dd}}{V_{dd}} \cdot t_d \qquad (2)$$

In (2), ΔV_{dd} refers to a DC offset to the nominal supply voltage. In case of AC supply noise, an alternative way to derive the alpha factor is to use a sinusoidal noise at a low frequency (noise period $\gg t_d$) for Spice simulation where the peak-to-peak values are used for ΔV_{dd} and Δt_d in (2). The resulting alpha factor should be very close to the value computed using a DC supply offset. Once the alpha factor is available, the average value of supply noise over the delay period can be used and the resulting jitter can be expressed as:

$$\Delta t_d(t) = \alpha \cdot \int_{t'=t}^{t+t_d} \frac{\Delta V_{dd}(t)}{V_{dd}} \, dt' \qquad (3)$$

B. Model Implemenrations in Matlab

The jitter model is implemented by Simulink tools in Matlab software due to its graphic view and easy scalability. In Simulink, any delay block is represented by an alpha factor and its nominal delay (t_d). Fig. 1 shows the detailed computation inside, essentially the implementation of (3). To simplify the computation, the power supply is pre-integrated so that the integration in (3) can be simply realized by delay and subtraction functions. The *in* and *out* are the input and output jitter in units of second and they both vary with time.

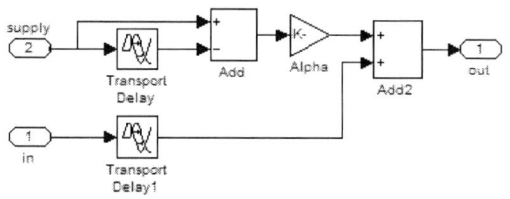

Fig. 1. The diagram shows the detailed jitter computation inside the model.

III. MODEL CORRELATION WITH SPICE SIMULATION

A. Circuit Without Biasing Voltages

To verify the modeling accuracy, an inverter chain with a 2GHz input clock is used as an example. An alpha factor of 1.3 and a nominal delay of 440ps are derived from Spice simulation. A sinusoidal noise with amplitude of 5% of nominal supply voltage is applied while its frequency is swept.

Fig. 2 plots the peak-peak absolute jitter for both Spice simulations and Simulink modeling. Both curves show low-pass characteristics and a good agreement is observed for the noise frequency up to 2GHz. The lower jitter at high frequencies is due to the AC averaging effect of the supply noise. A large difference is observed for some higher frequencies, primarily due to the non-linear effect starting to

dominate. Luckily, such inaccuracy is mitigated by the low jitter sensitivity and the low magnitude of the noise at high frequencies in many practical systems.

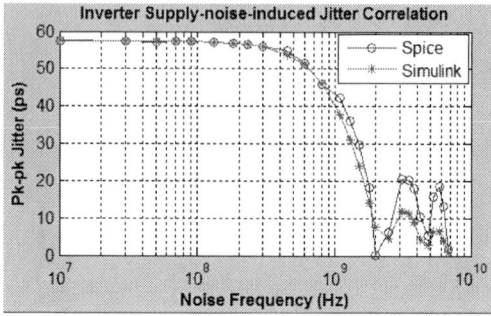

Fig. 2. The jitter spectrum shows a good correlation between simulation and Simulink modeling for an inverter chain with a 5% sinusoidal supply noise.

B. Circuits Requiring Biasing Voltages

To expand the application of the alpha model to the circuits that require biasing voltages, a ring oscillator consisting of current-starved CMOS inverters is studied. As shown in Fig. 3(A), the ring oscillator has eight stages. Each delay cell has two identical inverters with its schematics shown in Fig. 3(B). Under the nominal supply, the oscillator is running at 840MHz and the period jitter of the clock output is measured.

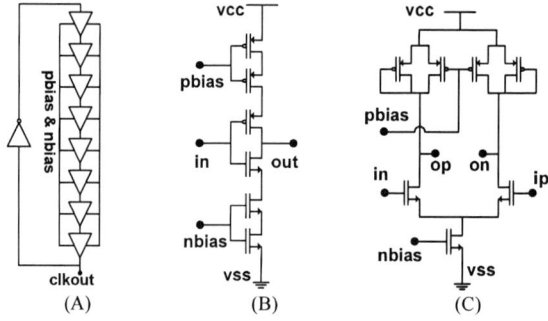

Fig. 3. (A) The ring oscillator has eight delay cells with each having two inverters. The schematics of a current-starved inverter (B) and a differential buffer (C) show two biasing voltages used for the delay control.

The modeling takes two steps. First, constant biasing voltages are assumed for *pbias* and *nbias* to isolate their impacts on delay. The ring oscillator is modeled as a regular delay block with a delay equivalent to its clock period, as shown in part C in Fig. 4. The second step is to include the impact of the coupling noise from the supply to the biasing voltages. Given the small magnitude of supply noise, the coupling noise is estimated using the transfer function, which is approximated by pole-zero systems in Simulink, as shown in part A & B in Fig. 4. The following integration, delay and subtraction function blocks are computing the average values of bias voltages over the delay period. The resulting jitter is the multiplication of the average value and the gain of the bias voltage. Here the supply output *raw_out* is the actual supply noise without the integration.

978-1-4799-0708-3/13 $31.00 © 2013 IEEE   148

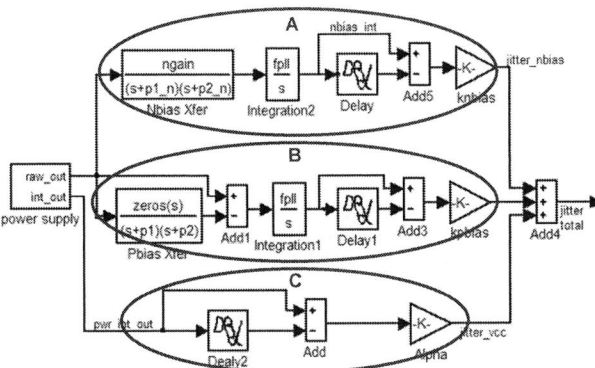

Fig. 4. The Simulink model of a ring oscillator includes the impacts of *pbias* and *nbias* voltages in addition to the basic alpha model for supply-induced jitter estimation.

Fig. 5 shows a good correlation of period jitter between the model and Spice simulation results for the ring oscillator. The difference between with and without the biasing voltages demonstrates that the modeling accuracy is greatly improved by including the bias network transfer function. The noise AC average effects also show up at high frequencies.

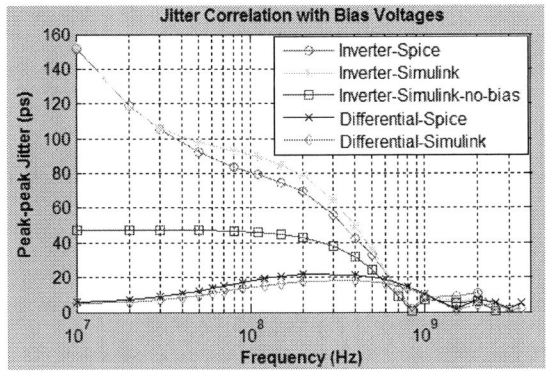

Fig. 5. The jitter spectrum shows good correlations between Spice and Simulink simulation for the period jitter for an inverter-based ring oscillator and the absolute jitter for a differential buffer-based delay line.

To prove the applicability of the proposed modeling method to differential circuits, the jitter correlation is performed for an eight-stage delay line with the differential delay cell as shown in Fig. 3(C). A close match of the absolute jitter is observed, as illustrated by the black and magenta curves in Fig. 5.

The examples listed above do not involve modeling the loop dynamics for a PLL or DLL. They can be easily added in Simulink as part of their complete behavioral model. For simplicity, their implementations are omitted here.

Here, we used an inverter chain and a ring oscillator as examples to illustrate how the alpha model is implemented. With the same methodology and principle, the alpha model can be applied to many other circuits. Afterwards, a model of full datum or clock path can be built by cascading individual circuit blocks. The jitter simulation can be carried out in Matlab at a much faster speed than the Spice simulation due to its simplicity of computation.

IV. MODEL CORRELATION WITH SILICON MEASUREMENT

A. Laboratory Setup

To further validate the accuracy of the proposed modeling technique, a ring oscillator described above is used from a testchip for silicon correlation. This testchip is for CPU product that includes many cores and high-speed IO interfaces, such as PCI-E, DDR, etc. Fig. 6 shows the high-level block diagram of the circuit under test from DDR IO function block.

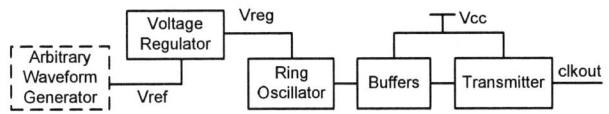

Fig. 6. The block diagram of the circuit under test shows the setup that enables introducing the AC noise to the ring oscillator power supply by varying the input reference voltage of the on-die voltage regulator.

The ring oscillator under test operates under an on-die linear voltage regulator that uses an off-chip reference voltage *vref*. Therefore, the supply noise for the oscillator can be introduced by varying the reference voltage in a controlled manner. The clock output of the oscillator goes to a transmitter which operates under a separate power supply. An arbitrary waveform generator is used to provide both DC and AC voltages to the reference voltage. The on-die supply *Vreg* is measured directly from silicon using pico-probes after Focused Ion Beam (FIB) process. The pico-probe is connected to an oscilloscope and pre-calibrated across a wide range of frequencies. The waveforms of the clock output and *Vreg* are captured simultaneously and saved for each measurement.

B. Measurement and Jitter Extraction

For the measurement in an actual system, it is impossible to de-embed the individual jitter contributions from all sources. To focus on the deterministic jitter from the added sinusoidal supply noise, jitter measurements are performed with and without such sinusoidal noise and its jitter contribution is evaluated by taking the difference between two measurements. This approach can be justified by the fact that the sinusoidal noise is uncorrelated with other jitter sources in the system. In addition, the same method is used between the measurement and the Simulink simulations. The jitter numbers shown below are the deterministic jitter after subtraction.

TABLE I
THE SINUSOIDAL OUTPUT SETTING OF ARBITRARY WAVEFORM GENERATOR

freq. (MHz)	10	10	20	20	50	50	80	80
Mag. (mv)	1	11	1	21	1	41	1	81

Table I shows the frequency and magnitude of the AC component of arbitrary waveform generator with a DC output of 0.7V. The minimum 1mV is used due to the limitation of the equipment. The magnitude at each frequency is chosen to ensure a 5-10mV AC noise at *Vreg*. Given the bandwidth of the voltage regulator, the frequency chosen is to avoid unrealistically large amplitude required from the generator. With a sampling rate of 40Gbps, a total of 1 million data points are captured for each channel, which includes approximately 20,000 clock cycles with a clock period of 1.2ns.

978-1-4799-0708-3/13 $31.00 © 2013 IEEE

The saved clock output waveform is loaded in Matlab and the period jitter is extracted. As shown in Fig. 7, the *N*-period peak-to-peak jitter increase linearly at each noise frequency when *N* goes up, which indicates the deterministic nature of the jitter source. It also can be observed that at higher noise frequency, the slope of the jitter increase is less steep, signaling the AC averaging effects starting to show up as *n* increases. The overlapping of the plots between 10 MHz and 20 MHz is coincidence due to the different amplitude of the AC noise at *Vreg* between two. The same applies to 50MHz and 80MHz.

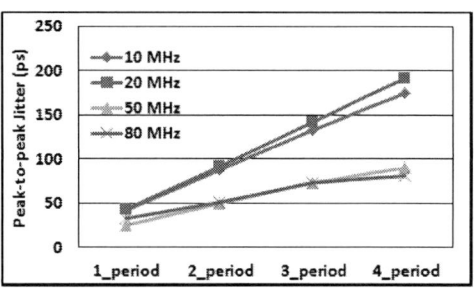

Fig. 7. N-period jitter from measurements shows a linear increase of jitter when the number of period increases at each supply noise frequency.

C. Modeling and Correlations

The AC noise of the measured *vreg* output is extracted and used for the supply noise in the Simulink model in Fig. 4. After simulation, the one-period jitter is imported back to Matlab where the *n*-period jitter is calculated.

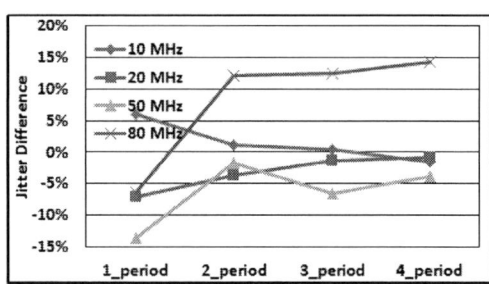

Fig. 8. The jitter differences between modeling and measured data are within ±15% range (normalized with respect to measured values).

Fig. 8 shows the difference between the simulated and the measured jitter after normalized with respect to measurement values. As shown, the simulated jitter numbers are closely correlated to the measured values, within ±15% difference.

V. SUMMARY

The proposed model carries a good balance among simplicity, accuracy and efficiency for supply-induced jitter behavioral modeling. The linear model yields the most accurate results for small supply variation (<10%) at low frequencies (below GHz). When the noise frequency goes higher, the model inaccuracy tends to rise for some circuits due to the increasing non-linear effects. In practical systems, sensitivity to non-linear effects is low due to the low magnitude of the noise and the low sensitivity of jitter at the high frequencies. The use of the frequency-independent alpha factor greatly simplifies the circuit characterization effort and computation required for the modeling, allowing fast simulation for interface-level I/O circuits. The Spice correlation work presented here focused on simple structures such as inverters and differential buffers. However, this technique has been successfully applied to many other circuit topologies, such as CML-CMOS converters, duty-cycle correction circuits and clock generators in memory IO designs. Further improvement includes the addition of other jitter sources and chip-to-chip interconnect effects.

ACKNOWLEDGMENT

The authors would like to gratefully acknowledge the help of the following individuals to this work: Zuoding Wang, Ravindran Mohanavelu and Dawson Kesling.

REFERENCES

[1] A. Demir, A. Mehrota, and J. Roychowdhury, "Phase noise in oscillator: A unifying theory and numerical methods for characterization", *IEEE Trans. On Circuits and Systems – I: Fundamental Theory and Applications*, vol. 47, no. 5, pp. 655-674, May. 2000.

[2] A. Hajimiri, and T. H. Lee, "A general theory of phase noise in electrical oscillators", *IEEE J. Solid-State Circuits*, vol. 33, no. 2, pp. 179-194, Feb. 1998.

[3] A. Hajimiri, S. Limotyrakis, and T. Lee, "Jitter and phase noise in ring oscillators", *IEEE J. Solid-State Circuits*, vol. 34, no. 6, pp. 790-804, Jun. 1999.

[4] N. Barton, D. Ozis, T. Fiez and K. Mayaram, "The effect of supply and substrate noise on jitter in ring oscillator", *Proc. IEEE Custom Integrated Circuits Conf. (CICC)*, pp. 505-508, May. 2002.

[5] M. Ierssel, and H. Yamaguchi, A. Sheikholeslami, H. Tamura, and W. Walker, "Event-driven Modeling of CDR Jitter Induced by power-supply noise, finite decision-circuit bandwidth, and channel ISI", *IEEE Trans. On Circuits and Systems – I: Regular Papers*, vol. 55, no. 5, pp. 1306-1315, Jun. 2008.

[6] J. Kim, Y. Lu, and R. Dutton, "Modeling and simulation of jitter in phase-locked loops due to substrate noise", *Proc. IEEE Int. Behav. Model. Simul. Workshop (BMAS)*, pp. 25-30, Sept. 2005.

[7] X. Lai, and J. Roychowdhury, "Fast, accurate prediction of PLL jitter induced by power grid noise", *Proc. IEEE Custom Integrated Circuits Conf. (CICC)*, pp. 121-124, Oct. 2004.

[8] F. Herzel, and B. Razavi, "A study of oscillator jitter due to supply and substrate noise", *IEEE Trans. On Circuits and Systems – II: Analog and Signal Processing*, vol. 46, no. 1, pp. 56-62, Jan. 1999.

[9] P. Heydari, and M. Pedram, "Analysis of jitter due to power-supply noise in Phase-locked loops", *Proc. IEEE Custom Integrated Circuits Conf. (CICC)*, pp. 443-446, May. 2000.

[10] A. Strak and H. Tenhunen, "Analysis of timing jitter in inverters induced by power-supply noise", *Int. Conf. Design and Test of Integrated Systems in Nanoscale Technology (DTIS)*, pp. 53-56, Sept. 2006.

[11] F. Yuan, "Power sensitivity of low-voltage CMOS current-mode circuits", *Canadian Conf. Electrical and Computer Engineering*, pp. 1741-1744, May. 2004.

[12] H. Lan, R. Schmitt and C. Yuan, "Prediction and measurement of supply noise induced jitter in high-speed IO interfaces", *DesignCon*, Feb. 2009.

[13] H. Lan, M. Han and R. Schmitt, "Modeling and measurement of supply noise induced jitter in a 12.8Gbps single-ended memory interface," *EPEP'12*, pp. 43-46, Oct. 2012.

[14] T. Sakurai and A. R. Newton, "Alpha-power law MOSFET model and its applications to CMOS inverter delay and other formulas", *IEEE J. Solid-State Circuits*, vol. 25, no. 2, pp. 584–594, Apr. 1990.

[15] L. H. Chen, M. Marek-Sadowska, and F. Brewer, "Buffer delay change in the presence of power and ground noise", *IEEE Trans. VLSI Syst.*, vol. 11, no. 3, pp. 461–473, June 2003.

978-1-4799-0708-3/13 $31.00 © 2013 IEEE

Mitigating the Impact of Sinusoidal Jitter and Duty Cycle Distortion on Random Jitter estimation by Tailfit Algorithm

Nitin Kumar Chhabra
PIMDS(TRnD)
STMicroelectronics Pvt. Ltd(India)
nitin.chhabra@st.com

Kushagra Bhatheja
EEE Department
BITS Pilani(India)
kushagrabhatheja@gmail.com

Jai Narayan Tripathi, Rajkumar Nagpal,Rakesh Malik
PIMDS(TRnD)
STMicroelectronics Pvt. Ltd(India)
{jainarayan.tripathi,rajkumar.nagpal,rakesh.malik}@st.com

Abstract —Random jitter (RJ) estimation based on Tail Fit algorithm are generally inaccurate in presence of deterministic components like the presence of sinusoidal jitter (SJ) and duty cycle distortion (DCD). Addition of deterministic jitter changes the standard deviation of the tail region of resulting jitter probability density function. A new methodology for random jitter estimation in presence of sinusoidal jitter and duty cycle distortion is presented. The method involves calculation of mathematical correction factor, which are derived and used to calculate the precise value of random jitter. The proposed methodology is validated by inducing known jitter sources in real channel simulator of ADS from Agilent.

Keywords – Tailfit, Random Jitter Estimation , Sinusoidal jitter , Duty cycle distortion .

I. INTRODUCTION

In data transmission, jitter is defined as the deviation of a data/clock transition edge from the ideal position. As the operating frequency of electronic systems is increasing the length of the bit interval continues to decrease. This reduces the time budget that is available for jitter. This has resulted in lots of design efforts to be directed towards containing jitter. In order to reduce jitter, the causes of jitter in a circuit need to be identified. It is here that decomposing jitter into its components proves useful. Another advantage of jitter segregation is that it can help to reduce the amount of time required to test a link for its bit error rate. Precise random jitter estimation is necessary as its impact in total jitter is very significant owning to multiplication factor coming from QBER factor which is BER dependent.

II. PRIOR ART AND LIMITATIONS

A. Dual Dirac Model

The dual Dirac model provides a quick estimation of the amount of jitter present for a given BER [1]. It is based on certain assumptions that are as mentioned below:

- Total Jitter (TJ) is composed of Deterministic Jitter (DJ) and Random Jitter (RJ).

- DJ has a bounded distribution.

- RJ has a Gaussian distribution.

- DJ distribution is composed of two Delta Dirac functions and the separation between these gives DJ.

- Jitter is stationary phenomenon.

B. Tail Fitting Algorithm

Since DJ is bounded, Tail portion of total pdf still represents the RJ distribution. Thus jitter segregation could be done through tail fitting [2].

C. Limitations of Dual Dirac for DJ measurements

It should be noted that there is a difference between Dual-Dirac DJ($\delta\delta$) and DJ(pk-pk). In fact,

$$DJ(\delta\delta) <= DJ(pk\text{-}pk) \qquad (1)$$

D. Limitations of Dual Dirac for RJ measurements

The application of Dual –Dirac approximation is based on the fundamental assumption that any deterministic jitter (DJ) distribution can be approximated by two delta functions separated by DJ($\delta\delta$). A jitter distribution that closely follows a dual-Dirac could result from pure duty cycle distortion (DCD) or square wave modulated jitter but in practical applications it is much more complicated. Figure 1 shows the convolution of a pdf formed by Sinusoidal Deterministic Jitter (DJ) with a Gaussian distribution, which represents the Random Jitter (RJ), to form the pdf of total jitter (TJ). The above figure clearly shows the deviation of the tail region of TJ pdf from the tail region of the Random Jitter Gaussian pdf.

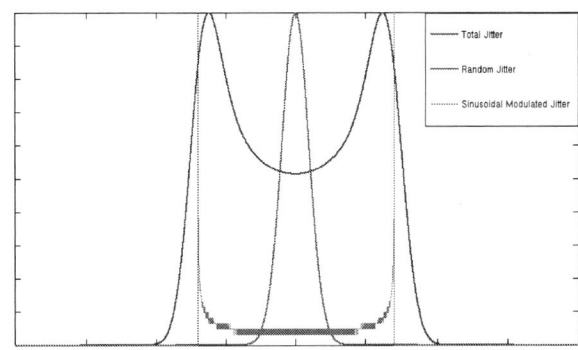

Figure 1 : Convolution of Gaussian with pdf of sinusoidal jitter

III. PROPOSED METHODOLOGY

In order to mitigate the impact of the sinusoidal jitter and duty cycle distortion on random jitter estimation by tail fit algorithm a new methodology is proposed.

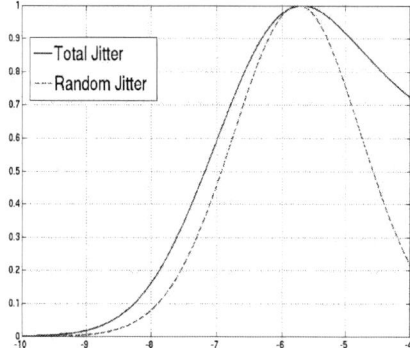

Figure 2 : Error introduced due to convolution

Convolution of random jitter probability density function (pdf) with the pdf of deterministic jitter gives the pdf of total jitter. Figure 2 clearly shows that there is a difference in the standard deviation of tail region of total jitter and that of original random jitter. The difference is due to the process of convolution which involves the overlap of one curve over the other. The methodology proposed in this paper computes this difference for various ratios of deterministic jitter to the values of random jitter and makes a lookup table for correction factor. This lookup table is then applied to the real scenarios to get the precise values of random jitter estimation.

Figure 3 gives the flow of the proposed methodology. Mathematical models of unit random jitter and deterministic jitter (sinusoidal jitter/duty cycle distortion) is taken and convolved to get the total jitter probability density function. Then initial guess of mean is computed and standard tail fit algorithm [2][4] is applied to get estimation of random jitter value. Error in random jitter estimation is computed and stored. Same method is repeated by varying the deterministic jitter and keeping the random jitter constant, to make a complete lookup table. Now this lookup table is basically representing the correction factor, which is use for precise estimation of random jitter in the real scenario. Detailed description of each block is as explained as below:

A. SJ/DCD Model

For modeling the sinusoidal jitter (SJ), a sinusoidal wave is taken with unit frequency and varying amplitude from 0 to 10 units in step of 0.1.The amplitude corresponds to half the value of Deterministic jitter peak to peak (DJpk-pk).

In order to model a duty cycle distortion (DCD), a vector with alternate positive and negative amplitude is taken. The amplitude is varied from 0 to 10 units in step of 0.1. The

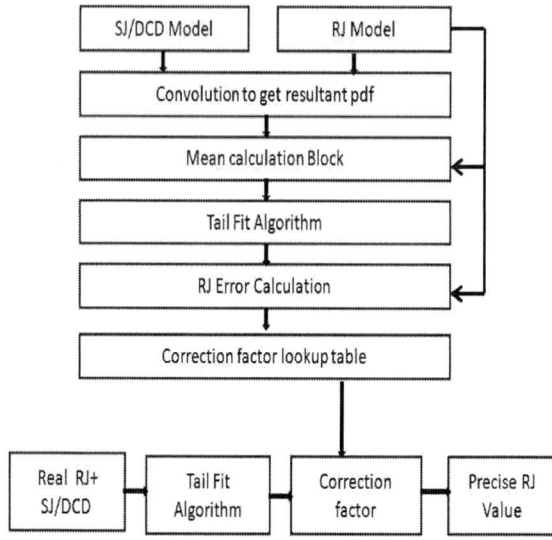

Figure 3: Proposed Methodology

amplitude corresponds to half the value of Deterministic jitter peak to peak (DJpk-pk).

B. RJ Model

Random jitter is modeled with a pure Gaussian curve with zero mean and unit standard deviation. The curve is normalized and the distribution is taken from -10 to +10 scale (+/- 10 sigma).

C. Convolution Block

The histogram of this deterministic jitter is made with the bins centered at same scale as that of Gaussian scale (i.e. from -10 to +10 scale). The two curves, Gaussian and Histogram of deterministic jitter value, represent the probability density function (pdf) of random jitter (RJ) and deterministic jitter (DJ) respectively. The convolution of these two pdf will result in the pdf of total jitter.

D. Mean Computation Block

Now tail region of the total jitter probability density function (1*sigma to 3*sigma) is taken and is matched with the tail of Gaussian curve (with unit sigma) by varying the mean to get the best match. The chi-squared test[6][2][4] is used to measure the degree of fitness between the Gaussian curve and the tail region of total jitter pdf. This comparison is done using the chi-square function which is defined as:

$$X^2 = \sum_{i=1}^{k} \frac{(x_i - E_i)^2}{E_i}$$

Here k is number of bins, x_i is the TJ pdf's y-axis value, whereas E_i is the y-axis value of the Gaussian curve. The Gaussian curve is created using the same x-axis data as the TJ pdf, allowing exact one to one comparison of the values.

The mean and chi-square value is stored. The same process is repeated for varying the amplitude of sinusoidal jitter/duty cycle distortion and the trend is noticed. Figure 3 give the variation of these chi-square values with respect to the magnitude of sinusoidal jitter amplitude and duty cycle distortion.

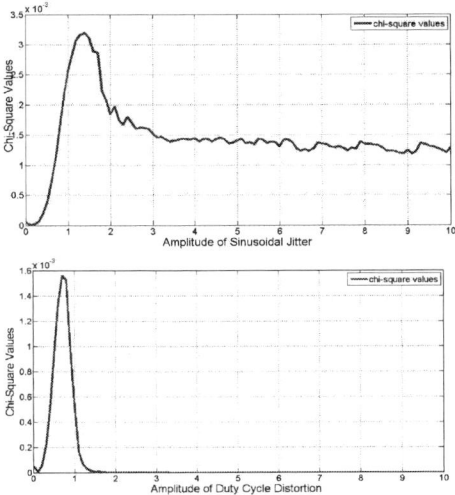

Figure 4: Variation of chi-square function with varying sinusoidal jitter and duty cycle distortion

E. Tail fit Algorithm

The chi-square value is relative, so in order to get the value of absolute error percentage, unit sigma and mean obtained from mean computation block is used as initial guess for the tail fitting algorithm[2][4].

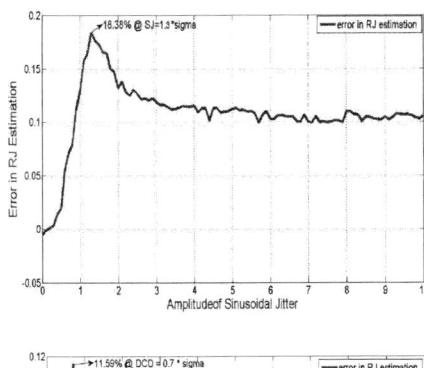

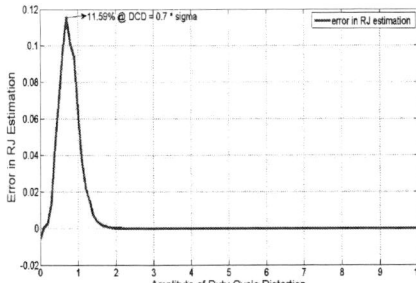

Figure 5: Variation of Error in random jitter with varying sinusoidal jitter and duty cycle distortion

F. RJ error computation Block

Comparing the random jitter results estimated by tail fit algorithm with the unit random jitter (as modeled in section III B) gives the absolute error percentage in random jitter estimation. The same process is repeated for varying the amplitude of sinusoidal jitter and the trend is noticed.

It is clear from the figure 5; error in estimation of RJ initially increases with the increase of SJ amplitude, till SJ amplitude becomes equal to 1.3 times RJ sigma. The error is maximum (18.38%) when SJ(pk-pk)=2*1.3*sigma, after which the error in estimation starts reducing. When SJ amplitude is more than 4 times of sigma value, error in estimation of RJ sigma becomes almost constant at 10%. However when duty cycle distortion amplitude is more than 2 times of RJ sigma value, error in estimation of RJ sigma becomes almost negligible.

G. Lookup Table for correction factor

The error trend for varying deterministic jitter values received from the error correction block is stored as a lookup table for the correction factor. The correction factor is stored for each ratio of the deterministic jitter to random jitter.

H. Application in real scenario

In case of real jittered signal, after applying the tail fit algorithm, the corresponding correction factor is applied to random jitter value obtained to get the precise value of random jitter.

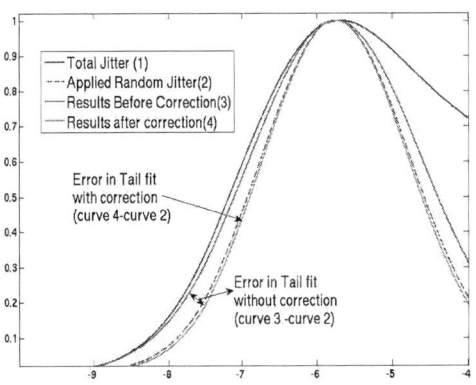

Figure 6: Improvement in random jitter estimation

Figure 6 clearly shows that the random jitter estimation using the tail fit algorithm follows standard deviation of the tail region of total jitter, which by default introduces an error. After the application of the correction factor on this estimation, the error is reduced and more precise value of random jitter can be estimated.

IV.   SIMULATION RESULTS

A. Random Jitter with Sinusoidal Jitter

In order to validate the above mathematical model, a simulation result of a channel simulator of ADS (Agilent) is taken. A pseudo random (PRBS9) source operating at the frequency of 5.0GHz is taken. A known amount of RJ sigma

(2.5% UI = 5ps) and SJ (@100MHz) is applied for 100000 bit patterns. The data is passed through a tail-fit algorithm [4] written in Matlab.

Figure 5 show the error percentage for various ratios of sinusoidal jitter amplitude to random jitter standard deviation in ideal mathematical model. A corresponding correction factor   is applied to the result obtained from standard tail fit algorithm to get the corrected results.

Table 1 : Comparison of Tail-fit Algorithm results with and without correction

Applied Values		CF	Results without correction factor		Results with correction factor	
RJ(ps)	SJ(ps)		RJ (ps)	Error	RJ (ps)	Error
5	5	1.1279	5.63	12.52%	4.99	-0.24%
5	6.5	1.1837	5.76	15.17%	4.86	-2.70%
5	10	1.1320	5.72	14.35%	5.05	1.02%
5	15	1.1181	5.55	10.93%	4.96	-0.79%
5	20	1.1149	5.48	9.69%	4.92	-1.61%
5	30	1.1029	5.48	9.68%	4.97	-0.55%
5	40	1.1095	5.50	10.07%	4.96	-0.80%

Table 1 shows the significance improvement of results obtained from standard Tail-fit algorithm when using the correction factor.

B. Random Jitter with Duty Cycle Distortion

In order to validate the above mathematical model, a simulation result of a channel simulator of ADS (Agilent) is taken. A pseudo random (PRBS9) source operating at the frequency of 5.0GHz is taken. A known amount of RJ sigma (2.5% UI = 5ps) and DCD is applied for 100000 bit patterns. The data is passed through a tail-fit algorithm [4] written in Matlab.

Table 2 : Comparison of Tail-fit Algorithm results with and without correction

Applied Values		CF	Results without correction factor		Results with correction factor	
RJ(ps)	DCD(ps)		RJ (ps)	Error	RJ (ps)	Error
5	1.5	1.0130	5.04	0.88%	4.98	-0.41%
5	2.5	1.0696	5.35	7.00%	5.00	0.03%
5	3.5	1.1159	5.59	11.76%	5.01	0.15%
5	4.5	1.0937	5.57	11.42%	5.09	1.14%
5	5.5	1.0368	5.28	5.51%	5.09	1.77%
5	6.5	1.0154	5.12	2.36%	5.04	0.81%
5	10	1.0001	5.09	1.83%	5.09	1.82%
5	30	1.0000	5.10	2.10%	5.10	2.10%

Figure 5 show the error percentage for various ratios of duty cycle distortion amplitude to random jitter standard deviation   in ideal mathematical model. A corresponding correction factor   is applied to the result obtained from standard tail fit algorithm to get the corrected results.

Table 2 shows the significance improvement of results obtained from standard Tail-fit algorithm when using the correction factor.

V. CONCLUSIONS

A new methodology is proposed to mitigate the impact of sinusoidal jitter and duty cycle distortion on random jitter estimation by tail fit algorithm. The methodology is validated by results obtained from a channel simulator of ADS (Agilent). Table 1 and 2 demonstrate the significant improvement in the random jitter estimation by the new methodology over the standard tail fit algorithm. It is a generic methodology which could be applied in software as well as hardware in instrumentation, which uses the Dual Dirac algorithm in real world scenario.

REFERENCES

[1] Agilent Technologies," Jitter Analysis: The dual Dirac Model, RJ/DJ, and Q-scale",December 31,2004

[2] M. P. Li, J. Wilstrup, R. Jessen; D. Petrich; "A new method for jitter decomposition through its distribution tail fitting", International Test Conference, 1999, pp 788-794

[3] Guy Foster; "Dual-Dirac, Scope Histograms and BERTScan Measurements - A Primer",

[4] McClure, Mark Scott," Digital Jitter Measurement and Separation" Texas Tech University Thesis(2005)

[5] Mike Peng Li ; "Jitter, Noise and Signal Integrity at High-Speed" – Prentice Hall, 2007.

[6] P. R. Belington, D. K. Robinson, "Data reduction and error analysis for the physical sciences", McGraw-Hill, Inc, 1992.

[7] Radhakrishnan, Nitin " Stressed Eye Analysis and Jitter Separation for High Speed Serial Links" Missouri University of Science and Tech. Thesis(2009)

[8] Agilent Technologies – "EZJIT and EZJIT Plus Jitter Analysis Software for Infiniium Series Oscilloscopes" – April 2006

[9] Agilent Technologies – "Analyzing Jitter Using Agilent EZJIT Plus Software" – October 2005

[10] Agilent Technologies, "Measuring jitter in digital systems", Application note 1448-1, June 1, 2003.

[11] Tecktronix Inc., "TDSJIT3: Jitter analysis and timing software", TDSJIT3 Product datasheet.

[12] Agilent Technologies, "Jitter fundamentals: Agilent 81250 ParBERT jitter injection and analysis capabilities", Application note 5988-9756EN, July 17, 2003

[13] Jitter Analysis Techniques for High Data Rates, Agilent Technologies Inc., Application Note 1432, February 2003

[14] Patrin, John, Li, Mike. "Characterizing jitter Histograms for Clock and DataCom Applications." DesignCon 2004. July 6, 2005

[15] Kuo Andy, Farahmand Touraj, Ou Nelson, Tabatabaei, Ivanov Andre, "Jitter Models and Measurement Methods for High Speed Serial Interconnects" IEEE ITC International Test Conference (2004).

978-1-4799-0708-3/13 $31.00 © 2013 IEEE

A Novel Flexible On-Die Decoupling Scheme Using Package Interconnects

Kundan Chand, Dan Oh, Hui Liu, and Hong Shi

Altera Corporation
101 Innovation Drive, San Jose, CA 95134
kchand@altera.com

Abstract—**Power supply noise induced jitter (PSIJ) is one of the critical bottlenecks for I/O signal performance. Good power distribution network (PDN) is a must for high-end system designs. Due to design limitation in die and package, providing sufficient on-die capacitor (ODC) or on-package capacitor (OPD) is a very challenging task. This paper presents a novel on-die decoupling scheme which places decoupling caps in core area and connects to I/O area by package interconnects. The presented method is applied to DDR interface in med-end FPGA devices. Excellent SSN improvement is achieved by implementing this scheme.**

Keywords-ISI, SSN, Crosstalk, Embedded IO, PDN, OPD, HDMiM

I. Introduction

Power distribution network design is a major challenge for DDR designs. The quality of the power supply, seen by the circuits on the die, is important for proper circuit performance and the ability to meet timing and jitter specifications. The power distribution network (PDN) for the package die combination is an important consideration in determining power supply quality. Through the use of on-package decoupling (OPD) capacitance and on-die capacitance (ODC), the PDN performance can be improved but it costs die or package area.

With continuous progress of semiconductor technology, hundreds of high speed fabric I/Os, sixty or more multi-gigabit data-rate transceivers, and a million logic elements (LEs) are incorporated on the same die. In order to reduce SSN noise, the supply pins continue to increase as shown in Figure 1. and Figure 2.

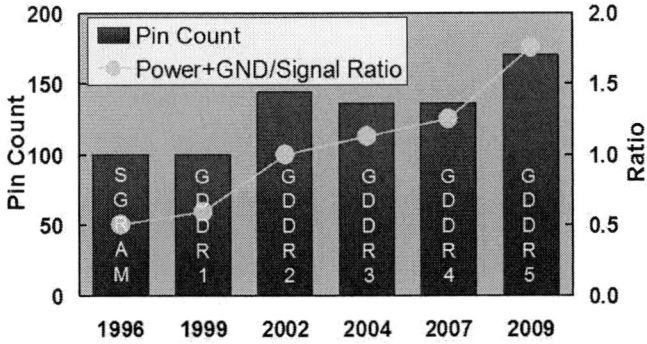

Figure 1.   Power distribution design challenges

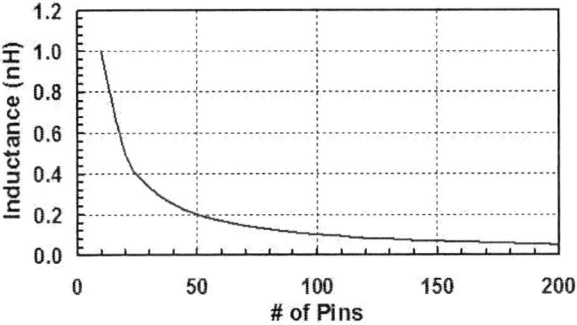

Figure 2.   Power supply inductance vs. number of pins

In this paper, a novel on-die decoupling scheme using package interconnects is presented which leads to excellent improvement in SSN performance. The scheme is applied to an FPGA monolithic package but it can be implemented in multi-die package applications as well as ASIC packages. The second section of the paper gives an overview of the PDN challenges for embedded vs. edge IO configurations used in FPGA applications. Section III describes the SSN issues in DDR3 designs and compares the performance between the two IO configurations. In Section IV and V, the novel ODC island scheme is explained and SSN improvement is quantified using relevant simulations. Finally, conclusions are given in Section VI.

II. Power Distribution Network Challenges in FPGA Applications

The integration of transceivers and DDR I/Os into FPGAs has greatly facilitated system designs in many applications. However, integrating them together has caused significant challenges in FPGA package designs. Figure 3. and Figure 4. display the die floorplan for FPGA test chip using embedded IO and edge IO columns. The embedded IO columns are located close to the center of the die whereas the edge IOs are located close to one edge. The embedded columns allow the FPGA designer to fit in more number of transceivers in the same die. While it is convenient to route out the PDN for the edge IOs, package routing becomes more challenging for embedded IO PDNs. The use of on-package decoupling capacitance (OPD) is also ineffective for embedded IOs due to large hook-up inductance.

978-1-4799-0708-3/13 $31.00 © 2013 IEEE   155

Figure 3. Figure 3. shows the VCCIO PDN impedance profile for embedded vs. edge IO for Altera FPGA test chip. The impedance peak is much higher for embedded IOs due to large VCCIO inductance caused by routing complexity. Since VCCIO is routed below the package core layer, OPD is ineffective due to large hook-up inductance. Therefore, the SSN performance for embedded IOs is much worse compared with edge IOs.

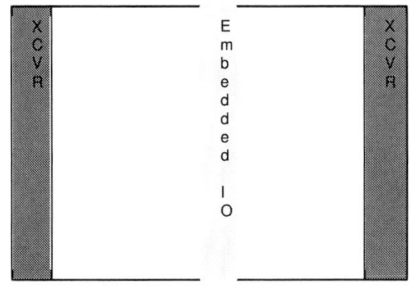

Figure 3.   Die floorplan showing edge IO column

Figure 4.   Die floorplan showing edge IO column

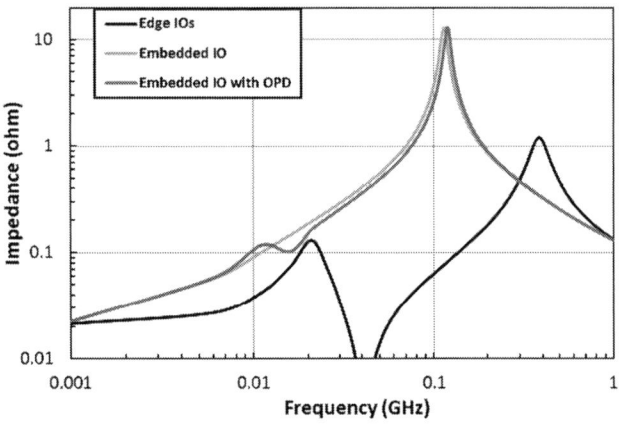

Figure 5.   VCCIO impedance profile

III.   SSN ISSUES IN DDR3 DESIGNS

In contrast to most chip-to-chip I/O interfaces that use differential signaling, the mainstream memory interface designs are based on single-ended signaling such as SSTL or PODL. Extending the data rate for single-ended signaling

beyond current data rate of a few Gb/s is becoming very difficult due to SSO noise. Interestingly, DDR4 has mitigated SSO noise compared with DDR3. This is due to two main reasons.

Firstly, the data bus inversion (DBI) coding technique introduced in GDDR4 reduces SSO noise significantly. DBI reduces the "weight", or total number of 1 or 0 states, of the bus by inverting the polarity of the bus values. Proper implementation of the encoding method will limit the "weight" of an N-bit bus to $N/2$. The choice of reducing the number of 1 or 0 states depends on the details of the bus termination. In GDDR4, which uses VCCIO referenced PODL signaling, power is saved by reducing the total number of 0 states, since they consume static current [3].

Secondly, DDR4 uses Pseudo-Open-Drain Logic (PODL) signaling which has only pull-up termination [1]. The presence of only a pull-up termination makes PODL signaling to behave similar to open-drain signaling which significantly reduces power consumption since there is no current flowing during the high state.

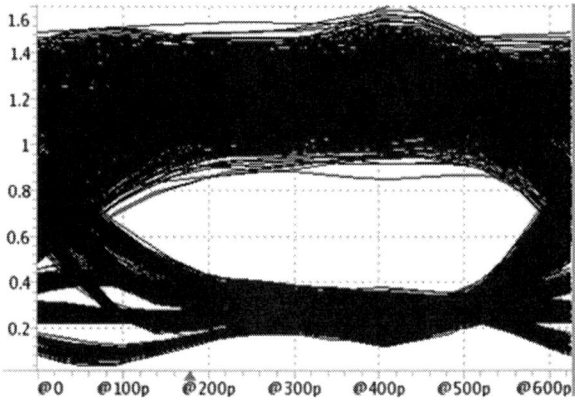

Figure 6.   DQ eye diagram for a typical DDR3 SSO simulation

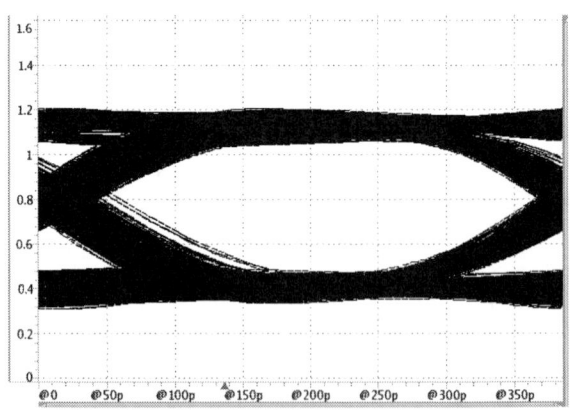

Figure 7.   DQ eye diagram for a typical DDR4 SSO simulation

Figure 6. and Figure 7. compare the DQ eye diagram for a typical DDR3 versus DDR4 setup under SSN condition.

Evidently, DDR3 has worse SSN performance compared to DDR4 due to above mentioned reasons. Therefore, for the purpose of this paper we focus on mitigating SSO noise in DDR3 designs.

Figure 8. displays the setup for DDR3 SSN simulation. The eye diagram is simulated at the input of the DRAM receiver when 48 fabric IOs are toggled simultaneously. The victim is toggled using PRBS15 whereas the aggressors are using PRBS10. The simulation methodology used is explained in [1]. Figure 9. compares the victim eye diagram between embedded vs. edge IOs under SSN condition. As expected, the embedded IOs violate the DDR3 margin due to worse VCCIO PDN compared to edge IOs.

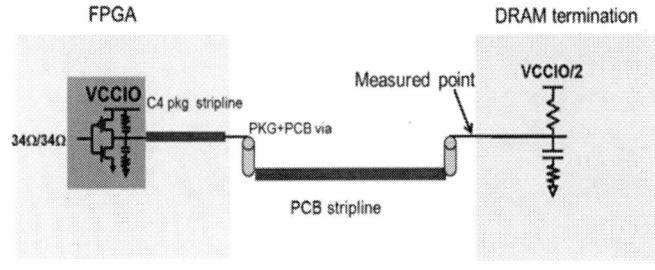

Figure 8.   DDR3 simulation setup

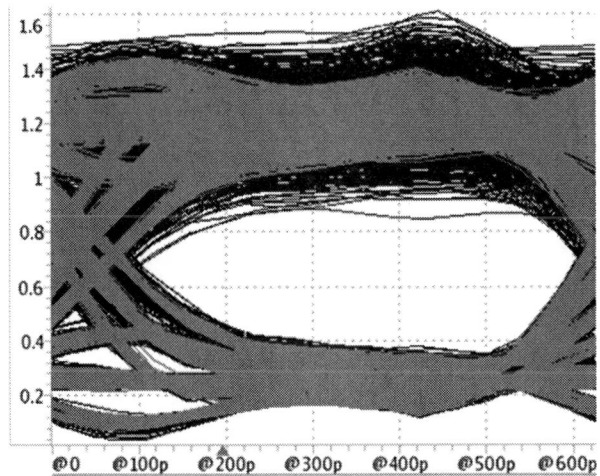

Figure 9.   DQ Eye-diagram for embedded (black) vs. edge IOs(red)

IV.   ODC ISLANDS

On-die decoupling cap (ODC) or on-package decoupling cap (OPD) can be used to improve VCCIO PDN for embedded IOs. ODC must be located near the current source in order to be effective. Increasing ODC generally leads to increase in die area. Alternatively, connecting OPD closer to the die can be challenging due to routing difficulties in the package. As shown in Figure 10. , OPD becomes less effective for embedded IOs due to large hook-up inductance.

HDMIM is a high density on-die metal cap where cap plates are located in between two of the upper die metal layers. It allows the re-use of die space for capacitance purposes,

which is also being used for regular circuitry. HDMiM can be built with low resistance connection to the PDN making it effective for high frequency. Figure 11. below shows a typical HDMiM cap implementation.

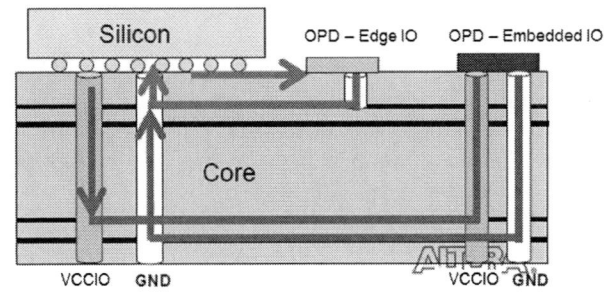

Figure 10. Package layout showing OPD hook-up connection

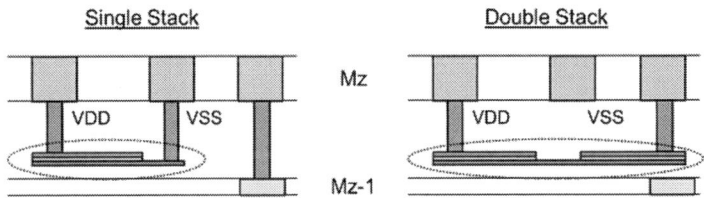

Figure 11. Typical HDMiM cap implementation

These HDMiM caps can be placed in different locations of the die (ODC islands) and then connected together in the package as shown Figure 12. . Actual usage of these ODC islands can be configured or altered using the package routing layer, instead of changing the die.

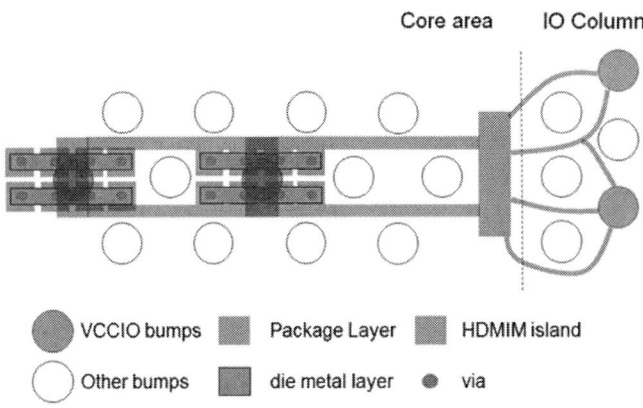

Figure 12. Novel on-die decoupling scheme using package layer

The above implementation is more effective than OPD due to less hook-up inductance and also easier to connect. It reduces the die cost and area by reducing ODC requirement. The same die can be configured to meet different application needs e.g. ODC can be assigned to different power rails just by modifying package design.

V. DDR3 SIMULATION RESULTS

Figure 13. is the VCCIO impedance profile showing improvement using ODC islands. There is significant PDN improvement in the mid frequency range of 50 - 300MHz by implementing this scheme. The multiple peaks in the red curve are due to the hook-up inductance resonating with ODC islands. The hook-up inductance value can be optimized further by changing the package design.

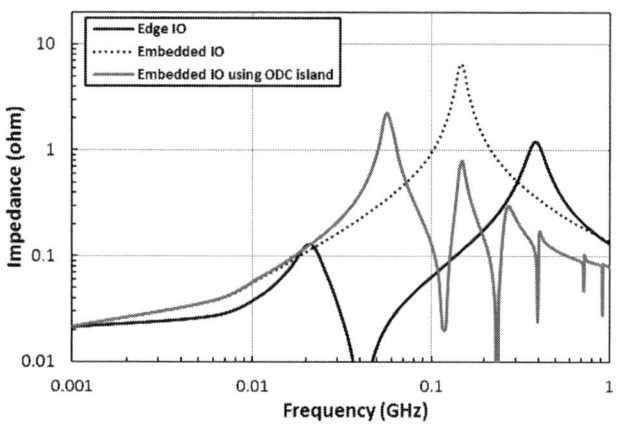

Figure 13. VCCIO impedance profile showing improvement using ODC island scheme

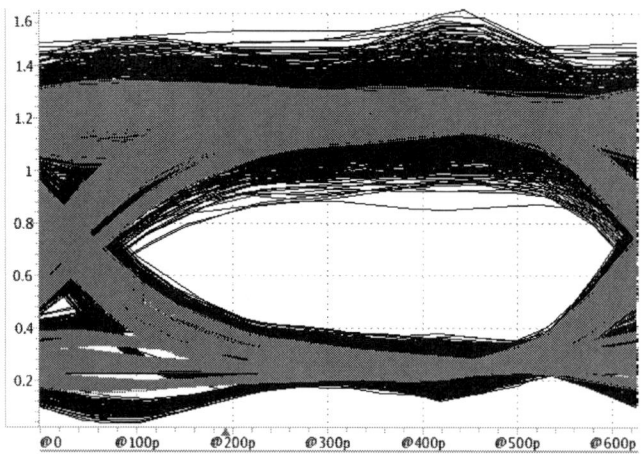

Figure 14. DQ eye improvement using ODC island

Figure 14. shows the victim eye improvement using the proposed scheme for embedded IOs. For this particular case

significant reduction in SSO noise is observed by implementing this scheme.

VI. SUMMARY

A novel on-die decoupling scheme is proposed which leads to significant reduction in SSO noise without utilizing extra die area. This scheme can also be implemented for multi-chip or multi-die applications where ODC in one chip can be used by other chips. Connection can be achieved by TSV, package layer, interposer trace, or RDL as shown in Figure 15. .

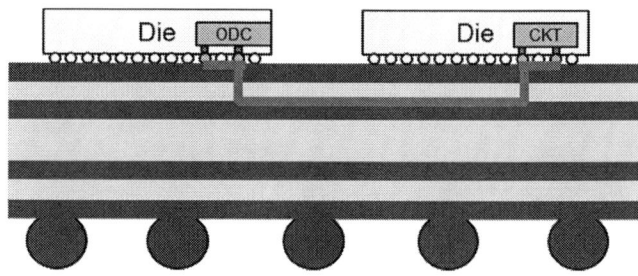

Figure 15. Other general applications for the decoupling scheme

This decoupling scheme implementation has various advantages. For certain designs, this technique can be more effective than OPD due to less hook-up inductance and is also easier to connect than OPD. Also, by utilizing HDMIM capacitors the overall system cost can be reduced due to less die area being used for ODC. Moreover, the same die can be configured to meet different package decoupling needs for multiple applications. However, this scheme will take up extra routing area on the package and hook-up inductance needs to be minimized for maximum benefits.

REFERENCES

[1] D. Oh and C. Yuan, *High-Speed Signaling: Jitter Modeling, Analysis and Budgeting*, Prentice Hall, 2011.

[2] L. D. Smith, R. E. Anderson, D. W. Forehand, T. J. Pelc, and T. Roy, "Power distribution system design methodology and capacitor selection for modern CMOS technology," *IEEE Transactions on Advanced Packaging*, pp. 284-291, Aug. 1999.

[3] D. Oh, W. Kim, et.al "Study of signal and power integrity challenges in high speed memory I/O designs using single-ended signaling schemes," *IEC DesignCon*, Santa Clara, CA, 2008.

[4] T. Granberg, *Handbook of Digital Techniques for High-Speed Design*, Upper Saddle River, NJ: Prentice Hall, NJ, 2004.

[5] M. R. Stan and W. P. Burleson, "Bus-invert coding for low-power I/O", *IEEE Transactions on VLSI Systems*, vol. 3, No. 1, pp. 49-58, March 1995.

Robust PoP Probing Solutions for High-Performance Application Processor Developments

Weiliang Yuan[1], SungJoo Kim[1], Woong Hwan Ryu[1]
[1]Design Technology, System LSI
Device Solutions, Samsung Electronics
Giheung-gu, Yongin-si, Gyeonggi-do, South Korea
wl06.yuan@samsung.com

SeongJae Moon[2] and Sangmin Lee[1]
[2]AP Development Team, System LSI
Device Solutions, Samsung Electronics
Giheung-gu, Yongin-si, Gyeonggi-do, South Korea

Abstract—**Probing solution is crucial for the designs and validation of high-speed PoP devices in high-performance mobile application processor (AP) development, where PoP assembly technology is used to achieve high data throughput & low cost. The paper has developed robust PoP probing solution for AP product development based on microwave network theory, including probe loading effect minimization, probing channel propagation removal, waveform reconstruction at die pad and interposer/ socket de-embedding, which are applied to the latest mobile application processor development. It is also straight forward and natural to extend the developed methods to TSV-based devices with WideIO interface.**

Keywords—package-on-package (PoP); application processor (AP); probing; waveform reconstruction; de-embedding

I. INTRODUCTION

With the requirement of higher data rate and smaller size in low-power portable applications such as smart phone, multiple chips are assembled with 3D stacking technologies such as package-on-package (PoP) technology, which achieve power saving and better signal integrity due to short signal length and less parasitics[1-2]. However, it has brought its own technical challenges, one of which is signal probing for device validation during the development. For example, LPDDR3 memory chip package is stacked on SOC package through PoP technology in the latest mobile AP devices, where IO signals are buried and inaccessible. As a solution, an interposer inserted between the top memory and bottom AP packages has been widely used, shown in Fig. 1, where memory package is assembled on the

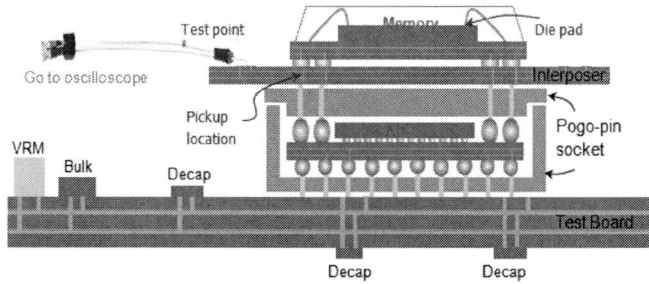

Fig. 1. Cross-section view of PoP device probing.

top of an interposer, which is connected with AP package through a pogo-pin socket, and AP package is then assembled

on test board through another socket. Parasitic effects on signal propagated between top memory and the bottom AP packages introduced by the interposer and sockets is negligible at low- and intermediate data-rate applications like LPDDR, LPDDR2, and test waveforms acquired at probe tip are considered its replica at IO pad of memory and AP devices. However, when data rate reaches multiple gigabit per second like LPDDR3 extension or LPDDR4, probe connector, interposer and socket bring enough parasitic, which causes two consequences. First, the waveform at probe tip cannot be considered as the replica of that on IO pad. Second, the waveform at IO pad acquired under test configuration is different from that in actual device operation, where neither interposer nor socket exists. Those challenge test and validation engineers. Because the waveform at the ball or the pad is desirable for compliance test or performance validation in device development, it is crucial to reconstruct signal waveform at the desired location from test data and to de-embed the effects from probe, interposer and socket, which is the focus of the paper. The paper intends to provide robust probing solution for the validation of high-speed AP device with PoP assembly, where interposer/socket effect is addresses and practical methods are developed, including interposer effect minimization, the method to remove probing channel propagation effect, the method to reconstruct the waveform at the pad, and the method to de-embed the probing interposer and sockets. Also, it is straightforward to extend the solution developed here to TSV-based devices, which is another novel 3D technology.

II. PROBE LOADING EFFECT & MINIMIZATION

The signal to be tested is probed with an interposer via a trace routed from signal pickup point to the top interposer layer, which is then connected to probe tip through a pair of wires. Hence, a stub is formed in high-speed signal path, which has significant impacts on high-speed signal to be tested. Probe loading effect is simulated with the simplified configuration, where the sockets are removed. Fig. 2 shows the simulated waveforms at signal pickup location inside the interposer with and without probe loading, that is, probing stub (remarked as R=0 & w/o probing). It is observed that probe loading has significant impact so that the probed waveforms are distorted by probing stub, which can mislead device validation and design. In order to minimize this effect, a resistor was used at the location as close to signal pickup location as possible to

978-1-4799-0708-3/13 $31.00 © 2013 IEEE

isolate the signal to be tested from the effect of probing stub, whose effectiveness is simulated with various isolating resistor (from 50ohms to 300ohms with the step of 50ohms). Fig. 2 shows the simulated waveforms, which indicates that the waveforms approach to the one without probe loading when

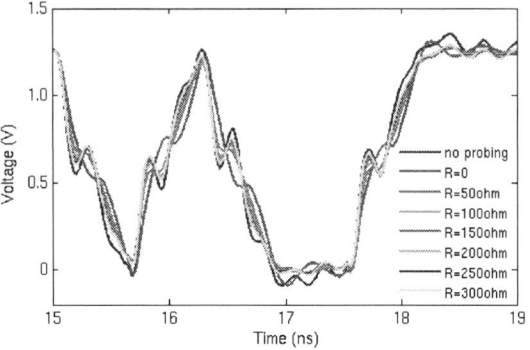

Fig. 2. The waveforms at signal pickup location with the isolating resistors.

the resistance increases and probe loading effect is minimized. It is suggested that an isolated resistor should be included at the location as close to signal pickup point as possible for high-performance PoP device validation.

III. PROBING CHANNEL PROPAGATION EFFECT & REMOVAL

However, the waveform is acquired at probe tip rather than signal pickup location. When high-speed signal to be tested is propagated from pickup location to probe tip, the waveform is distorted due to the parasitics from interposer, probe and the connection between them. The addition of an isolating resistor further exacerbates this distortion, a reverse effect when an isolating resistor is introduced. With HSPICE simulation, probing channel propagation effects can be observed through the comparison of voltage waveforms at probe tip, shown in Fig. 3 and Fig. 4. Fig. 3 is the result without the isolating

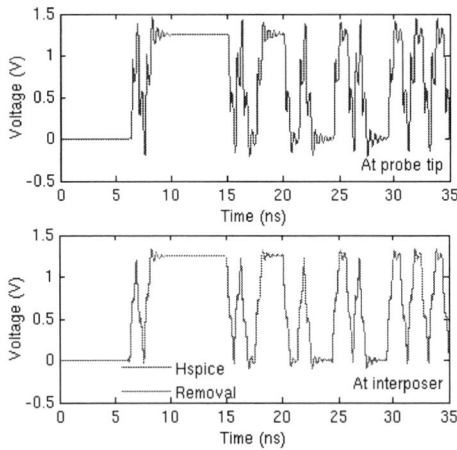

Fig. 3. The waveforms after probing channel propagation effect removal without an isolating resistor.

resistor while Fig. 4 is one with a 300-ohm resistor. The curves at probe tip are so different each other and also very different from one at pickup location. Hence, it is necessary to develop an approach to remove probing channel propagation effect from the parasitics between signal pickup location and probe

tip as well as the reverse effect from the isolating resistor and to recover the waveform at signal pickup location, which is achieved through the following formulation

$$V_{Pickup} = (A_p + B_p Y_{in-Probe})V_{probe} \tag{1}$$

where A_p, B_p are the elements in ABCD matrix of the network consisting of physical structures between pickup location and probe tip that can be extracted with electromagnetic solver or measured with vector network analyzer, and $Y_{in-probe}$ is input impedance of probe which usually has high impedance. With Eq. (1), the waveform of the signal at pickup location can be acquired from test waveform. In order to verify its accuracy, HSPICE simulation for one PoP test scenario is carried out from which the waveforms at probe tip and pickup location can

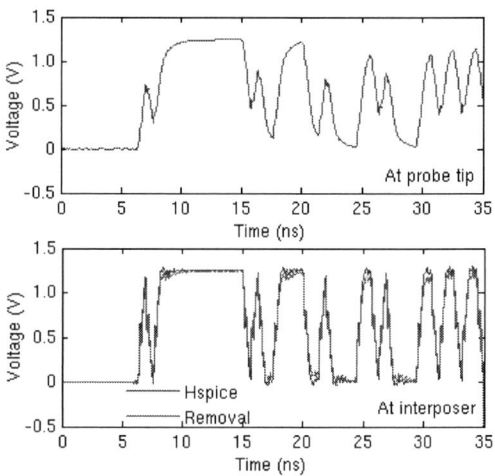

Fig. 4. The waveforms after of probing channel propagation effect removal with an isolating resistor (300ohms).

be obtained. The simulated waveform at probe tip is considered as test waveform, shown in the top plots of Fig. 3 & 4 for the cases without and with an isolating resistor respectively. The simulated ones at pickup location can be used as golden result. Then, the waveform at signal pickup location is recovered using Eq (1), shown in the bottom plots together with HSPICE simulation results. Their discrepancy indicates the accuracy of the developed method. It is observed that the waveform at probe tip is very different from that at signal pickup location, which implies that probing channel has significant effects on the signal to be tested so that its removal is important. On the other hand, the waveform at signal pickup location from HSPICE simulation is very close to that from Eq. (1), and its accuracy increases with less isolation from an isolating resistor, that is, the ability to accurately recover waveform increase when the isolating resistance decreases, which is natural. Optimal isolating resistor around 100-200ohms exists.

IV. WAVEFORM RECONSTRUCTION

Through probing loading effect minimization and probing channel propagation removal, the waveform was acquired at signal pickup location. However, the desired waveform is usually at device ball for compliance test or at die pad for device design & validation. Hence, it is crucial to reconstruct the waveform of high-speed signals at die pad or solder ball.

978-1-4799-0708-3/13 $31.00 © 2013 IEEE   160

With microwave network theory, waveform reconstruction at die pad can be achieved through the following formulation

$$V_{rvr} = \frac{A_{pp}D_{tt} - C_{pp}B_{tt} + (B_{pp}D_{tt} - D_{pp}B_{tt})Y_{in-Probe}}{A_{tt}D_{tt} - B_{tt}C_{tt}} V_p \quad (2)$$

Where A_{pp}, B_{pp}, C_{pp} D_{pp} are the elements in ABCD matrix of microwave network consisting of physical structures between signal pickup location and probe tip, and A_{tt}, B_{tt}, C_{tt} D_{tt} are that of microwave network consisting of physical structures between signal pickup location and receiver's die pad. Using the same way as that in last section, the waveform at receiver's die pad is reconstructed using the waveform at probe tip obtained from HSPICE simulation and compared with that from HSPICE simulation, shown in Fig. 5 (without an isolating resistor) and Fig. 6 (with 300ohm isolating resistor). Comparing them with the waveforms at signal pickup location

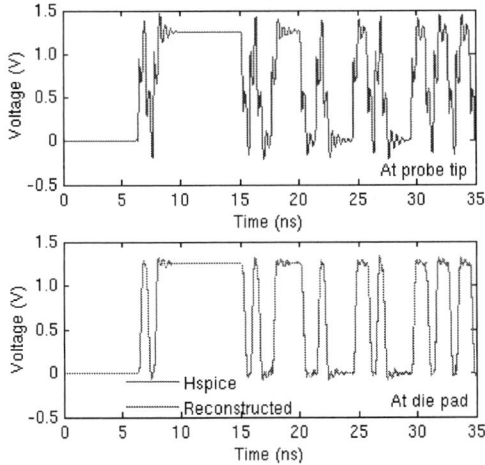

Fig. 5. The reconstructed waveforms at die pad without an isolating resistor.

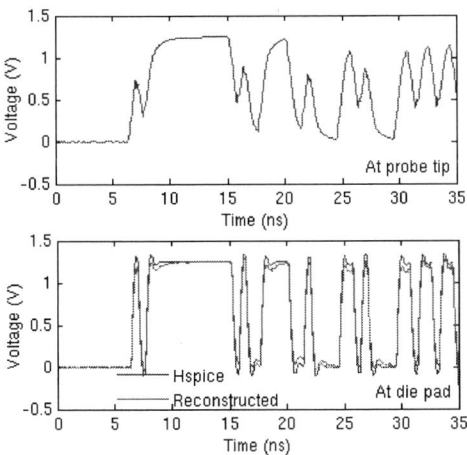

Fig. 6. The reconstructed waveforms at die pad with an isolating resistor (300ohms).

in Fig. 3 and 4, it is observed that the waveforms at die pad are also different from those at pickup location due to various parasitic effects along signal path. The comparisons (the bottom plot of Fig. 5 & 6) show high accuracy of waveform reconstruction method which, on the other hand, increases with the decrease of the isolation provided by an isolating resistor.

V. INTERPOSER/SOCKET DE-EMBEDDING

Up to now, the effects from the interposer and sockets for signal probing are still included, which does not present in actual device operation. In high- speed device validation, the effect from them has to be de-embedded in order to acquire real signal waveform, which can be achieved with the following formulation

$$V_{Rvr\_NIP} = \frac{Z_{Rvr}}{Z_{Prb}} \times$$

$$\frac{Z_{Prb}Z_{Drv}C_{Com\_IP} + Z_{Prb}A_{Com\_IP} + Z_{Drv}D_{Com\_IP} + B_{Com\_IP}}{Z_{Rvr}Z_{Drv}C_{Com\_NIP} + Z_{Rvr}A_{Com\_NIP} + Z_{Drv}D_{Com\_NIP} + B_{Com\_NIP}} V_{Prb} \quad (3)$$

where A_{com_IP}, B_{com_IP}, C_{com_IP}, D_{com_IP} are the elements in ABCD matrix of microwave network consisting of physical structures from driver's die pad to probe tip where the interposer and sockets are present, and A_{com_NIP}, B_{com_NIP}, C_{com_NIP}, D_{com_NIP} are those of microwave network consisting of physical structures from driver's die pad to receiver's pad

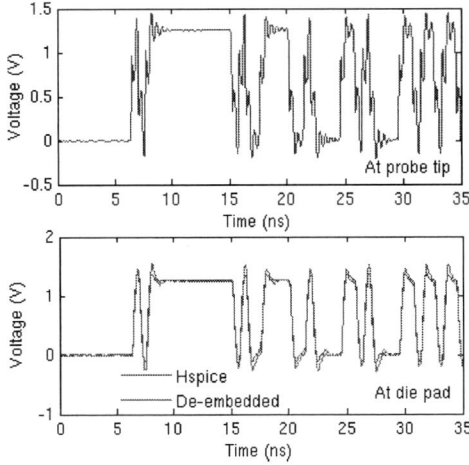

Fig. 7. The waveforms at die pad after the de-embedding of the interposer without an isolating resistor.

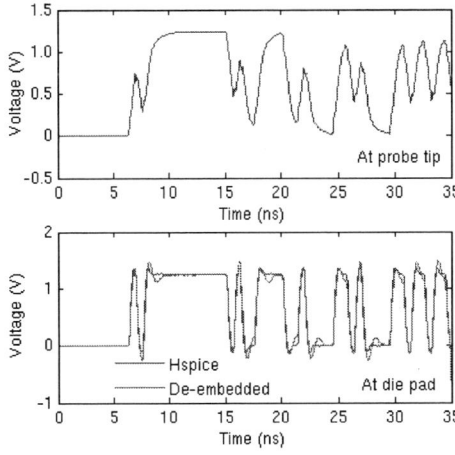

Fig. 8. The waveforms at die pad after the de-embedding of the interposer with an isolating resistor (300ohms).

under actual device operation scenario. Again, with Eq. (3), the waveform at die pad was acquired using the waveform at probe tip obtained from HSPICE simulation. They are shown in Fig.

7 and 8, which excludes the effects from test system such as interposer. Waveform comparisons show high accuracy of the developed de-embedding technique. The accuracy of the technique increases with the decrease of the isolation provided by an isolating resistor.

VI. MEASUREMENT

As discussed in previous sections, signal test for high-speed PoP device has to use an interposer between AP and memory packages for signal probing, two additional sockets with pogo-type pins might be accompanied for easy assembly and disassembly: one is for the connection of interposer PCB and AP package and the other is for the connection of AP and test PCB. Fig. 9 shows its actual configuration where test board and

Fig. 9. Top package, interposer PCB, socket, and bottom package inside.

probe attachment are removed for neatness. It is observed that the socket has long pogo pin, which brings inductive and capacitive parasitics. Also, signal routing inside the interposer is another source of such parastics, which worsens timing margin. The signals are probed and sensed at probe tip on the top side of the interposer with the following equipments:

- Digital signal analyzer with 8GHz bandwidth
- Probe with over- 8GHz bandwidth
- Solder tip with up to 20GHz bandwidth

Along with physical setup above, test pattern for signal

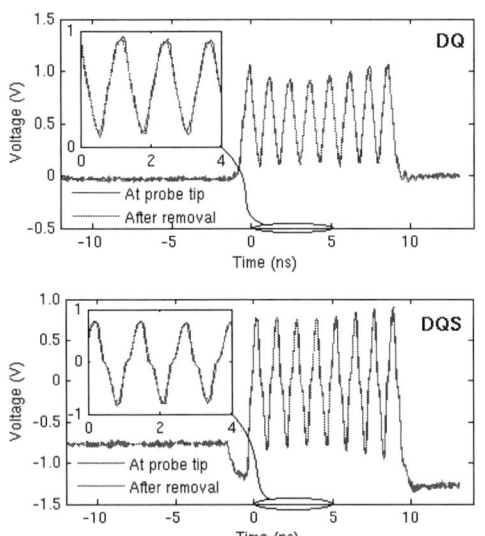

Fig. 10. DQ & DQS waveform at signal pickup location after probing channel propagation effect removal during writing operation.

measurement is generated by controlling register settings and signal waveform is acquired at probe tip. Network matrices are acquired through simulation. Fig. 10 shows the waveform of

one DQ and DQS respectively at probe tip and pickup location, which are recovered from the tested waveforms. It is observed that the waveforms at probe tip are close to the ones at pickup location because the stub for the nets in the interposer is short. However, probing channel effects quickly deteriorate the signals to be tested with longer stub or higher data rate, which will be shown later. Fig. 11 shows the waveforms at memory

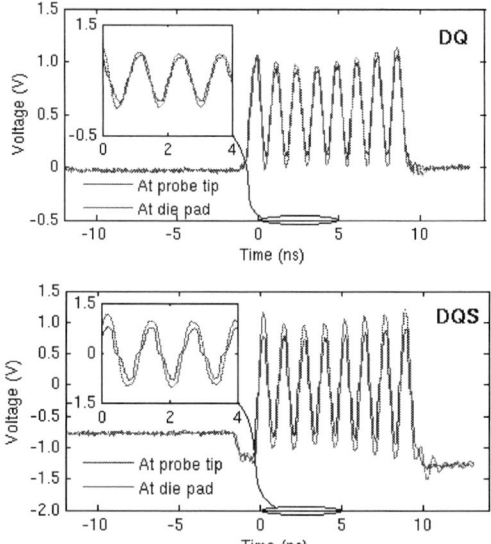

Fig. 11. DQ & DQS waveforms at memory die pad after waveform reconstruction during writing operation.

die pad reconstructed from from the tested ones at probe tip for one DQ and DQS respectively. Compared to the waveforms at probe tip, small notch disappears and temporal width in the waveform at die pad increases at the middle of voltage, which can be translated to wider eye width in eye diagram.

CONCLUSIONS

The paper has developed robust and complete probing solutions for high-speed PoP device testing & validation based microwave network theory, including probe loading minimization, probing channel propagation effect removal, waveform reconstruction at die pad, and interposer/socket de-embedding, which are crucial for high-performance mobile application processor development, where memory package (LPDDR3 extension and beyond) are stacked with AP package through PoP assembly technology. The developed techniques are verified and high accuracy is demonstrated through the comparison with HSPICE simulation. Finally, they are applied to actual AP test. It is straightforward for the developed techniques to apply them to TSV-based WideIO interface measurement although it is not demonstrated here.

REFERENCES

[1] W. H, Ryu and M. Wang, "A co-design methodology of signal integrity and power integrity," DesignCon, 2006.

[2] D. Oh and et al., "Design and characterization of a 12.8GB/s low power differential memory system for mobile applications," Electrical performance of Electronic Packaging and Systems (EPEPS'09), 2009, pp.33-36.

High Speed Links II

978-1-4799-0708-3/13 $31.00 © 2013 IEEE

High-speed DIMM-in-a-Package (DIAP) Memory Module

Zhuowen Sun, Kyongmo Bang, Kevin Chen, Richard Crisp
Invensas Corporation
2702 Orchard Parkway
San Jose, CA 95134
{zsun, kbang, kchen, rcrisp} @invensas.com

Abstract—**We presented a highly integrated quad-die memory module design using face-down wire bond assembly technology for DDR operations up to 2400 MT/s. We performed peak distortion analysis of the point-to-point DQ and fly-by Command/Address (C/A) memory buses with up to 9 aggressors on the DIAP memory module mounted on a Type-3 motherboard. We also discussed the tradeoffs when designing the package for better electrical performance and lower manufacturing cost.**

Keywords— DIAP, Soldered-down, On-board memory, Fly-by, DDR4

I. INTRODUCTION

Consumer electronic devices, such as notebooks and tablet PCs, continue to demand innovative packaging solutions for targeted IC performance in a smaller form factor than previous generations in order to deliver a slim and light-weight "mobile" device for good user experience [1]. The z-height requirement in particular is the most critical and challenging aspect of the system design. As an example, the next generation Ultrabook has a thickness target of only 12 mm [2]. Therefore, traditional SO-DIMM is facing an insurmountable hurdle while the so-called "soldered-down" memory package that directly soldered onto the motherboard is becoming more and more appealing.

An innovative soldered-down memory solution needs to meet different sets of requirements by the manufacturers and designers. In addition to the small form factor (x- and y-dimension) and low profile (z-height), OEMs also like to have the next-generation memory package delivered by a low-cost manufacturing process. From the chip and package designer's perspective, the resulting high density I/Os in a small memory module should have good signal integrity for operations at 1600 MT/s or faster. To satisfy all the needs noted above, we have previously introduced our DIMM-in-a-Package (DIAP) memory solutions implemented by a quad-die face down (QFD) wirebond assembly process [3-4]. The DIAP memory solution offers more aggressive form factors than traditional single-die package (SDP) or dual-die package (DDP). The DIAP's uniquely symmetric BGA ball-map enables the double-sided "clamshell" assembly on standard through-hole via (PTH) boards, rather than expensive high-density interconnect (HDI) boards, allowing denser and cheaper implementation of notebooks and tablet PCs [5]. We have already validated the performance of a 407-ball DIAP memory module consisting of four x16 DDR3L (1.35 V) dies using a balanced T-topology C/A memory bus at 1600 MT/s [6] on a Type-3 board, as shown in Fig. 1.

In this work, we employ the same design methodology as in [6] and try to explore the performance of quad-die DIAP with lower supply voltage (1.2V) at even higher speed (2400 MT/s) in a fly-by C/A memory bus. We highlight the analysis of the package design, including the performance and manufacturing cost tradeoffs and present some motherboard layout considerations for performance, before we conclude fly-by based DIAP memory solution is very promising for next generation memory, such as DDR4.

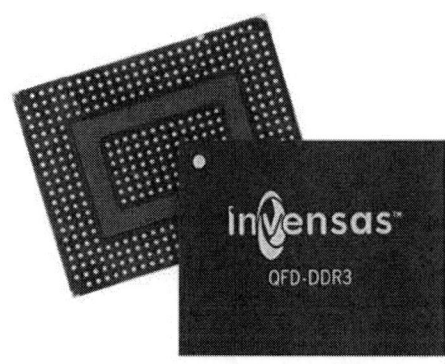

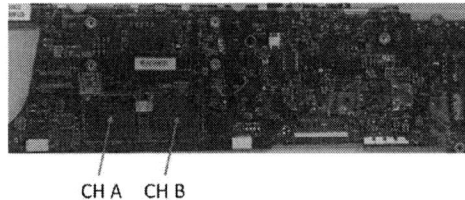

CH A   CH B

Fig. 1. Top and bottom side view of a 407-ball 1GB DIAP memory module and the finished motherboard assembly in a Dell XPS-12 Ultrabook.

II. FLY-BY DIAP MEMORY MODULE

DIAP is a multi-die solder-down memory module using face down wirebond assembly technology. Shown in Fig. 2 is an illustration of the quad-die placement inside the package. Fig. 3 shows the detailed X-Ray images of wirebond connection from the bottom layer of the package to the face-down dies through the substrate bonding window.

In this paper, we investigate the package design and cost/performance tradeoffs of DIAP memory modules when implementing a fly-by memory bus for up to 2400 MT/s operation. We keep the same DIAP package outline dimensions (22.5 x 17.5 x 1.2 mm) and QFD wire bond assembly as previously validated in the T-topology memory bus. The C/A signal group remains in the middle of the DIAP

978-1-4799-0708-3/13 $31.00 © 2013 IEEE

package and the DQ signal group is placed in the peripheral. In addition, we include four 4Gb dies (each x16 width) placed face-down without overlap in a single 2GB DIAP memory package (total x64 width), as opposed to the previously validated four 2 Gb dies in the 1 GB module [5-6].

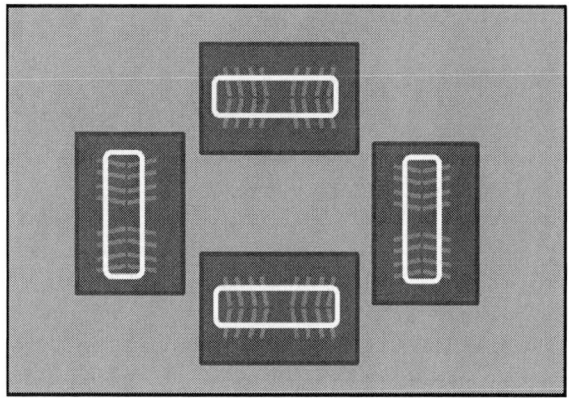

Fig. 2. An illustration of the four-die placement and the locations of wirebond and bonding windows inside the DIAP memory module.

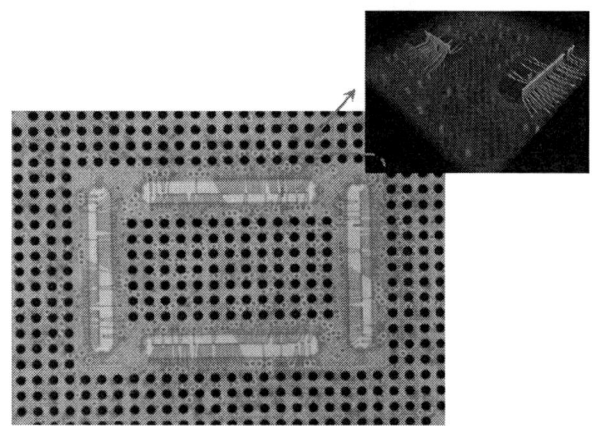

Fig. 3. The 2D X-ray photo of the DIAP package substrate and a detailed 3D X-ray photo details of the wirebond going through the bonding window.

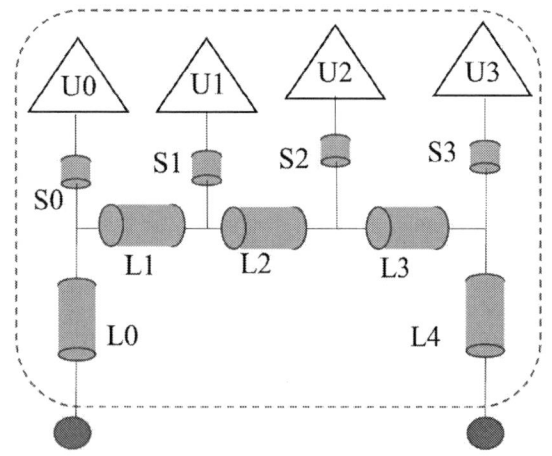

Fig. 4. An illustration of the fly-by memory bus inside the DIAP module.

In contrast to previous T-topology [6], the C/A net routing in the fly-by memory bus connects from one "fly-in" BGA ball

to another "fly-out" BGA ball, chaining all 4 memory dies and only tapping each of them along the way, as shown in Fig. 4. A signal layer routing inside the package substrate design is shown in Fig. 5. Unless carefully designed, the long traces running in parallel could degrade performance by adding significant crosstalk noises. This is different from the T-topology memory bus, where the two separate copies of the C/A pins give a little bit more routing flexibility and shorter neighboring traces running in parallel. Increasing the substrate layer counts would have alleviated routing density, but we limit our analysis to 6 layers, considering the aggressive DIAP package size and the goal to enable low-cost manufacturing.

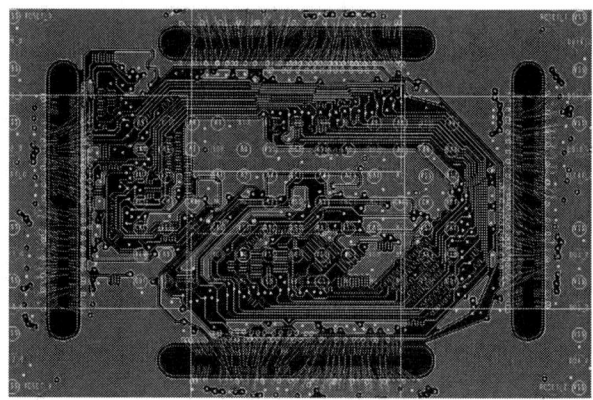

Fig. 5. Stripline layer routing for the fly-by C/A signal bus

III. PACKAGE AND BOARD CO-ANALYSIS RESULTS

Similar to our assessment of the T-topology memory bus [6], we conducted the HSPICE-based peak distortion analysis (PDA) of the full channel performance of the fly-by DIAP memory module. To showcase DIAP as a low-cost manufacturing enabling solution, we continue to assume our DIAP memory package is mounted on a 12-layer Type 3 plated-through-hole (PTH) motherboard with typical trace impedance of 46 Ohm, instead of more costly high-density interconnect (HDI) boards. The motherboard features, including via size, trace width, breakout sections are all based on production board requirements. With 0.8 mm ball pitch of the DIAP memory package, the channel routing from CPU to the DIAP module can be comfortably completed using the inner striplines of the 12-layer board. The outer surface layers are used only to break out CPU region and connect termination resistors to the mid-rail.

A. DQ Signal Group in DIAP

For the data signals, the CPU to DRAM connection is point-to-point. The DQs and their associated DQS routing inside the DIAP memory module have a maximum trace length of 10 mm (including the bondwire) and a maximum mismatch of 4 mils within the same byte inside the package. Using microstrip routing (65 Ohm) and with on-die termination (ODT) ON (120 Ohm), both READ and WRITE operations have good signal EYEs. Fig. 6 shows the EYE diagram of WRITE operation at 2400 MT/s.

978-1-4799-0708-3/13 $31.00 © 2013 IEEE

We also studied DQ signal sensitivity to the package trace length. As seen in Fig. 7, even with the longest trace length under consideration and at the highest operation speed, the EYE continues to have sizable openings. However, if more margins are needed, we could elect to route the DQ signals using stripline. Also shown in Fig. 7, the performance at 1866 MT/s using stripline (45 Ohm) has about 35 ps more timing margin than using microstrip of the same length. We observed similar timing margin gains at other speed bins. Because of stripline's higher manufacturing cost in production, microstrip is preferred for DQ routings, especially for 1866 MT/s and 2133 MT/s applications.

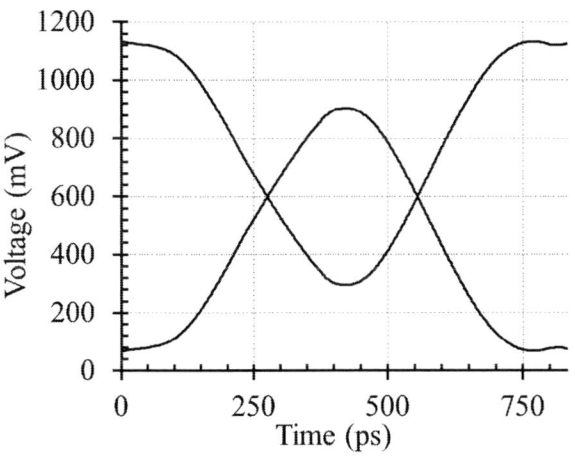

Fig. 6. EYE diagram of WRITE operation at 2400 MT/s (RON=26 Ohm, ODT=120 Ohm).

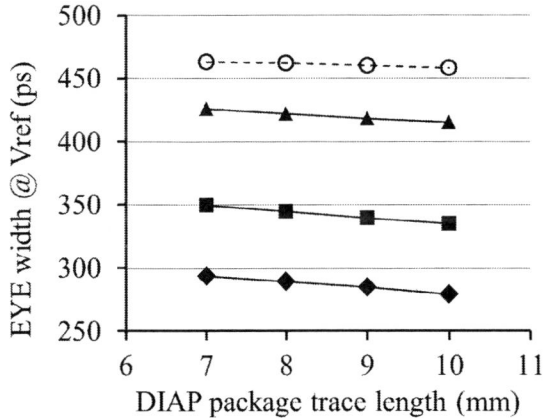

Fig. 7. EYE width dependence of DIAP package trace length at 1866 MT/s (triangle), 2133 MT/s (square), and 2400 MT/s (diamond) using microstrip routing. A case using stripline at 1866 MT/s (circle) is also shown for comparison.

B. C/A Signal Group in DIAP

For the package dimensions we are targeting, all fly-by C/A bus routing can be done using the 2 stripline layers in a 6-layer package substrate. The DIAP package's symmetric BGA ballmap design with respect to the central column enables one pin to "fly in" and another to "fly out". The total bus length (from "fly-in" pin to "fly-out" pin, as seen in Fig. 5) is approximately 50 mm and the stub to each memory die is about 2.3 mm including the bondwire length. To minimize the pin-to-pin

skew, the bus lengths are matched for all C/A signals with maximum mismatch of 4 mils. The channel connection is shown in Fig. 8. Notice the two motherboard vias (via2 and via3) for the termination resistor connection from the "fly-out" ball. As noted in Section II, the C/A balls are placed in the middle area of the package while DQs are at the peripheral. As such, the preferred C/A routing is to use motherboard vias to route to mid stripline layers, instead of breaking out on the surface layers using microstrip and going through the already dense DQ routing regions.

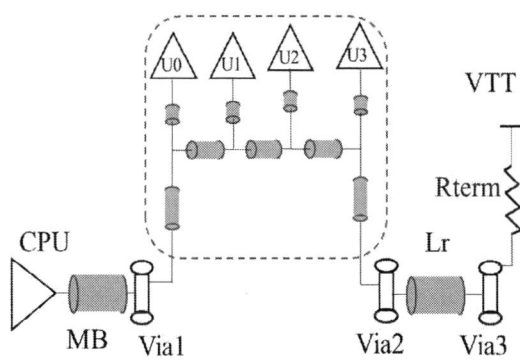

Fig. 8. Channel connection of the fly-by C/A memory bus.

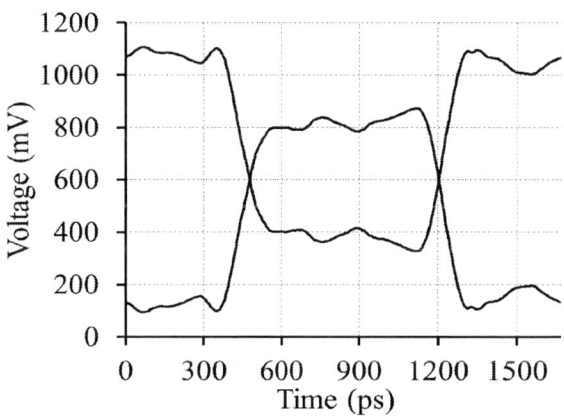

Fig. 9. EYE diagram of DRAM U0 at 2400 MT/s (UI=833 ps).

The C/A signals in the package have good voltage and timing margins. Fig. 9 shows the simulated performance of U0 at 2400 MT/s (UI=833 ps). Other DRAM dies have similar performances. To further illustrate the performance of the DIAP module, we summarized the EYE width (AC 135/ DC 90) at different speed bins in Fig. 10. As can be seen in Fig. 10, all DRAM dies have shown similar amount of openings, and the eye is consistently more than 80% of the UI at all speed bins. Memory package has long been identified to have the biggest influence on the overall signal EYE quality for traditional DIMM-based DDR3 applications [7]. Because the total trace length is fairly short in DIAP modules compared to conventional DIMMs, crosstalk control inside DIAP then becomes the key for improved SI, given the long lines running in parallel in the DIAP substrate layout and the requirement to use minimal layer count in our package substrate construction.

978-1-4799-0708-3/13 $31.00 © 2013 IEEE   167

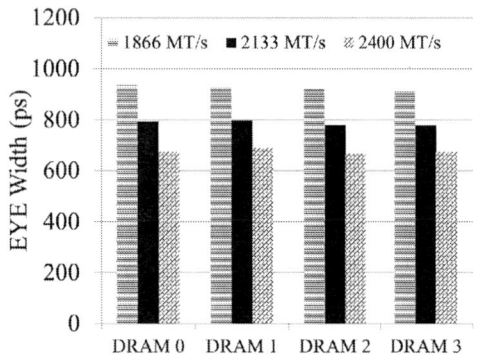

Fig. 10. EYE width of all four DRAM dies in DIAP at various speed bins.

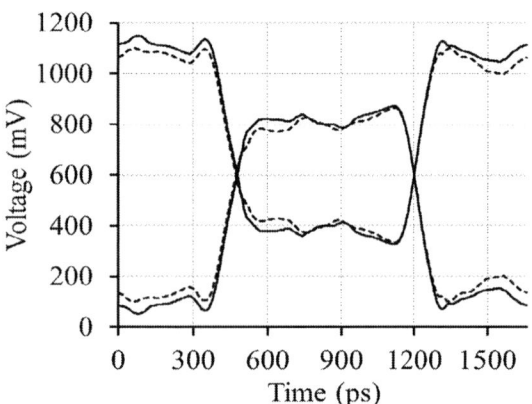

Fig. 11. Comparison of signal EYE using different trace impedances in DIAP routing: 56 Ohm (solid line) and 45 Ohm (dashed line).

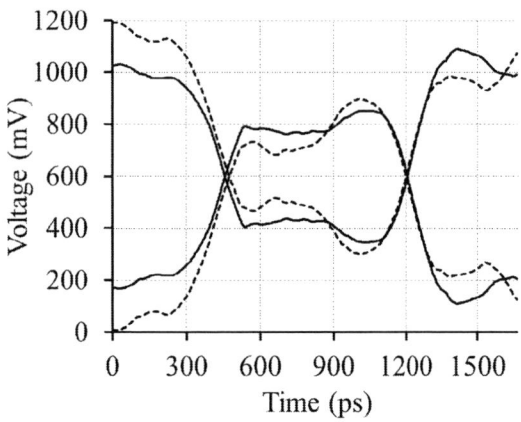

Fig. 12. Comparison of signal EYE using different termination resistor values: 35 Ohm (solid line) and 65 Ohm (dashed line)

To improve the voltage margin, narrower traces could be used to boost the performance further. Fig. 11 shows the performance comparison of two stripline traces of the same package trace length but different line impedance. As is clearly seen in the figure, the higher impedance gives rise to about 100 mV improvement to the eye opening. The improvement of the EYE is consistent with the known "neck-down" layout technique in fly-by DIMM designs and has been previously detailed in [8]. Using high-impedance lines also implies smaller trace width for the same substrate stackup and could further increase the trace spacing between neighboring lines for

reduced crosstalks. On the other hand, finer traces again would incur higher manufacturing cost in production and should only be used where more margins are truly needed.

The termination resistance is critical to minimize reflection. As seen in Fig. 12, we observe better signal EYE with lower termination resistance. To highlight the termination effect, Fig 13. plots the EYE opening for various termination resistances. We observe termination resistances closer to the line impedance (45 Ohm) give larger EYE openings. The fairly stable EYE openings for resistance between 30 to 45 Ohm imply a good manufacturing window for production.

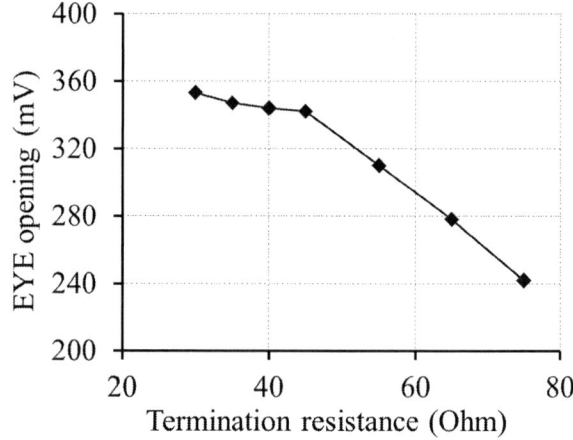

Fig. 13. EYE opening dependence of termination resistance at 2400 MT/s

ACKNOWLEDGMENT

We thank Gerald Pelissier at Dell, Becky Loop, Konika Ganguly, Bowen Liu and Rong Gao at Intel for their great help.

REFERENCES

[1] Samsung Semiconductor, "Low power transforming mobile PC industry," Intel Developer Forum, Sep 10-13 , 2011, San Francisco, CA

[2] S. Benn and J. Cheng, "Designing Ultrabook-engineering challenges and opportunities," Intel Developer Forum, April 2012, Beijing, China

[3] R. Crisp, W. Zohni, B. Haba, G. Pelissier and V. Bui, "A multi-die DRAM package for solder-down memory in Ultrabook and Tablet PC applications," International Conference on Electronics Packaging, Apr 17-20, 2012, Tokyo, Japan

[4] V. Solberg, S. McElrea, and W. Zohni: "xFD: A very thin two and four die package solution for high performance DDR SDRAM," 4th Electronics System Integration Technologies Conference (ESTC), Sep 17-20, 2012, Amsterdam, Netherland

[5] R. Crisp, K. Chen, Z. Sun and W. Zohni, "DIMM-in-a-PACKAGE memory module devices for ultra-thin client computing," International Conference on Electronics Packaging, Apr 10-12, 2013, Osaka, Japan

[6] Z. Sun, K. Chen and R. Crisp, "DIMM-in-a-PACKAGE (DIAP) signal integrity for high-performance on-board memory applications," 46th International Symposium on Microelectronics, Sep 29 - Oct3, 2013 Orlando, FL

[7] H. H. Chuang, et al. "Signal/Power integrity modeling of high-speed memory modules using chip-package-board co-analysis," IEEE Trans. Electronmagn. Compat.,vol. 2, no. 2, pp. 381–391, May 2010

[8] C. K. Chen, W. D. Guo, C. H. Yu, and W. R. Beei, "Signal integrity analysis of DDR3 high-speed memory module," Electrical Design of Advanced Packaging and Systems Symposium (EDAPS) 2008, 101-104, Dec. 10-12, 2008

Peak Distortion Analysis of Nonlinear Links

Wendemagegnehu T. Beyene

Rambus Inc., Sunnyvale, CA USA

ABSTRACT

This paper introduces statistical analysis of high-speed nonlinear links by extending the peak distortion analysis to include the impact of nonlinear drivers. By taking advantage of the unique characteristics of interconnect networks, the multilinear theory is applied to calculate the responses of the nonlinear networks by analyzing a series of linear networks.

First, the Volterra functional series is used to decompose the nonlinear link into multiple linear networks. Then, the peak distortion analysis of the high-speed nonlinear link is constructed from the linear combinations of sequences of peak distortion analyses of the decomposed linear networks. The analysis steps are derived analytically using an illustrative example. Finally, the accuracy of the method is verified for a high-speed memory link using time-domain simulation.

***Index Terms*—Multilinear theory, Nonlinear I/O driver, Peak distortion analysis, Statistical analysis, Volterra series.**

I. INTRODUCTION

AN interconnect-dominated system consists of a large number of linear, lumped and distributed elements and few nonlinear driver and termination networks. An estimation of the performance of chip-to-chip interconnect systems requires a transient analysis of nonlinear interface circuits and large frequency-dependent transmission-line systems. Consequently, these analyses can take significant CPU time because the link consists of a large number of distributed systems and the simulation time grows super-linearly with the size of the problem. Therefore, the simulation of large systems with few active devices using nonlinear time-domain simulators is too expensive in terms of time and memory requirements.

In order to quickly evaluate the performance of high-speed links, peak distortion analysis is applied to efficiently determine the worst-case eye diagram in [1]. Based on the linearity assumption, the channel response for arbitrary inputs as well as voltage and timing margins are derived from the single step response of the link. From the response and the assumption of the bit stream probability, the probability density function (PDF) of the channel intersymbol interference (ISI) is calculated. Based on these PDF's and any external noise and jitter distributions, statistical analysis are often used to study the impact of process variations, make architectural trade-off, and optimize the design of high-speed links.

Perhaps the key essential feature of statistical simulation is the linear time invariant (LTI) assumptions when calculating the ISI of the link. If the link is nonlinear or time variant, the result of the statistical simulation and the peak distortion analysis can be significantly incorrect. In current high-speed interface, a linear behavioral model cannot be used to represent the I/O driver in system simulations.

The block diagram, transmitter and the channel with PODL termination scheme, of a high-speed memory interface is shown in Fig. 1(a). Even though, the transistor I-V characteristics are voltage dependent as shown in Fig. 1(b), the driver is sized, biased and uses push-pull circuit with poly-transistor split, in order to provide a well-controlled driver impedance.

However, low-swing interfaces have lower supplies and do not use the linearization resistors, to save power and area of the drivers. Fig.2(a) shows the low voltage swing terminated logic (LVSTL) signaling for future advanced mobile memory interface [2]. This low-swing signaling has low IO power consumption and can operate beyond 4.3 *Gbps* in mobile and tablet channel environment. Fig.2(b) compares the eye diagrams of current and future mobile memory interfaces running at 1.6 *Gbps* and 3.2 *Gbps* respectively. The driver R_{on} in LVSTL signaling is not well controlled and thus it relies on the termination at the receiver to maintain the signal integrity. The pull-up and pull-down I-V characteristics are shown for PODL and LVSTL signaling in Fig.3(a) and Fig.3(b), respectively. Both pull-up and pull-down devices are highly nonlinear for LVSTL signaling when compared to PODL signaling.

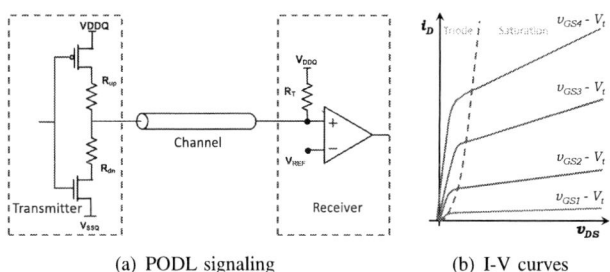

| (a) PODL signaling | (b) I-V curves |

Fig. 1.   PODL signaling for current high-speed memory interface.

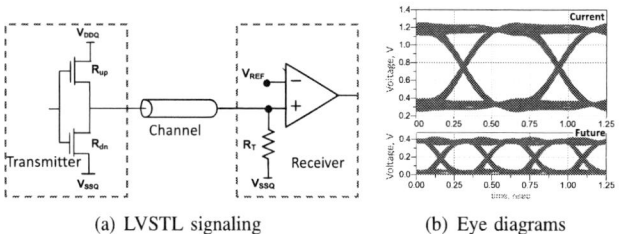

| (a) LVSTL signaling | (b) Eye diagrams |

Fig. 2.   LVSTL signaling for future high-speed memory interface.

In this paper, the multilinear theory is applied to perform peak distortion analysis of nonlinear links by decomposing them into linear subnetworks and analyzing using the standard peak distortion analysis method. Then, the final solutions are obtained by combining the solutions of the subnetworks.

In Section II, the peak distortion analysis is reviewed and the multilinear theory is introduced. First, the technique derived and then the analysis steps along the implementation details are presented. Simulation results of high-speed nonlinear links

978-1-4799-0708-3/13 $31.00 © 2013 IEEE

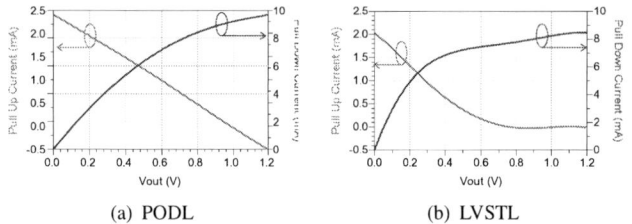

(a) PODL   (b) LVSTL

Fig. 3. I-V characteristics of the PODL and LVSTL I/O drivers.

are presented in Section III. The eye diagrams from the proposed methods are compared. Finally, the conclusion is given in Section IV.

II. PEAK DISTORTION ANALYSIS

If the high-speed link is linear and time-invariant, the transmitted signal can be modeled as a linear combination of step responses with shifted time. When the transmitted signals are bits, pulses with equal rise and fall times, the received waveforms can be calculated as a linear combination of single bit responses.

Based on this linear time-invariant (LTI) assumption, peak distortion analysis generates the worst-case eye diagrams based on the superposition of the bit responses. First, the single-bit response of the link is obtained by exciting the link with a pulse one unit interval wide. The single-bit response, $y(t)$, of a LTI system is given by the convolution of the transmitted single pulse, $x(t)$, and the impulse response of the link $h(t)$:

$$y(t) = x(t) \otimes h(t). \tag{1}$$

The worst-case data pattern and the worst-case eye diagram can be calculated by first folding this single-bit response, $y(t)$, into one unit-interval long and then finding the portions of the waveforms that degrade the received eye diagram assuming the transmitted bit is ONE or ZERO and finally adding up the segments. The worst-case upper eye shape when transmitting ONE is given by,

$$s_1(t) = y(t) + \sum_{\substack{k=-\infty \\ k \neq 0}}^{\infty} y(t-kT)|_{y(t-kT)<0} \tag{2}$$

$$+ \sum_{i=1}^{n} \sum_{k=-\infty}^{\infty} y^i(t-kT-t_i)|_{y^i(t-kT-t_i)<0},$$

where T is the symbol period, n is the number of crosstalk channels, t_i is the relative sampling point for each crosstalk single-bit responses, and y^i is the crosstalk single-bit response and the subscript '1' denotes the transmitted bit. Similarly, the worst-case lower eye shape when transmitting ZERO is given by,

$$s_0(t) = y(t) + \sum_{\substack{k=-\infty \\ k \neq 0}}^{\infty} y(t-kT)|_{y(t-kT)>0} \tag{3}$$

$$+ \sum_{i=1}^{n} \sum_{k=-\infty}^{\infty} y^i(t-kT-t_i)|_{y^i(t-kT-t_i)>0},$$

where the subscript '0' denotes the transmitted bit. The worst-case eye opening is given by the difference, $s_1(t) - s_0(t)$.

Assuming that the effects of the drivers and receivers in a link can be represented by linear model, the PDA provides very good approximations of the worst-case voltage and timing margins. However, when the nonlinearity effects of the drivers or receivers need to be consider in link analysis, extremely expensive time-domain simulation is needed to predict near worst-case performance prediction. One of the approaches to avoid time-domain simulation is to take advantage of the unique characteristics of the interconnect-dominated nonlinear links using multilinear techniques.

A. Multilinear Theory

In high-speed links, the passive interconnect systems integrate or smooth out the transmitted waveforms as it propagates along the signal path. The strength of the nonlinearities of the transmitters and receivers is reduced. Current high-speed and low-swing drivers show very mild nonlinearity around their operating points. This nonlinearity can accurately be captured via Volterra series expansion. Multilinear theory is used to solve the nonlinear networks by a sequential solution of a set of linear networks using efficient linear methods such as frequency-domain methods, reduced-order modeling, or peak distortion analysis techniques [3].

Volterra series expansion is used to describe the nonlinear network by combining the theory of convolution and Taylor series expansion [4]. The use of the multilinear theory or the Volterra series analysis has been used extensively in various applications of nonlinear analysis and optimization [4]- [6].

Volterra series analysis is the generalization of the convolution integral representation of linear systems. The input, $x(t)$, and output, $y(t)$, of any linear time-invariant system are related by its impulse response, $h(t)$, as in

$$y(t) = \int_{-\infty}^{\infty} h(\tau)x(t-\tau)d\tau. \tag{4}$$

The Fourier transform yields the frequency domain relationship as

$$Y(\omega) = H(\omega)X(\omega), \tag{5}$$

where $H(\omega)$ is the transfer function of the system, $X(\omega)$ and $Y(\omega)$ are the input and output signals. The response $y(t)$ of a weakly nonlinear system with small excitation is related to the input signal $x(t)$ by

$$y(t) = \sum_{n=1}^{\infty} y_n(t) \tag{6}$$

where

$$y_n(t) = h_0 + \int_{-\infty}^{\infty} h_1(\tau_1)x(t-\tau_1)d\tau_1$$

$$+ \int_{-\infty}^{\infty} \int_{-\infty}^{\infty} h_2(\tau_2,\tau_1)x(t-\tau_1)x(t-\tau_2)d\tau_1 d\tau_2$$

$$+ \int_{-\infty}^{\infty} \int_{-\infty}^{\infty} \int_{-\infty}^{\infty} h_3(\tau_1,\tau_2,\tau_3)x(t-\tau_1)$$

$$\times x(t-\tau_2)x(t-\tau_3)d\tau_1 d\tau_2 d\tau_3 + \cdots \tag{7}$$

978-1-4799-0708-3/13 $31.00 © 2013 IEEE

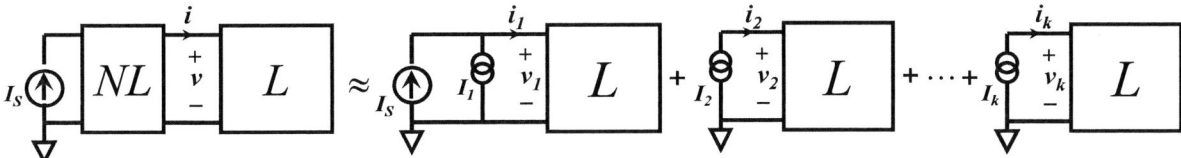

Fig. 4.   The original nonlinear network and the corresponding multiple linear networks.

where $h_n(\tau_1, \tau_2, \cdots, \tau_n)$ is called the $n-th$ order Volterra kernel or $n-th$ order nonlinear impulse response. The time domain Volterra Series response is related to the frequency-domain response by a multidimensional Fourier transformation, $H_n(\omega_1, \omega_2, \cdots, \omega_n)$.

In practice only finite number of Volterra transfer functions needs to be calculated. However, the determination of the Volterra kernel either in time domain or frequency domain of arbitrary nonlinear networks is generally difficult.

In multilinear theory, also known as the *method of nonlinear current*, the Volterra kernels are not explicitly calculated to determine the input-output relationship of nonlinear component [5]-[6]. Instead, the multilinear model decomposes the nonlinear network into multiple linear networks with sources that are linear, quadratic, cubic, and higher-order functions of currents and voltages that are derived from the lower order model as shown in Fig. 4. These current and voltage sources capture the nonlinearities in the original network. Therefore, the nonlinear system is decomposed into multiple linear networks that are sequentially and efficiently solved using linear models. This procedure is simple, fast, and computationally efficient.

Consider the simple low-swing signaling interconnect shown in Fig. 5 [7]. It consists of an open drain driver, an interconnect line, and a receiver. The current-voltage characteristic function, $I = f(V_G, V_D)$ of the driver, in Fig. 5, can be expanded in a two-dimensional Taylor series around V_G, V_D as,

$$
\begin{aligned}
I = & \left.\frac{\partial f}{\partial V_G}\right|_{V_G=V_1} v_g + \left.\frac{\partial f}{\partial V_D}\right|_{V_D=V_2} v_d \\
& + \frac{1}{2}\left.\frac{\partial f}{\partial V_G \partial V_D}\right|_{\substack{V_G = V_1 \\ V_D = V_2}} v_g v_d + \cdots
\end{aligned}
\tag{8}
$$

Then, the small signal nonlinear characteristic of the driver can be expressed as voltage-controlled conductance as

$$
i = g_0 v_g + g_1 v_d + g_2 v_g v_d + g_3 v_g^2 + g_4 v_d^2 + \cdots,
\tag{9}
$$

where $v_g \ll V_G$ and $v_d \ll V_D$.

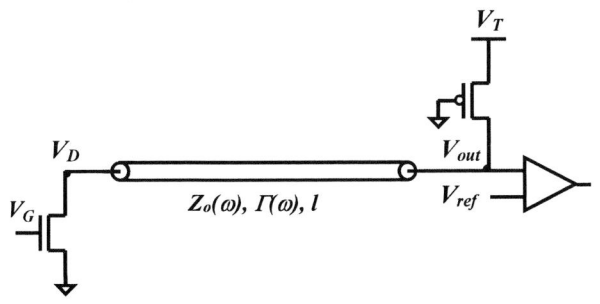

Fig. 5.   A simple reduced-swing interconnect system.

The current in the simplified driver model of Fig. 5 is controlled by the gate, V_G, and drain, V_D, voltages.

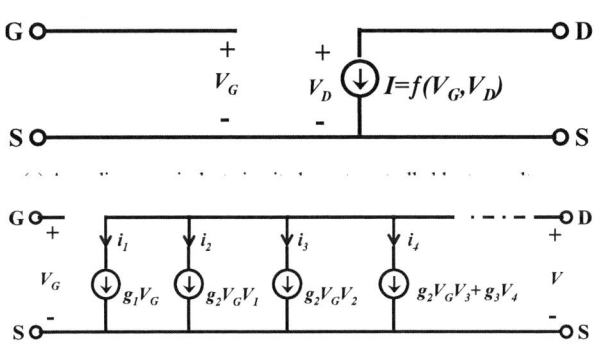

(b) The circuit in Fig 6(a) converted to linear resistor and a set of current sources.

Fig. 6.   A simple equivalent circuit for driver in Fig 5.

Thus, the voltage and current in any node and branch can be expressed in the series as

$$
v(t) = \sum_{n=0}^{\infty} v_n(t); \;\; i(t) = \sum_{n=0}^{\infty} i_n(t),
\tag{10}
$$

where $v_n(t)$ and $i_n(t)$ represent the sum of all nth-order mixing products of voltages and currents, respectively. Even though the current sources in Fig. 6(a) are nonlinear functions of node voltages, the higher-order networks are linear with independent sources which are known functions of the lower-order voltages and currents. Thus the analysis of the nonlinear circuit of Fig. 5 is reduced to the sequential analyses of linear networks as shown in Fig. 7.

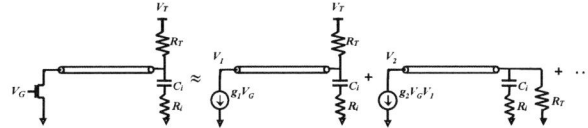

Fig. 7.   The equivalent nonlinear network of the interconnect system in Fig. 5 and a sequence of linear networks from multilinear theory.

The convergence rate of the method obtained depends on the level of nonlinearity. Consequently, the approach converges rapidly for networks with mild nonlinearities or for interconnect-dominated systems where the effects of nonlinear components do not strongly influence the response of the overall system [7].

The procedure for solving the nonlinear network using multilinear theory can be combined with peak distortion analysis to predict the worst-case eye diagrams of the high-speed nonlinear network. The expressions of peak distortion analysis of the upper and lower edges of the eye diagrams in

978-1-4799-0708-3/13 $31.00 © 2013 IEEE   171

(2) and (3) can be modified to handle nonlinear link as,

$$S_1(t) = s_1(t) + \sum_{r=1}^{m} s_1^r(t - t_r) \qquad (11)$$

$$S_0(t) = s_0(t) + \sum_{r=1}^{m} s_0^r(t - t_r), \qquad (12)$$

where t_r is the relative sampling point of the higher order responses, $s_1^r(t)$ and $s_0^r(t)$ represent the effects of the sum of all r^{th}-order nonlinearity when transmitting ONE and ZERO, respectively.

III. SIMULATION RESULTS

To illustrate the advantage and verify the accuracy of the proposed technique, the nonlinear driver model of a low voltage swing terminated logic interface is analyzed. The block diagram of the simplified signaling link, transmitter, the channel and the termination to ground, is shown in Fig. 8(a). The channel consists of several components between the I/O pads of the transmitter and receiver. This short channel is about 2-*in* long and consists of printed circuit board (PCB) traces in FR-4 substrate, a flip-chip and a wirebond packages for the transmitter and receiver chips, and vias, solder balls, as well as device capacitances at drivers and receivers. The impedances of the PCB and package traces are $40 \ \Omega$.

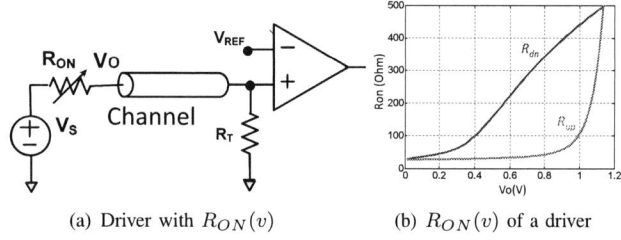

(a) Driver with $R_{ON}(v)$ (b) $R_{ON}(v)$ of a driver

Fig. 8. LVSTL signaling with $R_{ON}(v)$ and pull-up and pull-down drivers.

First, the output impedance of the driver, shown in Fig.8(b), is approximated by a low-order polynomial function and the nonlinear characteristic of the driver is expressed as voltage-controlled conductance as :

$$i = g_0 + g_1 v_o + g_2 v_o + g_3 v_o^2 + g_4 v_0^3 + \cdots . \qquad (13)$$

Then, the multilinear method is used to analyze the link. The single bit responses of the link with third-order nonlinearity are shown in Fig. 9. The zeroth, first, second, and third-order single bit responses of the link are shown in Fig. 9(a), Fig. 9(b),Fig. 9(c), and Fig. 9(d), respectively. Although the absolute magnitude of the high-order single-bit responses is reduced, the ratio of their pre or post-cursors to their main cursor increases.

The received eye diagrams of the link when transmitting a $100,000$ long PRBS at a data rate of $6.4 Gb/s$ is obtained using time domain simulations. Fig. 10 shows that prediction of the proposed method gives valid worst-case voltage and timing margins.

IV. CONCLUSIONS

In this paper, an approach to enable statistical channel simulation of nonlinear link by extending the peak distortion

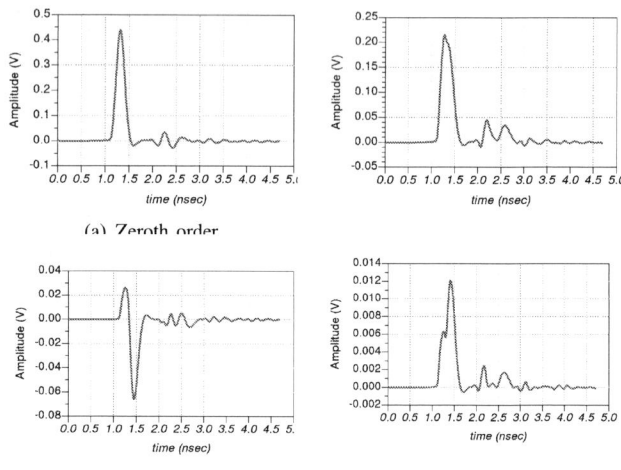

(a) Zeroth order

(c) Second order   (d) Third order

Fig. 9. Single-bit responses for a link with third-order nonlinearity.

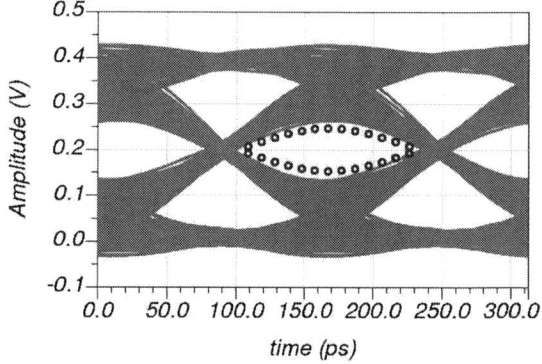

Fig. 10. An eye diagram from time-domain simulation of 100,000 bits with the edges of the upper and lower eyes from the proposed modified peak distortion analysis are shown in small circles, ○.

analysis to include the impact of nonlinear drivers is presented. First, the multilinear theory is used to decompose the link with nonlinear transmitters and receivers and solve using a sequence of linear solutions. Then, a series of peak distortion analyses of the linearized subnetworks are combined to predict the worst-case voltage and timing margins of high-speed nonlinear links. The worst-case margins obtained from the proposed method are compared favorably with results obtained from long time-domain simulation.

REFERENCES

[1] B.K. Casper et al, "An accurate and efficient analysis method for multi-Gb/s chip-to-chip signaling schemes," *IEEE Symposium on VLSI Circuits*, June 2002, pp. 54-57.

[2] J. Y. Choi, "Advanced mobile memory technology (LPDDR4)," *JEDEC, Mobile Forum*, May, 2013.

[3] M. Schetzen, *The Volterra & Wiener Theories of Nonlinear Systems*. Malabar, Floridia: Robert E. Krieger Publishing Company, Inc., 1989.

[4] V. Volterra, *Theory of Functionals and of Integral and Integro-Differential Equations*. New York, New York: Dover, 1959.

[5] M. Schetzen, " Multilinear theory of nonlinear networks," *Journal of the Franklin Institute*, Vol. 320, No. 5, 1985, pp. 221-247.

[6] S. Maas, *Nonlinear Microwave Circuits*. Norwood, Massachuset: Artech House Inc., 1988.

[7] W. T. Beyene, " Multilinear and waveform relaxation methods for efficient analysis of interconnect-dominated nonlinear networks," *IEEE Trans. Adv. Packag.*, Vol. 31, No.3, pp. 637-648, August 2008.

978-1-4799-0708-3/13 $31.00 © 2013 IEEE

A Low-Frequency Enhanced S-Parameter Handling Scheme for Time Domain Simulation of High Speed Interconnects

Chong Luo, Jingping Zhang, Jiayuan Fang
Sigrity R&D
Cadence Design Systems, Inc.
San Jose, CA 95134, USA.
cluo@cadence.com, zhangjp@cadence.com, fangj@cadence.com

Abstract—A multi-resolution scheme is proposed to efficiently handle the S parameter with rapid change at low frequency in time domain simulation. It is demonstrated by numerical results that, with the aid of the proposed scheme, the accuracy of time domain simulation based on aforementioned S parameter can be significantly improved without substantial increase of computation costs.

Keywords—S parameter; FFT; multi resolution method

I. INTRODUCTION

Capacitors are frequently used in high speed interconnects for DC blocking. One consequence of DC blocking capacitors is the rapid change of S parameters of interconnects at the low frequency. The magnitudes of S parameters can change from a value close to 0 to almost 1 in less than tens of KHz. For time domain simulation of interconnects, transient waveforms are often obtained through the inverse Fourier transforms of S parameter models. The rapid change of S parameters at the low frequency can cause serious problems in the time domain simulation as outlined in the following.

Conventional IFFT algorithm is based on an equidistance scheme. The samples of the input and output of IFFT are uniformly distributed in the frequency and the time domains. In order to maintain the accuracy of IFFT, the resolution has to be high enough to catch the fastest change of the data. It often requires very high resolution to capture the rapid change of S parameters caused by DC blocking capacitor at low frequency. For instance, the resolution may have to be as fine as KHz under some scenarios. Otherwise, the accuracy of IFFT is compromised.

If a uniformly high resolution scheme is used to compute the time domain wave from S parameters with local rapid change, the computation costs could be too high to be practical for any real applications. For example, if an S parameter of 8 GHz side bandwidth requires a resolution of 100 KHz, then it needs to store more than 100,000 samples for each of the elements of the S matrix. While conducting time marching, the computational cost for every impulse response would be 100,000 floating point multiplications at each time step.

In this paper, a multi resolution scheme is proposed. Based on the results of numerical simulations, it is found that, due to more efficient allocation of resources, the proposed method is able to handle the S parameter with rapid low frequency change accurately at a relatively low cost.

II. ALGORITHM

In order to use the S parameter for time domain simulation, the corresponding impulse response has to be computed by IFFT. In the single resolution scheme, the input for IFFT is the frequency domain data with a uniform resolution. Consequently, the time domain impulse also has a uniform resolution. The uniform resolutions employed in both the frequency and the time domains will lead to significant cost for handling S parameter that carries low frequency local rapid change for the following reasons. To capture the rapid change at low frequency, the resolution has to be sufficiently high. However, even though the S parameter carries rapid change at low frequency, the high frequency component usually does not change as fast as it does at the low frequency. The uniform high resolution is a waste of computational resource in the frequency domain. This conclusion holds true for the time domain also, since the high resolution employed to capture the fast changing early time response is also a waste for the smooth late time response.

A simple example is presented below to demonstrate the ineffectiveness of the single resolution scheme. As shown in Figure 1, the circuit consists of two transmission lines with DC blocking capacitors plugged in.

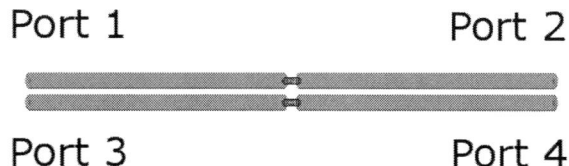

Figure 1. Structure of a 4-port network.

The width and length of those transmission lines are 1mm and 30mm, respectively. The capacitance of each of the two capacitors is 0.5uF. The magnitude of S11 from DC to 8GHz is presented in Figure 2.

978-1-4799-0708-3/13 $31.00 © 2013 IEEE

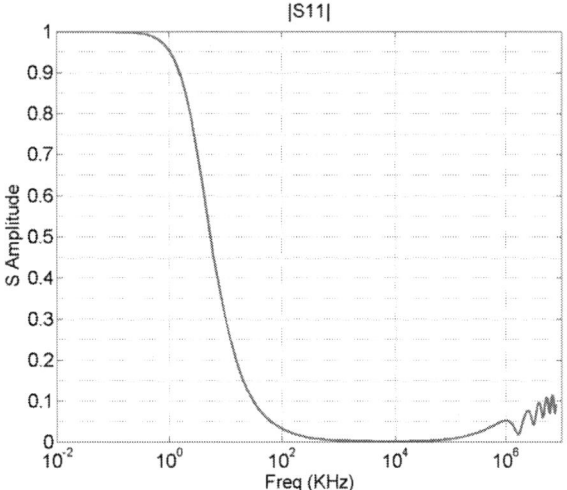

Figure 2. |S11| of the 4-port network.

It is found in Figure 2 that the magnitude of S11 drops from 0.95 to 0.05 in only 100 KHz. This observation suggests that the resolution of the IFFT has to be finer than 100 KHz in order to capture the change at the low frequency. Otherwise, late time error will be observed in the time domain. When such an error is significant, it can be observed almost immediately after the time domain simulation starts. The comparison of the voltages across port 1 computed in time domain simulation is presented in Figure 3. In this comparison, the reference is the voltage obtained directly by simulating the circuit shown in Figure 1 using transmission line and lumped circuit models (red line). The other two results are of simulations based on S parameter models shown in Figure 2. The difference between them is the resolution of IFFT. One is with the resolution of 78.125 KHz (blue dashed line), while the other one is 5 MHz (black dashed and dotted line).

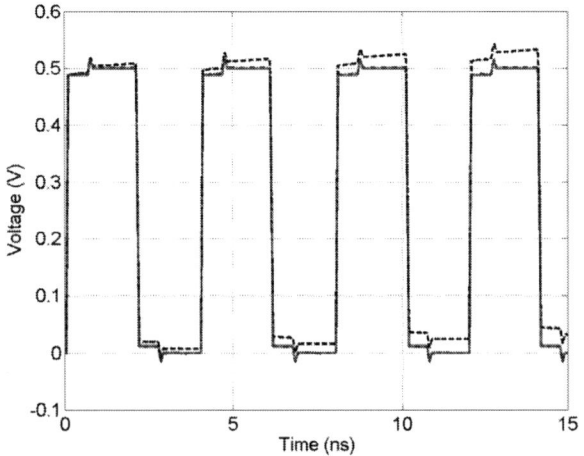

Figure 3. Comparison of voltages across port 1 computed using single resolution scheme.

It is observed in Figure 3 that the curve corresponding to 5 MHz resolution starts to deviate from the other two curves almost immediately after the simulation starts. This deviation is due to the loss of accuracy at the low frequency. By presenting in Figure 4 the comparison of the original S parameter (red line) and the S parameter recovered from the time domain impulse response of the simulation with 5 MHz resolution (blue dashed line), this conclusion can be demonstrated precisely.

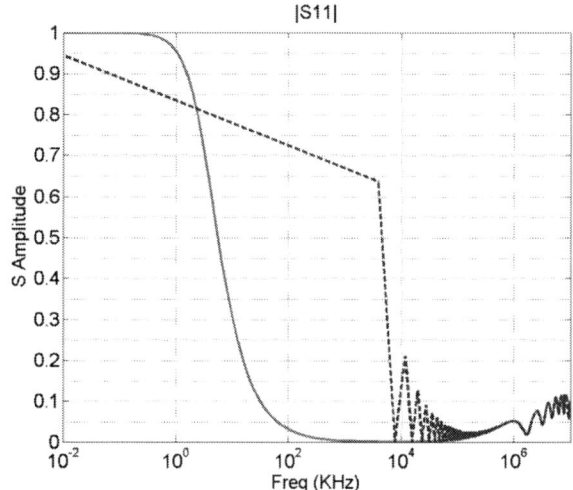

Figure 4. Comparison of the original and recovered S parameter

It is concluded from Figure 4 that, due to the inadequate IFFT resolution, significant error is introduced in the low frequency region.

In Figure 3, the curve corresponding to the simulation with resolution of 78.125 KHz is in good agreement with the reference. However, this accuracy is obtained with much higher cost. The CPU time and memory cost of the simulations based on 5 MHz resolution and 78.125 KHz resolution are presented in TABLE I.

TABLE I.   COST COMPARISON OF SINGLE RESOLUTION SIMULATIONS

Resolution	CPU time (s)	Memory (MB)
78.125 KHz	1001.57	421.36
5 MHz	34.18	52

As presented in TABLE I, the cost of the simulation with 78.125 KHz resolution is much higher than that of the simulation with 5 MHz resolution. Considering the model in this example is only a small 4 port network, the cost can be too high to be practical for many real applications with S parameters of such a nature.

Here we propose a multi resolution scheme as a solution to the high cost problem of single resolution scheme. Because this multi resolution scheme allocates more resource to the rapid changing low frequency region, it is therefore named as low frequency enhanced S parameter handling scheme. In this scheme, different resolutions are used for the low frequency and high frequency regions. For frequency lower than a preset boundary, high resolution is used. In order to suppress the error caused by the discontinuity in resolution, a transition frequency region is created between the high resolution and the low resolution regions. In the transition region, the high resolution data is multiplied by a raised cosine function which decreases with the frequency and the low resolution data is multiplied by a function that equals to one minuses the function multiplied to the high resolution data. Therefore, their magnitudes decrease to zero from one end of the transition region to another and the summation of them reproduces the original S parameter.

The scheme of multi resolution does cause artificial errors if not handled properly. It has been found that this error is controllable and it decreases linearly as the ratio of the width of the transition region and the resolution of low resolution data increases.

After the frequency domain data is split into the low frequency and high frequency components, the corresponding time domain impulse response can be computed from them by IFFTs of their corresponding resolutions. These two impulse responses will then be superimposed to obtain the total impulse response.

The impulse responses computed from the low frequency and high frequency components have different resolutions and lengths in time. Consequently, special care needs to be taken to find the total impulse response. To superimpose the responses of the low frequency and the high frequency components, interpolation is employed to raise the resolution of the low frequency data to that of the high frequency component. A transition region in time is also introduced at the end of the high frequency impulse response in order to avoid abrupt change in resolution. It is elaborated below on how the transition region in time is constructed.

Suppose the resolutions in the high and the low resolution regions are t_h and t_l, respectively, and the following relationship holds between them.

$$t_l = 2^n t_h \text{, where n is an integer.}$$

The transition region consists of n uniform resolution sub regions, with each of the sub regions of length t_l. The resolutions in this transition region decrease progressively from t_h in the first sub region towards $2^{n-1}t_h$ in the last sub region.

The structure of the transition region for the case n = 3 is illustrated in Figure 5.

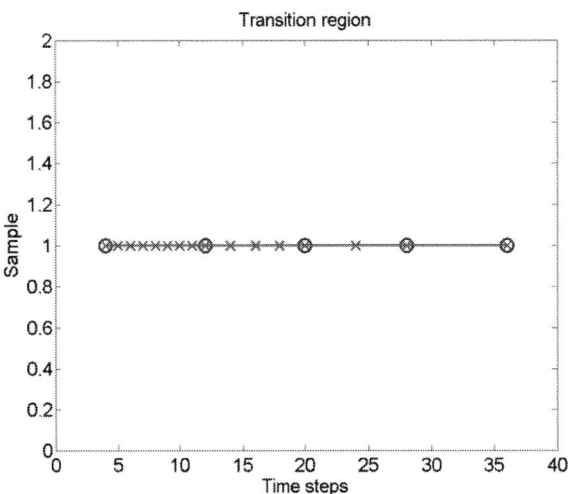

Figure 5 The structure of a transition region for n = 3

In Figure 5, the circles are corresponding to the original low resolution data from the low frequency component. The crosses are the data obtained from the interpolation of low resolution data. .

The simulation result based on the multi resolution scheme is presented in Figure 6 (blue dashed line). In this simulation, the resolution is 78.125 KHz for the low frequency component and 5 MHz for the high frequency part.

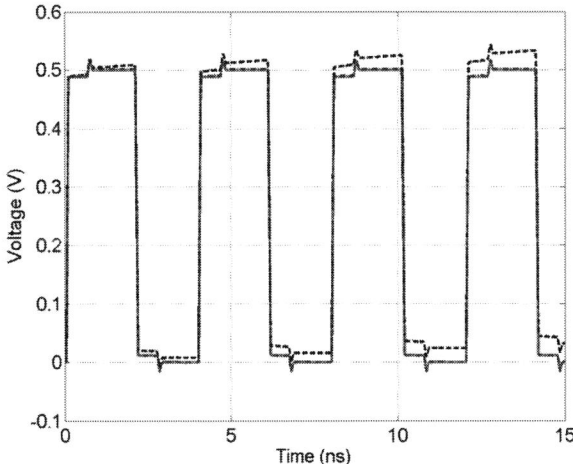

Figure 6. Comparison of voltages across port 1 computed using single resolution and multi resolution scheme.

As reference, the simulation results based on the single 5MHz resolution scheme (black dashed dotted line) and based on transmission line and lumped circuit models (red line) are also presented. From Figure 6, it can be seen that the curve obtained by using the low frequency enhanced multi resolution scheme agrees very well with that obtained directly from transmission line and lumped circuit models.

The comparison of the costs of the multi resolution and single resolution scheme is presented in TABLE II.

TABLE II.　COST COMPARISON OF SINGLE RESOLUTION AND MULTI RESOLTUION SCHEMES

Resolution	CPU time (s)	Memory (MB)
78.125 KHz (single)	1001.57	421.36
5 MHz (single)	34.18	52
Multi resolution	55.77	55.91

From the comparison of simulation results and the costs, the benefits of multi resolution scheme are clearly demonstrated. With a small amount increase of the cost, the multi resolution scheme yields a significantly higher accuracy.

III. NUMERICAL RESULTS

In this section, a more compelling example from a real application is used to demonstrate the effectiveness of the multi resolution scheme. In this case, the time domain response of a circuit to Gaussian pulse is computed with the proposed scheme. The result obtained from the vector fitted micromodel [1] is used as a reference. Figure 7 shows the S parameter used in this example.

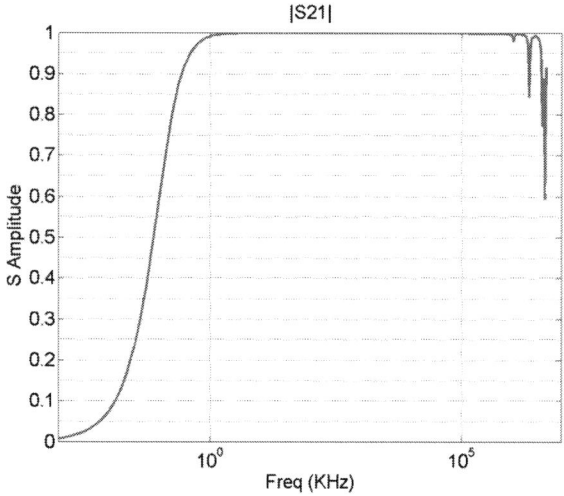

Figure 7. |S21| of a 4-port network.

Apparently, the S parameter of this network has a more steep change at low frequency. To obtain an accurate result, the resolution for the low frequency component has to be very high (488 Hz is used) and the resolution for the high frequency component is 2 MHz. The simulation results based on multi resolution and single resolution schemes are presented in Figure 8 together with the reference.

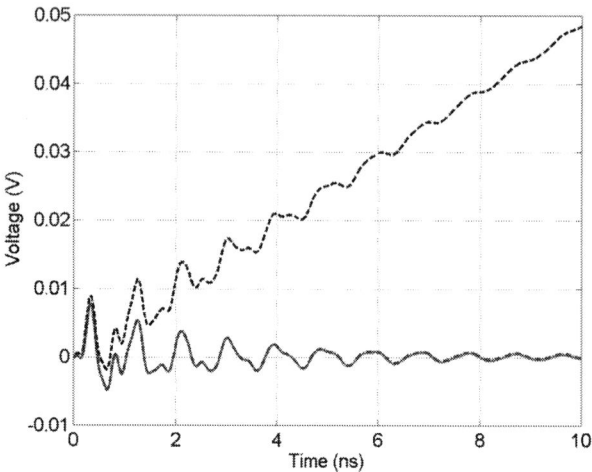

Figure 8. Comparison of voltage across port 2 computed using single resolution and multi resolution scheme

It can be clearly seen in Figure 8 that the single resolution scheme with 5MHz resolution generates an erroneous late time result (black dashed dotted line). On the other hand, the result of multi resolution scheme (blue dashed line) has a very good agreement with the reference (red line). It takes the multi resolution scheme 375 seconds to finish the simulation. If the single resolution scheme of 488Hz is used, it would take hours to complete the computation.

IV. CONCLUSION

A multi resolution scheme was proposed to handle the S parameter that carries rapid change at low frequency in time domain simulation. As demonstrated by numerical simulation, this scheme is able to significantly reduce the computation cost while maintain the accuracy of simulation.

REFERENCES

[1] B. Gustavsen and A. Semlyen, "Rational approximation of frequency domain responses by vector fitting", IEEE Trans. Power Delivery, vol. 14, no. 3, pp. 1052-1061, July 1999.

Poster Session

978-1-4799-0708-3/13 $31.00 © 2013 IEEE

A package-level implementation of traveling-wave switch using PIN-diodes

Saqib Kaleem, Sven Rentsch, Stefan Humbla, and Matthias Hein

RF and Microwave Research Laboratory, Ilmenau University of Technology, Ilmenau 98693, Germany

Phone: +49 3677 692695, email: saqib.kaleem@tu-ilmenau.de

Abstract—A single-pole single-throw switch constituting the transparent path in a reconfigurable switch matrix is presented, improving the availability of the switch matrix in case of an on-board power-fail. The design of the switch is based on the traveling-wave concept using packaged PIN-diodes on a thick-film ceramic substrate under the constraints on the number of PIN-diodes and power dissipation. The circuit design and analysis of the switch based on the concept of an artificial transmission-line are described. Unlike the traditional low-pass characteristics inherent to traveling-wave switches, high isolation is achieved by transforming the parasites of the PIN-diodes in the shunt arms to reveal band-pass characteristics besides assisting the assembly of the package. On-wafer measurements reveal an isolation $\geq$ 45 dB and an insertion loss $\leq$ 6 dB in the Ka-band downlink frequencies (17...22 GHz).

Keywords—Reconfigurable switch matrix, traveling-wave switch, low temperature co-fired ceramic (LTCC).

I. INTRODUCTION

A reconfigurable switch matrix (RSM) provides dynamic on-board transponder interconnectivity in satellite communications. A 4×4 reconfigurable switch matrix at Ka-band downlink frequencies (17...22 GHz) has been developed in a public research project related to the German geostationary satellite mission 'Heinrich Hertz' [1]. The PIN-diodes based multiple-throw switches are hybrid integrated on a multi-layer ceramic substrate using the low temperature co-fired ceramic (LTCC) technology. Thick-film substrates form the basis for high-density packages due to a high degree of integration. Due to the electrical and thermal characteristics of the ceramic substrate, thick-film circuits find their applications in harsh environment [2]. In an earlier research project, the LTCC technology was proven suitable for space applications in terms of robustness, reliability, lifetime and environmental compatibility [3].

The RSM module is comprised of eight single-pole four-throw (SP4T) microwave monolithic integrated circuit (MMIC) switches, placed in cavities of the ceramic substrate using double-side mounting. Besides MMIC, there are surface mount components and passive structures distributed in the eight-layered ceramic stack. The RSM module, therefore, represents a dense integration of functional components whilst possessing a compact size of 25 mm×25 mm as displayed in Fig. 1.

For the geostationary satellite mission, four transparent paths are introduced in the 4×4 RSM module in order to guarantee transmission in the case of on-board power-fail. The transparent paths become active only in the absence of power-supply and bypass the inner-core of the switch-matrix by connecting inputs directly to the corresponding outputs,

Figure 1. A camera view of the 4×4 RSM. The double-side mounted bare-die SP4T MMIC are encased in a Kovar lid and frame, resulting in a hermetically sealed ceramic package. The surface-mount components form the integrated bias circuitry for the PIN diodes.

demonstrated schematically in Fig. 2. A single transparent path is represented by a single-pole single-throw (SPST) switch.

This paper presents the implementation of the SPST switch based on the traveling-wave concept using packaged PIN-diodes on the ceramic substrate, where the PIN-diodes are used to detect the presence of on-board power-supply. Due to low parasites of the integrated switching device, the traveling-wave concept is a renowned approach at the integrated circuit level to realize low-loss and high-isolation switches from DC to millimeter-wave frequencies [4, 5]. However, using packaged PIN-diodes, parasites of the switching device and technological constraints limit the achievable insertion loss and isolation. In the following, the design and measurement of a high-isolation SPST switch are presented under a strict constraint on the number of PIN-diodes. On the packaging substrate, the through-hole via connections to ground and mounting pads degrade the intrinsic parasites of the PIN-diode. Therefore, an additional transmission line in the shunt-arm is used to

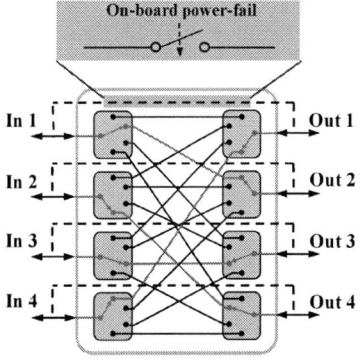

Figure 2. Functional representation of a 4x4 RSM. A particular state of the RSM is depicted; electrically active paths are marked in red while the black lines represent the isolated paths. A SPST switch forms the transparent path (dashed and highlighted) between inputs and outputs.

978-1-4799-0708-3/13 $31.00 © 2013 IEEE

restore the intrinsic parasites, however, revealing a band-pass characteristic of the traveling-wave SPST switch.

II. DESIGN AND ANALYSIS

A PIN-diode is represented by its equivalent model, a combination of passive lumped elements to reflect its intrinsic parasites; the ON-state is modeled by the on-resistance R_{on} while the OFF-state is represented by a parallel combination of the off-capacitance C_{off} and the isolation-resistance R_{off} [6]. Due to $\omega R_{\mathrm{off}} C_{\mathrm{off}} \gg 1$, the isolation-resistance R_{off} is ignored from the off-state equivalent model. R_{on} is a function of forward-current and increases with frequency. C_{off} depends on the reverse voltage and settles down to a constant value above a certain frequency limit (10 GHz for the PIN-diode used in this work), however, its reactance decreases with frequency. For a shunt-mounted PIN diode, this monotonous degradation of impedance implies reduced isolation and increased insertion loss at microwave frequencies.

The traveling-wave switch is composed of a chain of shunt-mounted PIN-diodes periodically loading a short transmission line of high impedance. A short transmission line ($l < \lambda/4$) is represented by its equivalent π-model as shown in Fig. 3 [7]. The equivalent inductance L_l and capacitance C_l are functions of the electrical length βl_l and characteristic impedance Z_l of the transmission line. Combining the equivalent models of the PIN-diode and high impedance transmission line, the equivalent models of the traveling-wave switch in the transmission and isolation states are shown in Fig 4. In the insertion-state, an artificial right-handed transmission-line is formed by $C_{\mathrm{off}} + 2C_l$ and L_l under the assumption $C_{\mathrm{off}} \gg C_l$. The impedance of the artificial transmission line Z_o depends on C_{off}, C_l and L_l. A detailed design procedure of the traveling-wave switch was outlined in [4]; the design parameters include the switching device periphery, the number of switching elements, and the impedance and length of the interconnect lines. However, using off-the-shelf PIN-diodes, the device periphery is not a design parameter. The number of PIN-diodes is constrained (in this work to 6) in order to limit the power dissipation and reduce the components count on the RSM module. Therefore, the design parameters left are C_l and L_l which can be expressed by Z_l and βl_l.

For the required characteristic impedance of the artificial transmission line Z_o, the impedance Z_l and the electrical length βl_l of the high impedance line are selected from Fig. 5. In the current work, for $Z_o = 50$ Ω, $Z_l = 70$ Ω and $\beta l_l = 0.23\pi = 42°$ are chosen, adhering to the LTCC design rules. The intermediate design parameters as well as the achievable theoretical insertion loss and isolation (computed by the set of equations in [4]) are enlisted in Table I. It can be observed that the capacitance associated with the high-impedance line C_l is of the order of C_{off}, therefore, the assumption of homogeneous artificial transmission line

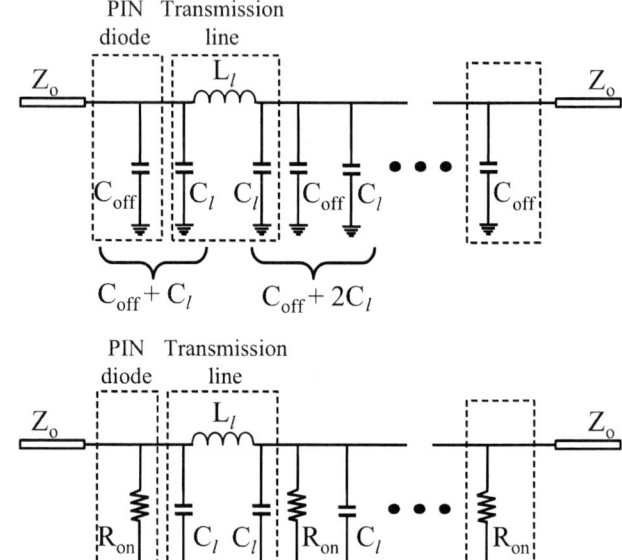

Figure 4. Equivalent models of a traveling-wave SPST switch represented by the equivalent model of a short transmission line and simplified PIN-diode models in the insertion-state (top) and isolation-state (bottom).

does not hold true in a strict sense. Consequently, the outer capacitance $C_{\mathrm{off}} + C_l$ in the equivalent circuit of Fig. 4 cannot be considered equal to the inner capacitance $C_{\mathrm{off}} + 2C_l$ when $C_{\mathrm{off}} \approx C_l$. Based on the design values mentioned in Table I, a low-pass characteristic results.

In the isolation-state, the PIN-diodes in their ON-states form a very low-impedance transmission path between input and output as shown in Fig. 4, exhibiting reflective characteristic. The equivalent circuit, in this case, is a shunt resistance of $\sim 0.01 R_{\mathrm{on}}$ with estimated isolation of 50 dB at 20 GHz.

The implementation of the traveling-wave switch on a substrate is represented by the equivalent circuit depicted in Fig. 6, in which the non-ideal effects are included, e.g. the bond-wires ($L_{\mathrm{bond-wire}}$) at the anode and cathode of the PIN-diode, through-hole via (L_{via}) and the mounting pad (Z_{pad} and βl_{pad}) to connect PIN-diodes to the back-side ground metalization. This via and the contact-pad form a non-ideal ground at high-frequencies. To meet the LTCC design rules, the impedance and the electrical length of the contact pads are

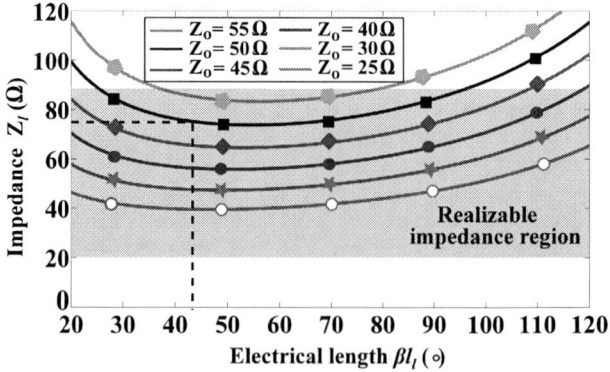

Figure 5. The dependency of impedance Z_l and electrical length βl_l of the high-impedance transmission line on the characteristic impedance of the artificial transmission line Z_o.

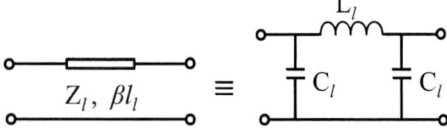

Figure 3. Equivalent π-model of an ideal transmission line of short length ($\beta l_l < \pi/4$).

978-1-4799-0708-3/13 $31.00 © 2013 IEEE

Table I. Design Values of Traveling-Wave Switch at 20 GHz

Design parameters	Values
C_{off} and R_{on}	0.06 pF and 8 Ω
Number of PIN diodes (n)	6
Impedance of interconnect line (Z_l)	73.6 Ω
Electrical length (βl_l)	42°
Equivalent inductance (L_l)	0.4 nH
Equivalent capacitance (C_l)	0.045 pF
Impedance of artificial line (Z_o)	50 Ω
Estimated return loss and insertion loss	15 dB and 2 dB
Estimated isolation	50 dB

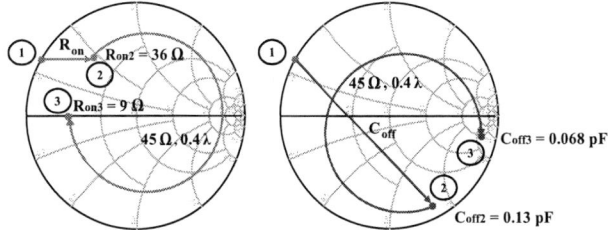

Figure 7. Impedance transformation at points 1, 2 and 3 in Fig. 6 in the isolation (left) and insertion states (right) of the SPST switch at 20 GHz. A single transmission line in the shunt arm restores the intrinsic parasites of the PIN-diode (mentioned in Table I).

chosen as $Z_{pad} < 50$ Ω and $\beta l_{pad} = 0.125\pi = 22°$. Due to significant length $\beta l_{pad} + L_{via} \approx 0.19\pi = 33°$, the isolation and insertion loss deviate significantly from the theoretical estimates mentioned in Table I.

The impedance transformation at different points marked in the equivalent circuit of Fig. 6 at 20 GHz is shown in Fig. 7. Due to non-ideal ground effects, the R_{on} and C_{off} values of the diodes are transformed to equivalent values of R_{on2} and C_{off2} respectively, at Point 2. A transformation by a single transmission line with impedance Z_x and electrical length βl_x shifts the parasites to point 3. On the Smith chart, the transformed parasites R_{on3} and C_{off3} lie close to the intrinsic values of the PIN-diodes in both isolation and transmission states. Therefore, the earlier computed structure of traveling-wave switch can still be utilized, however, revealing a band-pass characteristic.

III. EXPERIMENTAL RESULTS

Two versions of the manufactured SPST switches based on the traveling-wave concept are shown in Fig. 8. The PIN-diodes (MACOM MA4SPS552) were epoxy glued to the microstrip lines on the DuPont 951PX substrate (thickness ~203 μm). The inductance $L_{bond-wire}$ is, therefore, quite low. Even though DuPont 951PX possesses reasonable dielectric losses compared to other thick-film substrates, this particular tape system is selected because of its reliable manufacturing, wide range of available pastes, and its hermetic packaging capabilities. On-wafer measurements were carried out by using microstrip-to-coplanar waveguide transitions and the S-parameters were extracted after de-embedding.

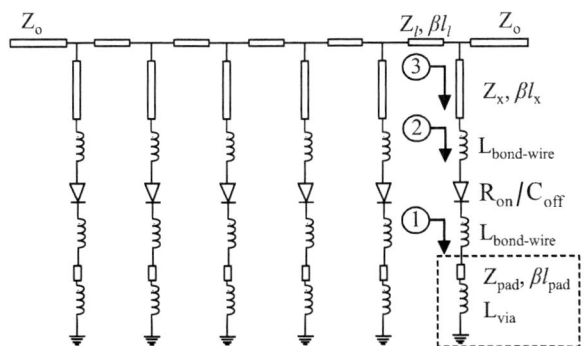

Figure 6. Equivalent circuit of the traveling-wave switch under realistic packaging influence like through-hole ground-via, mounting pads, bond-wires and intrinsic parasites of the PIN-diode. The relevant positions to observe the impedance transformation are marked at positions 1, 2 and 3 in the shunt arm.

A. PIN-diode characteristics

In the first step, R_{on} and C_{off} of the PIN-diode were extracted from measured S-parameters. For this purpose, a PIN-diode was mounted in series with a microstrip line. The PIN-diode was biased by a Programmable Network Analyzer (PNA) eliminating external bias-tees; a power supply connected to the rear-end of the PNA provided the required bias at the probing ports. In the presented scenario, the PIN-diode is used to detect the presence of on-board power-supply; the bias conditions are selected as +5 V, 5 mA (ON) and 0 V, 0 mA (OFF). This choice of forward current is sufficient to achieve high isolation whilst limiting the power dissipation in the SPST switch to 150 mW. For the shunt-mounted PIN-diode, increasing the forward current lowers R_{on} and improves the isolation at the expense of higher DC power dissipation. The expressions to extract R_{on} and C_{off} from the S-parameters as well as the extracted values are given in Table II [8]. The frequency dependencies of R_{on} and C_{off} are shown in Fig. 9. The figure-of-merit of the switching device is expressed in terms of its quality factor Q_s defined as the ratio of the off-state impedance to the on-state impedance [9]. This parameter is found to be 16.5 for the selected PIN-diode.

B. SPST Switch

The S-parameters of the SPST switch were measured at two bias levels, i.e. isolation-state at +5 V, 30 mA and insertion-state at no bias. The DC bias is provided at the open stub connected to the center conductor by a $\lambda/4$-line. Bias-free on-wafer contacts were established by addition of two DC-block single-layer capacitors (ATC113CHC1R2TT) at the input and output. In a separate measurement, the DC block capacitors were measured up to 30 GHz, exhibiting no parallel nor

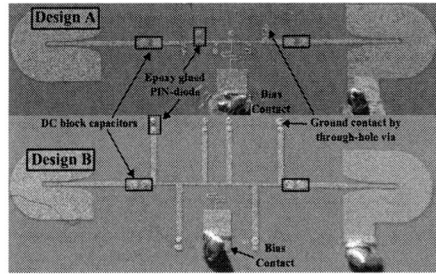

Figure 8. Manufactured traveling-wave SPST switches; 6 PIN-diodes are epoxy glued to the microstrip lines with their cathodes grounded by minimum length of contact-pad and through-hole via (top—Design-A). A transmission line in the shunt arm of each PIN diode transforms non-ideal effects due to packaging constraints (bottom—Design-B).

978-1-4799-0708-3/13 $31.00 © 2013 IEEE

Table II. EXTRACTED PARAMETERS OF SERIES PIN-DIODE AT 20 GHZ

Parameter	Extraction formula	Value
R_{on}	$2Z_o[10^{S_{21}/20}-1]$	8 Ω
C_{off}	$1/4\pi f_o Z_o\sqrt{10^{S_{43}/10}-1}$	0.06 pF
Q_s	$1/2\pi R_{\mathrm{on}}C_{\mathrm{off}}$	16.5

series resonances in the operational band. The measured S-parameters of the capacitor revealed an insertion loss of 0.8 dB at 20 GHz.

Two substrates of the SPST switches were assembled for comparison; one with a conventional traveling-wave SPST switch (Design-A) while the other is a SPST switch employing a transformation in the shunt arms (Design-B). The measured de-embedded S-parameters are displayed in Fig. 10. It is observed that the Design-A reveals a low-pass characteristic, as expected. The measured insertion loss and isolation for Design-A are ≤7 dB and ≥ 38 dB, respectively, over the frequency range 17...22 GHz. The reflection coefficients are ≤10 dB and the 3-dB cut-off frequency is 20.75 GHz. For Design-B, a band-pass behavior is evident with an insertion loss of ≤7 dB, return loss ≥8 dB and an isolation ≥45 dB in the required frequency band. It is clear that the restoration of the intrinsic parasites of the shunt-mounted PIN-diodes improves the isolation significantly without increasing the number of PIN-diodes. The poor results for return loss and insertion loss in Design-B are expected to improve (to ≥15 dB and ≤6 dB respectively) by adjusting the high-impedance interconnect lines.

IV. CONCLUSIONS

This paper presented the implementation of the traveling-wave switch concept using off-the-shelf PIN-diodes on the packaging substrate. By using a single transmission line in the shunt arm, the non-ideal effects of ground-via, mounting-pads and the parasites of the PIN-diodes are transformed to form an artificial transmission-line of 50 Ω with bandpass characteristic. On-wafer measurements reveal high isolation >45 dB and an insertion loss <6 dB by using only 6 PIN-diodes in the switching chain. Two DC-block capacitors in the transmission path increase the insertion loss by 1.6 dB. The SPST switch constitutes the transparent-path on the 4×4 reconfigurable switch-matrix module. The DC power consumption of a transparent path is 150 mW limiting the power dissipation of the entire switch matrix module < 2 W.

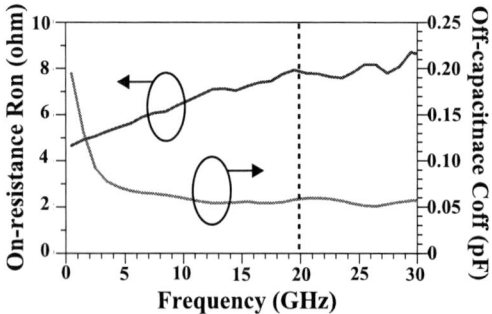

Figure 9. Frequency dependency of extracted parasites of the PIN-diode. R_{on} demonstrate monotonically increase with frequency while C_{off} settles down to almost a constant value.

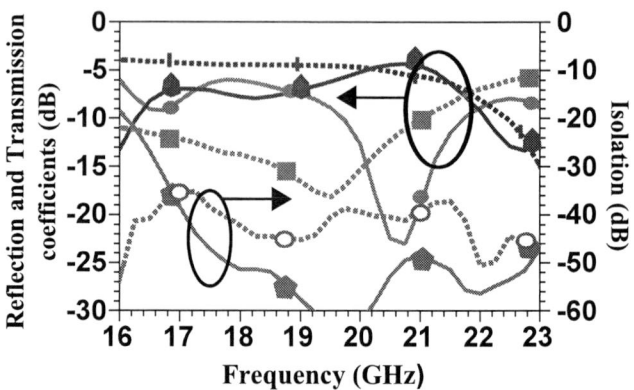

	Design-A	Design-B
Isolation	-○-	◆
Reflection Coefficients	-■-	◆
Transmission Coefficients	-+-	◆

Figure 10. Measured S-parameters of the SPST switches. Compared to Design-A, an improved isolation of ~10 dB is achieved by transforming the packaging effects in Design-B.

ACKNOWLEDGMENT

The authors would like to appreciate the efforts of Dirk Stöpel, Center for Micro- and Nanotechnologies to manufacture the ceramic substrates and Naveel Shaukat for characterizing PIN-diodes. Scientific support from Jens Müller and Ralf Stephan are highly acknowledged. Technical assistance from Michael Huhn and Matthias Zocher are appreciated.

REFERENCES

[1] S. Kaleem, S. Humbla, S. Rentsch, D. Stöpel and M.A. Hein, "Reconfigurable 4×4 switch matrix for Ka-band geostationary satellite mission", *Frequenz Journal of RF-Engineering and Telecommunications*, vol. 66, issue 11-12, pp. 347-354, December 2012.

[2] D.R. Schröder and L. J. Rexing, "LTCC MCM technology for Military environements", *Proceedings of the International Conference on Multichip Modules (ICMCM)*, pp. 612-617, April 1994.

[3] S. Humbla, R. Stephan, D. Stöpel, J. Müller and M. A. Hein, "Low-Earth-Orbit verification of a Reconfigurable 4×4 switch matrix and potential applications for satellite communications", *International Conference on Electromagnetics in Advanecd Applications (ICEAA)*,September 2013.

[4] Kun-You Lin, Wen-Hua Tu, Chen, Ping-Yu, Hong-Yeh Chang, Huei Wang, Wu and Ruey-Beei, "Millimeter-wave MMIC passive HEMT switches using traveling-wave concept", *IEEE Trans. Microwave Theory and Tech.*, vol. 52, issue 8, pp. 1798-1808, August 2004.

[5] Ruei-Bin Lai, Shih-Fong Chao, Zuo-Min Tsai, Jeffrey Lee and Huei Wang, "Topology analysis and design of passive HEMT millimeter-wave multiple-port switches", *IEEE Trans. Microwave Theory and Tech.*, vol. 56, issue 7, pp. 1545-1554, July 2008.

[6] H. Takasu, "Estimation of equivalent circuit parameters for a millimeter-wave GaAs PIN diode switch", *IEEE proceedings of the Circuits, Devices and Systems* , vol. 150, issue 2, pp. 92-94, April 2003.

[7] Yuan Liao and Thai Nguyen "Alternative way to derive equivalent π circuit model of transmission lines", 39th *Southeastern Symposium on System theory (SSST)* , pp. 186-188, March 2007.

[8] Microsemi Corp.-Watertown, "The PIN diode circuit desiners handbook", Doc.98=WPD-RDJ007, 1998.

[9] K. Kurokawa and W.O. Schlosser, "Quality factor of switching diodes for digital modulation", *Proc. IEEE*, vol. 58, issue 1, pp. 180-181, January 1970.

Implementation of a RF Front-end Module by Embedding ICs in Molding Package

Jong-In Ryu[#1], Se Hoon Park[#], Jong Min Yook[#], Dongsu Kim[#], and Jong Chul Park[#]

[#]Package Research Center
Korea Electronics Technology Institute
Seongnam-si, South Korea
[1]aceryu@keti.re.kr

Abstract— This paper presents a front-end module (FEM) for Wireless Personal Area Network (WPAN). This module is composed of a switch IC for Tx/Rx, a Tx Balun, and Rx band-pass filter (BPF). These components are embedded in molding package by employing the proposed package technology. Molding package process is to add lamination and pattern in contrast to normal package of ICs. This process, therefore, has advantage of short connection from chip to chip, compact size, and low cost in comparison to package using wire-bonding. Embedded components are connected by copper line on molding package. In order to implement a proposed FEM, commercial switch IC, a Balun, and a BPF by using integrated passive device (IPD) technology are embedded. The size of implemented FEM module is given as 3.8 mm × 2.6 mm × 0.6 mm. A FEM is, at first, measured in view of S-parameter and then it is tested in test environment for WPAN system. Tx path and Rx path include a switch and there are a Tx Balun and a Rx BPF in each path. Measured insertion loss and return loss in Rx path are better than 1.7 dB and 15.4 dB, respectively. In Tx path which include Balun, measured insertion loss, return loss, and phase unbalance are almost 1.7 dB, 16.3 dB, and 2 degree, respectively. Measured Tx power and Rx sensitivity in overall WPAN system by applying implemented module are 18.6 dBm and -74 dBm at error rate 0.4 %. The WPAN system has good performance by using implemented module.

Keywords—front-end module, molding package, embedding chip component, embedded module, WPAN

I. INTRODUCTION

Wireless communication is widely used in various fields [1]. Wireless communication is classified with wireless personal area network (WPAN), wireless local area network (WLAN), wireless metropolitan network (WMAN), and wireless wide area network (WWAN) in corresponding to distance. Bluetooth, zigbee, and UWB are employed as WPAN technology. WAPN is focused on this paper. Especially, 2.4 GHz ISM band is focused on and proposed module can be available for this solution. The embedding components technology and modules has been research in view of compact and low cost [2]-[5]. ICs and discrete components are embedded in various material substrates using a lot of embedding technology.

Molding package process is proposed in this paper to implement a front-end module (FEM) as low cost and compact. The proposed FEM is composed of a SPDT IC, Tx balun, and a Rx band-pass filter (BPF) as shown in fig. 1 (a). A SPDT IC is given as die as commercial product and both a balun and a BPF are made from silicon substrate as integrated passive device (IPD). Fig. 1(b) shows the overall test RF system. RF board consists of a proposed FEM, a RF Transceiver, regulator, and a TCXO. This board is connected with a baseband board. Data communication between a RF IC and baseband is accomplished with Rx/Tx data lines.

This paper is organized as follows. Procedure of molding package, process to embed components, applied IPDs, and implementation are described and discussed in section II. Section III describes the design, schematic, and CAD design. A fabricated module and a RF system are measured and summarized in section IV. Section V concludes the embedded wireless module in molding package.

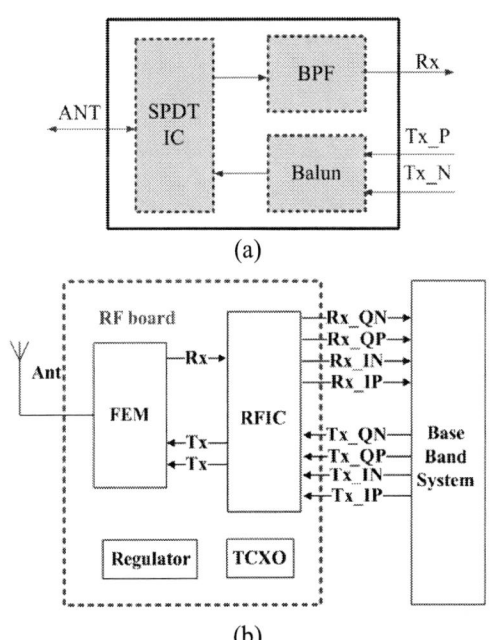

(a)

(b)

Fig. 1 A block diagram for a proposed : (a) a FEM and (b) test system for WPAN

II. PROCESS

A proposed front-end module employs molding and sequential build up process using organic materials which are adopted as substrates. Carrier film, Ajinomoto-bonding-film (ABF), and epoxy for molding package are mainly processed. ABF is very compatible with attaching ICs due to good adhesive to silicon and with forming fine line. ABF is suitable thin substrate by using semi additive process. Interconnection between formed line pattern and the pad of a embedded IC is carried out by UV laser drilling and Cu pattern plating processes. A planar surface and void free cavity filling are processed by vacuum lamination. After lamination and laser drilling, Cu plating and etching processes are performed.

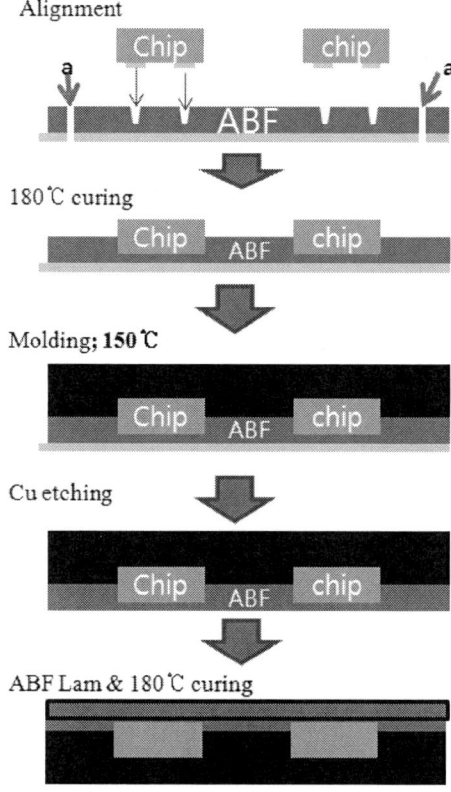

Fig. 2 A process for molding package by embedding ICs

The detail fabrication process for a module with embedded components in molding is demonstrated in Fig.2 and explained in follow as:

- ABF film with via hole for aligning the position of embedded ICs and ICs are prepared. These component are mounted in ABF film in accordance with position,
- This is cured at 180 degree temperature,
- Molding liquid is injected in module at 150 degree temperature,
- Copper is removed by etching,
- One ABF in top-side are laminated as 1st process, and this is cured at 180 degree temperature,
- UV laser drills into the Pads of IC and through via and Cu is plated and line pattern is formed on ABF.

Fig.3 (a) and (b) illustrate a Rx BPF and a Tx Balun. These components were designed and fabricated by IPD technology. The shape of coil and rectangular is inductor and capacitor. Proposed IPDs has Ti/Au layer for capacitor, Su-8 layer for inductor, and passivation layer as depicted in fig.3 (b). The high resistive silicon is given as substrate.

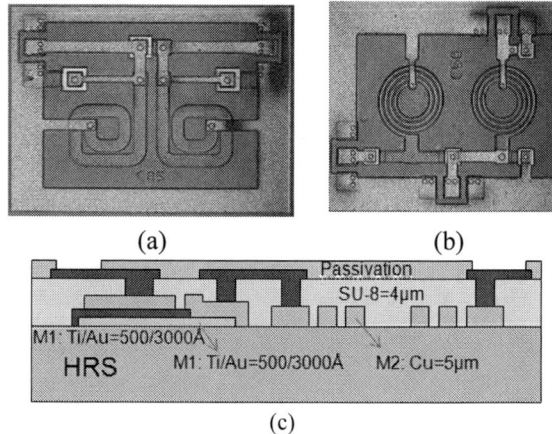

Fig.3 A photograph for (a) Rx BPF and (b) Tx balun and (b) the cross section of IPD

A switch IC, a Tx Balun, and a Rx BPF are embedded in epoxy molding package and each component is connected with copper line on ABF film as shown in fig. 4. The dimension of a FEM is 3.8 mm × 2.6 mm × 0.3 mm. The module has ball grid array (BGA) and 20 balls. The size of ball is given as 300 um. As a result, the total of 0.6 mm thickness is obtained in this paper. A FEM and a RF transceiver for ISM are mounted in printed-circuit-board (PCB). This RF board is composed of a FEM, a RF IC, a TCXO, regulators, and etc. RF transceiver is made from Airoha company and the model name is AL2236.

Fig.4 A cross section of the proposed FEM module

III. DESIGN FOR A MODULE

In this paper, the design procedure of proposed module is as follows: first design is carried out that the feasibility test for embedding only one switch IC in package, switch module is processed and test, a FEM is designed and fabricated, insertion loss and return loss of a FEM are measured, and, at last, the full WPAN system is tested for Tx power and Rx sensitivity after mounting a FEM.

The dielectric constant of FR4 and ABF in PCB substrate is 4.5 and 3.3, respectively. On basis of dielectric constant and loss, a RF transmission line is simulated. Signal crosstalk and

978-1-4799-0708-3/13 $31.00 © 2013 IEEE

interference are analyzed. An embedded component and lines are designed in molding package by using CAD artwork tool from Autodesk company [6]. An embedded component is designed by foaming the pads of an IC at inner layer.

Fig.5 (a) shows the schematic for a proposed RF board. A FEM includes the switch, a BPF, and Balun and a RF IC is connected to it. Fig. 5 (b) shows the layout of a FEM. A Balun, a BPF, and a switch IC are located at left side, right side, and middle-top side.

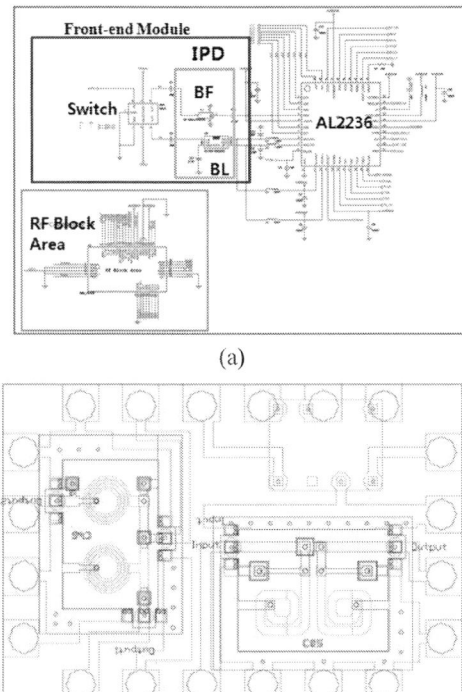

(a)

(b)

Fig. 5 (a) the schematic for a RF board, and (b) layout for a FEM

In this paper, the design by embedding ICs has the following issues : (1) open and short circuit test ; (2) tests for connection between IC pads and pads of module ; (3) adjustment for verified points to perform an embedded IC ; (4) confirmation for the improvement of performance.

A input and output pins of switch is short in each other under no DC supply, input/output pins of a BPF is short to GND, and an input/output pins of a Balun are open each other. This is first test point to via connection.

IV. FABRICATION AND MEASUREMENT

Fig.6 shows photographs for fabricated samples embedding ICs. A bare die for a switch and fabricated shape is located at left and right. Photo solder resistor is employed to be convenient for ball bumping. An implemented FEM is depicted in fig.6 (a) and (b). The ball bumps are at all side edge and brown color lines are fabricated copper lines in molding module. The cross section of a FEM is depicted in fig.6 (c). Bright yellow line depicts plated-via by copper between IC pads and PCB patterns. After process 5 step as

mention as previous section II, and the top view of a fabricated is shown in fig.6 (d)

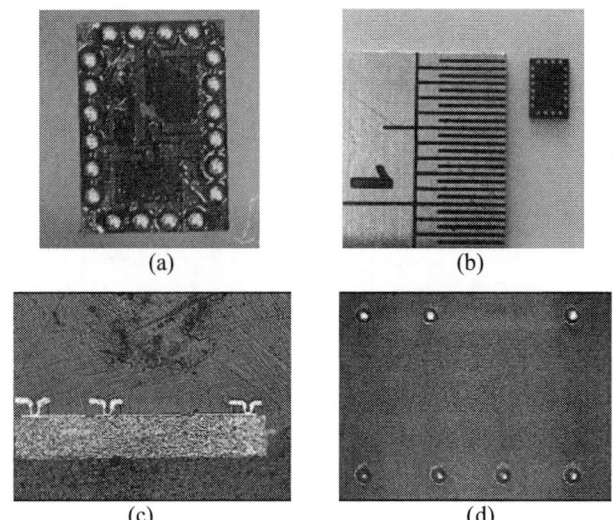

Fig. 6 photographs for (a) a fabricated FEM, (b) the size of a FEM, (c) the cross section of module, and (d) top-view in the middle of process

The implemented test boards for the module are shown in fig.7. Insertion loss and return loss are measured by using test board for a FEM as depicted in fig.7 (a). Fig,7 (b) shows a RF board that includes a FEM, a RF IC, a TCXO, and etc. WPAN test system is depicted in fig.7 (c) and baseband system is under a RF board.

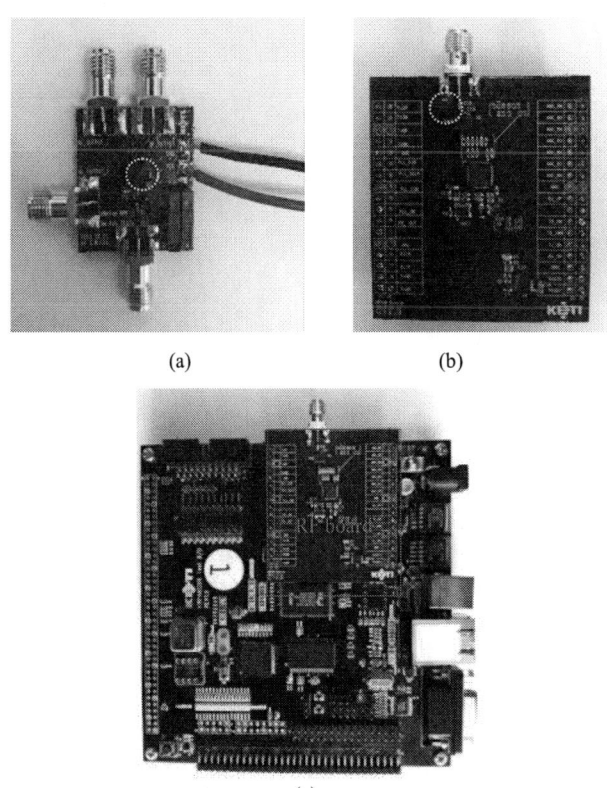

Fig. 7 Implemented (a) test board for a FEM , (b) a RF board, and (c) WPAN test system

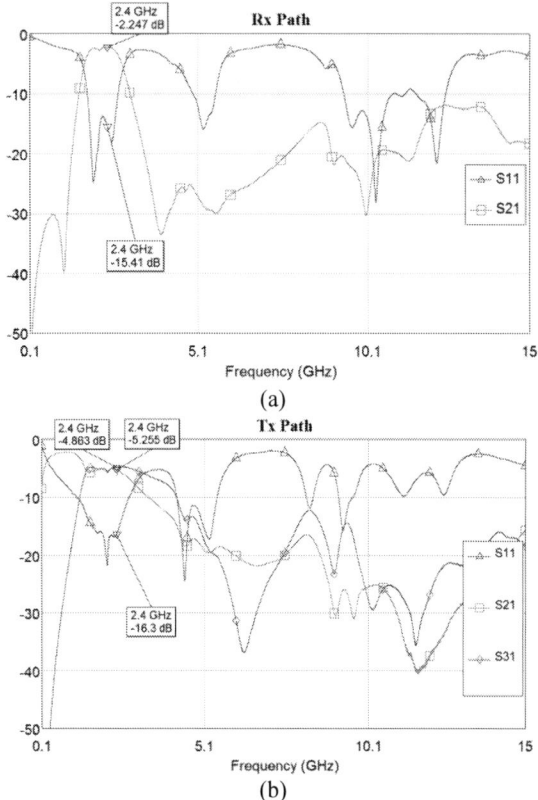

Fig. 8 Measured data for (a) Rx path and (b) Tx path

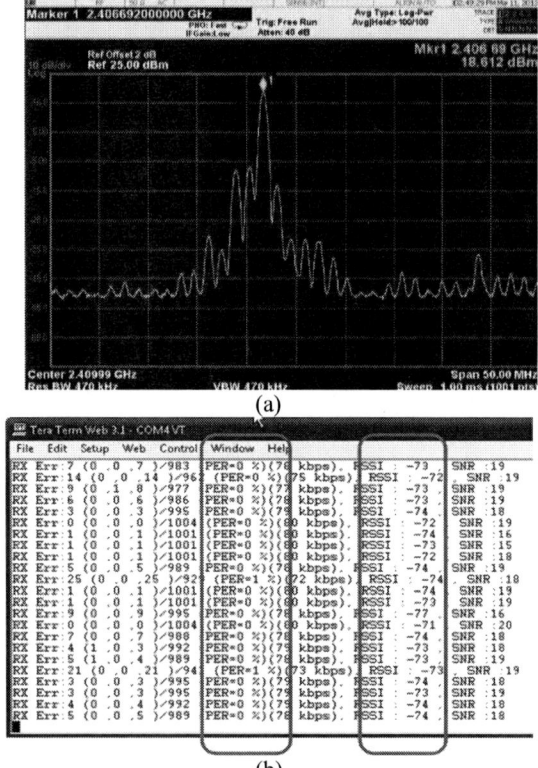

Fig. 9 Measured data for (a) Tx power and (b) Rx sensitivity

A presented compact module was tested by using Agilent's N5230 network analyser for S-parameters and Tx power and

Rx sensitivity is obtained by using self test set-up. Test set-up is inevitable to measure new device by adopting new communication technology. Rx sensitivity is main problem because any instrument doesn't supply wanted stream signal and standard. Two test devices are communicated with each other to measure sensitivity by adopting non-signalling method. Two devices are connected through attenuator as air interface. The value of attenuator is raised as if distance is increased because distance means the RF loss between antennas in view of RF system. RF sensitivity is measured as -75 dBm if Tx power and attenuator are given as 10 dBm and 85 dB, respectively.

A FEM is, at first, measured in S-parameter and then it is test in test environment for WPAN system. Rx path includes a switch and a BPF. The Measured insertion loss and return loss as shown in fig.8 (a) are better than 2.2 dB and 15.4 dB, respectively. Tx path included a Balun and a switch. Tx Path has a Balun, so port is separated positive and negative. Measured insertion loss and return loss are shown in fig.8 (b). Measured insertion loss, return loss, and phase unbalance are almost 5.2 dB, 16.3 dB, and 2 degree. In the Tx path case, insertion loss includes 3 dB of the Balun and 0.5 dB of test board. Therefore, Rx loss and Tx loss are 1.7 dB and 1.7 dB, respectively. By using test set-up as mentioned as previous paragraph, measured Tx power and Rx sensitivity in overall WPAN system by applying implemented module are 18.6 dBm and -74 dBm at error rate 0.4 % as shown in fig.9.

V. CONCLUSION

A compact FEM for WPAN was designed and implemented by embedding components with molding process. A FEM and WPAN system were set-up and each test was accomplish. This paper presented that process and test results for embedded module. The WPAN system has good performance by using implemented module.

REFERENCES

[1] T. G. Lenihan, and E. J. Vardaman, "Worldwide Perspectives on SiP Market: Technology Trends and Challenges," *in Proc 7th Int. Conf. on Electronics Packaging Technology*, 2006.

[2] R.R. Tummala, M. Swaminathan, M.M. Tentzeris, J. Laskar, Gee-Kung Chang, S. Sitaraman, D. Keezer, D. Guidotti, Zhaoran Huang, Kyutae Lim, Lixi Wan, S.K. Bhattacharya, V. Su ndaram, Fuhan Liu, and P.M. Raj, "The SOP for miniaturized, mixed-signal computing, communication, and consumer systems of the next decade," *IEEE Trans. on Advanced Packaging*, vol.27, no.2, pp.250-267, 2004.

[3] R. Ulrich, "Embedded Resistors and Capacitors for Organic-based SOP," *IEEE Trans. on Advanced Packaging*, vol.27, no.2, pp.326-331, 2004.

[4] J. Ryu, D. Kim, S.H Park, J.C Park and J. C. Kim, "Investigation of an Embedded RF Switch IC in Printed Circuit Board," *in Proceeding of the Electrical Design of Advanced Packaging and Systems Sym.*, pp. 81-84, 2008.

[5] J. Ryu, S. Park, J. Moon, D. Kim, J.C Kim and N. Kang, "Implementation of a Front-End-Module by Embedding a RF Switch IC and a Power Amplifier in Printed-Circuit-board," *in Proceeding of the Electric Components and Technology Conf.*, pp. 1920-1923, 2009.

[6] Autodesk Inc., User's Manual, AutoCAD 2007, 2007.

Accurate Characterization of Lossy Interconnects from TDR Waveforms

Ping Liu
Sigrity R&D
Cadence Design System, Inc
Shanghai, China
liuping@cadence.com

Jingping Zhang, Jiayuan Fang
Sigrity R&D
Cadence Design System, Inc
San Jose, USA
zhangjp@cadence.com, fangj@cadence.com

Abstract—**This paper proposes an accurate solution to characterize the lossy interconnects with discontinuities from TDR waveforms. A new lossy peeling algorithm is proposed to determine the interconnect impedance of lossy structures. An efficient fitting algorithm is applied to extract the W-element model of a lossy structure from an initial impedance profile extracted by a lossless peeling algorithm. The proposed method also improves the accuracy of late time response which has been difficult to achieve by previous algorithms. The approach presented in this paper is able to recover the impedance profiles of complicated lossy structures.**

Keywords—TDR measurements; lossless peeling algorithm; lossy peeling algorithm; W-element model; TDR impedance profile; curve fitting

I. INTRODUCTION

In high-speed system designs, it is important to determine the impedances of passive interconnects and extract their models along the entire channel length. The waveforms from the Time Domain Reflectometry (TDR) [1] are widely used for impedance measurements due to their intuitiveness.

In recent years, several techniques [2]-[5] based on TDR measurements have been proposed to characterize and model interconnects with discontinuities such as traces, bends, vias and lumped elements. The lossless peeling algorithm [2]-[4] has been developed to calculate the impedance and extract the equivalent circuit of an interconnect channel with discontinuities. However, the method is based on the approximation that the interconnect is lossless. For higher frequency and longer interconnect length, such an approximation leads to more substantial error. The impedance profile recovered by the lossless peeling algorithm has an upward tilt due to the loss of the interconnect [6][7]. Such an impedance profile fails to represent the actual impedance of the interconnect. A lossy peeling algorithm has been proposed to extract the impedance profile of the lossy structures [8]. Unfortunately, with this method, a constant resistive loss of the structure needs to be pre-known, and the accuracy deteriorates when the skin effect loss becomes more significant.

In this paper, we first propose a modified lossy peeling method to extract the impedances of lossy structures in Section II. This method extracts the resistive loss from the impedance profile recovered by the lossless peeling algorithm and applies different resistive loss element in different time step. With this approach, the accuracy of the impedances of lossy interconnects is improved. In Section III, we analyze the impedance profile of lossy structures extracted by the lossless peeling algorithm and illustrate some interesting findings, which are useful for our proposed solution. A new solution flow is proposed in detail for characterizing the lossy interconnects in Section IV. Section V gives an example to verify the solution. Finally a conclusion is drawn in Section VI.

II. NEW LOSSY PEELING ALGORITHM

A. Lossless Peeling Algorithm

The lossless peeling algorithm has been widely used to recover the impedance profile along the whole length of an interconnect channel with discontinuities. Consider an interconnect shown in Figure 1, represented by a series of N+1 impedance sections. Each section has the same time delay, but maybe of different characteristic impedance. With the initial input TDR voltage and the relationship between the reflected wave and the transmitted wave at each junction, the real value of the characteristic impedance of each section can be calculated by the peeling algorithm [2][3]. The lossless peeling algorithm is based on the approximation that the interconnect is lossless. For a uniform section of lossy line, the impedance recovered by this approach is not a constant, but of a positive tilt, as will be shown later in this paper. In addition, the error in the impedance profile extracted for an early section of lossy interconnect further impacts the accuracy of the impedance profile extracted for a later section of the lossy line. With such shortcomings, the lossless peeling algorithm fails to accurately extract long lossy interconnects.

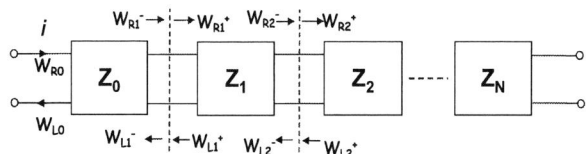

Fig. 1. The interconnect is discretized into N+1 sections of constant impedance in each section.

B. New Lossy Peeling Algorithm

J.M. Jong presented a lossy peeling algorithm, in which the model of an interconnect is constructed by cascaded resistive sections as shown in Figure 2 [8]. Each section includes a constant impedance section and a resistor R_k which represents the resistive loss of the kth section. This model can well separate the resistive loss part from the total impedance.

978-1-4799-0708-3/13 $31.00 © 2013 IEEE

However, with this method, the resistive element R_k in each section should be pre-calculated for the lossy interconnect structure [8]. In general, as we get the TDR/T data, the interconnect structure is not pre-known. Therefore such a lossy peeling algorithm is not feasible in real applications.

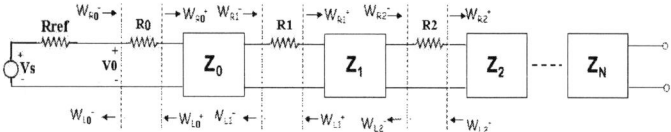

Fig. 2. The interconnect is discretized into N+1 sections of constant impedance plus a resistor in each section.

In this paper, we extract the resistive element R_k from the impedance profile recovered by the lossless peeling algorithm. It has been observed that it is the resistive loss that creates the positive tilt on the impedance profile and the resistive loss can be estimated from the impedance increment in the impedance profile. Based on that, we use the impedance increment between neighboring sections as the resistive element R_k of the kth section. Figure 3 shows the TDR impedance result of a single lossy microstrip line recovered by this new lossy peeling algorithm. The impedance profile is flat and much more accurate than that recovered by the lossless peeling algorithm. For the case that multiple lines are cascaded, the new lossy peeling algorithm can also generate more accurate impedance profile than the lossless peeling algorithm, but its impedance profile corresponding to the late TDR response is still not accurate enough. Therefore we need to look for more viable solutions.

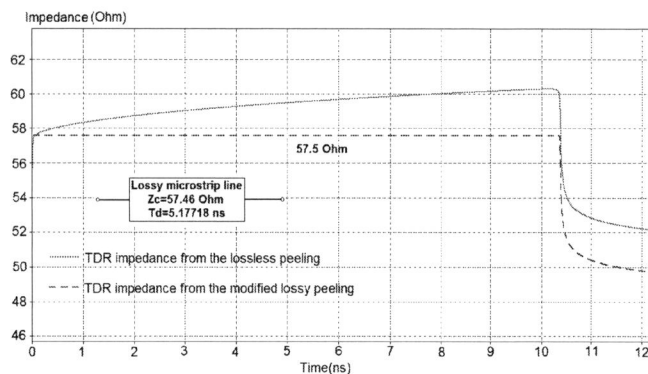

Fig. 3. The TDR impedance comparison between the lossless peeling result and the modified lossy peeling result.

III. SOME FINDINGS ON IMPEDANCE PROFILES

As an example, we first use a W element to model a lossy transmission line, and compute the TDR waveform of the lossy transmission line.

A. W-element model extraction from the impedance profile

The dash curve of Figure 4 shows the impedance profile of the single transmission line recovered by the lossless peeling algorithm. From the impedance curve, the time delay T_d of the transmission line can be obtained. For a W element with given parameters L_0, C_0, R_{DC} and R_s, its step function response can be found analytically as Eq.(1), see Appendix. The TDR

impedance curve can be fitted by the 2nd order least square method with the curve of Eq.(1).

$$z(t)=\left(1+\frac{R_s}{\pi L}\sqrt{t}+\frac{R_{DC}}{2L}t\right)\cdot Z_0 \tag{1}$$

$$Z_0=\sqrt{L_0/C_0} \tag{2}$$

$$T_d=\sqrt{L_0 C_0} \tag{3}$$

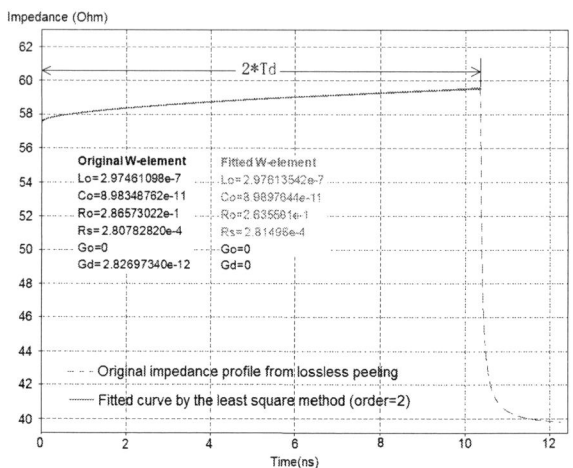

Fig. 4. Lossless peeling impedance profile and its fitting curve of a single transmission line represented by W element.

The solid curve of Figure 4 is the fitted curve by Eq.(1), which is in very good agreement with the original impedance profile from TDR waveform. The original W-element parameters and the extracted W-element parameters are also in very good agreement as shown in the figure. This illustrates that, from TDR waveforms, one can extract a W-element model which includes the frequency-dependent skin-effect loss rather than a constant resistive loss. In this illustration, we assume G_d which represents the dielectric loss is zero.

B. Loss effects of the previous block on the latter block

Figure 5 shows the impedance profiles of two cases recovered from the lossless peeling algorithm. Both cases include two cascaded transmission lines. Case 1 has two cascaded lossless transmission lines, whose impedance profile is the curve marked by triangles. Case 2 includes a lossy transmission line cascaded with a lossless transmission line, whose impedance profile is the curve marked by circles.

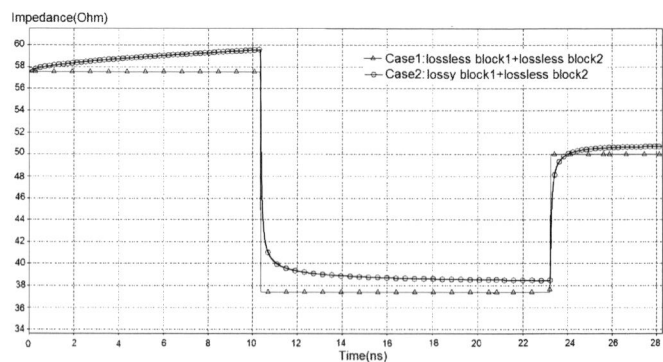

Fig. 5. Lossless peeling impedance profiles of two cases with two cascaded transmission lines.

From the curve marked by circles, it is noted that, although the second block is lossless, the impedance profile of the second block is affected by the loss of the first block. That is, the impedance profile of block 2 includes not only the loss effect of block 2 itself but also the loss effect from block 1.

Figure 6 adds two more impedance curves. The curve marked by squares, Z3, is for the case that block 1 is lossless and the cascaded block 2 is lossy. The curve marked by crosses, Z0, is for the case of two cascaded lossy blocks. Let's examine the impedance profiles of block 2 in Figure 6. The curve Z3 only includes the loss of block 2; the curve Z1 only includes the loss of block 1; the curve Z0 includes both block 1 loss and block 2 loss; while the curve Z2 is a base line for no loss in both blocks. It can be found that there is a relationship among the four curves as shown in Eq.(4).

$$Z3 = Z0 - (Z1 - Z2) \qquad (4)$$

In Eq. (4), (Z1- Z2) of block 2 is the loss effect from block 1, which is found to have little dependence on the impedance of block 2. This observation is useful for the proposed solution in the next section.

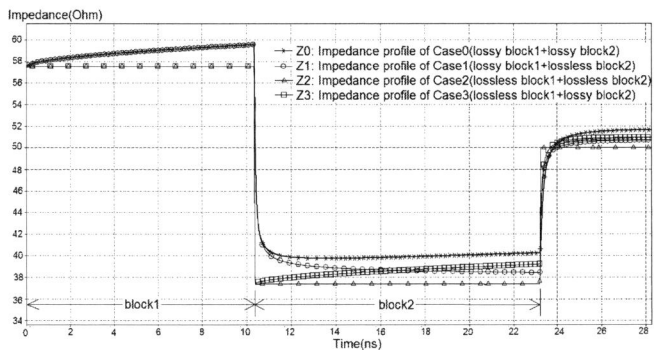

Fig. 6. Lossless peeling impedance profiles of four cases with two cascaded transmission lines.

IV. PROPOSED SOLUTION FLOW

Assuming the DUT (device under test) includes two cascaded lossy transmission lines and we need to determine its impedance profile from TDR measurements.

Firstly, we can get the impedance profile Z0 marked by stars in Figure 7 using the lossless peeling algorithm. The W-element model of block 1, W_1, can be obtained by fitting the impedance curve Z0 of block 1 using the method described in Section III.A. An approximate transmission line model (a lossless transmission line model with two parameters-- impedance and time delay), T_2', of block2 can be obtained by the new lossy peeling algorithm in Section II.B. Secondly, construct DUT1 with the fitted W-element model of block 1 (W_1) and the approximated transmission line model of block 2 (T_2'). By the lossless peeling on DUT1, we can generate the curve Z1 marked by circles in Figure 7. A lossless transmission line model of block 1 (T_1) can be obtained easily from its W-element model (W_1). Construct DUT2 as the transmission line model of block 1 (T_1) cascaded with the estimated transmission line model of block2 (T_2'). By the lossless peeling on DUT2, we can obtain the curve Z2 marked by triangles in Figure 7. Finally, the curve Z3 of block 2 marked by squares can be

found by Eq. (4). By performing the fitting on Z3 of block 2, we can obtain the W-element model of block 2 (W_2).

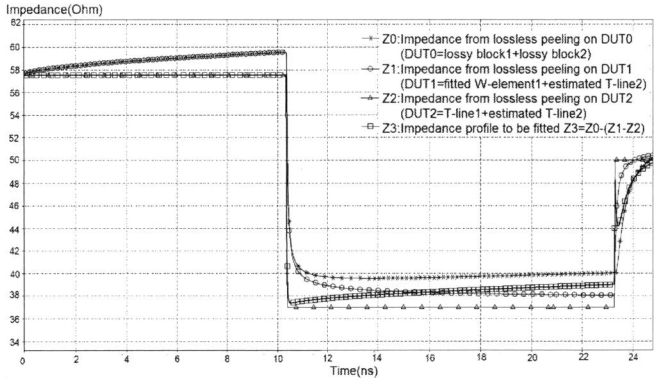

Fig. 7. Lossless peeling impedance profiles generated during the calculation process of the new solution

The above process can be extended to cases with multiple cascaded lossy transmission lines. With this solution, the W-element model of every block can be extracted and the whole impedance profile can be recovered thereby. The complete flow of the proposed solution can be summarized as follows in Figure 8.

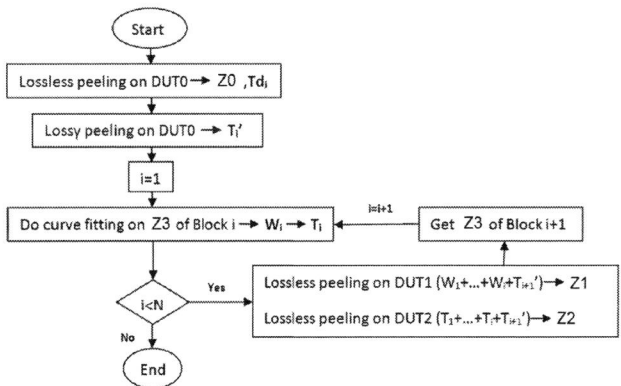

Fig. 8. The flow of the new solution

1. *Do lossless peeling on DUT0 (original structure)*

    1) Get the impedance profile --- curve Z0;

    2) Do block partition according to the curve discontinuities (assume the block number is N) and get the time delay Td_i of every block;

2. *Do curve fitting on the impedance profile of block1 recovered by the lossless peeling algorithm to get its W-element model (W_1) and its T-line model (T_1);*

3. *Do the new lossy peeling on DUT0 and get the estimated characteristic impedance Zc_i' of blocks except block1. Assume every block with estimated parameters is a T-line T_i' (Zc_i', Td_i);*

4. *Create DUT1 and DUT2 and do lossless peeling on them:*

    DUT1: $W_1 + W_2 + ... + W_{i-1} + T_i'$--- obtain the curve Z1;

    DUT2: $T_1 + T_2 + ... + T_{i-1} + T_i'$ --- obtain the curve Z2.

5. *Get the curve Z3 of block i by Eq. (4);*

6. *Do the curve fitting on the Z3 curve of block i by the 2-order least square method to get the W-element parameters of Block i. Record the model as W_i;*

7. *Use the accurate Zc_i and Td_i of block i to create a T-line T_i;*

If i=N, stop; else i=i+1, return step 4.

V. NUMERICAL EXAMPLES

An example is used to demonstrate the effectiveness of the proposed solution. This case is a structure including four cascaded lossy transmission lines as shown in Figure 9. The rise time Tr of the input step pulse is 30 ps and the time step dt is 5 ps. The real W-element parameters of the lossy microstrip lines and the lossy strip lines are also presented in the figure.

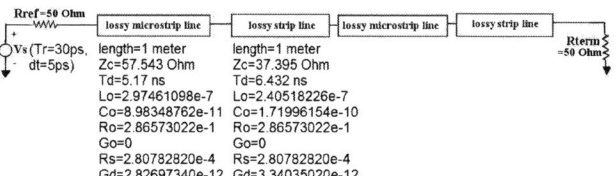

Fig. 9. Four cascaded lossy lines

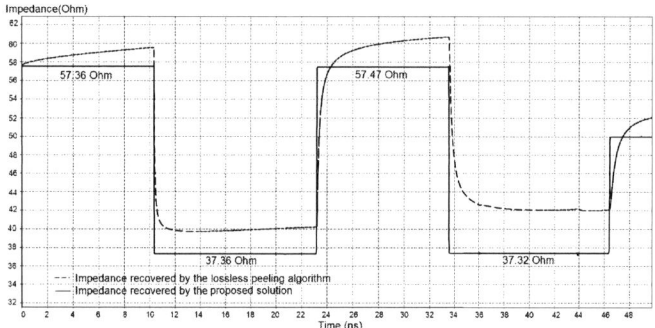

Fig. 10. Impedance profile comparison between the lossless peeling algorithm and the proposed solution

As shown in Figure 10, the dash curve is the impedance profile recovered by the lossless peeling algorithm on TDR waveform of the structure. Obviously it has a large deviation from the true impedance of the components shown in Figure 9. The solid curve is the impedance profile recovered by the proposed solution. Compared with the true characteristic impedance of every block, the error of the final result is within 0.2%, which is much more accurate than the lossless peeling result.

VI. CONCLUSION

This paper proposes a new solution to accurately determine the impedance profiles of lossy interconnects from TDR waveforms. The TDR waveforms can be from actual instrumentation measurements or numerical simulations. In this paper, we only present the solution for cascaded lossy transmission lines. In fact, the method can be extended to more complicated structures including vias and lumped elements, which will be presented in future papers. In addition, this paper assumes the metal loss dominates over the dielectric loss. How to extract impedance profiles under significant dielectric loss can be our future work.

APPENDIX

Derivation Process of Eq.(1)

Assume the dielectric loss G can be ignored, so the characteristic impedance of a transmission line in the frequency domain is [9]

$$Z_c(\omega) = \sqrt{\frac{R(\omega) + j\omega L}{j\omega C}} \qquad (5)$$

Since

$$R(\omega) = R_{DC} + R_{AC}$$

$$R_{AC} = (1+j)R_s\sqrt{f}^{[10]}$$

$$Z_0 = \sqrt{L/C}$$

Eq.(5) becomes

$$Z_c(\omega) = Z_0\sqrt{1 + \frac{jR_s\sqrt{f} + R_s\sqrt{f} + R_{DC}}{j\omega L}} \qquad (6)$$

Using a 1st order Taylor-series approximation, we can obtain

$$Z_c(\omega) \approx Z_0\left(1 + \frac{1}{2}\frac{jR_s\sqrt{f} + R_s\sqrt{f} + R_{DC}}{j\omega L}\right) \qquad (7)$$

Perform the inverse Fourier transform on Eq. (7) and get a time-domain impedance $Z_c(t)$. The step response of the characteristic impedance can be obtained by the convolution of $Z_c(t)$ and the step function u(t) as follows

$$z(t) = z_c(t) * u(t) = Z_0\left(1 + \frac{R_s}{\pi L}\sqrt{t} + \frac{R_{DC}}{2L}t\right) \qquad (8)$$

REFERENCES

[1] IPC-TM-650 Test methods manual 2.5.5.7 "Characteristic Impedance of Lines on Printed Boards by TDR".

[2] Scott C. Burkhart, "Arbitrary pulse shape synthesis via nonuniform transmission lines," IEEE Transactions on Microwave Theory and Techniques, vol. 38, pp. 1514-1518, No. 10, October 1990.

[3] J.M.Jong and V.K.Tripathi, "Time-domain characterization of interconnect discontinuities in high-speed circuis," IEEE Transactions on Components, hybrids, and manufacturing Technology, vol.15, pp.497-504, No. 4, August 1992.

[4] SPEED2000 TDR/TDT Simulation Tutorial, Product Version 16.6, Cadence Design System, December, 2012.

[5] Signal Integrity Modeling of Gigabit Backplanes, Cables and Connectors Using TDR. Application Notes, TDA Systems, 2001.

[6] Luis Navarro, Eugene mayevskiy, Timothy Chairet, "Application of launch point extrapolation technique to measure characteristic impedance of high frequency cables with TDR," DesignCon 2009.

[7] Istvan Novak, etc., "Determining PCB trace impedance by TDR: challenges and possible solutions," DesignCon 2013.

[8] J.M.Jong, etc., "Lossy interconnect modeling from TDR/T measurements," Proceedings of EPEP 1994.

[9] Howard Johnson and Martin Graham, High-speed Signal Propagation: Advanced Black Magic, Prentice Hall PTR, 2003.

[10] HSPICE User Guide: Signal Integrity, Version E-2010.12, December 2010.

978-1-4799-0708-3/13 $31.00 © 2013 IEEE

Per-Unit-Length Parameter Extraction for Lossy Multi-conductor Power Cables

Bernhard Wunsch*, Ivica Stevanović*†, and Stanislav Skibin*

*ABB Switzerland Ltd., Corporate Research, Segelhofstrasse 1K, CH-5405 Baden-Dättwil, Switzerland

{bernhard.wunsch, stanislav.skibin}@ch.abb.com

†Federal Office of Communications OFCOM, Zukunftstrasse 44, CH-2501 Biel/Bienne, Switzerland

ivica.stevanovic@bakom.admin.ch

Abstract—**In this paper, a reliable and accurate extraction of per-unit-length parameters of multiconductor cables from measured admittance or scattering parameters is presented. Once the per-unit-length parameters are available, the admittance for any other length of the cable of the same type can be calculated, which allows creating broadband models for arbitrary cable length. The proposed approach exploits the symmetry of the cable and has been validated against measured admittances of industrial multi-conductor power cables of different lengths.**

I. INTRODUCTION

Multi-conductor transmission lines can be described by telegraph equations [1]. An accurate numerical simulation or direct measurement of the per-unit-length (p.u.l.) parameters is, however, difficult [2]. In contrast, accurate characterization of the cable by s-parameter measurements is possible and based on such measurement data, accurate behavioral models can be created [3]. However, a short-coming of behavioral models is that they are typically linked to the measured cable type and even length. In contrast, the p.u.l. parameters characterize the same cable type for any cable length. A reliable and accurate conversion from measured y- or s-parameters to the p.u.l. parameters is, therefore, desirable. Inverting the well-known equations for y-parameters in order to extract the p.u.l. parameters is computationally involved and highly sensitive to measurement errors [4] and better methods are necessary.

In this paper, we show that by exploiting symmetries of the cable, the recovery of the p.u.l. parameters from measurements can be simplified significantly. For uniform cables with rotational symmetry in their cross section, the inversion of matrix equations needed for the recovery of the p.u.l. parameters can be reduced to scalar equations and these can be solved analytically. The accuracy of the extracted p.u.l. parameters is shown by reproducing the measurement data from the extracted parameters of a shielded four-conductor power cable of 150 m length. Finally, the p.u.l. parameters extracted from the 150 m cable are used to calculate the admittances of the same type of cable but of 5 m length, which again agree closely with measurements. The good accuracy between predicted and measured admittances also shows that the quasi-transverse field approximation underlying the telegraph equations is well justified for the frequency region of interest (conducted noise emission of up to 30 MHz). The s-parameters for power cables have been measured accurately using vector network analyzer (VNA) with appropriate deembeding of fixtures [2], [3].

II. RECOVERING P.U.L. PARAMETERS FROM MEASURED ADMITTANCES

In microwave engineering, it is a common task to describe the propagation of radio frequency signals on transmission lines [1], [5]. The same methods can be applied to obtain high-frequency models of long power cables. The main assumption in this approach is that the electromagnetic waves produced by the currents are quasi-TEM modes, which is a good approximation as long as the cable diameter is much smaller than the wavelength, and the ohmic losses of the conductors are small. The propagation of voltages and currents along uniform transmission lines/cables is then described by means of telegraph equations [1]

$$\partial_x \mathbf{V}(x, \omega) = -\mathbf{Z}'(\omega)\mathbf{I}(x, \omega) \tag{1}$$

$$\partial_x \mathbf{I}(x, \omega) = -\mathbf{Y}'(\omega)\mathbf{Y}(x, \omega) \tag{2}$$

Here, x denotes the position along the line, $\mathbf{Z}'(\omega) = \mathbf{R}'(\omega) + j\omega \mathbf{L}'(\omega)$ is the p.u.l. impedance matrix and $\mathbf{Y}'(\omega) = j\omega \mathbf{C}'(\omega)$ is the p.u.l. admittance matrix, which are determined by the matrices of resistance, $\mathbf{R}'(\omega)$, capacitance, $\mathbf{C}'(\omega)$, and inductance, $\mathbf{L}'(\omega)$. We denote the length of the line by d so that $0 \leq x \leq d$. We follow common practice and assume the electric insulation of the dielectric to be perfect and neglect any leakage conductance.

Current and voltage vectors $\mathbf{I}(x, \omega)$ and $\mathbf{V}(x, \omega)$ in (1), (2) have N_C components, one for each conductor. Correspondingly, $\mathbf{Z}'$ and $\mathbf{Y}'$ are of dimension $N_C \times N_C$. The currents in the positive x-direction are counted positive. The current in the shield is assumed to be the negative sum of all currents in the conductors, so that there is zero total current at each cross section. Each component of the voltage vector represents the voltage between the corresponding conductor and the shield at a given cross section defined by x.

Solving the telegraph equations (1), (2) for each frequency, one can express the multiport admittance matrix of the cable in terms of the p.u.l. parameters [1]

$$\begin{bmatrix} \mathbf{I}_L \\ \mathbf{I}_R \end{bmatrix} = \begin{bmatrix} \mathbf{Y}_S & \mathbf{Y}_M \\ \mathbf{Y}_M & \mathbf{Y}_S \end{bmatrix} \begin{bmatrix} \mathbf{V}_L \\ \mathbf{V}_R \end{bmatrix} \tag{3}$$

978-1-4799-0708-3/13 $31.00 © 2013 IEEE

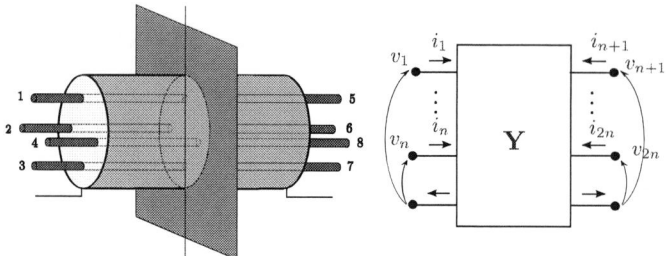

Fig. 1. Left: Schematic of shielded four-conductor power cable showing 4-fold rotational symmetry of cross section and reflection symmetry with respect to indicated mirror plane. Right: Description of cable as a multiport network. The admittance matrix relates voltages between signal conductors and shield (reference conductor) to the currents entering the signal conductors.

$$\mathbf{Y}_{\mathrm{M}}(d,\omega) = -\frac{\sqrt{d\mathbf{Y}'(\omega)d\mathbf{Z}'(\omega)}}{\sinh\left(d\mathbf{Y}'(\omega)d\mathbf{Z}'(\omega)\right)d\mathbf{Z}'} \tag{4}$$

$$\mathbf{Y}_{\mathrm{S}}(d,\omega) = -\mathbf{Y}_{\mathrm{M}}(d,\omega)\cosh\left(d\mathbf{Y}'(\omega)d\mathbf{Z}'(\omega)\right) \tag{5}$$

with $\mathbf{I}_{\mathrm{L}}(\omega) = \mathbf{I}(x=0,\omega)$ and $\mathbf{V}_{\mathrm{L}}(\omega) = \mathbf{V}(x=0,\omega)$ denoting currents and voltages on the left terminal of the cable and $\mathbf{I}_{\mathrm{R}}(\omega) = -\mathbf{I}(x=d,\omega)$ and $\mathbf{V}_{\mathrm{R}}(\omega) = \mathbf{V}(x=d,\omega)$ denoting currents and voltages on the right terminal, respectively. The minus sign in the definition of $\mathbf{I}_{\mathrm{R}}(\omega)$ is needed since the positive current direction for the multiport corresponds to currents directed towards the cable (see Fig. 1).

The equations (3), (4), and (5) have the attractive feature, that the p.u.l. parameters of a uniform cable only depend on frequency and the cross section of the cable, but are length independent, so that for the known p.u.l. parameters the multiport admittance matrix can be calculated for any cable length.

A. Exploiting symmetries of cable

For cables with rotationally symmetric cross sections, the matrix equations (3), (4), and (5) can be reduced to scalar equations. As an example, take the shielded four-conductor cable schematically shown in Fig. 1. The cross section of the cable is invariant under a rotation of 90° around its symmetry axis. Equivalently, a symmetric shielded N_{C}-conductor cable is invariant under $360°/N_{\mathrm{C}}$ rotations. The rotational symmetry of the cable causes a special structure of the $N_{\mathrm{C}} \times N_{\mathrm{C}}$ matrices describing the cable. For example, the p.u.l. parameters, mutual, or self admittances are circulant matrices [6]–[8], so that each subsequent row is given by the previous row shifted by one to the right. Additionally, these matrices have only $\lfloor N_{\mathrm{C}}/2 \rfloor + 1$, distinct entries and eigenvalues, where $\lfloor x \rfloor$ denotes the floor function of x. Furthermore, they commute with each other and can all be diagonalized by the same orthogonal matrix, $\mathbf{T}$, that only depends on the number of conductors N_{C}. In the case of the four-conductor cable, a possible orthogonal matrix is given

by

$$\mathbf{T} = \left[\frac{1}{2}\begin{pmatrix} 1 \\ 1 \\ 1 \\ 1 \end{pmatrix}, \frac{1}{\sqrt{2}}\begin{pmatrix} 1 \\ 0 \\ -1 \\ 0 \end{pmatrix}, \frac{1}{2}\begin{pmatrix} 1 \\ -1 \\ 1 \\ -1 \end{pmatrix}, \frac{1}{\sqrt{2}}\begin{pmatrix} 0 \\ 1 \\ 0 \\ -1 \end{pmatrix} \right] \tag{6}$$

The cable has also a mirror symmetry with respect to its symmetry plane located in the middle between the two endings of the cable (see Fig. 1). Consequently, the full admittance matrix can be block diagonalized by introducing open circuit and short circuit currents and voltages, which are the even and odd superpositions of the terminal currents and voltages $\mathbf{I}_{\mathrm{OC}} = \mathbf{I}_{\mathrm{L}} + \mathbf{I}_{\mathrm{R}}$, $\mathbf{I}_{\mathrm{SC}} = \mathbf{I}_{\mathrm{L}} - \mathbf{I}_{\mathrm{R}}$, and analogously for $\mathbf{V}_{\mathrm{OC}}$ and $\mathbf{V}_{\mathrm{SC}}$ [9]

$$\begin{bmatrix} \mathbf{I}_{\mathrm{OC}} \\ \mathbf{I}_{\mathrm{SC}} \end{bmatrix} = \begin{bmatrix} \mathbf{Y}_{\mathrm{OC}} & \mathbf{0} \\ \mathbf{0} & \mathbf{Y}_{\mathrm{SC}} \end{bmatrix} \begin{bmatrix} \mathbf{V}_{\mathrm{OC}} \\ \mathbf{V}_{\mathrm{SC}} \end{bmatrix} \tag{7}$$

where $\mathbf{Y}_{\mathrm{OC}} = \mathbf{Y}_{\mathrm{S}} + \mathbf{Y}_{\mathrm{M}}$ and $\mathbf{Y}_{\mathrm{SC}} = \mathbf{Y}_{\mathrm{S}} - \mathbf{Y}_{\mathrm{M}}$. Due to rotational symmetry the N_{C} eigenmodes of both admittances $\mathbf{Y}_{\mathrm{OC}}$ and $\mathbf{Y}_{\mathrm{SC}}$ can again be obtained by the unitary transformation $\mathbf{T}$ introduced above.

Applying the transformation $\mathbf{T}$ to (3), (4), and (5) results in scalar equations for each of the N_{C} eigenvalues

$$Y_{\mathrm{M}i} = -\frac{d\gamma_i}{dZ_i' \sinh(d\gamma_i)} \tag{8}$$

$$Y_{\mathrm{S}i} = -Y_{\mathrm{M}i}\cosh(d\gamma_i) \tag{9}$$

Here $\gamma_i = \alpha_i + \mathrm{j}\beta_i = \sqrt{Y_i' Z_i'}$ denotes the propagation parameter and the index $i = 1, \dots, N_{\mathrm{C}}$ numerates the eigenvalues, e.g., $Y_{\mathrm{M}i}$ is determined by $\mathrm{diag}(Y_{\mathrm{M}1}, \dots, Y_{\mathrm{M}N_{\mathrm{C}}}) = \mathbf{T}^{-1}\mathbf{Y}_{\mathrm{M}}\mathbf{T}$.

The frequency dependence is not indicated explicitly in (8) and (9), to improve readability, although except for the constant length d and the constant $\mathrm{j} = \sqrt{-1}$ all the other quantities are frequency dependent. The attenuation parameter α_i is given by the real part of the propagation parameter and characterizes the damping of the waves. The phase parameter β_i is the imaginary part of the propagation parameter and determines the phase velocity of wave propagation via $\beta_i(\omega) = \omega/v_{\mathrm{phase}\,i}(\omega)$. Since the phase velocity is typically weakly frequency dependent, the phase parameter grows approximately linearly with frequency.

III. RESULTS

We note that recovering the p.u.l. parameters by inverting the scalar equations (8) and (9) for each eigenvalue separately is much simpler than inverting the original matrix equations (3), (4), and (5). One remaining difficulty is that for complex numbers, the inverse of hyperbolic functions are multi-valued, since the phase of a complex number is only determined up to multiples of 2π, i.e., $\exp(z) = \exp(z + \mathrm{j}2\pi i)$ for any integer i. The multiple branches for complex functions lead to a whole family of solutions for each eigenvalue, which

978-1-4799-0708-3/13 $31.00 © 2013 IEEE

can be distinguished by i:

$$d\gamma_i = d\alpha_i + \mathrm{j}d\beta_i = \cosh^{-1}(Y_{Si}/Y_{Mi}) + \mathrm{j}2\pi i \qquad (10)$$

$$Y_i' = -Y_{Mi}\gamma_i \sinh(d\gamma_i) \; ; \; Z_i' = \frac{\gamma_i^2}{Y_i'} \qquad (11)$$

The integer i can be inferred from physical reasoning. At DC the admittances are real and the phase parameter β_i has to be zero implying $i = 0$. With increasing frequency, the phase parameter grows continuously. In order to ensure continuity of β_i, discontinuities stemming from the term $\cosh^{-1}(x)$ with $x = \frac{Y_{Si}}{Y_{Mi}}$ that occur for $\mathrm{Im}(x) = 0$ and $\mathrm{Re}(x) < 0$ have to be compensated by increasing i by one.

With this insight, the propagation parameters γ_i can be unambiguously extracted from the measurement data. Their real parts α_i are shown in Fig. 2. The phase velocity that determines the phase parameter $\beta_i = \omega/v_{\text{phase } i}$ is shown in Fig. 3.

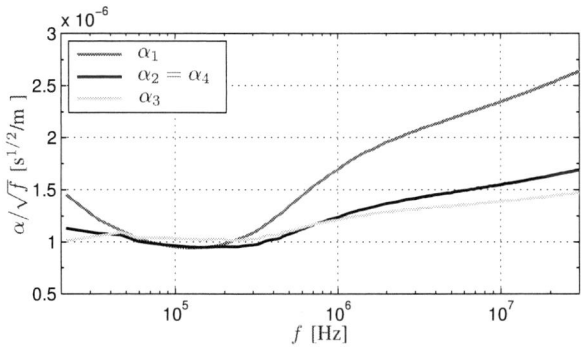

Fig. 2. Frequency dependence of attenuation parameter α for different eigenmodes. At high frequencies $\alpha \approx \frac{R'}{2}\sqrt{\frac{L}{C}}$ and $\alpha \propto \sqrt{f}$ due to p.u.l. resistance R', see Fig. 6.

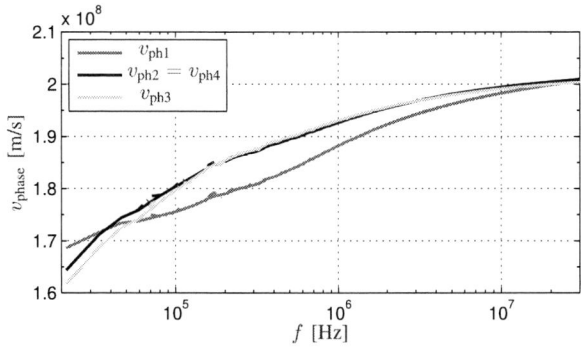

Fig. 3. Frequency dependence of phase velocity $v_{\text{phase}} = \omega/\beta$ for different eigenmodes. The phase velocity increases with increasing frequency approximately by $v_{\text{phase}} \approx \frac{1}{\sqrt{L'C'}}\left(1 - \frac{R'^2}{8\omega^2 L'^2}\right)$.

In order to extract C_i', L_i', and R_i', we first determine the p.u.l. admittance $Y_i' = G_i' + \mathrm{j}\omega C_i'$ by (11). In principle, its real part would allow calculating the p.u.l. leakage conductivity through the dielectric insulation between the conductors G_i'. However, the original s-parameter measurement using VNA

does not provide sufficient precision for the very low values of G'. In fact, the wire coating dielectrics like PVC are very good insulators and that is why $G' = 0$ is used. The experimental verification for neglecting the leakage conductivity is that the terminal behavior of cables of different lengths are accurately predicted from the extracted p.u.l. parameters with zero leakage conductance. The extracted p.u.l. capacitance $C_i' = \mathrm{Im}(Y_i')/\omega$ can well be fitted by a frequency independent constant as shown in Fig. 4. We averaged the capacitances extracted for frequencies $0 \le f \le 2\,\mathrm{MHz}$ thus putting more weight to lower frequencies. Having determined Y_i', the real and imaginary part of the p.u.l. impedance $Z_i' = R_i' + \mathrm{j}\omega L_i'$ given by (11) can be written as $R_i' = 2\alpha_i\beta_i/(C_i'\omega)$ and $L_i' = (\beta_i^2 - \alpha_i^2)/(C_i'\omega)$. The results of the extracted p.u.l. parameters are shown in Figs. 4, 5, and 6.

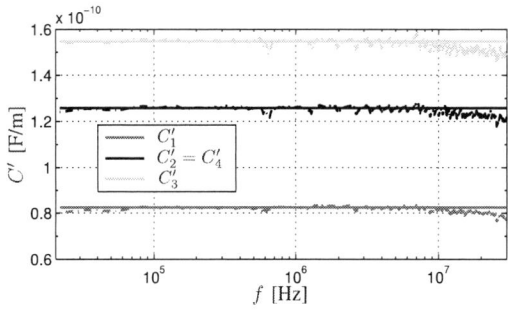

Fig. 4. Per-unit-length capacitance of different eigenmodes fitted by constant values obtained by averaging the extracted capacitances.

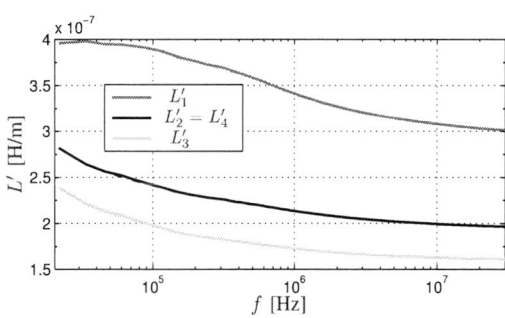

Fig. 5. Per-unit-length inductance decreases with increasing frequency due to skin and proximity effect that decrease the internal inductance of conductors.

As consistency check, we used the extracted p.u.l. parameters to again calculate the admittances using (8) and (9) and the agreement is very good, as shown in Figs. 7 and 8. As a further verification, the admittance matrix for a 5 m long cable of the same type is calculated using the same p.u.l. parameters. The results of predicted and measured admittances are shown in Figs. 9 and 10. The agreement is good, particularly when noting that the 5 m cable is not just a 5 m part cut from the 150 m cable but is a physically distinct cable of the same type (manufacturer).

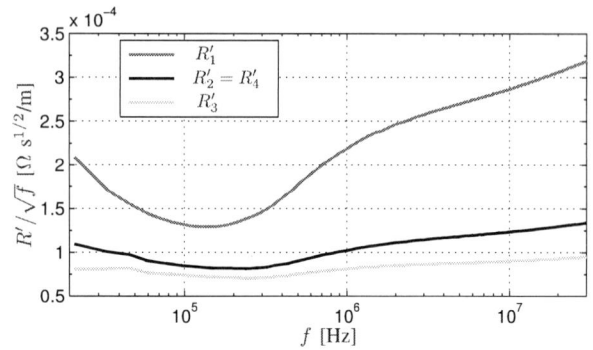

Fig. 6. Per-unit-length resistance is scaled with $1/\sqrt{f}$ due to expected skin effect.

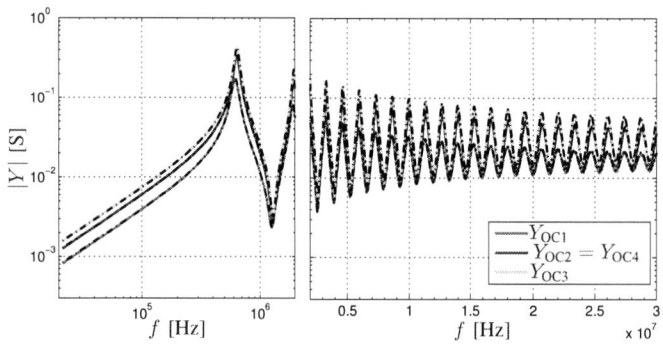

Fig. 7. Agreement between measured eigenmodes of open-circuit admittances of 150 m cable shown as colored solid lines and results calculated by extracted p.u.l. parameters shown as dashed lines. Per-unit-length parameters were extracted from this measurement data.

IV. Summary and Conclusion

An accurate and reliable approach for the recovery of the p.u.l. parameters from measured y- or s-parameters was presented. Exploiting symmetries of typical power cables, analytical formulas for p.u.l. parameters as function of measured admittances were derived and verified against measurement for different cable lengths. The recovery of the p.u.l. parameters, links behavioral models with physical models and allows the creation of stable, accurate, and efficient high-frequency

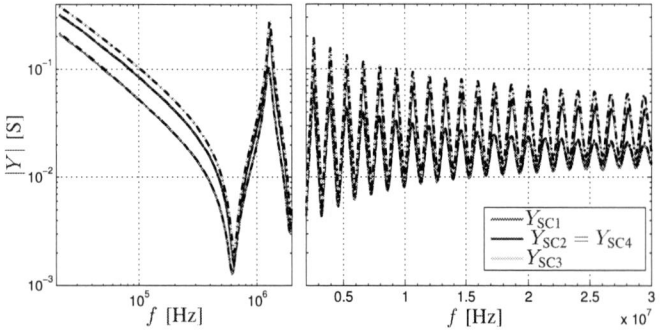

Fig. 8. Same as Fig. 7 for short-circuit admittances.

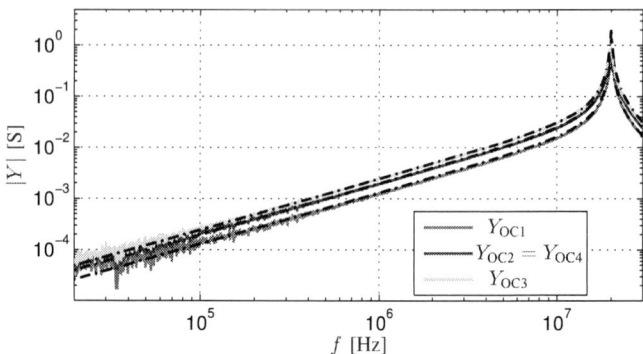

Fig. 9. Comparison between calculated and measured open-circuit admittances of 5 m cable. Per-unit-length parameters were extracted from 150 m cable. The good agreement proves the possibility to accurately predict measurements for different cable lengths based on the extracted line parameters.

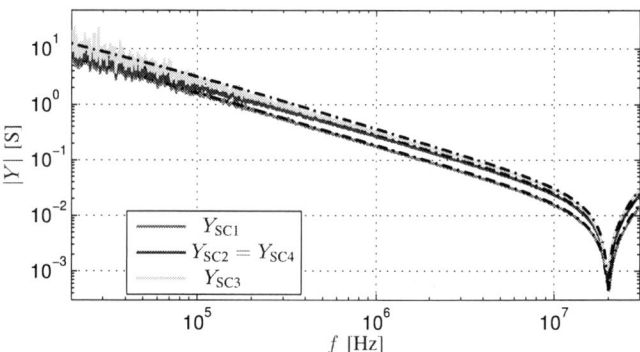

Fig. 10. Same as Fig. 9 for short-circuit admittances.

equivalent circuits for shielded multi-conductor cables [2], [3].

References

[1] C. Paul, *Analysis of Multiconductor Transmission Lines*. Hoboken, New Jersey: Wiley, 2008.

[2] S. Skibin, B. Wunsch, I. Stevanović, and B. Gustavsen, "High frequency cable models for system level simulations in power electronics applications," in *Int. Symp. on Electromagnetic Compatibility EMC Europe 2012*, Rome, Italy, Sep. 17-21, 2012, pp. 1–6.

[3] I. Stevanović, B. Wunsch, G.-L. Madonna, M.-F. Vancu, and S. Skibin, "Multiconductor cable modeling for EMI simulations in power electronics," in *38th Annual Conference of the IEEE Industrial Electronics Society IECON 2012*, Montreal, Canada, Oct. 25-28, 2012, pp. 5366–5371.

[4] L. Knockaert, D. De Zutter, F. Olyslager, E. Laermans, and J. De Geest, "Recovering lossy multiconductor transmission line parameters from impedance or scattering representations," *IEEE Transactions on Advanced Packaging*, vol. 25, no. 2, pp. 200–205, May 2002.

[5] R. Achar and M. Nakhla, "Simulation of high-speed interconnects," *Proc. IEEE*, vol. 89, no. 5, pp. 693–728, May 2001.

[6] M. AbuShaaban and S. Scanlan, "Modal circuit decomposition of lossy multiconductor transmission lines," *IEEE Trans. Microw. Theory Tech.*, vol. 44, no. 7, pp. 1046–1056, Jul. 1996.

[7] C. Paul, "Decoupling the multiconductor transmission line equations," *IEEE Trans. Microw. Theory Tech.*, vol. 44, no. 8, pp. 1429–1440, 1996.

[8] B. Wunsch, I. Stevanović, and S. Skibin, "Improved accurate high frequency models of multiconductor power cables exploiting symmetries," in *Int. Symp. on Electromagnetic Compatibility EMC Europe 2013*, 2013.

[9] B. Gustavsen and A. Semlyen, "Admittance-based modeling of transmission lines by a folded line equivalent," *IEEE Trans. Power Del.*, vol. 24, no. 1, pp. 231–239, Jan. 2009.

Differential Through-Silicon-Vias Modeling and Design Optimization to Benefit 3D IC Performance

Yang Yi
Department of Computer Science & Electrical Engineering
University of Missouri - Kansas City
Kansas City, MO 64110, USA
cindyyi@umkc.edu

Yaping Zhou
NVidia Corporation.
Santa Clara, CA, 95050, USA
yapingz@nvidia.com

Abstract — **Through Silicon Vias (TSVs) constitute key components interconnecting adjacent dies vertically to form three dimensional integrated circuit (3D IC). In this paper, we present an accurate electrical circuit model for differential through silicon vias (TSVs) considering the metal oxide semiconductor capacitance effects and study the effect of differential TSVs on the signal integrity with high data rate signals (up to 25Gbps) using eye diagram approach. Furthermore, we find the nonlinear TSV capacitance has the most predominant impact for the 3D IC performance, and thus, its negative effect to the system performance should be minimized. We optimize the parameters of TSVs architecture and manufacturing process to obtain the minimum depletion capacitance in the desired operating voltage region based on the nature of the TSVs C-V characteristics. Our study shows minimizing the TSVs capacitance could significantly improve the 3D IC performance, which help in developing effective design guidelines for TSVs in 3D IC.**

Keywords — *Through Silicon Vias (TSVs); Capacitance; Modeling; Design Optimization.*

I. Introduction

Three dimensional integrated circuit (3D IC) has gained widespread recognition as it offers promising option to stack dies vertically, which reduces the interconnect length and increases the performance in terms of delay, power and form factor [1].

3DIC can be achieved by a combination of bonding and through silicon vias (TSVs). TSVs are most commonly fabricated by high aspect ratio deep silicon etching lining made of dielectric layer for electrical isolation, and super conformal filling with conductor such as copper. A dielectric layer, usually SiO_2 is deposited on the hole walls to isolate the metals from the substrate. TSVs are embedded in Si, and they are electrically isolated from the silicon layer with a dielectric liner, which form metal oxide semiconductor (MOS) structure. The conductive metal core acts as the gate in a MOS capacitor.

There have been several publications focusing on the electrical modeling and simulations of TSVs. Most publications [2-7] related to TSVs modeling and simulations consider the Si substrate as a lossy dielectric material with low conductivity. Ignoring the semiconductor properties of the substrate and the resulting MOS capacitance introduces significant inaccuracies in the modeling of the capacitance in these structures. Ref. [8-9] studied the impact of a bias voltage on TSVs characteristics (mostly capacitance) theoretically and experimentally. However, it requires solving both the full Maxwell equations and Boltzmann transport equations at the same time. Therefore, it is computationally intensive and difficult to apply to the channel simulations.

Recently, we made an advance in proposing a novel circuit model for multiple TSVs [10]. The compact lumped circuit model accounts for wide frequency range, high frequency skin effect, eddy currents in substrate, and metal-oxide-semiconductor (MOS) effect. The circuit model also includes a closed form expression for the large signal and nonlinear capacitance which is dependent on the biasing of the TSVs with respect to the substrate. The model's compactness and compatibility with SPICE simulators allows the electrical modeling of various TSVs arrangements without the need for computationally expensive field solvers, which significantly reduces the simulation running time. The proposed circuit model could be a promising solution to both academic and industrial need for broadband electrical modeling of TSVs in 3D IC.

Through the channel simulations with eye diagram approach, we found that the nonlinear TSV capacitance has the most predominant impact for the 3D IC performance, and thus, its negative effect to the system performance should be minimized. Therefore, it is advantageous to have a small TSV capacitance to obtain a faster signal response and lower signal distortion. The nature of TSV C-V characteristics depends on the TSV process and the TSV architecture, and both these factors ought to be tuned to achieve minimum TSV capacitance in the desired operating voltage region.

In this paper, we extend our work in [10] and optimize the parameters of TSVs architecture and manufacturing process to obtain the minimum depletion capacitance in the desired operating voltage region based on the nature of the TSVs C-V characteristics. Our study will help in developing design guidelines for TSVs in 3D IC. Furthermore, TSVs can be used as useful variable capacitors in 3D IC design by utilizing the MOS capacitance effect.

978-1-4799-0708-3/13 $31.00 © 2013 IEEE

II. DIFFERENTIAL TSVs MODELING AND SIMULATION

Fig. 1 shows the self-biasing of differential TSVs carrying a differential signal. The differential mode has a positive voltage on one TSV and a negative one on the other. There is a virtual electrical ground in the silicon substrate in order satisfy the voltage boundary conditions. In a silicon substrate, a hole is etched out, a thin silicon dioxide layer (TSV liner) is formed, and a conductive via is plated. A TSV is essentially a metal insulator semiconductor (MIS) device.

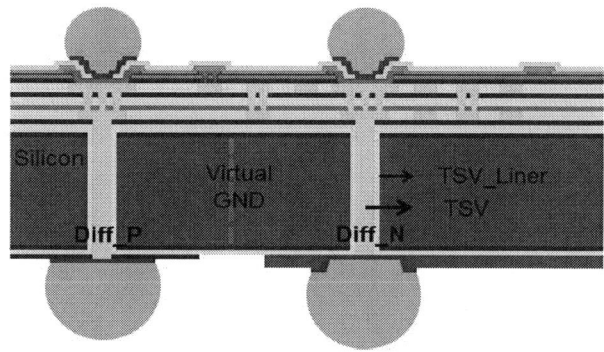

Figure 1: Self-biasing of Differential TSVs Structure

The equivalent circuit model for differential signal TSVs is presented in [10]. The proposed circuit model contains resistance (R), capacitance (C), and inductance (L). Since frequent switching of high speed signals can dynamically bias TSV metal insulator semiconductor (MIS) interface and allocate TSV MIS into accumulation or depletion regions, the TSV capacitance is nonlinear and dependent on the biasing of the TSVs. The capacitive effect of each TSV with the differential signal is modeled as a nonlinear capacitance $C_{TSV}(V(t))$, which is dependent on the biasing of the TSVs with respect to the substrate. The TSV capacitance $C_{TSV}(V(t))$ is the series combination of the oxide capacitance C_{ox} and the depletion capacitance C_{dep}. Note that there is no capacitive coupling between the two differential TSVs since they are shielded by the virtual ground. The circuit model accurately captures all the parasitic elements of various TSVs arrangements and accounts for wide frequency range, high frequency skin effect, eddy currents in substrate, and metal oxide semiconductor (MOS) effect.

We simulated the differential TSVs with high data-rate signals using eye diagram approach. Note that the nonlinear and voltage dependent TSV capacitance $C_{TSV}(V(t))$ can be modeled in HSPICE simulator by using the Verilog A block or using the equation-based capacitance. Fig. 2 shows the simulate eye diagrams of pseudo-random data transmitted over the wafer level TSV, chip level TSV, and interposer TSV, which are running at 10Gbps and 25Gbps. For chip-level bonding, low density TSVs having diameters approximately between 5 and 30 *um* are mostly used. High density TSVs for 3D IC applications typically have diameters less than 5 *um* and are used for wafer bonding processes. TSVs in Si interposers usually have diameters larger than 50 *um*. As shown in Fig. 2, the eye diagram of interposer TSV is closed

when its data rate reach 25Gbps, which enables us to see the imperfection of the signal. As jitter closes the eye horizontally and noise closes the eye vertically, the sampling area gets squeezed, increasing the chance to sample incorrectly and get bit errors. The nonlinear TSV capacitance has the most predominant impact for the 3D IC performance [5], and thus, its negative effect to the system performance should be minimized.

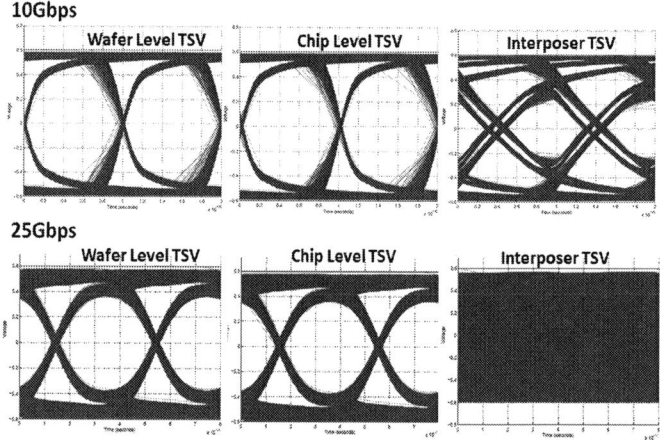

Figure 2: Eye Diagram of TSVs

III. TSVs CAPACITANCE REDUCTION TO IMPROVE 3D IC PERFORMANCE

As discussed in section II, it is advantageous to have a small TSV capacitance to obtain a faster signal response and lower signal distortion. The nature of TSV C-V characteristics depends on the TSV process and the TSV architecture, and both these factors ought to be tune to achieve minimum TSV capacitance in the desired operating voltage region.

Since the Si substrate is doped, a typical TSV C-V curve has accumulation and depletion regions. When a voltage V is applied to a TSV, if $V < V_{FB}$ (flat band voltage), the positively charged holes in silicon are dragged to Si-SiO2 interface, and an accumulation layer is formed; if $V > V_{FB}$, holes are pushed away and a depletion region is formed. Note that the flat band voltage (V_{FB}) is defined as:

$$V_{FB} = \varphi_m - \varphi_{si} - \frac{Q_S}{C_{ox}}, \tag{1}$$

where φ_m and φ_{si} are the work functions of TSV metal and silicon, respectively, Q_s is the space charge, and C_{ox} is the silicon liner capacitance. V_{FB} depends on material properties as well as fabrication processes (doping, etc.).

When the TSV MIS is in the accumulation region, the electrical field is confined in the SiO2 liner. The TSV capacitance is equal to C_{ox}, which is the capacitance of the SiO2 liner. C_{ox} can be calculated through equation (2):

$$C_{ox} = 2\pi\varepsilon_{ox} \frac{h}{\ln(\frac{r_{ox}}{r_{TSV}})}, \tag{2}$$

where h is the TSV height, ε_{ox} is the dielectric constant of the SiO_2 liner , r_{TSV} is the radius of the TSV, and r_{ox} is the radius of a TSV with SiO2 liner in the accumulation region.

When TSV MIS is in the depletion region, the electrical field can penetrate into the substrate, and the TSV capacitance is equal to C_{ox} in series with C_{dep}, which is defined as the capacitance of the depletion region. To calculate the depletion capacitance C_{dep}, we first need to figure out how to calculate depletion width w_{dep} since the large digital signal swing makes the depletion region to change its depletion width dynamically. w_{dep} is defined as:

$$w_{dep} = r_{dep} - r_{ox}. \tag{3}$$

The depletion width w_{dep} can be calculated by solving a Poisson's equations of scalar potential in the depletion region and a Laplace's equation in the SiO_2 liner as shown in equation (4):

$$V - V_{FB} = \frac{qN_a r_{ox}^2}{2\varepsilon_{ox}}(1 - \frac{r_{dep}^2}{r_{ox}^2})\ln(\frac{r_{TSV}}{r_{ox}}) - \frac{qN_a}{2\varepsilon_S}(\frac{r_{dep}^2 - r_{ox}^2}{2} + r_{dep}^2 \ln(\frac{r_{ox}}{r_{dep}})) \tag{4}$$

Depletion width w_{dep} is usually sub-micron in silicon substrates [5, 11-12]. The SiO_2 liner thickness (t_{ox}) which is denoted as:

$$t_{ox} = r_{ox} - r_{TSV}, \tag{5}$$

is also submicron [2] [3]. Therefore, we have the following equations for the relationship between w_{dep}, t_{ox} and r_{ox}

$$\frac{w_{dep}}{r_{ox}} << 1, \text{ and } \frac{t_{ox}}{r_{ox}} << 1.$$

Therefore, the equation (4) can be simplified accordingly with the following steps:

$$\left(1 - \frac{r_{dep}^2}{r_{ox}^2}\right) = 1 - \left(\frac{r_{ox} + w_{dep}}{r_{ox}}\right)^2 = 1 - \left(1 + \frac{w_{dep}}{r_{ox}}\right)^2$$

$$\doteq 1 - \left(1 + \frac{2w_{dep}}{r_{ox}}\right) = -\frac{2w_{dep}}{r_{ox}} \tag{6}$$

$$\ln\frac{r_{TSV}}{r_{ox}} = \ln\left(\frac{r_{ox} - t_{ox}}{r_{ox}}\right) = \ln\left(1 - \frac{t_{ox}}{r_{ox}}\right) = -\frac{t_{ox}}{r_{ox}} \tag{7}$$

$$\frac{r_{dep}^2 - r_{ox}^2}{2} = \frac{(r_{ox} + w_{dep})^2 - r_{ox}^2}{2} = \frac{2w_{dep}r_{ox} + w_{dep}^2}{2} \tag{8}$$

$$r_{dep}^2 \ln\frac{r_{ox}}{r_{dep}} = (r_{ox} + w_{dep})^2 \ln\left(\frac{r_{ox}}{r_{ox} + w_{dep}}\right)$$

$$\approx (r_{ox} + w_{dep})^2 \ln\left(1 - \frac{w_{dep}}{r_{ox}}\right) \tag{9}$$

$$\approx (r_{ox} + w_{dep})^2 \left(-\frac{w_{dep}}{r_{ox}}\right)$$

Based on equation (6-9), equation (4) could be transformed to:

$$V - V_{FB} = qN_a\left(\frac{w_{dep}t_{ox}}{\varepsilon_{ox}} + \frac{3w_{dep}^2}{4\varepsilon_s}\right) \tag{10}$$

Accordingly, we can calculate the depletion width w_{dep} using equation (11):

$$\frac{3}{4\varepsilon_s}w_{dep}^2 + \frac{t_{ox}}{\varepsilon_{ox}}w_{dep} + \frac{(-V + V_{FB})}{qN_a} = 0 \tag{11}$$

The depletion width w_{dep} can be solved when the TSV is in depletion region and expressed in equation (12):

$$w_{dep} = \frac{2\varepsilon_s}{3\varepsilon_{ox}}\left(-t_{ox} + \sqrt{t_{ox}^2 + \frac{3\varepsilon_{ox}^2}{\varepsilon_S}\frac{V - V_{FB}}{qN_a}}\right). \tag{12}$$

As shown in equation (12), the large digital signal swing makes the depletion region to change its depletion width dynamically. Fig. 3 shows one example of w_{dep} changes with the supply voltage V.

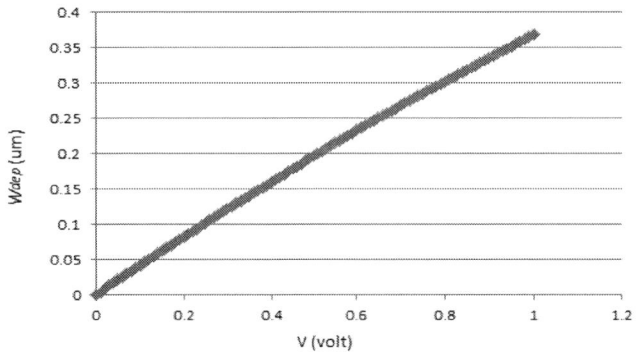

Figure 3: Dynamic change of w_{dep} with the supply voltage

The C-V characteristics of TSVs are dependent on their geometrical parameters (such as the TSV diameter, height, pitch, and oxide layer thickness) and electrical parameters (such as the metal conductivity, oxide permittivity, and the silicon substrate resistivity).

In the accumulation region when $V < V_{FB}$, TSV capacitance is oxide capacitance. For high frequency signals, the TSV capacitance is equal the series of oxide capacitance and depletion capacitance in the depletion region when $V > V_{FB}$. In order to reduce the TSV capacitance, the TSV must be biased in the deep depletion region. This condition could be satisfied by applying an appropriate negative bias voltage on a p-type Si substrate or an appropriate positive bias voltage on an n-type Si substrate. Alternatively, the TSV can be filled with a low work function metal for p-type Si (high work function metal for n-type Si) such that it is biased in the deep depletion region by the given signal range.

Furthermore, the diameter of the TSV and the thickness of the Si substrate should be reduced to minimize the TSV capacitance. A thick liner should be used to reduce the liner capacitance. An appropriate material should be selected with a low dielectric constant will be helpful. Furthermore, a high resistivity Si substrate is helpful to reduce the depletion capacitance.

IV. SIMULTATION RESULTS

Fig. 4-6 show the horizontal eye opening and vertical eye opening of the wafer level TSV, chip level TSV, and interposer TSV at accumulation region (high Cap.), depletion region (medium Cap.), and deep depletion region (low Cap.). As shown in Fig. 4-6, optimizing the parameters of TSVs architecture and manufacturing process to obtain the minimum depletion capacitance could drastically enhance the TSV electrical performance across wide frequency range.

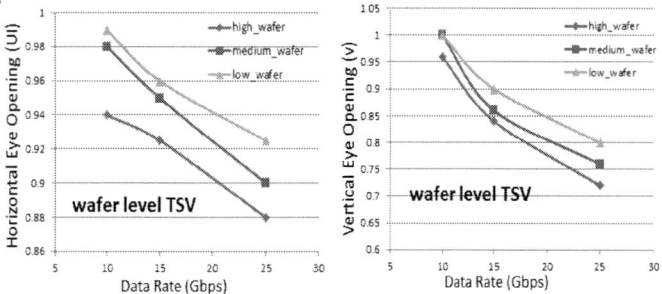

Figure 4: Horizontal & Vertical Eye Opening of Wafer Level TSV

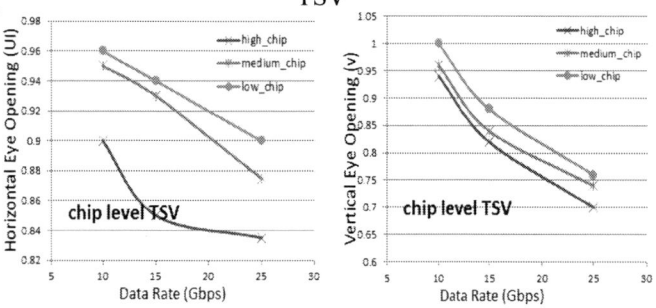

Figure 5: Horizontal & Vertical Eye Opening of Chip Level TSV

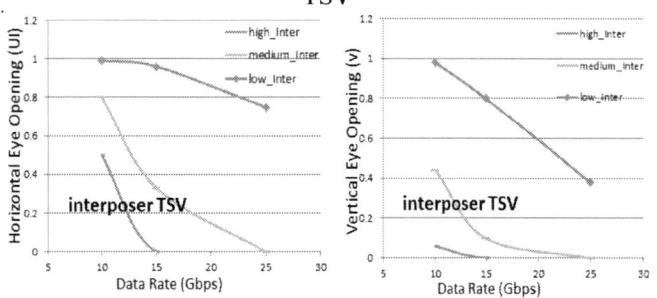

Figure 6: Horizontal & Vertical Eye Opening of Interposer TSV

Fig. 7 shows the eye diagram of interposer TSV with optimized depletion capacitance when the signal data rate at 25Gbps. The optimized depletion capacitance is achieved by biasing the TSV into the deep depletion region with low work function metal for p-type Si. Compared with the eye diagram shown in Fig. 2, we can find the performance of interposer TSV significantly improved by reducing the TSV capacitance. Our study helps in developing design guidelines for TSVs in 3D IC. Furthermore, TSVs can be used as useful variable capacitors in 3D IC design by utilizing the MOS capacitance effect.

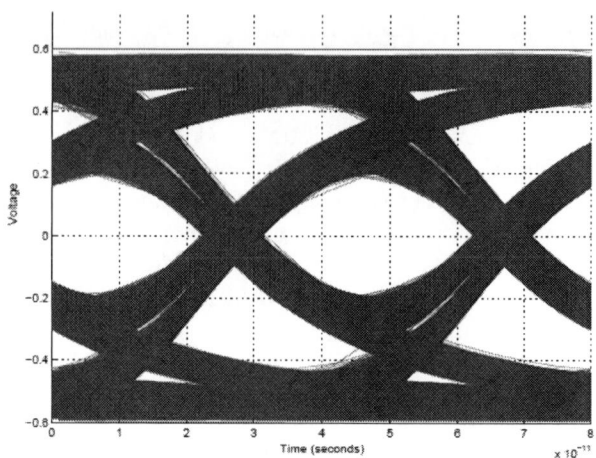

Figure 7: Eye diagram of Interposer TSV with optimized depletion capacitance at 25Gbps

REFERENCES

[1] J. Jiang, Q. Xu, and B. Eklow, "On effective TSV repair for 3D-stacked ICs," In the procedings of *IEEE Design, Automation & Test in Europe Conference & Exhibition*, pp.793-798, 2012.

[2] I. Ndip, B. Curran, K. Lobbicke, S. Guttowski, H. Reichl, K. Lang, and H. Henke, "High-frequency modeling of TSVs for 3D chip integration and silicon interposers considering skin-effect, dielectric quasi-TEM and slow wave modes", *IEEE Trans. CPMT*, Vol. 1, No. 10, pp. 1627, 2011

[3] K. Han, M. Swaminathan, and T. Bandyopadhyay, "Electromagnetic modeling of TSV interconnections using cylindrical modal basis functions", *IEEE Trans. Adv. Pkg*, Vol. 33, No. 4, pp. 804, 2010

[4] J. Cho, E. Song, K. Yoon, J.Pak, J. Kim, W. Lee, T. Song, K. Kim, J. Lee, H. Lee, K. Park, S. Yang, M. Suh, K. Byun, and J. Kim, "Modeling and analysis of through-silicon-via noise coupling and suppression using guard ring", *IEEE Trans. CPMT*, Vol. 1, No. 2, pp. 220, 2011

[5] L. L. W. Leung and K. J. Chen, "Microwave characterization and modeling of high aspect ratio through-wafer interconnect vias in silicon substrates," *IEEE Trans. Microw. Theory Tech.*, vol. 53, no. 8, pp. 2472–2480, Aug. 2005.

[6] P. Luo, T. Wang, C. Wey, L. Cheng, B. Sheu, Y. Shi, "Reliable Power Delivery System Design for Three-Dimensional Integrated Circuits," In the procedings of *IEEE Computer Society Annual Symposium on VLSI*, pp.356-361, 2012.

[7] R. Weerasekera, L.-R. Zheng, D. Pamunuwa, and H. Tenhunen "Extending systems-on-chip to the third dimension: performance, cost and technological tradeoffs," In the procedings of *IEEE/ACM Int. Conf. on Comp.-Aided Design*, pp. 212 ‑ 219, 2007.

[8] G. Katti, M. Stucchi, K. de Meyer, and W. Dehaene, "Electrical Modeling and characterization of through-silicon-via for three dimensional ICs", *IEEE Trans. Elec. Dev.*, Vol. 57, No. 1, pp. 256, 2011.

[9] T. Bandyopadhya, K. Han, D. Chung, R. Chatterjee, M. Swaminathan, and R. Tummala, "Rigorous Electrical Modeling of Through Silicon Vias (TSVs) with MOS Capacitance Effects", *IEEE Trans. CPMT*, Vol. 1, No. 6, pp. 893, 2011.

[10] Y. Yi and Y. Zhou, ``A Novel Circuit Model for Multiple Through-Silicon-Vias (TSVs)", In Proceedings of *IEEE International Conference on 3D System Integration*, Oct. 2013.

[11] Z. Guo and G. Pan, "On simplified fast modal analysis for through silicon vias in layered media based upon full-wave solutions," *IEEE Trans. Adv. Packag.*, vol. 33, no. 2, pp. 517–523, May 2010.

[12] J. S. Pak, C. Ryu, and J. Kim, "Electrical characterization of trough silicon via (TSV) depending on structural and material parameters based on 3-D full wave simulation," In the procedings of *Interational Conference of Electronic Materials and Packaging*, pp. 1–6, Nov. 2012.

978-1-4799-0708-3/13 $31.00 © 2013 IEEE

On-line Real-time Temperature and Power Estimation of an IC Using Time-domain Thermal Filters

Jaeha Kung, Minki Cho, Sudhakar Yalamanchili, and Saibal Mukhopadhyay

School of ECE, Georgia Institute of Technology, Atlanta, GA 30332, USA

Email: <jhkung, mcho8>@gatech.edu, <sudha, saibal>@ece.gatech.edu; Ph: 404-894-2688

Abstract— **A methodology is presented for on-line and real-time estimation of transient variations in temperature and average power of a chip after fabrication and packaging. Measurements from a 130nm CMOS test chip demonstrate the accuracy of the proposed approach.**

I. INTRODUCTION

The dynamic and leakage power of ICs have increased significantly with scaling resulting in elevated chip temperature. High temperature degrades reliability and performance of ICs, increases leakage power due to positive feedback between leakage and temperature, and results in higher cooling and packaging cost [1]. Hence, it is crucial to estimate transient variations in on-chip temperature and power to drive on-line power/energy management techniques for reliable system design and operation.

Accurate, real-time and on-line estimation of transient temperature of a packaged IC (i.e. post-silicon) is crucial for efficient power/thermal management. The fine-grain design-time thermal simulation tools based on finite element, finite volume, or distributed RC networks, although accurate, but require accurate estimate of the thermal properties of materials (like thermal conductivity and heat capacity) [2-3]. However, these properties can have chip-to-chip or time-dependent variations. Moreover, due to the positive feedback between leakage and temperature, the chip-to-chip process variation (hence, leakage) can result in chip-to-chip variation in temperature for same power pattern [4-5]; the pre-silicon estimates do not capture this variation. Likewise real-time and on-line characterization of transient variations in average power dissipation is also challenging. The design time average power estimation tools are significantly slow for use in real-time. The dynamic variation of instantaneous power can be monitored using on-line current sensors [6] or by sensing voltage variation across a sleep transistor [7]. However, designing current or power sensors are challenging at low operating voltages. Moreover, when an average or a slowly varying power pattern is needed (for thermal design/ management), these approaches require additional filtering or incur significant memory overhead.

To address the challenge of post-silicon thermal analysis, Cho et al. have proposed Thermal System Identification (TSI), a methodology to characterize the relation between power dissipation and temperature of a packaged IC in the frequency domain using on-chip power/thermal measurements [4] (Fig. 1). Once the relation, referred to as the *thermal filter*, is characterized, authors have proposed a Fast Fourier Transform (FFT) and Inverse FFT (IFFT) based approach for temperature prediction [4]. While TSI is suitable for post-silicon thermal analysis, it requires complex FFT/IFFT computation and does not provide real-time temperature estimation.

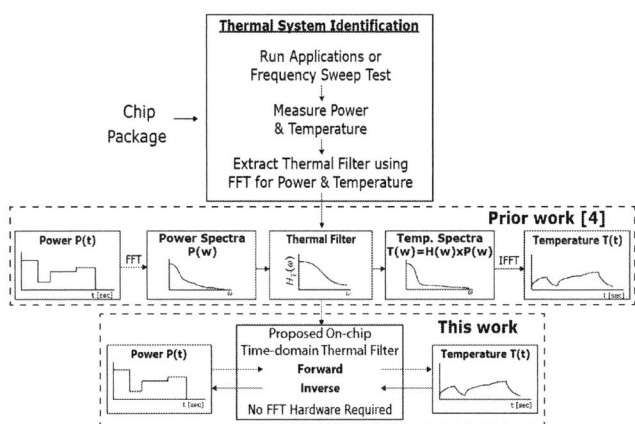

Fig. 1. The basic concept of *Thermal System Identification* (TSI) method [4] and the proposed time-domain thermal filter.

This paper presents a methodology for on-line and real-time estimation of transient variations in temperature and average power of individual ICs after fabrication/packaging. The proposed approach is based on TSI, but instead of using FFT/IFFT, we propose an on-chip time-domain thermal filter to represent the thermal system (Fig. 1). The filter is tuned post-silicon to represent the frequency response of the thermal system of an IC extracted by TSI. The time-domain filter takes the time-domain power variation as input and directly estimates time-domain temperature variation. Moreover, it is proposed that the inverse of the extracted thermal filter can be used to estimate power dissipation from a measured temperature response. A time-domain inverse thermal filter is designed to capture this inverse relation and estimate time-domain power variations from a measured transient temperature pattern. The proposed time-domain temperature and power estimation method is validated through measurement of a 130nm CMOS test chip.

II. DESIGN OF TIME-DOMAIN THERMAL FILTERS

A. Background – Thermal System Identification

The basic approach of TSI, as proposed by Cho et al., is to extract the frequency response of a thermal system after fabrication and packaging through power and temperature measurements [4]. Assume a single point (x, y) on a chip where a temperature sensor is located. The power consumption, $P(t)$, at location (x, y) is known. Temperature, $T(t)$, is then obtained by performing a convolution between power and the impulse response of the thermal system, $h_T(t)$. It can be written as

$$T(t) = P(t) * h_T(t). \qquad (1)$$

The convolution operation is simply a multiplication in the frequency domain. Consider Fourier Transform, we obtain

$$T(\omega) = P(\omega) \times H_T(\omega), \qquad (2)$$

978-1-4799-0708-3/13 $31.00 © 2013 IEEE

where $H_T(\omega)$ is the frequency response of the thermal system between the power generation and temperature observation point. Thus, we can define TSI as a method to identify $H_T(\omega)$ using the relation (2). Once $H_T(\omega)$ is extracted, the temperature for a given power pattern is estimated using FFT and IFFT as shown in Fig. 1.

The thermal filter extraction is performed only once (or very few times) for an IC in frequency domain (see [4] for details, also summarized in Section III). But temperature prediction using the extracted filter is required during real time operation. During the temperature estimation the FFT/IFFT based approach incurs the complexity associated with transformations between time and frequency domain including computation of FFT/IFFT. This paper proposes to design a time-domain filter having the same frequency response as $H_T(\omega)$ to eliminate the complex FFT/IFFT computations. In the following sections, we discuss a forward $[H_T(\omega)]$ and an inverse $[H_T^{-1}(\omega)]$ thermal filter design to obtain real-time estimates of temperature or power variations of the chip.

B. Time-Domain Thermal Filter to Estimate Temperatue

We first explore a mathematical realization of the thermal filter having a flexible slope in gain and phase response (Fig. 2). As it will be noticed in Section III in the frequency response of the thermal system obtained by TSI, the filter gain decreases slowly with frequency (absolute value of the slope less than 20dB/decade). It is difficult to satisfy a smoothly decreasing frequency response by simply controlling poles and zeros. However, there is a methodology making it possible; a low pass filter using a fractional-order capacitor (FOC) [8]. Conceptually, a FOC is an element whose current $i(t)$ and voltage $v(t)$ are related by

$$i(t) = \frac{C_\alpha \cdot d^\alpha v(t)}{dt^\alpha}, \tag{3}$$

where C_α is the capacitance of FOC and α is a non-integer exponent ($0 < \alpha < 1$). The forward thermal filter response can be designed by a fractional-order low pass filter (FLPF) having a transfer function as follows

$$H_{FLPF}(s) = \frac{A_{DC} \cdot \omega_f}{s^\alpha + \omega_f}, \tag{4}$$

where ω_f is a fractional cutoff frequency and A_{DC} is a DC gain. Then, the gain and phase response of the thermal filter, equivalently FLPF, will become

$$Gain = |H(j\omega)| = \frac{A_{DC} \cdot \omega_f}{\sqrt{\omega^{2\alpha} + 2\omega_f \omega^\alpha \cos\left(\frac{\alpha\pi}{2}\right) + \omega_f^2}},$$

$$Phase(rad) = \angle H(j\omega) = -\tan^{-1}\frac{\omega^\alpha \sin\left(\frac{\alpha\pi}{2}\right)}{\omega^\alpha \cos\left(\frac{\alpha\pi}{2}\right) + \omega_f}. \tag{5}$$

When $\alpha = 1$, (4) becomes a conventional first-order low pass filter amplified by A_{DC}. By controlling α, the gain and phase response can be adjusted to fit an extracted thermal filter response from the TSI method (Fig. 3, ω_f is set to 20π rad).

The realization of a FOC is being studied for a decade [9]. Using an actual FOC, the thermal filter can be designed by a simple RC_α circuit (Fig. 2(a)). However, these processes are yet commercialized in industry. Therefore, we present a functional filter that can be implemented using digital system

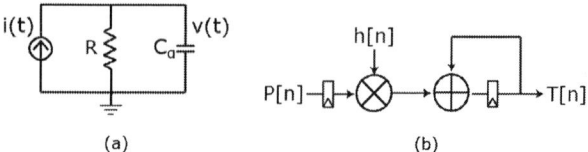

Fig. 2. (a) Simple RC_α circuit and (b) a digital system for the time-domain thermal filter design.

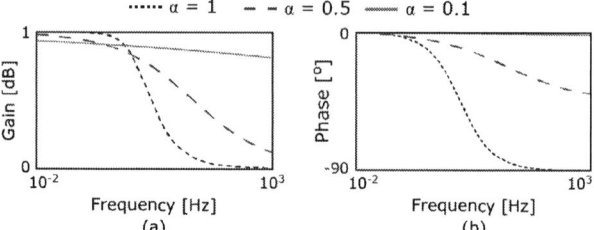

Fig. 3. Frequency response of FLPF: (a) gain response and (b) phase response with different α value (0.1, 0.5, and 1).

(Fig. 2(b)). Performing the inverse Laplace Transform on (4) when $\alpha = 0.5$ gives a time-domain impulse response $h(t)$ as

$$h(t) = A_{DC} \cdot \omega_f \left[\frac{1}{\sqrt{\pi t}} - \omega_f \cdot e^{\omega_f^2 t} \cdot erfc\left(\omega_f \sqrt{t}\right)\right], \tag{6}$$

where $erfc(\cdot)$ is the complementary error function.

In the time-domain approach, the convolution operation is required to predict temperature with a given power pattern. Assume $h[k \cdot T_s]$ is stored in memory where k is an integer and T_s is the sampling time. Consider temperature prediction for 1s with a given power pattern sampled at $T_s = 1ms$ (the number of sample n is 1000). The convolution requires $n^2/2$ of multiplication of *real* numbers. If the multiplication is performed at 1GHz clock, temperature is predicted in 0.5ms with a much simpler circuit than FFT/IFFT where multiplications of complex numbers are involved.

C. Time-Domain Inverse Thermal Filter to Estimate Power

In this section, we discuss methodology to characterize power variation over a finite duration from the measured temperature and using the inverse of the thermal filter extracted from TSI. From (2), we obtain

$$P(\omega) = T(\omega) \times H_T^{-1}(\omega). \tag{7}$$

The problem becomes how to implement a filter circuit which behaves similarly as $[H_T^{-1}(\omega)]$. For a given location (x, y), $H_T^{-1}(\omega)$ simply becomes $1/H_T(\omega)$; dividing (2) by $H_T(\omega)$ has to be equal to (7). The gain and phase response of the inverse thermal filter can be obtained by modifying (5) as

$$Gain = |H^{-1}(j\omega)| = \frac{\sqrt{\omega^{2\alpha} + 2\omega_f \omega^\alpha \cos\left(\frac{\alpha\pi}{2}\right) + \omega_f^2}}{A_{DC} \cdot \omega_f} \tag{8a}$$

$$Phase\,(°) = \angle H^{-1}(j\omega) = \tan^{-1}\frac{\omega^\alpha \sin\left(\frac{\alpha\pi}{2}\right)}{\omega^\alpha \cos\left(\frac{\alpha\pi}{2}\right) + \omega_f}, \tag{8b}$$

Although the frequency response of the inverse thermal filter is similar to a high pass filter (phase response is quite different from a conventional high pass filter; phase increases as ω increases), high frequency components of power could not be completely restored due to low pass nature of the thermal filter. Thus, power characterization with the inverse thermal filter is useful for estimating average or slowly varying power.

From (8a), it shows that the gain goes to infinity as the ω

978-1-4799-0708-3/13 $31.00 © 2013 IEEE

increases. Since we are dealing with the estimation of slowly varying power pattern, we can ignore high frequency components. Thus, we use a low pass filter with a cutoff frequency at 1kHz after the inverse thermal filter to make the system converge. Then, the transfer function of fractional-order high pass filter (FHPF) becomes

$$H_{FHPF}(s) = \frac{s^\alpha + \omega_f}{A_{DC} \cdot \omega_f} \cdot \frac{\omega_c}{s + \omega_c}, \quad (9)$$

where ω_c is the cutoff frequency of the low pass filter connected to the inverse thermal filter. Similar to forward filter case, the time-domain impulse response, $h^{-1}(t)$, can be obtained by the inverse Laplace Transform of $H_{FHPF}(s)$ which is

$$h^{-1}(t) = \frac{\omega_c \cdot [1/\sqrt{\pi t} + e^{-\omega_c t}(\omega_f + \sqrt{-\omega_c} \cdot \mathrm{erf}(\sqrt{-\omega_c t}))]}{A_{DC} \cdot \omega_f}. \quad (10)$$

Then, we can use the same digital system as in Fig. 2(b) using $h^{-1}[k \cdot T_s]$ as $h[k]$ and use temperature as an input.

III. EXPERIMENTAL RESULTS

A. Test Chip and Thermal Filter Extraction

For the verification of TSI method described in Section II, a test chip (Fig. 4(a)) has been designed [5]. The design of the test-chip and its application to verify TSI using FFT/IFFT (as proposed in [4]) is presented in [5]. This paper uses the test-chip to verify the accuracy of the time-domain filter. There are three important components; a poly-based heater, an external PMOS, and a BJT-based analog thermal sensor as in Fig. 4(b). The bias voltage is externally applied at PMOS gate to generate a desired power pattern. Power density of the heater is estimated by dividing the dissipated total power of the heater ($= V^2_{DD,heater}/R$) by the heater area. Temperature of the chip is measured from thermal sensors.

Using this test chip, the gain and phase response of a thermal system can be extracted at a given frequency [4-5]. During an AC analysis, a square wave with a fixed fundamental frequency f is applied at the gate of external PMOS. By performing Fourier Transform for both power and temperature, the gain and phase information of thermal filter response at f can be obtained by relation (2). Similar experiment is done at each frequency [in the range 0Hz to

1kHz (47 data points)] to obtain the amplitude and the phase response of the thermal filter at all frequencies. For a DC analysis, 0V is applied to the gate of an external PMOS so that to provide ($V_{DD,heater}-V_{th}$) to the heater. To get an accurate DC gain, voltage is applied for five minutes. After the steady state is reached, DC gain is achievable by dividing temperature (T - $T_{ambient}$) by the heater power. Fig. 5 shows the thermal filter behaves like a low-pass filter as expected. As shown with black dotted line in Fig. 5(a), the gain smoothly decreases in logarithmic scale which is about -6dB/decade.

B. Temperature Prediction with the Time-Domain Filter

As mentioned in Section II, time-domain thermal filter is designed to predict temperature in real-time. To verify the method with the test chip, variables in (5) is set to match the extracted thermal filter. The half-power frequency, ω_h, is used as a reference which satisfies

$$|H(j\omega_h)| = \frac{1}{\sqrt{2}} \cdot |H(j\omega_{passband})|. \quad (11)$$

Now, ω_h can be obtained by

$$\omega_h = \omega_f^{1/\alpha} \cdot \left(\sqrt{1 + cos^2\left(\frac{\alpha\pi}{2}\right)} - \cos\left(\frac{\alpha\pi}{2}\right) \right)^{1/\alpha}, \quad (12)$$

where $\alpha = 0.5$ to fit the slope of the gain response, but also to simplify the inverse Laplace Transform for computing the time-domain impulse response $h(t)$ as (6). From the extracted thermal filter response, the gain reached half-power at ~0.13Hz. Using selected ω_h and α, ω_f is computed which is 0.7Hz. With these values for a time-domain thermal filter, frequency response of the designed time-domain filter (red solid line) is compared to the extracted thermal filter response (Fig. 5).

To verify the accuracy of the time-domain thermal filter, arbitrary power patterns are used. The input voltage of the external PMOS of the test circuit is generated by the National Instrument PXIe-6363 to generate the power patterns shown in Fig. 6(a). Temperature for each power pattern is measured using the thermal sensor (the one closest to the heater). Fig. 6 shows that the time-domain thermal filter can accurately estimate temperature. The error is computed by ($T_{time\_domain} - T_{TSI}$) and was within ±2.5°C. Note the very high-frequency fluctuations are mostly measurement and sensor noise and as evident from Fig. 6, the time-domain thermal filter helps reduce their effect.

C. Power Estimation with the Time-Domain Inverse Filter

Assuming that the thermal filter response is extracted by

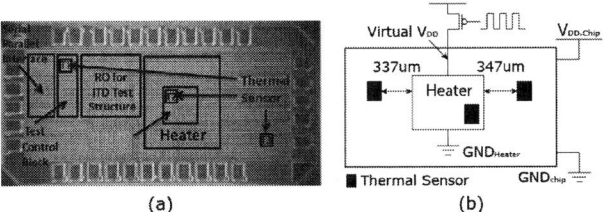

(a)

Fig. 4. Test structure for the experiment: (a) a die photo of the test chip and (b) a schematic of the test circuit.

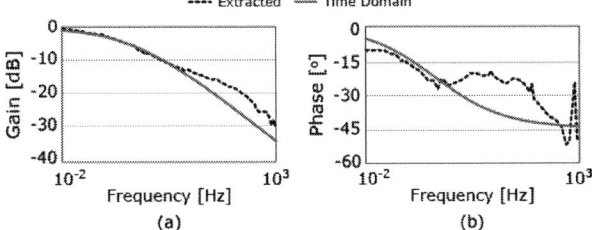

(a)   (b)

Fig. 5. (a) The amplitude response and (b) the phase response of extracted (black) and time-domain (red) thermal filter.

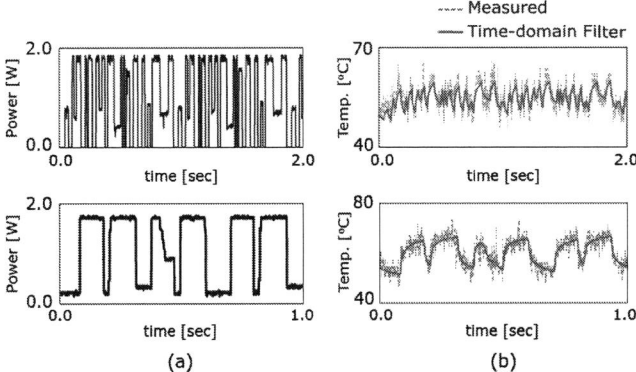

(a)   (b)

Fig. 6. Temperature prediction result: measured temperature (green, dotted) and temperature obtained by using the time-domain filter (blue, solid).

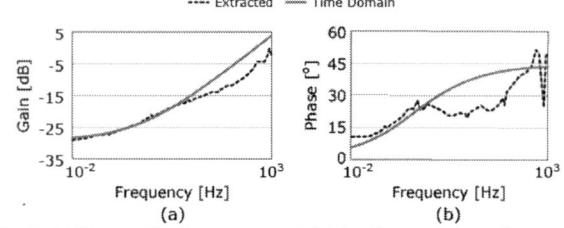

Fig. 7. (a) The amplitude response and (b) the phase response of extracted (black) and time-domain (red) inverse thermal filter.

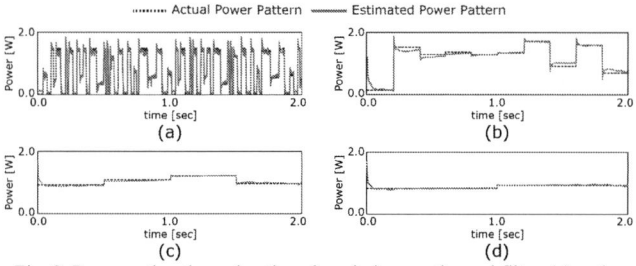

Fig. 8. Power estimation using time-domain inverse thermal filter: (a) entire power pattern, (b) average power for each 0.2 seconds interval, (c) for each 0.5 seconds, and (d) for each 1 second.

the TSI method, an inverse thermal filter can be constructed. Fig. 7 shows the extracted inverse thermal filter (black line) and its time-domain approximation based on (8). The potential of the inverse thermal filter on power estimation is verified in Fig. 8 with the power pattern shown with black dotted line in Fig. 8(a). First, the power pattern is applied to the test chip and the temperature sensor output is noted. The sensed temperature is filtered to eliminate the high-frequency noise (as observed in Fig. 6) and used as input to the time-domain inverse thermal filter. The output of the inverse thermal filter is the estimated power which is reasonably accurate as shown in Fig. 8. To stress the efficacy of characterizing a slowly varying power pattern, estimating average power for certain time interval is demonstrated. Three different time intervals are chosen for the experiment; 0.2, 0.5, and 1 second. As the time interval becomes larger, power estimation gets more accurate. Estimation error on average power over 2 seconds ranges from 0.22% (d) to 3.85% (a). Therefore, estimation accuracy increases for average power.

D. Example of Application of the Time-domain Filters

The real-time and on-line prediction of temperature can allow efficient thermal and power aware task scheduling. One such example is shown in Fig. 9(a) where at a given time instant a scheduler needs to estimate the temperature consequences of scheduling various tasks in the IC. The expected power patterns of the alternative tasks (future power) and the current temperature are known. The proposed filters

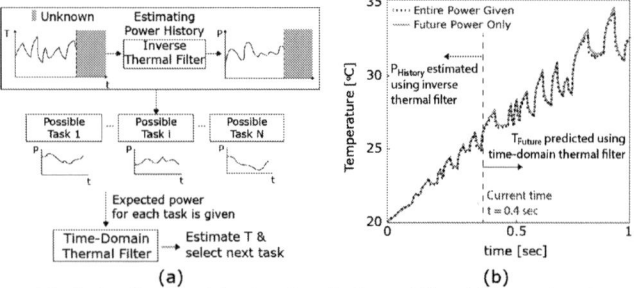

Fig. 9. Application of the time-domain thermal filter: (a) dynamic task scheduling and (b) accuracy of temperature prediction.

can be used to address this problem. First, the inverse thermal filter is used to estimate the past power based on the temperature history. The estimated power history is augmented with expected future power for various tasks and passed through the forward thermal filter to predict future temperature. As an example, consider the power/temperature pattern in Fig. 6 (top). Assume (i) power is unknown prior to 0.4 second (power history), (ii) the temperature history till 0.4 second is known, (iii) power after 0.4 second is known (future power). Fig. 9(b) illustrates that the future temperature predicted using the time-domain filters are close to the actual measured temperature over the entire duration.

IV. CONCLUSION

We have presented an on-line temperature and power estimation approach using time-domain forward and inverse thermal filters, respectively. Accuracy of proposed time-domain filter is validated experimentally using a 130nm CMOS test-chip. The developed approach can estimate transient variations in temperature and power in real time considering the chip-to-chip variations in thermal properties of materials as well as device characteristics. The filters define the unique thermal signature of each IC which can be used during run-time dynamic thermal management. The proactive power/thermal estimates enable fine-grained on-line power management techniques that can maximize the performance by operating close to thermal capacity but with fewer thermal emergencies and performance throttling events. For example, predicted temperature variations can be used to estimate available thermal capacity for turbo boosting specific cores avoiding the premature throttling and relative inefficiencies that occur due to thermal coupling between cores on a die [10].

Acknowledgment: This paper is based on work supported by Semiconductor Research Corporation (#2084.001).

REFERENCES

[1] H. F. Hamann, et al., "Hotspot-limited microprocessors: direct temperature and power distribution measurements," *IEEE Journal of Solid-State Circuits*, vol. 42, no. 1, pp. 56-65, Jan. 2007.

[2] W. Huang, et al. "HotSpot: a compact thermal modeling methodology for early-stage VLSI design," *IEEE Tran. on Very Large Scale Integration Systems*, vol. 14, no. 5, pp. 501-513, May 2006.

[3] K. Sankaranarayanan, et al., "A case for thermal-aware floorplanning at the microarchitectural level," *Journal of Instruction-Level Parallelism*, vol. 7, Oct. 2005.

[4] M. Cho, et al., "Thermal System Identification (TSI): a methodology for post-silicon characterization and prediction of the transient thermal field in multicore chips," in *IEEE SEMI-THERM*, pp.118-124, Mar. 2012.

[5] M. Cho, et al., "Methods and circuits for post-silicon characterization and prediction of transient thermal field in integrated circuits," accepted for publication in *IEEE Tran. on Components, Packaging and Manufacturing Technology*, Available: http://www.ece.gatech.edu/research/labs/GREEN/index_files/cho_thermal_system_identification_TCPMT.pdf.

[6] C. Y. Leung "An integrated CMOS current-sensing circuit for low-voltage current-mode buck regulator," *IEEE Tran. Circuits and Systems II*, vol. 52, no. 7, pp. 394-397, July 2005.

[7] N. Mehta, "Dynamic supply and threshold voltage scaling for CMOS digital circuits using in-situ power monitor," *IEEE Tran. on Very Large Scale Integration*, vol. 20, no. 5, pp. 892-901, May 2012.

[8] A. G. Radwan, et al, "First-order filters generalized to the fractional domain," *Journal of Circuits, Systems, and Computers*, vol. 17, no. 1, pp. 55-66, Feb. 2008.

[9] T. C. Haba, et al., "Influence of the electrical parameters on the input impedance of a fractal structure realised on silicon," *Journal of Chaos, Solitons and Fractals*, vol. 24, no. 2, pp. 479-490, Apr. 2005.

[10] I. Paul, et al., "Cooperative boosting: needy versus greedy power management," in *IEEE/ACM ISCA*, pp. 285-296, June 2013.

Modeling Broadband Equivalent Circuit of Interconnects with Full Wave Electromagnetic Solver

Jun Wei Wu

School of Electronics and Information
WuHan University, WuHan, HuBei 430072, China
antennawu@gmail.com

José E. Schutt-Ainé, Zhi-Guo Qian* and Weng Cho Chew

Department of Electrical and Computer Engineering
University of Illinois at Urbana-Champaign
Urbanda, IL 61801, USA.
jesa@illinois.edu, zhiguo.qian@intel.com, w-chew@uiuc.edu

Abstract—**This paper presents a method to find the broadband equivalent circuit of multi-port interconnects with full wave electromagnetic solver. The interconnects are simulated by augmented electric field integral equation (A-EFIE). Stable and passive pole-residue model is generated by vector fitting and passivity enforcement. Multi-port equivalent circuit is synthesized from the pole-residue model, and can be easily incorporated into SPICE-like simulator for both frequency domain and time domain simulation. Numerical example is given to verify the method.**

I. INTRODUCTION

As the operating frequency and packaging density increases higher and higher, interconnects can significantly affect the electrical performance of the packaging. So interconnects analysis becomes an essential part of the design process. Due to the nonuniform geometry structure and the wide bandwidth of the signal, quasi-static analysis tools are inadequate. It is imperative to analyse the interconnects over broadband using full wave electromagnetic solver.

The electric field integral equation (EFIE) solved by method of moments using Rao-Wilton-Gillson (RWG) basis function [1] is a powerful full wave electromagnetic solver. However, when applied to analyse packaging structure, it suffers from the low frequency breakdown problem [2], as the dimensions of most interconnects are just a tiny fraction of wavelength, even for operating frequencies as high as several gigahertz. The partial element equivalent circuit (PEEC) method is used to solve low frequency EM-CKT problems in [3]. A-EFIE [4] [5] [6] separates the contribution of the vector potential and the scalar potential. With proper frequency scaling, it is free of low-frequency breakdown problem. It captures both the wave physics and circuit physics, and can simulate interconnects over broadband. .

For time domain simulation, a problem is interfacing the output of frequency domain electromagnetic solver with the time domain models of nonlinear devices. One direct way is the time domain convolution. If the frequency domain responses can be approximated by low order rational function using Padé approximation or vector fitting [7], the convolution

can be evaluated in a recursive manner which has $O(N)$ computational complexity [8].

Another way is the equivalent circuit synthesis approach. Equivalent circuits are generated by network synthesis, and can be incorporated into existing SPICE-like simulators easily. Moreover, the model can be used for both frequency domain and time domain simulation.

In this work, we show how to find the broadband equivalent circuit of interconnects with full wave EM solver. A-EFIE and vector fitting are reviewed briefly. Then multi-port equivalent circuit synthesis method is presented. Finally an application example is given to verify the method.

II. AUGMENTED EFIE FORMULATION

For the geometry structure to be analysed, we mesh it into p triangular patches, with e inner edges. We denote the RWG basis function as $\mathbf{\Lambda}_i(\mathbf{r})$, which is normalized by removing the edge length, and the pulse basis function as $h_i(\mathbf{r})$. We define three matrices as [4] [5] [6]

$$\left[\overline{b}\right]_m = -\int_{S_m} \mathbf{\Lambda}_m(\mathbf{r}) \cdot \mathbf{E}_{inc}(\mathbf{r}) \, dS \tag{1}$$

$$\left[\overline{V}\right]_{m,n} = \mu_r \cdot \int_{S_m} \mathbf{\Lambda}_m(\mathbf{r}) \cdot \int_{S_n} g(\mathbf{r}, \mathbf{r}') \mathbf{\Lambda}_n(\mathbf{r}') \, dS' dS \tag{2}$$

$$\left[\overline{P}\right]_{m,n} = \varepsilon_r^{-1} \cdot \int_{T_m} h_m(\mathbf{r}) \cdot \int_{T_n} g(\mathbf{r}, \mathbf{r}') h_n(\mathbf{r}') \, dS' dS \tag{3}$$

A-EFIE is written as

$$\begin{bmatrix} \overline{\mathbf{V}} & \overline{\mathbf{D}}^T \cdot \overline{\mathbf{P}} \cdot \overline{\mathbf{B}} \\ \overline{\mathbf{F}} \cdot \overline{\mathbf{D}} & k_0^2 \overline{\mathbf{I}}_r \end{bmatrix} \cdot \begin{bmatrix} ik_0 \mathbf{J} \\ c_0 \boldsymbol{\rho}_r \end{bmatrix} = \begin{bmatrix} \eta_0^{-1} \mathbf{b} \\ \mathbf{0} \end{bmatrix} \tag{4}$$

t is the number of discontinuous parts of the geometry. $\mathbf{J} \in \mathbb{C}^{e \times 1}$ is the current coefficient vector, $\boldsymbol{\rho}_r \in \mathbb{C}^{(p-t) \times 1}$ is the reduced charge coefficient vector. Matrix $\overline{\mathbf{D}} \in \mathbb{R}^{p \times e}$ describes the relationship between the triangles and common edges of the mesh. Matrix $\overline{\mathbf{F}} \in \mathbb{R}^{(p-t) \times p}$ maps the full vector forward to the reduced one, and matrix $\overline{\mathbf{B}} \in \mathbb{R}^{p \times (p-t)}$ projects the reduced vector backward to the full one.

Formulation (4) is immune from the low-frequency breakdown problem, and it has been used for interconnects and packaging

*Zhi-Guo Qian is currently at Intel Corporation.

analysis in [4] [5] [6]. In such applications, admittance at the exciting port is deduced from $\mathbf{J}$. In next section, we show how to convert the admittance at discrete frequency points into pole-residue model.

III. POLE-RESIDUE MODELING USING VECTOR FITTING

Given the admittance matrix $\mathbf{Y}(s)$ generated by either simulation or measurement, vector fitting [7] finds a pole-residue model (5) which approximates the original data in the least square sense.

$$\mathbf{Y}(s) \approx \sum_{l=1}^{L} \frac{\mathbf{res}_l}{s - p_l} + \mathbf{F} + s\mathbf{E} \tag{5}$$

Since the original data is admittance, several physics constraints need to be satisfied.

1) The poles are the characteristic frequencies of the system, common poles are used to fit all the elements of $\mathbf{Y}$ matrix.
2) For stable system, the poles should be in the left half s plane.
3) For causal system, the poles and residues should be either real or come in complex conjugate pairs, and $\mathbf{F}$ and $\mathbf{E}$ should be real.
4) For passive system, $Re\left(\mathbf{V}^\dagger \mathbf{Y} \mathbf{V}\right)$ is non-negative for any exciting vector $\mathbf{V}$. For reciprocal network, this means that the real part of the $\mathbf{Y}$ is positive definite, i.e.,

$$eig\left(Re\{\mathbf{Y}\}\right) > 0 \tag{6}$$

The vector fitting is based on pole relocation technique. Within a few steps of iteration, optimal solution can be obtained, and the stability is also guaranteed. Although the algorithm fits the original data accurately at the sample points, violation of passivity condition can occur between sample intervals or out of the sample range. Passivity is assessed by computing the eigenvalue of the singular test matrix. Passivity enforcement is achieved by perturbing the eigenvalues of the residue matrices while minimize the influence on the fitting error. Details about vector fitting and passivity enforcement can be found in [7] [9] [10].

IV. EQUIVALENT CIRCUIT OF MULTI-PORT NETWORK

As the accuracy, stability and passivity of pole-residue model have been guaranteed by the vector fitting and passivity enforcement, an equivalent circuit can be derived from the pole-residue model directly with the knowledge of network synthesis. Firstly, we consider the equivalent circuit of one-port network.

A. Equivalent Circuit of One-Port Network

For convenience we rewrite one matrix element of (5) as

$$Y(s) = \sum_{m=1}^{m=M} \frac{res'_m}{s - p'_m} + \sum_{n=1}^{n=N} \left[\frac{res''_n}{s - p''_n} + \frac{res''^*_n}{s - p''^*_n} \right] + F + sE \tag{7}$$

M is the number of real poles, N is the number of complex conjugate pole pairs, and $M + N = L$. The equivalent circuit of (7) is given Fig. 1 with the following relations [11]

$$L = \frac{1}{res'_m} \tag{8}$$

$$R = -\frac{p'_m}{res'_m} \tag{9}$$

$$R''_n = -\frac{p''_n + p''^*_n}{res''_n + res''^*_n} \tag{10}$$

$$L''_n = \frac{1}{res''_n + res''^*_n} \tag{11}$$

$$C''_n = \frac{res''_n + res''^*_n}{p''_n p''^*_n} \tag{12}$$

$$I''_n = Y''_n V_{Cn} \tag{13}$$

$$Y''_n = -\frac{res''_n p''^*_n + res''^*_n p''_n}{p''_n p''^*_n} \tag{14}$$

$$R = \frac{1}{F} \tag{15}$$

$$C = E \tag{16}$$

The resistors, inductors, and capacitors determined by (8)-(16)

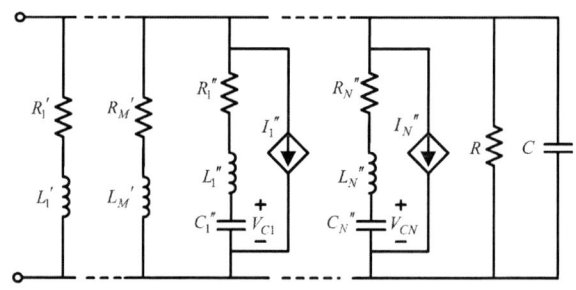

Fig. 1. Equivalent circuit of pole-residue model (7).

may have negative values, however, as the passivity is guaranteed at the mathematical level by the passivity enforcement process, the circuit is still stable and passive.

B. Consideration about Multi-Port Reciprocal Network

From the definition of Y parameter, 2-port reciprocal network has a Π-type equivalent circuit in Fig. 2(a) with $Y_1 = Y_{11} + Y_{12}$ and $Y_2 = Y_{21} + Y_{22}$. This concept is also valid for general n-port reciprocal network [12], which can be fully characterized by

$$\begin{cases} Y_i = \sum_{j=1}^{n} Y_{ij} & i = 1, \cdots, n \\ -Y_{ij} & i = 1, \cdots, n; j = i+1, \cdots, n \end{cases} \tag{17}$$

This is a degenerate description of reciprocal n-port network. For example, a 4-port reciprocal network can be described by Fig. 2(b) which has ten elements. The equivalent circuits of the ten elements are extracted using the method of the previous subsection.

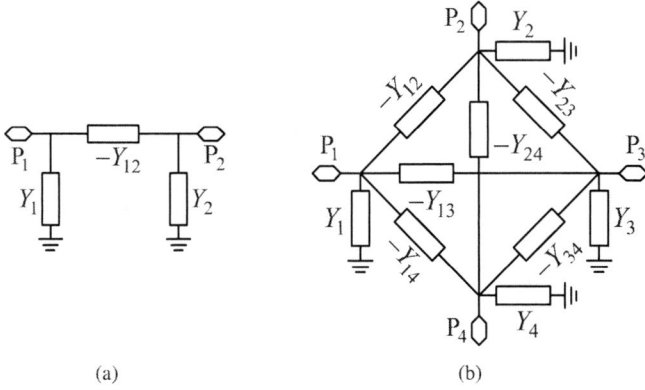

(a)   (b)

Fig. 2. Equivalent circuits of 2-port and 4-port reciprocal networks.

V. APPLICATION EXAMPLE

In this section we extract the broadband equivalent circuit of a 4-port interconnects shown in Fig. 3. It is cut out of a realistic package board. The overall size is 8 mm, and the triangular mesh has 3691 patches. The highest frequency supported by the mesh is about 40 GHz.

Firstly, A-EFIE solver is used to compute the admittance matrix from 1 Hz to 40 GHz. Y_{11} obtained by the solver is shown in Fig. 4. As can be seen, the magnitude of Y_{11} changes linearly in the low frequency region, implying that the solver does not break down in the low frequency case.

Vector fitting and the passivity enforcement are used to fit the

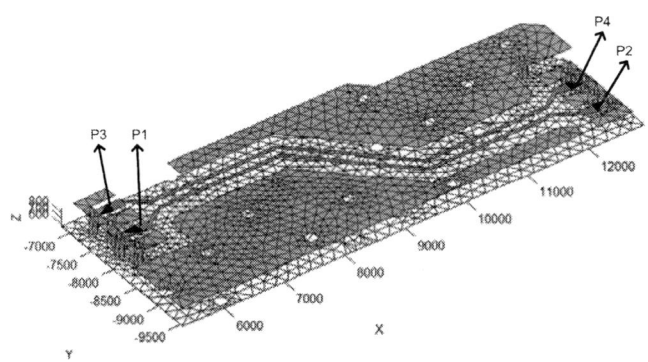

Fig. 3. Geometry of interconnects. Triangular mesh with four ports marked. Unit: 1 micron.

output of the A-EFIE solver. As the network is reciprocal, only the diagonal and upper triangular part of $\mathbf{Y}$ matrix need to be fitted. With $L = 25$ and $\mathbf{E} = \mathbf{0}$ in (5), satisfactory pole-residue is obtained. We compare Y_{11} of the pole-residue model with the original one of A-EFIE output in Fig. 4. The deviation is very small. Fig. 5 plots the distribution of the poles. All of the poles are located in the left half plane. The four eigenvalues of $Re(\mathbf{Y})$ over the whole frequency band, shown in Fig. 6, are always positive. Thus the pole-residue model is stable and passive. Ten admittance elements, i.e., Y_1, Y_2, Y_3, Y_4, $-Y_{12}$, $-Y_{13}$, $-Y_{14}$, $-Y_{23}$, $-Y_{24}$, $-Y_{34}$, are computed by combining pole-residue model following (17). Then they are converted

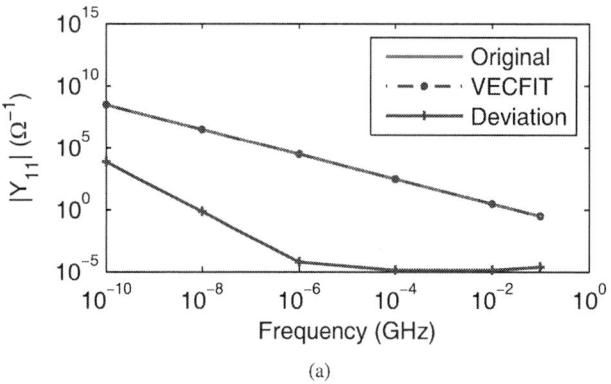

(a)

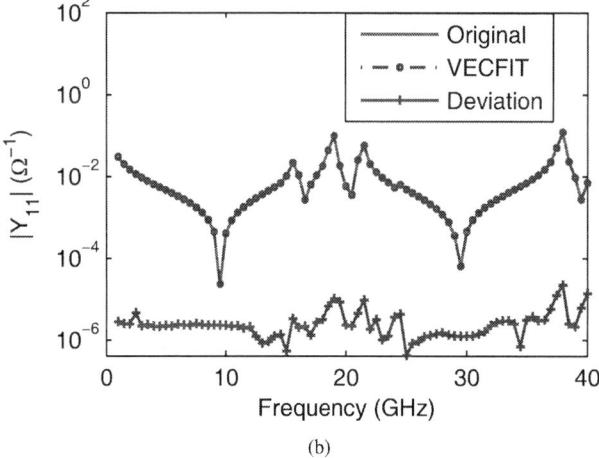

(b)

Fig. 4. Comparison of Y_{11} between the original output of A-EFIE and the vector fitting with passivity enforcement. (a) 10^{-10} GHz - 10^{-1} GHz. (b) 1 GHz - 40 GHz.

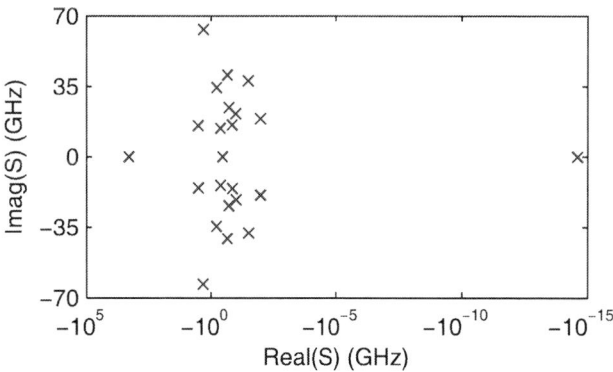

Fig. 5. Poles of the pole-residue model.

into ten elementary circuits using (8)-(16). Connecting the ten elementary circuits according to Fig. 2(b), a final 4-port equivalent circuit is obtained. The final equivalent circuit is then imported into circuit simulator for frequency domain simulation. All the 16 admittance parameters of the circuit simulator agree well with the original output of the A-EFIE solver. For space limitation, we only show the comparison of Y_{11} in Fig. 7 and Fig. 8. The differences are indistinguishable, and the extracted circuits are equivalent to the interconnects

978-1-4799-0708-3/13 $31.00 © 2013 IEEE   205

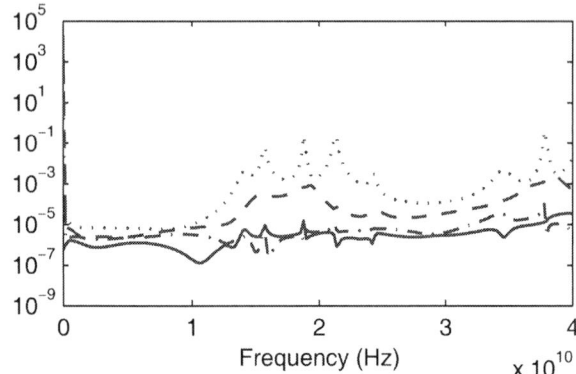

Fig. 6. Eigenvalues of $Re\,(\mathbf{Y})$ over frequency.

over the whole frequency range.

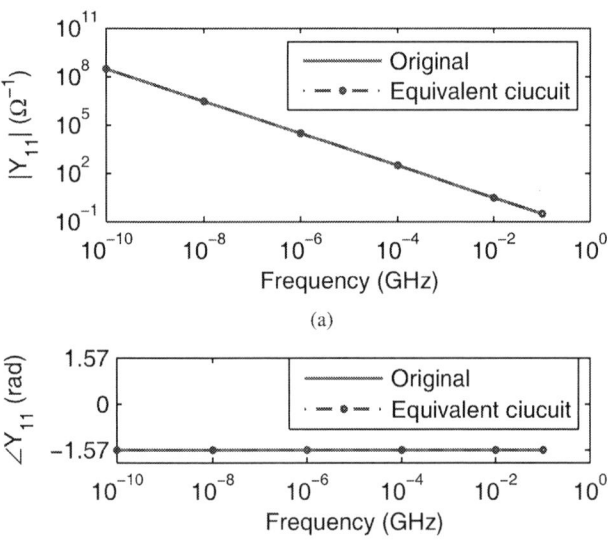

Fig. 7. Magnitude and phase of Y_{11} between 10^{-10} GHz and 10^{-1} GHz.

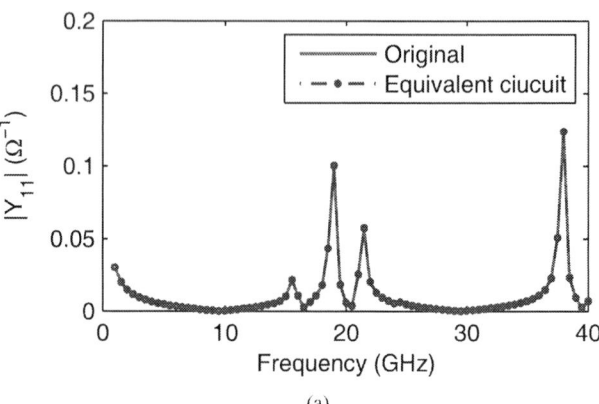

(a)

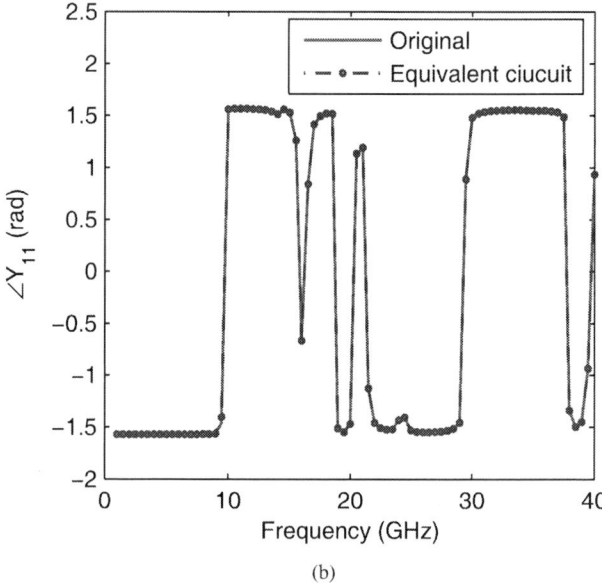

(b)

Fig. 8. Magnitude and phase of Y_{11} between 1 GHz and 40 GHz.

VI. CONCLUSION

In this paper, a method to find the broadband equivalent circuit of multi-port interconnects using full wave broadband electromagnetic solver is presented. The equivalent circuit is accurate, stable and passive over broad bandwidth. Numerical example shows that the method is valid and exact.

REFERENCES

[1] S. M. Rao, D. R. Wilton, and A. W. Glisson, "Eelectromagnetic scattering by surface of arbitrary shape," *IEEE Trans. Antennas Propag.*, vol. 30, pp. 409–418, May 1982.

[2] Z. G. Qian and W. C. Chew, "A quantitative study on the low frequency breakdown of EFIE," *Micro. Opt. Technol. Lett*, vol. 50, no. 5, pp. 1159–1162, May 2008.

[3] D. Gope, A. Ruehli, and V. Jandhyala, "Solving low-frequency EM-CKT problems using the PEEC method," *IEEE Trans. Adv. Packag.*, vol. 30, pp. 313–320, May 2007.

[4] Z. G. Qian and W. C. Chew, "An augmented electric field integral equation for high-speed interconnect analysis," *Micro. Opt. Technol. Lett*, vol. 50, no. 10, pp. 2658–2662, Oct. 2008.

[5] ——, "Packaging modeling using fast broadband surface integral equation method," in *Proc. IEEE 17th Topical Meeting on Electrical Performance of Electron. Packag.*, Oct. 2008, pp. 347–350.

[6] ——, "Fast full-wave surface integral equation solver for multiscale structure modeling," *IEEE Trans. Antennas Propag.*, vol. 57, pp. 3594–3601, Nov. 2009.

[7] B. Gustavsen and A. Semlyen, "Rational approximation of frequency domain responses by vector fitting," *IEEE Trans. Power Del.*, vol. 14, pp. 1052–1061, Jul. 1999.

[8] A. Semlyen and A. Dabuleanu, "Fast and accurate switching transient calculations on transmission lines with grond return using recursive convolutions," *IEEE Trans. Power App. Syst.*, vol. 94, pp. 561–571, Mar./Apr. 1975.

[9] B. Gustavsen, "Fast passivity enforcement for pole-residue models by perturbation of residue matrix eigenvalues," *IEEE Trans. Power Del.*, vol. 23, pp. 2278–2285, Oct. 2008.

[10] A. Semlyen and B. Gustavsen, "A half-size singularity test matrix for fast and reliable passivity assessment of rational models," *IEEE Trans. Power Del.*, vol. 24, pp. 345–351, Jan. 2009.

[11] G. Antonini, "SPICE equivalent circuits of frequency-domain responses," *IEEE Trans. Electromagn. Compat.*, vol. 45, pp. 502–512, Aug. 2003.

[12] Z. Qi, H. Yu, P. Liu, S. X.-D. Tan, and L. He, "Wideband passive multiport model order reduction and realization of RLCM circuits," *IEEE Trans. Comput.-Aided Design Integr. Circuits Syst.*, vol. 25, pp. 1496–1509, Aug. 2006.

Efficient Performance Evaluation of High-Speed Differential Interconnect Lines with Via Discontinuities

Hyewon Kim, Junghyun Lee, and Yungseon Eo
Dept. Electronics and Communication Engineering
Hanyang University, Ansan, South Korea
e-mail: eo@giga.hanyang.ac.kr

Abstract—**Vias in transmission lines cause significant waveform distortion due to electromagnetic wave reflections and vias may behavior like a band rejection filter due to resonances in between reference planes. These undesirable effects significantly exacerbate the signal integrity in high-speed and high-frequency integrated circuits, packaged modules, or printed circuit board (PCB). In this work, discontinuous interconnect lines including vias are experimentally characterized with S-parameter measurements in the frequency range of 10 MHz to 40 GHz. Then base on the experimental characterizations, an accurate circuit model to represent electrical characteristics of the vias is developed. It is shown that with the proposed technique, the high-performance integrated system can be efficiently designed by avoiding the conventional strict layout design rules for the discontinuous lines.**

Keywords—Circuit model, resonance, scattering parameters, signal integrity, vias.

I. INTRODUCTION

Over the past several decades, the ceaseless performance improvement of integrated electronic system is mainly due to transistor miniaturization. Now, a considerable number of the reliability issues and bandwidth limitation of such high-performance integrated electronic circuits or systems are concerned with interconnect lines between circuit blocks [1]. Therefore, accurate experimental characterization and efficient performance evaluation for data links become an integral part of todays high-speed system design.

In order to reduce layout area and to increase the system performance, integrated electronic system design using vertical interconnections (i.e., vias) are nowadays indispensable. Moreover, as the circuit switching speed drastically increases, differential signaling is very usual in many interface signaling such as chip-to-chip interconnects or global interconnect lines within a chip (i.e., RF I-signal and Q-signal, clock signal, and high-speed data paths between memory and processor). In general, the differential signaling requires at least two times wider layout spacing than the single ended signaling. Furthermore, since discontinuous effects have a significant effect on the signal integrity, high-speed signal lines are constrained with very strict layout design rules that may not allow to include vias or discontinuities. In practice, the strict layout design rules may cause many crucial design conflicts due to routing congestion, timing malfunction, and manufacturing cost increase. In order to relax the routing congestion problems, flexible layout design with bending or multi-layered layout

designs with vias are necessary and occasionally inevitable in both single-ended and differential signaling. Thus, the accurate signal integrity verification of discontinuous interconnect lines becomes crucial.

Up to date, many circuit designers model the vias as simple LC or RLC circuit [2]. However, since simple LC/RLC via model does not accurately reflect band rejection characteristics of transmission lines, the model has a fundamental limitation to be applied to high-speed circuit performance evaluation. Recently, hybrid model that exploits the advantages of both the analytical and numerical techniques is proposed [3-4]. However, these models may not accurately reflect process variations and non-ideal effects without experimental characterizations. In this work, discontinuous transmission lines with vias are experimentally characterized. Then based on the experimental characterizations, a more accurate via model that reflects resonance effect is developed. Thereby, it is shown that the signal integrity for discontinuous interconnect lines can be efficiently verified.

II. EXPERIMENTAL CHARACTERIZATION AND MODELING OF DISCONTINUOUS INTERCONNECT LINES

A. Experimental setup

For experimental characterization, various test patterns were designed and fabricated by using a typical 4-layer package process. Top and bottom layers are assigned as signal traces and inner layers are reserved for power/ground reference planes. The dimensions of a test pattern are shown in Fig. 1(a). The radii of the via hole, clearance hole, and via pad are 100, 380, and 250 μm, respectively and the width of the signal line is 80 μm. The vendor-provided dielectric constant for the inter-metal-dielectric (IMD) material is 4.2 to 4.3, while the loss tangent of the dielectrics is given as 0.011 at 1 GHz. The S-parameters for the test patterns were measured from 10 MHz to 40 GHz using a vector network analyzer (Agilent E8361A) in which each port was connected to a probe tip (Cascade ACP40-A-GSG-150 for single line and GGB 40A-GS-150/40A-SG-150-D-150 for differential lines). The VNA was calibrated by employing a short-open-load-thru (SOLT) calibration method with an impedance standard substrate. The pad size for probing is 100 μm by 100 μm and the pitch between the pads is 150 μm. The parasitic effects of the pads were de-embedded using a Y-parameter de-embedding technique [5].

As shown in Fig. 1(b), unlike a uniform transmission line (i.e., a straight transmission line without any via), the return

978-1-4799-0708-3/13 $31.00 © 2013 IEEE

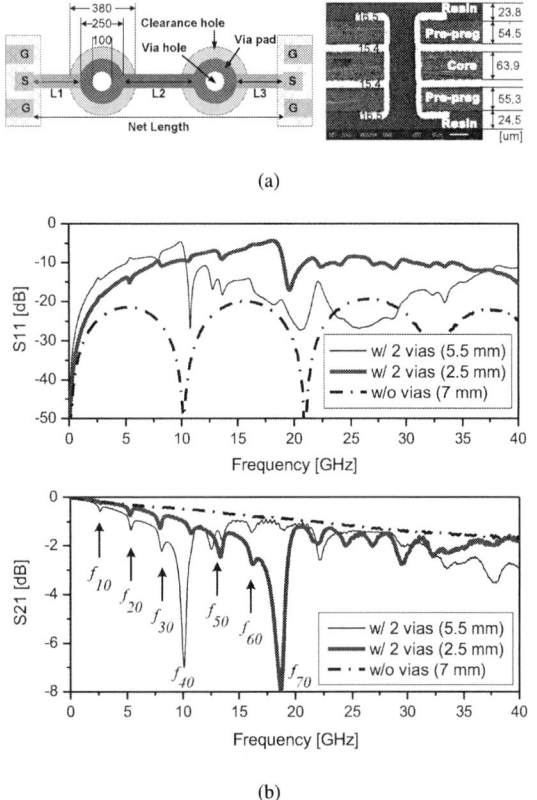

(a)

(b)

Figure 1. Test pattern. (a) top- and cross-section view (b) measurement result.

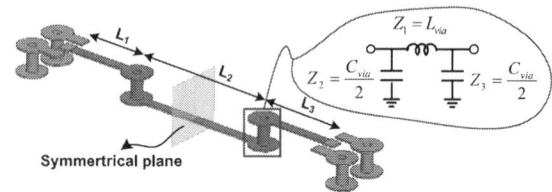

Figure 2. Symmetrical (i.e., $L_1 = L_3 = L_2/2$) test structure of a pair of vias (power/ground planes are not shown) and its equivalent circuit model.

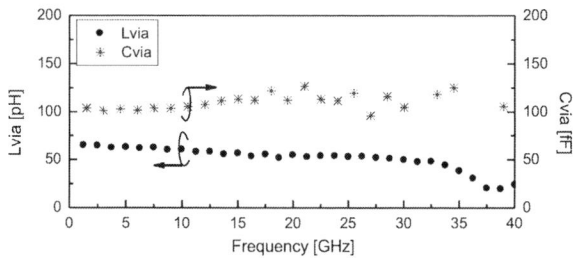

Figure 3. Via inductance and capacitance. The capacitance is about 100 fF and inductance is about 60 pH.

$$T_{22} = \frac{T_{m22} + 2}{\sqrt{T_{m11} + T_{m22} + 2}}, \quad (4)$$

$$T_{12} = \frac{T_{m12}}{T_{11} + T_{22}} = \frac{T_{m12}}{2}\left(\frac{\sqrt{T_{m11} + T_{m22} + 2}}{(T_{m11} + T_{m22} + 4)}\right) = -T_{21}. \quad (5)$$

Thus, from the network parameters for a single via, the circuit model parameters of Fig. 2 can be determined as shown in Fig. 3.

C. Resonance effects

The circuit model of Fig. 2 is not enough to represent the electromagnetic phenomena since the resonance effect is not taken into account. When a signal propagates on transmission lines with vias, the current flowing in the vias induces undesirable electromagnetic waves between the reference planes (ground and power plane). These waves are reflected back and forth, consequently forming resonances. If the resonance frequencies due to the parallel-plates are concerned with the signal trace, they may have considerable effects on signal propagation. For two transmission lines with 2 vias that have different line lengths, the resonance effects are investigated using the measured S-parameters, as shown in Fig. 1(b).

Assuming the reference planes as a rectangular-shaped wave-guide, the resonance frequencies can be approximately estimated using the wave-guide theory as

$$f_{mn} = \frac{c}{2\sqrt{\varepsilon_r \mu_r}}\sqrt{\left(\frac{m}{a}\right)^2 + \left(\frac{n}{b}\right)^2}, \quad (6)$$

where c is the speed of light, ε_r is the dielectric constant, μ_r is the relative permeability, and a and b are the size of the planes in meters ($a \geq b$). The parallel-plate resonances are also related to the electrical length of signal path. While the small resonance frequencies (e.g., $f_{10}, f_{20}, f_{30}, ...$) can be estimated using (6), the largest resonance frequencies can be estimated by determining the electrical length of the layout pattern. The resonances can be reflected by adding a parallel RLC resonance circuit as shown in Fig. 4. The circuit model

loss and signal propagation characteristics of the discontinuous transmission line are substantially deteriorated due to the reactive characteristics of the vias, even though their physical size is quite small. Thus, the reactive via effects must be accurately taken into account in the discontinuous transmission line circuit model. In order to model a via as a circuit, the capacitance and the inductance due to the physical length of the via barrel have to be determined.

B. Via circuit model and model parameter determination

The 2-port network parameters of a via can be determined by using its symmetrical properties. As shown in Fig. 2, the structure can be considered approximately symmetrical. Thus, the measured network parameters for two symmetrical vias can be represented with the cascaded network parameters of two single vias

$$\begin{bmatrix} T_{m11} & T_{m12} \\ T_{m21} & T_{m22} \end{bmatrix}_{measured} \approx \begin{bmatrix} T_{11} & T_{12} \\ T_{21} & T_{22} \end{bmatrix}\begin{bmatrix} T_{11} & T_{12} \\ T_{21} & T_{22} \end{bmatrix}. \quad (1)$$

Note, for the reciprocal network, T-parameters satisfy the following relationship,

$$T_{11}T_{22} - T_{12}T_{21} = 1. \quad (2)$$

Thus, following T-parameters for a single via can be derived

$$T_{11} = \frac{T_{m11} + 2}{\sqrt{T_{m11} + T_{m22} + 2}}, \quad (3)$$

978-1-4799-0708-3/13 $31.00 © 2013 IEEE

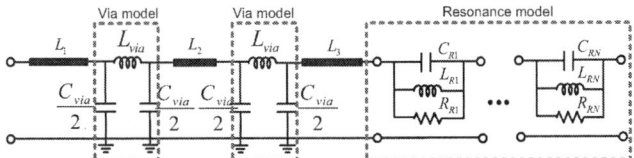

Figure 4. Proposed circuit model for transmission line with 2 vias

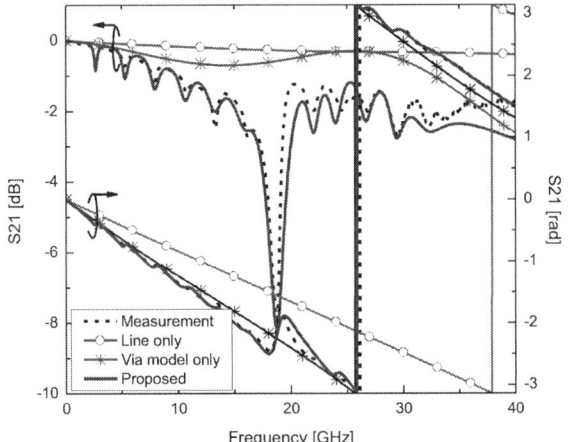

Figure 5. S-parameter data comparison.

parameters can be readily determined with layout dimensions or measurement data.

D. Via model verification

In the transmission line with vias as shown in Fig. 4, the number of parallel-RLC circuits depends on the size of the planes and the interested frequency range. Respective resonance frequencies require corresponding parallel-RLC circuits. With the model, a SPICE simulation was performed. The simulation results were compared with the measurement data. The circuit model shows excellent agreement with the measurement data, as shown in Fig. 5. With the proposed circuit model, the signal integrity deterioration due to the vias can be accurately investigated. It is noteworthy that the via inductance and capacitance contribute to the phase delay, whereas the resonance effects distort both the phase and magnitude in the vicinity of the resonance frequency.

III. EVALUATION OF DIFFERENTIAL LINES WITH VIAS

Differential signaling requires two signal lines as a pair. One line is driven by positive signal while the other line is the exactly opposite signal with the same magnitude. In differential signaling, the phase difference between two signals has to be maintained with 180 degree all along the line. The line discontinuities may significantly shift the phase back and forth, whereas the reflected waves due to the discontinuities significantly distort the signal wave-shape. The phase shift increases the random jitter and the reflected waves decrease the eye-height.

In order to investigate the signal integrity deterioration for the differential lines, 4-port S-parameters for four test patterns as shown in Fig. 6 are measured in the frequency

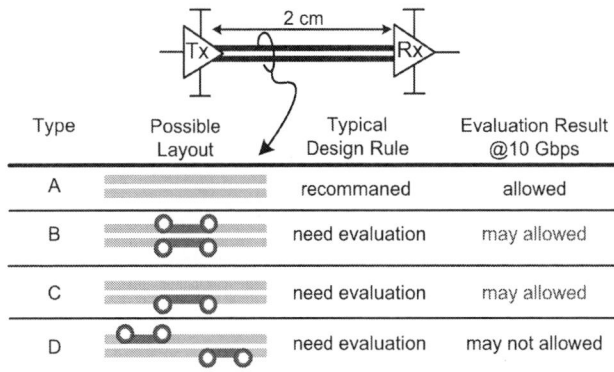

Figure 6. Various types of layout examples and evaluation result of differential lines with vias.

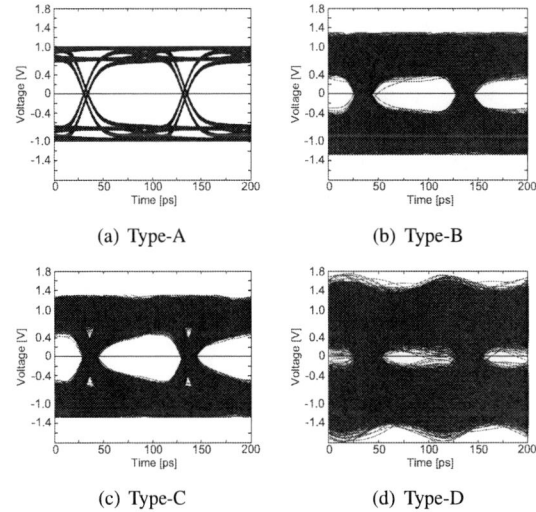

Figure 7. Eye-diagrams for the four test patterns: HSPICE simulation using the measured S-parameters with 10 Gbps 30,000 bits PRBS.

range of 10 MHz to 40 GHz. Eye-diagrams were determined with the measured S-parameters. Note that in practice, the signal integrity verification of the discontinuity lines based on the measurement data or 3-dimensional numerical simulations is neither efficient in computation time nor versatile to be applied to numerous discontinuity lines. In contrast, the proposed model can be much more efficiently and conveniently employed with the conventional transmission line models for the signal integrity verification of the diverse discontinuity lines.

A. Measurement-based time-domain observation

The eye diagrams for the 4 test patterns are determined as shown in Fig. 7. Clearly, discontinuities exacerbate the signal integrity. The signal integrity deteriorations are due to three major factors (i.e., reflection, resonance, and signal coupling). It is noteworthy that although the signal integrity is considerably deteriorated, the type-B and type-C may be allowed for practical circuits, whereas type-D is definitely not the case. Thus, although the discontinuous lines show worse performance than the uniform straight lines, too much strict

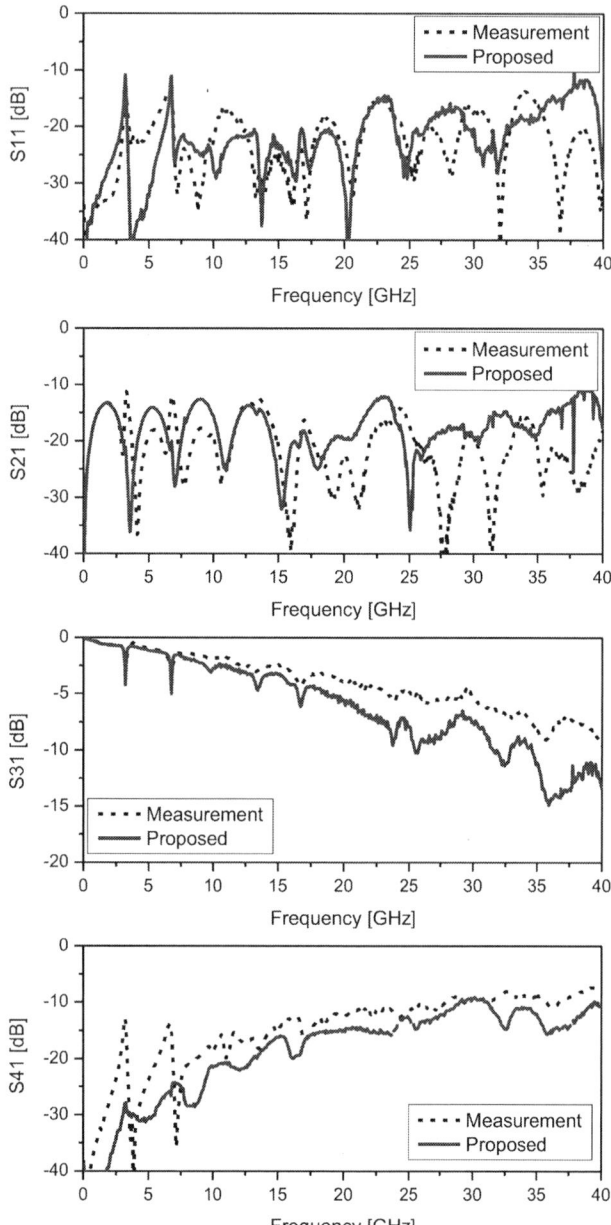

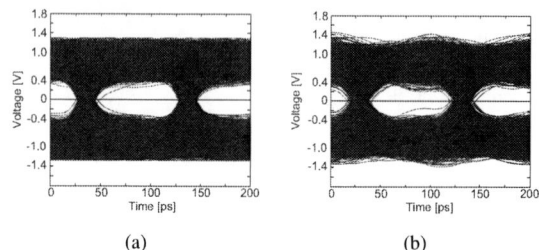

Figure 9. Eye-diagram comparison for the type-B. (a) S-parameter-measurement-based simulation. (b) simulation using the proposed model.

model, the simulation using the proposed model shows minor difference with the S-parameter measurement-based one. Note that since the most of the discontinuity effects can be captured with the simulation using the model, the various discontinuous lines may be much more efficiently evaluated with the proposed model.

IV. CONCLUSION

In this work, discontinuous interconnect lines are experimentally characterized with high-frequency S-parameter measurement data in the frequency range of 10 MHz to 40 GHz. In the transmission lines with vias, resonances cannot be neglected any more. Discontinuous interconnect lines for both single and differential lines are investigated using a novel via model with resonance effects. It is shown that if the signal integrity for discontinuous lines is accurately evaluated, high-performance integrated system can be much more efficiently designed by avoiding the conventional strict layout design rules for the discontinuous lines.

ACKNOWLEDGMENT

This work was supported by the project, Development of Technologies for Next-generation Electromagnetic Wave Measurement Standards, of the Korea Research Institute of Standards and Science under Grant 12011016.

REFERENCES

[1] J. D. Meindl, Beyond Moore's Law: the interconnect era, *Computing in Science and Engineering*, vol. 5, no. 1, pp. 20–24, Jan.-Feb. 2003.

[2] S. H. Hall and H. L. Heck, *Advanced signal integrity for high-speed digital designs.* Hoboken, New Jersey: John Wiley and Sons, Inc., 2009.

[3] R. Rimolo-Donadio, X. Gu, Y. H. Kwark, M. B. Ritter, B. Archambeault, F. de Paulis, Y. Zhang, J. Fan, H.-D. Bürns, and C. Schuster, "Physics-based via and trace models for efficient link simulation on multilayer structures up to 40 ghz," *IEEE Transactions on Microwave Theory and Techniques*, vol. 57, no. 8, pp. 2072–2083, Aug. 2009.

[4] Z. Z. Oo, E.-X. Liu, X. C. Wei, Y. Zhang, and E.-P. Li, "Cascaded microwave network approach for power and signal integrity analysis of multilayer electronic packages," *IEEE Transactions on Components, Packaging and Manufacturing Technology*, vol. 1, no. 9, pp. 1428–1437, Sep. 2011.

[5] P. J. van Wijnen, H. R. Classen, and E. A. Wolsheimer, "A new straightforward calibration and correction procedure for on wafer high frequency s-parameter measurements (45 mhz-18 ghz)," in *Proc. IEEE BCTM*, Sep. 1987, pp. 70–73.

Figure 8. Comparison of the measured S-parameters for the type-B with those using the model.

layout design rules for the discontinuous lines may not be necessary.

B. Evaluation based on proposed circuit model

The via model that takes the resonance effect into account can be employed for the efficient signal integrity verification of discontinuous differential signal lines. The proposed model shows good agreements with the measured S-parameters for discontinuous differential lines as shown in Fig. 8. The eye-diagram based on the proposed model is compared with the S-parameter-based SPICE simulation (see Fig. 9). Since the coupling effect between vias is not included in the circuit

In-Depth Analysis of Power Noise Coupling Between Core and Periphery Power Rails

Guang Chen, WernShin Choo, Shishuang Sun, and Dan Oh

Altera Corporation
101 Innovation Drive, San Jose, CA 95134
guchen@altera.com

Abstract— **Core and periphery digital blocks often use the same voltage level. Due to the large current drawn by core, shielding periphery power from core power noise is highly desirable. This paper presents various separation schemes and compares pros and cons. Fundamental power noise coupling mechanisms between core and periphery powers are described in detail. Two major sources of power noise coupling are studied: inductive noise coupling and direct current draw through a common share point. Based on this analysis, we propose cost optimized board decoupling schemes.**

Keywords-core and periphery power coupling, ferrite bead, power separation, power integrity, power transfer noise

I. INTRODUCTION

Modern integrated circuit devices (ICs) including conventional microprocessors or FPGA devices incorporate numerous digital blocks or cores. Due to slow advance in packaging technology relative to ICs, large power consumption, and reduction of supply voltage, providing a stable power supply voltage for modern integrated devices is an ever challenging task [1] [2]. Having different power supply rails for each core or different periphery blocks is not practical in terms of cost or not even possible due to limited decoupling area. Special attention is required when designing a power distribution network for core and associated periphery blocks [3].

Core and periphery power (or different core power rails) often shares the same power domain. Since core and periphery current activities can be significantly different, it is highly desirable to separate two power domains. Separation of these two power domains can be done in various ways. First, connecting them on chip is the simplest approach and it eases power distribution network (PDN) designs. This approach can be used when noise characteristics for core and periphery blocks are well known and controlled. Two different connection points are available in package substrates: above and below package core. Connection above package core is not recommended for separation as it has a least inductance path between the two power rails. However, this connection is useful for tying multiple periphery blocks. Connecting below the package core provides isolation which is closer to tying at the PCB side. However, it does not provide an option to implement additional shielding that can be implemented in PCB.

The simplest way to handle two power domains on PCB is tying them to the same PCB power stripe/plane. The parasitic inductance associated with power delivery network helps reducing the noise coupling between core and periphery power. In general, placing a shared power plane near the bottom of the PCB stack-up improves isolation between shared rails as this increases parasitic inductance. This approach does not raise routing complexity or BOM cost. But it provides only limited noise isolation.

Two power rails have to be physically separated if better isolation is required. Two power rails are then connected to a common point near the voltage regulator module (VRM). One of the power stripes can be connected to the VRM via a ferrite bead to further increase the isolation as the bead helps filter out any medium to high frequency noise. Physical separation increases the complexity of the PCB layout task due to the physical separation requirements. The ferrite bead also introduces additional IR drop and limits the maximum power can be delivered to the isolated power rail. Another drawback is the increased BOM cost. The isolated power rail may require individual PCB decoupling network since separation also blocks the PCB decoupling network sharing.

Good noise isolation improves FPGA device performance as coupled noise can cause issues such as clock jitter increase or power supply voltage being out of spec. As demonstrated in the following section, noise can propagate to victim power rail via multiple coupling mechanisms. It is important to understand how noise propagates under these mechanisms in various power sharing scenarios in order to design effective PCB decoupling schemes.

In this paper, we present in depth study of the noise coupling between FPGA power rails in different shared PCB power delivery network designs. We first examine the two power noise coupling mechanisms. Then, we review the performance of different decoupling schemes and propose cost optimized board decoupling schemes.

II. SOURCE OF POWER NOISE COUPLING

Core activity creates a wide spectrum of current draw. The high frequency current component is supplied by core on-die decoupling capacitor (ODC). The remaining current component has to come from external sources, such as package and PCB decoupling capacitors, and voltage regulator module (VRM). In the core and peripheral power sharing case, the core also draws current from periphery ODC causing significant coupling noise to periphery power rail.

Figure 1 illustrates a typical PCB power sharing configuration. The FPGA device is mounted on top of a motherboard. The package power balls of two power rails are connected at the shared power plane located at the bottom of the motherboard stack up via PTH vias since top motherboard layers are typically assigned for high-speed signal routing. This shared plane area is important for noise isolation as most of the noise coupling happens within this region. There are two noise coupling mechanisms in the illustrated scheme: direct current draw and inductive current coupling.

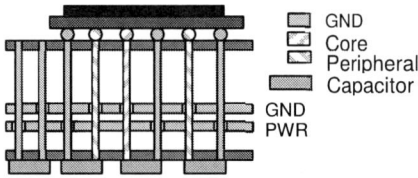

Figure 1. PCB power sharing layout

Noise coupling between two rails can be studied in both time and frequency domains. Time domain data provides direct measurement of the coupled noise magnitude that can be checked against device noise spec requirement. Frequency domain data provides an overview of the noise coupling profile over the frequency. The high impedance peaks in the noise profile indicate the frequencies that strong noise coupling happens. This information is helpful in identifying proper noise reduction solution.

A. Noise Due toDirect Current Draw

The core draws current from peripheral rail through shared power delivery structure when two rails are physically connected and causes noise coupling. Figure 2 illustrates a transfer impedance profile along with self-impedance profiles of core and peripheral rails. The locations of the two impedance peaks of the transfer impedance profile are aligned with the impedance peaks of the core and peripheral power rails. This suggests that there is a strong correlation between core activity and coupled noise to periphery power rail.

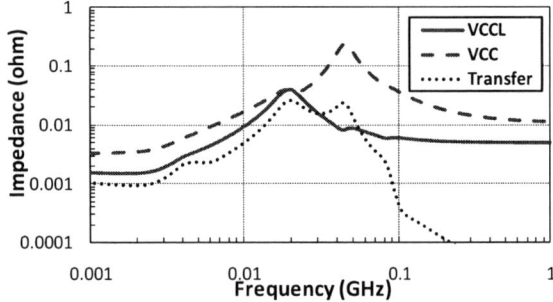

Figure 2. Core, peripheral, and transfer (direct current draw) PDN impedance profiles

Small current draw around the peripheral rail resonance peak frequency can excite peripheral rail resonance and the magnitude of resonance voltage is largely determined by the Q factor of the resonance network. Therefore, core activity impact can still be big when the two peak frequencies are far apart. It

amplifies the transfer impedance hence the coupling and tends to be the worst case when two peaks aligned.

B. Noise due to Inductive Current Coupling

The core current draw can also excite current draw change on peripheral rail through conventional inductive coupling which occurs regardless of power sharing. Coupling happens at both package and PCB. But the majority of the noise coupling is within a via pin field underneath the FPGA device. The pin field has a large amount of closely arranged long PCB vias. The mutual inductance between nearby vias is high because the structure does not have good ground shielding comparing to the planar structure.

Figure 3 plots the transfer impedance profile along with PDN impedance of physically separated core and peripheral power rails. Similar to the transfer impedance profile of direct current draw, the two impedance peaks in the transfer impedance profile of inductive coupling are aligned with the core and peripheral PDN peak frequency and the coupled noise magnitude is the biggest near the peripheral PDN impedance peak frequency. The inductive coupled noise magnitude is also affected by the mutual inductance between core and peripheral power vias, which is determined by the coupling section length and relative position of core and peripheral power vias.

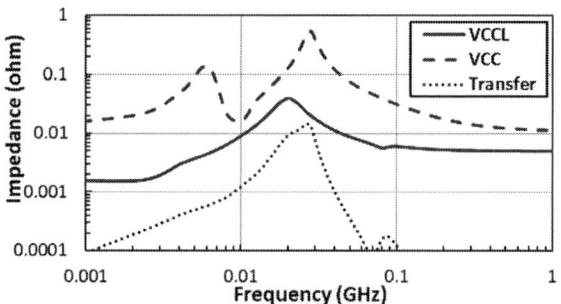

Figure 3. Core, peripheral and transfer (inductive coupling) PDN impedance profiles

III. PERFORMANCE OF DIFFERENT DECOUPLING SCHEMES

Core noise is attenuated by the PDN parasitic inductance as it propagates to PCB. The amount of attenuation depends on where the component sits in the noise spectrum. Low frequency noise is barely attenuated by PDN parasitic inductance. The noise reduction is achieved via PCB decoupling. High frequency noise is greatly attenuated by PDN parasitic inductance and its impact is limited within local die area. Effort to reduce noise coupling mainly focuses on the mid frequency noise component. As explained previously, there are two noise coupling mechanisms, direct current draw and inductive coupling and the method to reduce noise coupled via two mechanisms is different.

Ideally, powering each power rail with individual VRM can eliminates noise coupling due to direct current draw. Inductive power noise coupling can also be greatly reduced if the location of those power stripes can be further optimized. But this approach raises PCB cost significantly. Power sharing among power rails with same voltage is desired to reduce PCB cost.

978-1-4799-0708-3/13 $31.00 © 2013 IEEE   212

When power rail is shared, it is best if the sharing point is located further away from the die to minimize the direct coupling noise. For instance, moving the shared power plane to the bottom of the PCB stack up extends the PCB via length and increases the parasitic inductance. This improves noise isolation between sharing rails. Two commonly used PCB separation designs, power stripes and ferrite bead, can further increase the parasitic inductance and improve noise isolation.

A. Separation Using Power Stripes

In this approach, the two power rails are connected to different power stripes on PCB. The two power stripes are then connected together somewhere outside via pin field (Figure 4). Two power stripes are usually located at different layers of the PCB. This reduces the length of via coupling section. As the result, the magnitude of the noise due to inductive coupling is reduced. Besides inductive coupling noise reduction, the approach provides slightly better isolation to the noise due to direct current draw as the noise has to travel longer distance.

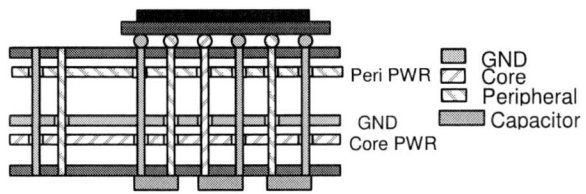

Figure 4.   Noise isolation using power stripe

Routing complexity is increased as the approach now requires two power stripes instead of a single power plane. The number of PCB decoupling capacitors required may also increase. PCB spreading inductance increases because of the separation. The effectiveness of shared PCB bulk capacitors are not affected since these capacitors are not sensitive to the PCB spreading inductance increase. But high frequency capacitors are sensitive to the inductance increase and each separated power rail needs its own high frequency decoupling capacitors as the result.

B. Separation Using Ferrite Beads

High isolation (30dB ~ 60dB) between core and peripheral power rail is required if peripheral power rail is extremely noise sensitive. Noise coupling due to direct current draw consists of the major portion of the noise seen at peripheral power rail. Parasitic inductance associated with PDN structure simply cannot provide attenuation required to achieve such a big isolation and ferrite bead has to be used. Figure 5 shows the equivalent circuit of a ferrite bead and its simulated transfer impedance profile [4]. A ferrite bead delivers DC power to the isolated power rail while blocks the noise coupling within its stop band. PCB routing for the approach is similar to that of the separated power stripes approach except that the one power rail now connects to the other power rail via ferrite bead.

Routing complexity increase for the ferrite bead approach is the same comparing to the power stripe approach. But the BOM cost will increase. Ferrite bead inductance increases the mounting inductance of PCB decoupling capacitors isolated by

the bead and reduces their effectiveness in controlling voltage ripple for both high frequency and bulk capacitors. More capacitors have to be placed after the bead to provide complete PDN frequency coverage for isolated power rail. Ferrite bead also introduces extra DC IR drop and this may result in power supply being out of spec. It is not a concern for power rails with small current consumption. But it requires multiple ferrite beads in parallel to meet the IR drop requirement if the current consumption of the isolated power rail is very large.

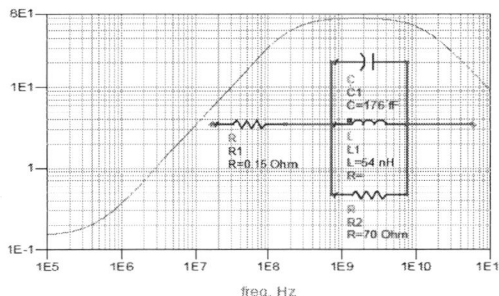

Figure 5.   Equivalent circuit and frequency response of a ferrite bead

C. Improving PCB Decoupling

As illustrated in Figures 2 and 3, the noise coupling is very large around the peripheral rail resonance peak frequency. The noise coupled can be reduced if the peripheral rail resonance peak impedance value is reduced. Commonly used method is using PCB decoupling capacitors.

The equivalent circuit model of a capacitor consists of a resistor, a capacitor, and an inductor in shunt. The impedance profile of the circuit is a "V" shaped curve. The low impedance area around the capacitor resonance frequency is where the capacitor is effective in controlling PDN impedance. If PCB decoupling capacitors has sufficiently low impedance compared to periphery power PDN impedance, the couple noise will be minimized.

Resonance frequency of a capacitor is determined by its inductance and capacitance. Reducing inductance increases effective decoupling range of the capacitor. Commonly used methods to reduce inductance include increasing number of capacitors, choosing capacitor form factor with smaller parasitic inductance, and optimizing mounting and placement of the capacitors [5].

IV. CASE STUDY

The transient noise on peripheral rail for different core and peripheral connection schemes in a typical power sharing scenario are simulated to demonstrate the noise isolation effectiveness of the previous schemes. The thickness of the PCB board studied is 114mil. FPGA device is mounted on top of the PCB board. The PCB decoupling capacitors are placed at the bottom of the PCB board. FPGA core and peripheral power rail share a 1.1V power supply.

Four power sharing schemes are studied. Case 1: Core and peripheral rail are connected near the bottom of the PCB. Via length is 114mil. Case 2: Core and peripheral rail are directly connected near the top of the PCB and via length is 9mil. Case

3: Core and peripheral rail are isolate with power stripe. Via length is 114mil for core power plane and 53mil for peripheral power stripe. Case 4: Core and peripheral rail are isolated with ferrite bead. Via length is 114mil for core power plane and 53mil for peripheral power stripe.

Figure 6 compares simulated transfer impedance between core and peripheral rail. Figure 7 compares simulated transient waveform on Peripheral rail. Table I summarizes simulated core and peripheral noise magnitude for each case.

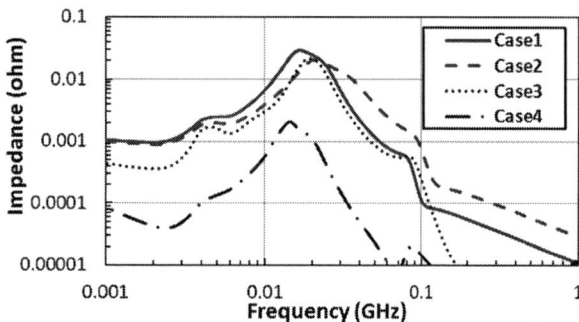

Figure 6. Transfer impedance profiles

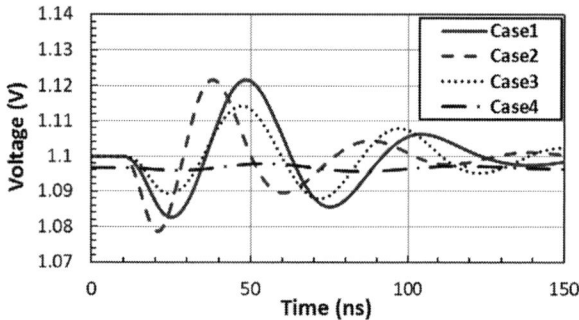

Figure 7. Coupled noise waveform

Table I. Comparison of simulated transient noise

	Core noise V_{p-p} (mV)	Peripheral noise V_{p-p} (mV)	Noise isolation (dB)
Case 1	67	39	-4.7
Case 2	53	43	-1.8
Case 3	72	25	-9.2
Case 4	74	2	-31.4

Case 1 represents a typical PCB PDN design and results are the base line for the study. Case 2 eliminates inductive coupling by moving core and peripheral connection to the top of the PCB stack up. Core noise magnitude reduction also indicates resonance peak impedance of the core and peripheral PDN is reduced as the result of the change. But noise isolation is reduced from -4.7dB in Case 1 to -1.8dB in Case 2 due to increase in noise due to direct current draw. This illustrates that noise from direct current draw contributes most to the power noise coupling.

Separating core and peripheral power rail and moving peripheral rail to higher layer reduces peripheral rail PDN

impedance peak and inductive coupling (Case 3). As the result, noise isolation increases to -9.2dB. Increasing peripheral power plane inductance further improves noise isolation, but results in higher resonance peak impedance. Peripheral power rail sees higher power noise due to its own activity as the result.

Ferrite bead provides noise isolation solution for rails very sensitive to the noise. Simulation shows -31.4dB isolation to the core power noise with selected ferrite bead (Case 4). But DC resistance of the selected bead is 0.15Ohm and this causes an extra 150mV voltage drop for every 1A current draw, which makes it difficult to meet power rail IR drop requirement when isolated rail current demand is high.

V. Conclusions

Noise generated at core power rail can propagate to peripheral power rail when two rails share same power delivery system on PCB. The coupled noise has impact on device performance. In this paper, two major power noise coupling mechanisms, direct current draw and inductive noise coupling, are studied. Results suggest that noise due to direct current draw is the major contributor of the noise coupling.

Optimizing PCB power delivery network design helps reducing the power noise coupling. Commonly used PDN design optimizations include: PCB stack up optimization, isolation using power stripe and isolation using ferrite beads. This study examines the effectiveness of these methods in reducing power noise coupling. Case study suggests that:

- Pushing shared power plane to the bottom of the PCB stack up increases noise isolation, but the noise isolation in shared case is limited.

- Isolation using power stripe design along stack up optimization can achieve 10dB noise isolation without power limit.

- Ferrite bead has good noise isolation. But it limits the maximum power due to high ESR of the bead, which limits its application to low power rails only

It is important to fully understand the power consumption and the noise sensitivity of the peripheral power rail so we can choose proper isolation solution based on performance requirement.

References

[1] L. D. Smith, R. E. Anderson, D. W. Forehand, T. J. Pelc, and T. Roy, "Power distribution system design methodology and capacitor selection for modern CMOS technology," *IEEE Transactions on Advanced Packaging*, pp. 284-291, Aug. 1999.

[2] S. Sun, L. Smith, and P. Boyle, "On-chip PDN noise characterization and modeling," presented at the *IEC DesignCon*, Santa Clara, CA. 2010.

[3] Istvan Novak. (2008). *Power Distribution Network Design Methodologies*. International Engineering Consortium

[4] Altera application notes AN583: Designing Power Isolation Filters with Ferrite Beads for Altera FPGAs

[5] Altera application notes AN574: Printed Circuit Board (PCB) Power Delivery Network (PDN) Design

Analysis and Verification of Board Power Delivery Network Impact on DDR3L Memory Interface in ARM SoC Application

Dinh T. Tran, GaWon Kim, Max (Sunghwan) Min,
Harold Bautista, and Nian Zhou

System LSI SoC Bay Area R&D (SBR)
Samsung Semiconductor Inc.
San Jose, California, USA
dinh.tran@ssi.samsung.com

Baekkyu Choi*, Seungyong Cha**, Se-ho You**

*Memory Division DRAM Solution Team
**Package Development Team, Semiconductor R&D
Center
Samsung Electronics Corporation (SEC)
Hwasung-City, Gyeonggi-Do, South Korea

Abstract— *In this paper, a DDR3L simulation topology for ARM SoC application is presented and the impact of the board power delivery network (PDN) on a DDR3L memory interface is simulated and analyzed. The analysis of the DDR3L PDN of the package and board will be discussed in the frequency-domain while the DDR power noise and DDR3L data signals are analyzed in the time-domain. A timing jitter comparison of the measured and simulated eye-diagram is presented. Finally, the analysis verifies the importance of including the board PDN to accurately predict the performance of the DDR3L memory interface in the simulation and modeling environment.*

Keywords— ARM SoC; DDR3L; board PDN

I. INTRODUCTION

In recent years, social network service (SNS) has increased in popularity and have enabled users to seamlessly share their content online. Thus, this trend makes the IT companies need huge data storage and fast data transfer. ARM SoC application is aiming to solve this dramatic increase demand for servers in data centers and to process low intensive workloads for media storage applications, data center applications, and clouding computing applications.

One of the main memory interfaces of ARM SoC applications is DDR3L. The DDR3L interface is designed to operate at the maximum speed of 1600Mbps with one Dual Ranks DIMM (Dual Inline Memory Module) topology. The DIMM can be UDIMM (Un-buffer DIMM) or SODIMM (Small Outline DIMM). In order to ensure the DDR3L system meets the performance requirements, early pre-layout and post-layout simulations are significantly important before the final implementation of chip, package and board. However, it's difficult to have accurate board models with power delivery network (PDN) at the early stage of ARM SoC development. Therefore, the simulation topology in Fig. 1(a) is commonly used for modeling and simulation of DDR memory interface. Usually, it includes one DDR3L byte lane model which have DQ (data), DQS (data strobe), and transistor SoC and DDR I/O models with package PDN.

In this paper, the simulation topology of DDR3L memory interface with the board PDN for ARM SoC application is introduced as shown in Fig. 1(b). Simulations with and without the board PDN are performed, which makes it possible to quantify and verify its impact on DDR3L memory interface. The analysis of DDR PDN of package and board will be discussed in frequency-domain. Also, DDR power noise and DDR3L DQ signals will be analyzed with/without the board PDN in time-domain. The measured and simulated eye-diagrams and timing jitter will be compared to verify the board PDN impact on DDR3L memory interface. It will show that including the board PDN is critical to predict accurately the performance of the DDR3L memory interface in modeling and the simulation.

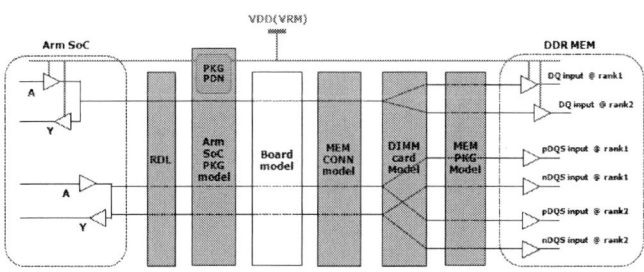

(a) Simulation Topology of DDR3L Interface without Board PDN

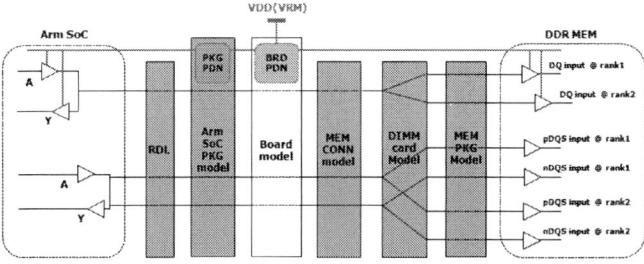

(b) Simulation Topology of DDR3L Interface with Board PDN

Fig. 1. Simulation Topology of DDR3L Memory Interface for ARM SoC Application

978-1-4799-0708-3/13 $31.00 © 2013 IEEE

II. ANALYSIS OF DDR POWER DELIVERY NETWORK IN FREQUENCY-DOMAIN

In order to analyze the DDR PDN of an ARM SoC application, the self-impedances (Z11 parameters) of DDR PDN on SoC package and SoC board were modeled and simulated using commercial tools. The frequency sweep was from DC up to 10GHz.

Fig. 2(a) shows the simulated self-impedance of the DDR PDN at the bump in SoC package. Up to 2.7GHz, the impedance plot shows the plane capacitance of SoC package PDN. After 2.7GHz, several plane resonances of SoC package PDN are depicted. Fig. 2(b) shows the simulated self-impedance of SoC board PDN at the ball port with VRM (voltage regulator module) short termination. Due to the VRM short termination, the impedance curve shows the inductance up to 1.5GHz. This board PDN model included decoupling capacitors and other discrete components. The dominant resonance peak in DDR board PDN is 3.6GHz.

In Fig. 3(a), the simulated self-impedances of SoC package PDN and board PDN are plotted together. The plane capacitance of the SoC package and the inductance of the SoC board are crossing at 1.95GHz. The combined SoC package PDN and SoC board PDN has the simulated self-impedance as plotted in Fig. 3(b). The first big resonance, at 1.95GHz, comes from the LC resonance, the plane capacitance of the SoC package PDN and the inductance of the SoC board PDN. Before the 1.95GHz resonance, there is the inductance curve which represents the summation of the package inductance and SoC board inductance. Resonances after 1.95GHz are mainly attributed to the SoC package PDN but their impedances have

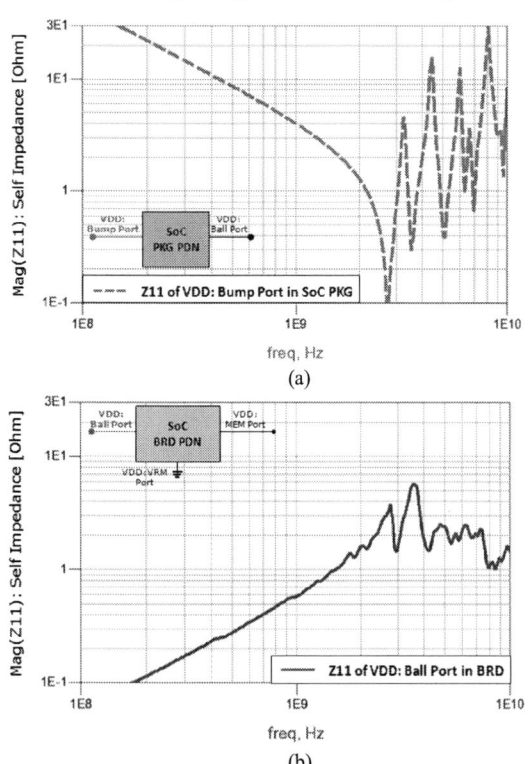

Fig. 2. (a) Simulated Self-impedance of DDR PDN at bump port in SoC package (b) Simulated Self-impedance of DDR PDN at ball port in SoC board including board decaps with VRM short

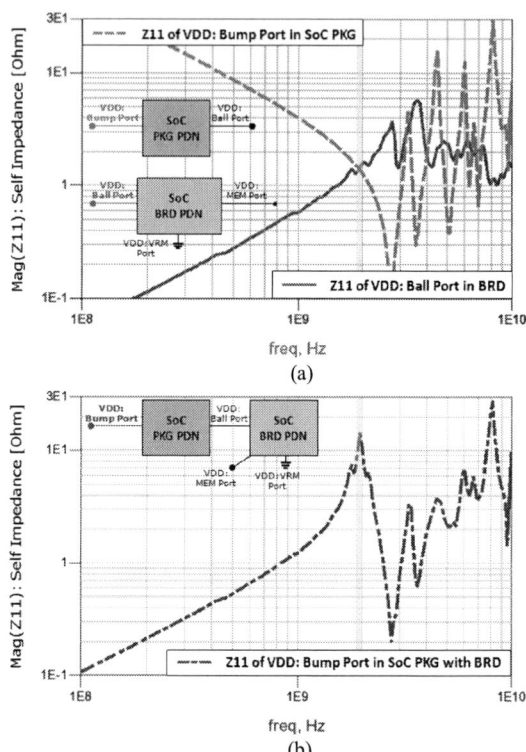

Fig. 3. (a) Simulated Self-impedances of DDR at bump port in SoC package PDN and at ball port in SoC board PDN, separately (b) Simulated Self-impedance of DDR at bump port in SoC package PDN with SoC board PDN

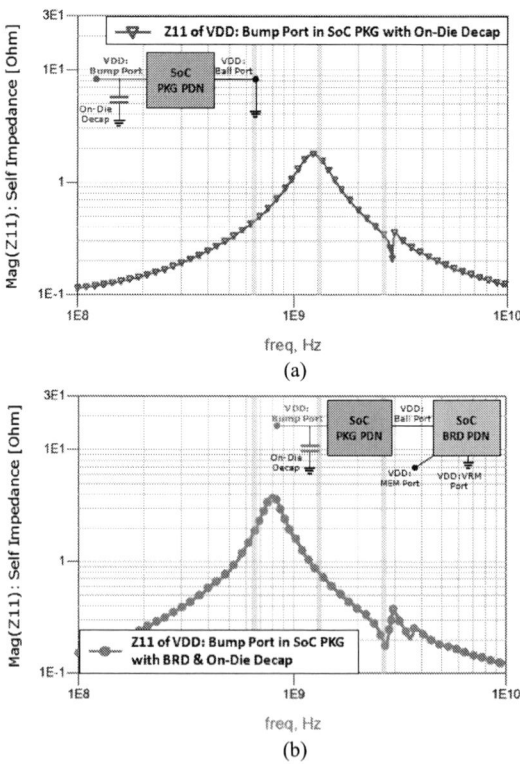

Fig. 4. (a) Simulated Self-impedance of DDR at bump port in SoC package PDN with SoC board PDN and On-Die decoupling capacitor (b) Simulated Self-impedance of DDR at bump port in SoC package PDN with On-Die decoupling capacitor

decreased values due to the board PDN.

Fig. 4 depicts the simulated self-impedances of the SoC DDR PDN at bump port with/without the board PDN including on-die decoupling capacitor model. In order to isolate the effect of the board PDN, the self-impedance of the SoC package PDN at the bump port with short termination and on-die decoupling capacitor model was simulated as displayed in Fig 4(a). Due to the on-die decoupling capacitor model, the high frequency resonances are all removed and one big resonance remains in each plot. The main resonance occurs at 1.23GHz, as shown in Fig 4(a), comes from the inductance of the SoC package and the on-die decoupling capacitor. Comparing the impedance graph in Fig. 3(b) and Fig. 4(b) shows that the original resonance shifted from 1.95GHz to 794MHz, which is caused by the on-die decoupling capacitor model. Also, the impedance at the resonance frequency in Fig. 4(b) is higher than that in Fig. 4(a), which means the SSN (Simultaneous Switching Noise) voltage of DDR power in the simulation topology with board PDN in Fig. 1(b) will be bigger than the noise voltage from the topology without board PDN.

Fig. 4 has three frequency indicators, one at the DDR3L operating frequency (667MHz) and the 1st and 2nd harmonics of it (1.33GHz and 2.66GHz). The 1.23GHz resonance in Fig. 4(a) is close to the 1st harmonics (1.33GHz) and the resonance at 794MHz in Fig. 4(b) is close to the original DDR3L operating frequency (667MHz). Frequency-domain analysis says that the 667MHz SSN is the dominant noise in the DDR PDN since the combined DDR PDN has biggest self-impedance at 794MHz. One important conclusion from this analysis is that the simulation topology without the board PDN, the 794MHz resonance in Fig. 4(b) will not be captured properly. The 1.23GHz resonance will be considered the main resonance instead of the 794MHz resonance. In that case, the analysis of SSN in the DDR PDN will be compromised. Also, the timing jitter of the DDR signals which is affected by the SSN will not correlate accurately with the measured timing jitter data. Chapter III will discuss time-domain waveforms to verify this prediction.

III. ANALYSIS OF DDR BOARD PDN IMPACT ON DDR3L MEMORY INTERFACE IN TIME-DOMAIN

To analyze the board DDR PDN impact on the DDR3L memory interface, the DDR3L DQ write signal at memory package and DDR power at SoC bump port were simulated with/without SoC board PDN, as explained in Fig. 1. Fig. 5(a) shows the simulated waveforms without the board PDN. The DDR power has 120mV peak-to-peak SSN voltage and this

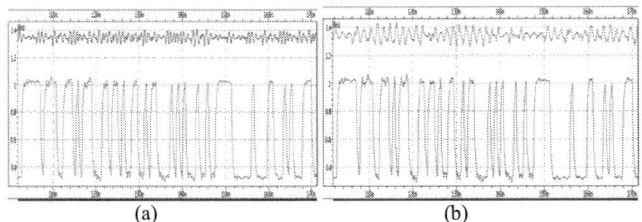

Fig. 5. (a) Simulated DDR3L DQ write signal and DDR SSN voltage waveform without SoC board PDN (b) Simulated DDR3L DQ write signal and DDR SSN voltage waveform with SoC board PDN

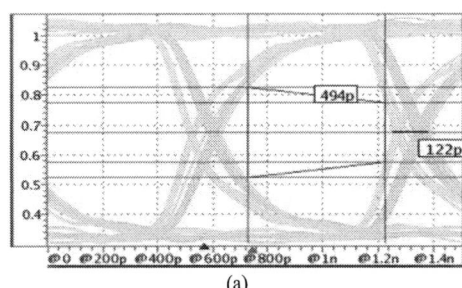

(a)

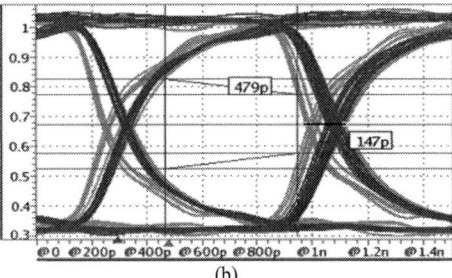

(b)

Fig. 6. (a) Simulated eye-diagram without SoC board PDN: 1333Mbps SSON (Simultaneous Switching Output Noise) pattern for DDR3L DQ write signal (b) Simulated eye-diagram with SoC board PDN: 1333Mbps SSON pattern for DDR3L DQ write signal

SSN has 746ps period which equates to 1.34GHz frequency component. Since the DQ write signal is affected by this SSN, it has the glitches on the DDR DQ write signal waveform. The SSN of DDR power is measured as 180mV peak-to-peak voltage as displayed in Fig. 5(b) and it has 1.46ns period that represents to a 685MHz frequency component.

These SSN simulation results show excellent correlations with the analysis in the frequency-domain, as mentioned in section II. Without the board PDN, the SSN has less peak-to-peak voltage and the frequency of the SSN is very close to the prediction. This 60mV SSN voltage difference between Fig 5(a) and (b) is caused by the board PDN. Also, it can be demonstrated that the pattern transitions of DQ write signals are causing the SSN voltage in DDR power in Fig. 5. The big SSN voltage which is probed at the SoC bump pad could be aligned to the frequent transition range of data pattern in DQ write signal with some delay.

Fig. 6 shows the simulated eye-diagrams of DDR3L DQ write signal using the simulation topology with/without the board PDN. The data pattern is 1333Mbps SSON (Simultaneous Switching Output Noise) pattern which is intentionally made for generation of the worst timing jitter. The eye-diagram in Fig. 6(a) is similar to one in Fig. 6(b) in terms of waveform shapes and ISI response. But, the timing jitter is different. Without the board PDN, the timing jitter is 122ps. With the board PDN, the timing jitter is 147ps in Fig. 6(b). There is 25ps timing jitter in correspond to the 60mV increase in the SSN voltage ripple introduced by the board PDN. This is a significant impact to the DDR3 timing budget. Therefore, the conclusion is the board PDN should be included in the simulation topology.

IV. VERIFICATION OF BOARD PDN IMPACT ON DDR3L MEMORY INTERFACE

In this section, the measurement results will be compared to the simulation results to verify the board PDN impact on the DDR3L memory interface. To measure the time-domain waveforms at the memory interface side, a BGA interposer made by Samsung was used for the measurement between the DDR3L memory and the UDIMM under test which seats in DIMM socket on the ARM SoC board. This interposer makes possible to directly measure the time-domain waveforms as close as possible to the load/source but also adds some parasitic loading and ISI effect on the measured waveforms.

In Fig. 7(a) and (b), the simulated eye-diagrams of DDR3L DQ write signal are described using the simulation topology with/without the SoC board PDN and with interposer model. The interposer model was included in the simulation topology for the correlation with the measurement data. The used data pattern is the same as the 1333Mbps SSON pattern which generates the worst timing jitter. Two simulated eye-diagrams in Fig. 7(a) and (b) also show very similar shape and same ISI response, but they have bigger reflection effect caused by the interposer model. Without the SoC board PDN, the simulated eye-diagram in Fig. 7(a) has 132ps timing jitter. And, the

simulated eye-diagram including the SoC board PDN in Fig. 7(b) shows 154ps of timing jitter. This 22ps difference of timing jitter is definitely caused by the simulation topology including the SoC board PDN. One thing can be noticed here is that the interposer model generates additional 10ps of timing jitter.

The measured eye-diagram of DDR3L DQ write signal using the same 1333Mbps SSON pattern is plotted in Fig. 7(c). The measured eye-diagram shape has perfect agreement with the simulated one. And, it has 224.5ps timing jitter. There is a 34.5ps difference between the simulated and measured timing jitter. This difference could be generated from the on-chip timing jitter portions such as PLL clock source jitter, clock distribution noise jitter, and several on-chip physical layout skews, which are not included in the simulation topology. In spite of the 34.5ps timing jitter difference, it can be definitely said that the simulated eye-diagram and timing jitter including the SoC board PDN show excellent correlation with the measured data. Without the SoC board PDN, the timing jitter correlation will not be accurate enough.

V. CONCLUSIONS

In this paper, the simulation topology of a DDR3L for an ARM SoC application was introduced and the importance of including the board PDN for accurate predictions was verified. It was also demonstrated that the impact of the board PDN on the DDR3L memory interface did, not only affect the SSN voltage level but also the timing jitter. The analysis of DDR PDN in package and board with/without on-die decoupling capacitor was discussed in the frequency-domain. Through this analysis, the SSN voltage and frequency component was predicted. The time-domain waveforms of DDR power noise and DDR3L DQ write signals were displayed to confirm this prediction. Consequently, the comparison of the measured and simulated eye-diagrams and timing jitter were compared to verify the board PDN impact on DDR3L memory interface for ARM SoC application. The simulated eye-diagrams without interposer model display consistent results with the frequency-domain plots in previous analysis. Also, the simulated eye-diagrams with the interposer model show excellent correlation with the measured eye-diagram. Including the board PDN is critical to accurately predict timing jitter in the simulation and modeling of a DDR3L memory interface.

ACKNOWLEDGMENT

The authors would like to give appreciation to Sungjoo Kim and Hyun Joon Kang in Next Generation AP team who helped to bring up the system for this project.

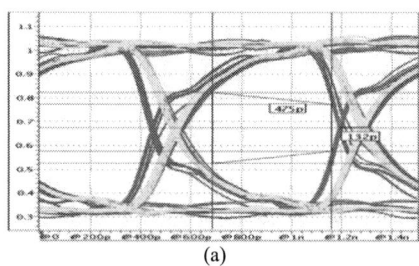

(a)

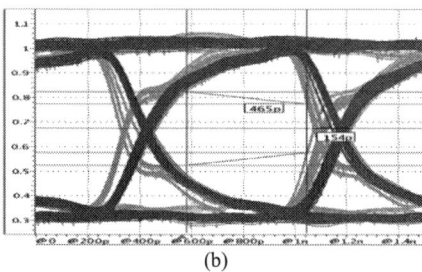

(b)

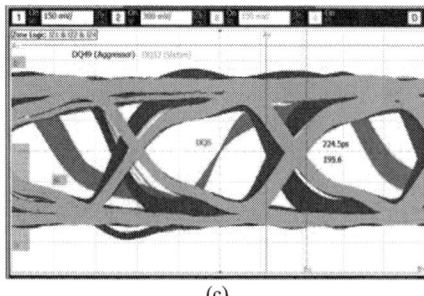

(c)

Fig. 7. (a) Simulated eye-diagram without SoC board PDN and with Interposer: 1333Mbps SSON (Simultaneous Switching Output Noise) pattern for DDR3L DQ write signal (b) Simulated eye-diagram with SoC board PDN and Interposer: 1333Mbps SSON pattern for DDR3L DQ write signal (c) Measured eye-diagram: 1333Mbps SSON pattern for DDR3L DQ write signal

REFERENCES

[1] Laddha, V. and Swaminathan, M., *"Correlation of PDN impedance with jitter and voltage margin for high speed channels"* in Proceedings of 2008 Electrical Performance of Electronic Packaging (EPEP), pp. 73-76.

[2] Swaminathan, M. and Eugin, A. E., *Power Integrity Modeling and Design for Semiconductors and Systems*, Prentice hall, 2007.

[3] Kyung Suk (Dan) Oh and Xing Chao (Chuck) Yuan, *High-Speed Signaling: Jitter Modeling, Analysis, and Budgeting*, Prentice hall, 2011.

[4] Hall, S. H. & Heck, H. L., *Advanced Signal Integrity for High-Speed Digital Designs*, IEEE: Wiley, 2008.

978-1-4799-0708-3/13 $31.00 © 2013 IEEE

Minimal-Order Circuit Model Based Fast Electromagnetic Simulation

Qing He, Duo Chen, Jianfang Zhu, and Dan Jiao

School of Electrical and Computer Engineering, Purdue University
West Lafayette, IN 47907, USA, djiao@purdue.edu

Abstract— State-of-the-art electromagnetics-based circuit simulators perform circuit simulation on the physical layout of an integrated circuit, the size of which can be large even for an optimal-complexity simulator to analyze it efficiently. We find a frequency- and time-independent minimal-order model of the integrated circuit from DC to high frequencies for any prescribed accuracy based on electromagnetics. The model size is only two for RC circuits and minimal such as tens for RLC and fullwave circuits, regardless of the original circuit size and the number of input/output ports. Numerical experiments demonstrate the superior performance of the proposed minimal-order model based fast electromagnetic simulation.

Keywords—Circuit modeling; circuit simulation; electromagnetic simulation; reduced order model

I. INTRODUCTION

In existing electromagnetics (EM)-based circuit simulation frameworks, the linear network of a circuit is analyzed directly in time domain by an electromagnetics-based approach, thereby bypassing the circuit extraction step. Moreover, an electromagnetics-based circuit simulation approach possesses first-principles based accuracy, thus allowing for an integrated circuit to be accurately analyzed from zero to high frequencies. A state-of-the-art EM-based circuit simulator now is capable of simulating a circuit including both nonlinear devices and the linear network in linear (optimal) complexity [1]. However, the problem size handled by state-of-the-art electromagnetics-based circuit simulators is the size of circuit layout. The entire physical layout of an IC is simulated once and once again at each time instant. As a consequence, even though an electromagnetics-based circuit simulator has an optimal complexity, it is not feasible for such a simulator to perform a system-level circuit analysis in real time since the layout size of a large-scale IC such as a global power delivery network and a full-chip mixed-signal IC is too large even for an optimal-complexity simulator to analyze it efficiently.

The contribution of this work is an electromagnetics-based minimal-order circuit modeling and simulation method, which is different from existing electromagnetic and circuit simulation algorithms. In this new modeling and simulation method, we find a minimal-order model of the integrated circuits from DC to high frequencies for prescribed accuracy based on electromagnetics, which has not been found by existing model order reduction methods [2-4]. For circuits dominated by RC effects, the order of the proposed model is only 2 regardless of the original circuit size and the number of inputs and outputs involved in the circuit. For circuits dominated by RLC or full-wave effects, the order of the

proposed model is also minimal for achieving the prescribed accuracy. As a result, the EM-based simulation of large-scale circuits can be accurately and rapidly performed on the proposed minimal-order model, the size of which is many orders of magnitude smaller than the original layout size as well as the size of reduced order models currently available. From the proposed minimal-order model, one can recover circuit solutions at any node of interest. Furthermore, the proposed minimal-order model is both frequency and time independent, and hence it can be used to characterize the broadband response of an integrated circuit and conveniently integrated into other circuit simulators.

In addition, we propose fast algorithms to generate the proposed minimal-order model efficiently, and hence the computational overhead due to model generation is small. Moreover, we do not incur additional computation in preserving the passivity and stability of the proposed model as the way the model is constructed naturally ensures the passivity and stability of the proposed model based simulation. The proposed circuit simulation method retains the strength of an EM-based circuit simulator in bypassing circuit extraction and possessing first-principles based accuracy, while eliminating its disadvantage in performing circuit simulation on the original layout, the size of which is many orders of magnitude larger than the size of the proposed minimal-order model. Numerical experiments have demonstrated the superior performance of the proposed minimal-order circuit model based electromagnetic simulation.

II. ELECTROMAGNETICS-BASED MINIMAL-ORDER CIRCUIT SIMULATION FRAMEWORK

Consider an integrated circuit consisting of a linear network and nonlinear circuits. The physical phenomena in the linear network from DC to light are governed by Maxwell's equations

$$\nabla \times [\mu_r^{-1} \nabla \times \mathbf{E}(\mathbf{r},t)] + \mu_0 \varepsilon \partial_t^2 \mathbf{E}(\mathbf{r},t) + $$
$$\mu_0 \sigma \partial_t \mathbf{E}(\mathbf{r},t) = -\mu_0 \partial_t \mathbf{J}(\mathbf{r},t) \qquad (1)$$

where $\mathbf{E}$ is electric field, μ_0 is free-space permeability, μ_r is relative permeability, ε is permittivity, σ is conductivity, $\mathbf{J}$ is current density, and $\mathbf{r}$ denotes a 3-D point in the circuit layout.

We discretize (1) in time and space by a finite element method, resulting in the following second-order linear system of equations [1]

$$\mathbf{T}\ddot{u}(t) + \mathbf{G}\dot{u}(t) + \mathbf{S}u(t) = \dot{I}(t) , \qquad (2)$$

in which $\mathbf{T}$, $\mathbf{G}$, $\mathbf{S}$ are *sparse* matrices, u is the vector of the voltage drop on each edge of a discretized layout, and I is a

978-1-4799-0708-3/13 $31.00 © 2013 IEEE   219

vector of current sources. The $\ddot{u}(t)$ denotes the second-order time derivative of $u(t)$, $\dot{u}(t)$ is the first-order derivative, and $\dot{I}(t)$ is the first-order time derivative of $I(t)$. Matrices $\mathbf{T}$, $\mathbf{G}$, $\mathbf{S}$, and vector I are assembled from their elemental contributions as the following

$$\mathbf{T}_{ij} = \mu_0 \varepsilon < \mathbf{N}_i, \mathbf{N}_j >_V , \qquad \mathbf{G}_{ij} = \mu_0 \sigma < \mathbf{N}_i, \mathbf{N}_j >_V$$

$$\mathbf{S}_{ij} = \mu_r^{-1} < \nabla \times \mathbf{N}_i, \nabla \times \mathbf{N}_j >_V , \quad I = -\mu_0 < \mathbf{N}_i, \partial_t \mathbf{J} >_V , \qquad (3)$$

where $\mathbf{N}_i$ and $\mathbf{N}_j$ are the vector basis functions used to expand unknown field $\mathbf{E}$ in each element of the discretized layout and $<.,.>_V$ denotes a volume integration. $\mathbf{T}$, $\mathbf{G}$ and $\mathbf{S}$ are all sparse, and hence can be assembled in linear time and storage. In addition, $\mathbf{T}$ is symmetric and positive definite, $\mathbf{S}$ as well as $\mathbf{G}$ is symmetric semi-positive definite. As a result, (2) is guaranteed to be stable and passive.

The size of (2), denoted by N, is proportional to the size of the physical layout. When performing circuit simulation on (2), one has to simulate a huge linear system of layout size once and once again in time. To overcome this bottleneck problem, in this work, we propose to find a minimal-order model of the linear network that is frequency and time independent to represent the solution of (2). The detail of this model and its fast generation will be elaborated in Section III. Now, assuming this model is available for use, denoting it by $\mathbf{V}$, the order of which is k. The solution of (2) can then be represented as

$$u(t) = \mathbf{V}_{N \times k} u_r(t) , \qquad (4)$$

where $u_r(t)$ is the unknown coefficient vector to be solved. Substituting (4) into (2) and multiplying both sides of (2) by $\mathbf{V}^T$, we obtain

$$\mathbf{T}_r \ddot{u}_r(t) + \mathbf{G}_r \dot{u}_r(t) + \mathbf{S}_r u_r(t) = \dot{I}_r(t) , \qquad (5)$$

where the reduced matrices and vectors are

$$\mathbf{T}_r = \mathbf{V}_{N \times k}^T \mathbf{T} \mathbf{V}_{N \times k}, \ \mathbf{G}_r = \mathbf{V}_{N \times k}^T \mathbf{G} \mathbf{V}_{N \times k}, \ \mathbf{S}_r = \mathbf{V}_{N \times k}^T \mathbf{S} \mathbf{V}_{N \times k}, \ \dot{I}_r = \mathbf{V}_{N \times k}^T \dot{I} \ (6)$$

each of which is of size k. The matrices are also frequency and time independent. As a result, the circuit simulation can be performed on (5), the size of which is orders of magnitude smaller than the size of the original system (2). After simulating (5), the solution of (2), $u(t)$, can be obtained from (4). If only m pins are of interest, only m rows of $\mathbf{V}$ need to be stored and computed, the cost of which is $O(m)$. In addition, the proposed minimal-order simulation preserves passivity and stability because in the minimal-order system, $\mathbf{T}_r$ remains positive definite, $\mathbf{G}_r$ and $\mathbf{S}_r$ remain semi-positive definite.

To simulate the combined nonlinear-linear network, we co-simulate the following three systems of equations:

$$\mathbf{T}_r \ddot{u}_r(t) + \mathbf{G}_r \dot{u}_r(t) + \mathbf{S}_r u_r(t) = \dot{I}_r(t), \ I_r(t) = \mathbf{V}_{N \times k}^T (I_s(t) + I_{nl}(t)) \quad (7a)$$

$$\tilde{\mathbf{G}}(\cdot)x + \tilde{\mathbf{C}}(\cdot)\dot{x} = b, \text{ where } x = [V_c, I_c, v_c, i_c]^T \qquad (7b)$$

$$I_{nl} + I_c = 0, \ V_c = V_c , \qquad (7c)$$

where equation (7a) is the same as (5) which has reduced the huge linear system (2) from order N to order k; equation (7b) is the nonlinear system of equations obtained from a modified nodal analysis, and (7c) is the equation enforcing the current and voltage continuity condition at the interface between the linear network and the nonlinear network.

III. PROPOSED MINIMAL-ORDER CIRCUIT MODEL AND ITS FAST GENERATION

A. Proposed Minimal Order Model

The model $\mathbf{V}$ is used to represent the *solution* of the linear system instead of the linear system itself. Hence, we propose to derive a theoretical model of the inverse of the linear system from DC to high frequencies, from which a frequency-independent minimal-order model of the linear network can be identified. This model is valid for general circuits the layout of which contains arbitrarily shaped lossy conductors embedded in inhomogeneous and dispersive dielectrics.

To facilitate an analytical derivation, we transform (2) to frequency domain, obtaining

$$\mathbf{A}(\omega)u = -j\omega I(\omega) , \qquad (8)$$

where

$$\mathbf{A}(\omega) = \mathbf{S} - \omega^2 \mathbf{T} + j\omega \mathbf{G} . \qquad (9)$$

In (8), we divide unknowns into two groups: unknowns outside conductors u_o and unknowns inside conductors u_i. For unknowns that reside on the conducting surface, we categorize them into u_i. The system matrix $\mathbf{A}(\omega)$ shown in (8) is correspondingly cast into the following form:

$$\mathbf{A}(\omega) = \begin{bmatrix} \mathbf{A}_{oo} & \mathbf{A}_{oi} \\ \mathbf{A}_{io} & \mathbf{A}_{ii} \end{bmatrix}, \qquad (10)$$

the inverse of which can be written as

$$\mathbf{A}(\omega)^{-1} = \begin{bmatrix} \mathbf{A}_{oo}^{-1} + \mathbf{A}_{oo}^{-1} \mathbf{A}_{oi} \tilde{\mathbf{A}}_{ii}^{-1} \mathbf{A}_{io} \mathbf{A}_{oo}^{-1} & -\mathbf{A}_{oo}^{-1} \mathbf{A}_{oi} \tilde{\mathbf{A}}_{ii}^{-1} \\ -\tilde{\mathbf{A}}_{ii}^{-1} \mathbf{A}_{io} \mathbf{A}_{oo}^{-1} & \tilde{\mathbf{A}}_{ii}^{-1} \end{bmatrix} \quad (11)$$

where

$$\tilde{\mathbf{A}}_{ii} = \mathbf{A}_{ii} - \mathbf{A}_{io} \mathbf{A}_{oo}^{-1} \mathbf{A}_{oi} \qquad (12)$$

is the Schur complement of $\mathbf{A}_{ii}$, which captures the coupling from what is outside conductors to what is inside. The current excitation is launched outside conductors. Thus, the right hand side of (8) can be written as $\{-j\omega I \ \ 0\}^T$. Multiplying it with (11), we have

$$u_o = (\mathbf{A}_{oo}^{-1} + \mathbf{A}_{oo}^{-1} \mathbf{A}_{oi} \tilde{\mathbf{A}}_{ii}^{-1} \mathbf{A}_{io} \mathbf{A}_{oo}^{-1})(-j\omega I)$$

$$u_i = -\tilde{\mathbf{A}}_{ii}^{-1} \mathbf{A}_{io} \mathbf{A}_{oo}^{-1}(-j\omega I) \qquad (13)$$

After a careful derivation, we obtain

$$u_o = -V_0 V_0^T I / (j\omega) + V_h [\Lambda_h - \omega^2 \mathbf{I}]^{-1} V_h^T I / (j\omega) - \mathbf{Q} u_i$$

$$u_i = V_{ii,0} V_{ii,0}^T \mathbf{Q}^T I + V_{ii,h} (\Lambda_{ii,h} + j\omega \mathbf{I})^{-1} V_{ii,h}^T \mathbf{Q}^T (-j\omega I) \qquad (14)$$

where $\mathbf{Q} = V_0 V_0^T \mathbf{T}_{oi} + V_h (\Lambda_h - \omega^2 \mathbf{I})^{-1} V_h^T (\mathbf{S}_{oi} - \omega^2 \mathbf{T}_{oi})$, boldface $\mathbf{I}$ denotes an identity matrix, V_0 and V_h are, respectively, the eigenvectors corresponding to the zero and nonzero eigenvalues of the following frequency-independent generalized eigenvalue problem

$$\mathbf{S}_{oo} v = \lambda \mathbf{T}_{oo} v ; \qquad (15)$$

and Λ_h is the diagonal matrix formed by nonzero eigenvalues. The $V_{ii,0}$ and $V_{ii,h}$ are, respectively, the eigenvectors corresponding to the zero and nonzero eigenvalues of the generalized eigenvalue problem formulated inside conductors:

$$\tilde{\mathbf{S}}_{ii} x = \lambda_{ii} \mathbf{G}_{ii} x \qquad (16)$$

where

$$\tilde{\mathbf{S}}_{ii} = \mathbf{S}_{ii} - \mathbf{A}_{io}\mathbf{A}_{oo}^{-1}\mathbf{A}_{oi}. \tag{17}$$

Similarly, $\Lambda_{ii,h}$ is a diagonal matrix consisting of all the nonzero eigenvalues of (16).

The solution shown in (14) is a broadband solution of the voltage distribution in the entire circuit from zero to high frequencies where Maxwell's equations hold true. At relatively low frequencies where the contribution from higher-order eigenvectors (modes) V_h and $V_{ii,h}$ can be neglected in (14), the voltage solution inside conductors u_i and that outside conductors u_o can be explicitly written as [5]

$$\begin{aligned} u_o &= -V_0 V_0^T I / (j\omega) - \mathbf{Q}_l V_{ii,0} V_{ii,0}^T \mathbf{Q}_l^T I \\ u_i &= V_{ii,0} V_{ii,0}^T \mathbf{Q}_l^T I, \quad \text{where } f \in (0, f_{ref,\max}) \end{aligned}, \tag{18}$$

in which $\mathbf{Q}_l = V_0 V_0^T \mathbf{T}_{oi} + V_h(\Lambda_h)^{-1}V_h^T \mathbf{S}_{oi}$ is the low-frequency counterpart of $\mathbf{Q}$, and $f_{ref,\max}$ denotes the maximum frequency at which (18) can yield a desired accuracy. The $f_{ref,\max}$ can be quantitatively determined from the weights of higher-order modes in the circuit solution based on (14) and a required accuracy. In (18), except for ω, all the other terms are frequency independent.

A careful examination of (18) reveals that the solution can be expanded by two groups of vectors

$$\Phi_1 = \begin{pmatrix} -\mathbf{Q}_l V_{ii,0} \\ V_{ii,0} \end{pmatrix} \text{ and } \Phi_2 = \begin{pmatrix} V_0 \\ 0 \end{pmatrix}, \tag{19}$$

which span the real, and imaginary part of the solution, respectively. It can be proven that $\mathbf{S}[\Phi_1, \Phi_2] = 0$, and hence $[\Phi_1, \Phi_2]$ constitutes a complete nullspace of $\mathbf{S}$. The size of the nullspace could be large. Here, we reduce the size of nullspace to 2 by utilizing the fact that all the nullspace eigenvectors share the same zero eigenvalue in common, and hence for any given right hand side vector I, all the nullspace vectors in Φ_1 are grouped together, yielding a single vector wr-based representation of the real part of u, u_{re}, as the following:

$$u_{re} = \begin{pmatrix} re(u_o) \\ re(u_i) \end{pmatrix} = \begin{pmatrix} -\mathbf{Q}V_{ii,0}V_{ii,0}^T \mathbf{Q}^T I \\ V_{ii,0}V_{ii,0}^T \mathbf{Q}^T I \end{pmatrix} = wr. \tag{20-a}$$

Similarly, a single vector wi-based representation of u_{im} is obtained from

$$u_{im} = \begin{pmatrix} im(u_o) \\ im(u_i) \end{pmatrix} = -\frac{1}{j\omega}\begin{pmatrix} V_0 V_0^T I \\ 0 \end{pmatrix} = -\frac{1}{j\omega}wi. \tag{20-b}$$

The two frequency-independent vectors wr and wi form a complete and accurate space for representing the circuit solution of (2) for $f \in (0, f_{ref,\max})$. As a result, we obtain a circuit model of order 2 for the contribution from all nullspace modes as the following:

$$\mathbf{V}_0 = [wr, wi], \tag{21}$$

which is termed static modes in the sequel.

At frequencies where not only static modes but also higher-order modes (termed *fullwave modes*) make important contributions, the solution of the linear system (8) for a right hand side b, as can be seen from (14), can be written as:

$$\text{For } \forall \omega, \quad [\mathbf{A}(\omega)]^{-1}b = \tag{22}$$

$$\begin{bmatrix} V_0 & -\mathbf{Q}V_{ii,0} & V_h & * \\ 0 & V_{ii,0} & 0 & * \end{bmatrix} \cdots \begin{bmatrix} (-\omega^2\mathbf{I})^{-1} & & & \\ & (j\omega\mathbf{I})^{-1} & & \\ & & (\Lambda_h - \omega^2\mathbf{I})^{-1} & \\ & & & * \end{bmatrix} \begin{bmatrix} V_0 & -\mathbf{Q}V_{ii,0} & V_h & * \\ 0 & V_{ii,0} & 0 & * \end{bmatrix}^T b$$

where * denotes terms associated with additional higher-order eigenvectors. The solution resulting from the first two groups of vectors for a given right hand side I is nothing but $\mathbf{V}_0$ shown in (21), the order of which is 2. For circuits operating at higher frequencies, as can be seen from (22), the weights of higher-order modes become larger and larger. Given an accuracy requirement, the circuit model $\mathbf{V}$ should include some higher-order modes $\mathbf{V}_h$ in addition to $\mathbf{V}_0$ for achieving the required accuracy. Thus, we have

$$\mathbf{V} = [\mathbf{V}_0, \mathbf{V}_h]. \tag{23}$$

In this case, the inductance and full-wave effects become important in the circuit response. But the resultant model order, k, is still minimal for the prescribed accuracy based on the following Theorem: Given an accuracy requirement ε, the rank-r representation ($\mathbf{R}$) generated from singular value decomposition (SVD) is a minimal rank approximation of the original matrix $\mathbf{M}$ that fulfills $\|\mathbf{M} - \mathbf{R}\|_2 \leq \varepsilon$ [6]. It is clear that the singular values of $[\mathbf{A}(\omega)]^{-1}$ correspond to the entries of the center diagonal matrix of (22).

B. Fast Generation of the Proposed Model

To obtain the $\mathbf{V}_0$ of order 2 shown in (21), apparently, we need to solve generalized eigenvalue problems in (15) and (16). This can be computationally expensive when the circuit size is large. In fact, this is not necessary. Since we have already found the theoretical model of the circuit solution as shown in (18) in a wide range of frequencies from DC to $f_{ref,\max}$, solving (8) at any frequency in this range will provide us with accurate two vectors in (21). To be specific, at one frequency f_{ref} between DC and $f_{ref,\max}$, we solve the original system (8) to obtain the field solution u_{ref}. We then separate this solution into two vectors, namely, the real part u_{re} and the imaginary part u_{im}. The $\mathbf{V}_0$ can then be obtained as

$$\mathbf{V}_0 = [u_{re}, u_{im}]. \tag{24}$$

Although the eigenvalue solution is bypassed, solving (8) at a relatively low frequency is still a computational challenge when the circuit size is large. This challenge can be overcome by a linear-complexity direct matrix solution based divide-and-conquer algorithm we recently developed in [7]. In this algorithm, we decompose the original large-scale system matrix shown in (9) rigorously into small sparse matrices that are fully decoupled, we then synthesize the solution of (9) from the nullspace of the small sparse matrices. We also developed an efficient linear-complexity algorithm to extract the higher-order modes $\mathbf{V}_h$, the detail of which is described in [8] and omitted here due to space limit.

IV. SIMULATION RESULTS

First, we simulated two M6-M8 power grids backed by package planes. The first one had 7, 10 and 5 pairs of VCCs and VSSs respectively in M6, M7, and M8. The dielectric constant was 3.16, 3.13, 4.3, and 6.5 respectively from M6 to the top. The wire width, thickness, and pitch were 1, 0.5, and 4 μm in M6 and M7. The fourth pair of the VCC/VSS on M6 was excited at the middle point by a Gaussian derivative current source with $t_0 = 4\tau$ and $\tau = 3 \times 10^{-10}$ s. The time cost in model generation was only 87 s. The voltages between the first pair of M6 VCC/VSS and the top package plane at the middle of the power grid are plotted in Fig. 1. Excellent agreement between the proposed method and the reference method [1] is observed. We also generated a simple resistance (R)-based netlist with 840 resistances for SPICE simulation. The SPICE

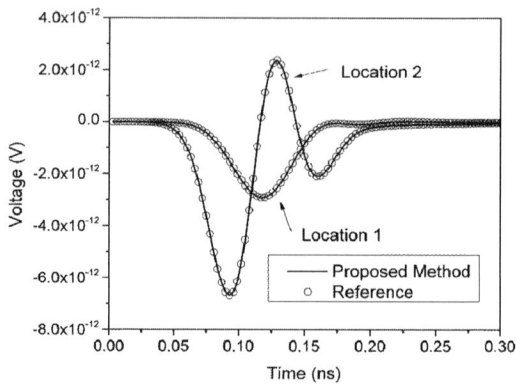

Fig. 2. Time-domain voltage waveforms simulated by the proposed method in comparison with a reference method [1].

holes, a center via layer consisting of both power and ground vias, and a top power plane with via holes, occupying a chip area of 2500 μm × 2500 μm. The on-chip part consists of two metal layers. VCC and VSS vias traverse via holes and contacted package planes through bumps. The on-chip power rails are connected through on-chip vias. The total unknown number of (2) is 754,397. The size of the proposed model determined by the method in Section III is 78, which is almost 10,000 times smaller than the original layout size. Fig. 2 shows the time domain waveforms of two voltages in the combined die and package circuit. Accurate results are obtained.

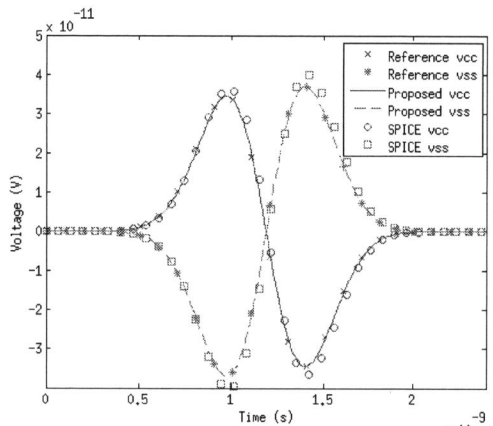

Fig. 1. Simulation of an M6-M8 power grid.

Table I: Relative error of entire solution vector $\|u - u_{\text{ref}}\| / \|u_{\text{ref}}\|$.

Time (ns)	0.1	0.9	1	1.4	1.5	1.6
Error (%)	1.78	0.97	0.14	0.2	1.01	2.04

result shows a slight difference with the other two methods, which can be attributed to the inaccuracy of the R-based model. In Table I, we list the relative error of the entire solution vector of the proposed method, which has circuit solutions at 83,039 points, in comparison with the reference method. It is worth mentioning that although the model size for this example is only 2, after $u_r(t)$ is found from the reduced system (5) of order 2; the circuit solution at any discretized point in the physical layout can be recovered from (4). Because of the reduced $O(2)$ system, the proposed method only cost 0.01 s for simulating the entire time window of 2.5 ns, whereas the reference method used 139.05 s and SPICE with a netlist of 840 resistances used 0.9 s in simulation. We then simulated a larger M6-M8 power grid with 100 pairs of VCC/VSSs in M7. The size of the original system was increased to 0.81 million, while the proposed model order remains 2. The reference method failed to complete the simulation of 2.5 ns time in 10 hours, while the proposed method only used 0.01 s, and SPICE with an R-based netlist cost 10.6 s.

In the example above, the order of the proposed model is only 2 due to a relatively small electric size of the underlying circuit and hence RC-dominated circuit behavior. Next, we simulate a combined die-package power delivery structure provided by Intel. The package structure involves irregular geometry shapes. It contains a bottom ground plane with via

V. CONCLUSIONS

An electromagnetics-based circuit simulator with a minimal-order circuit model is developed for fast simulation of integrated circuits and packages from zero to high frequencies. The proposed time- and frequency-independent minimal-order circuit model can also be readily used by other circuit simulators to significantly accelerate circuit simulation.

ACKNOWLEDGMENT

This work was supported by a grant from Intel Corporation, a grant from Office of Naval Research under award N00014-10-1-0482, and a grant from NSF under award No. 1065318.

REFERENCES

[1] Q. He, D. Chen, and D. Jiao, "From Layout Directly to Simulation: A First-Principle Guided Circuit Simulator of Linear Complexity and Its Efficient Parallelization," *IEEE Trans. on Components, Packaging, and Manufacturing Tech.* 2(4): 687-699, April 2012.

[2] A. Odabasioglu, M. Celik, and L. T. Pileggi, "PRIMA: Passive reduced-order interconnect macromodeling algorithm," *TCAD*, 17(8), 1998.

[3] J. R. Phillips, L. Daniel, and L. M. Silveira, "Guaranteed passive balanced transformation for model order reduction," *TCAD*, 22(8), 2003.

[4] H. Wu and A. C. Cangellaris, "Model-Order Reduction of Finite-Element Approximations of Passive Electromagnetic Devices Including Lumped Electrical-Circuit Models," *IEEE Trans. MTT*, 52(9), 2004.

[5] J. Zhu and D. Jiao, "A rigorous solution to the low-frequency breakdown in full-wave finite-element-based analysis of general problems involving inhomogeneous lossless/lossy dielectrics and non-ideal conductors," *IEEE Trans MTT*, 59(12):3294–3306, 2011.

[6] C. F. Van Loan and G. H. Golub, Matrix Computations, Johns Hopkins University Press, London, 1996.

[7] Q. He and D. Jiao, "A Rigorous Divide-and-Conquer Algorithm for Fast DC-Mode Extraction," *IEEE Int. Symp. Ant. and Propagat.*, July 2013.

[8] D. Chen, "Time-domain orthogonal finite-element reduction-recovery (OrFE-RR) method for electromagnetics-based analysis of very large scale integrated circuit and package problems," Doctoral dissertation, Purdue University, 2012.

Characterization and Analysis of Vertical Coupling Impact on Receiver Performance in High Speed Serial Interface

Yujeong Shim, Dan Oh, Shishuang Sun, Jianmin Zhang, Jenny Jiang, Cuong Nguyen, Janani Chandrasekhar, and Yuri Tretiakov

Altera Corporation
101 Innovation DR, San Jose, CA, 95051, USA
yshim@altera.com

Abstract— **As data rate increases, crosstalk becomes a significant source of high jitter. Although many techniques have been investigated to reduce crosstalk, it is not possible to fully eliminate coupling. In particular, near-end coupling between transmitter (TX) to receiver (RX) occurs at the interface of chip and PCB is one of the main sources of crosstalk. This TX to RX coupling is detrimental compared to other TX to TX or RX to TX coupling since the aggressor TX swing is large and the victim RX signal has slow edge. Thus, RX jitter is sensitive to coupling noise from TX. Therefore, it is important to investigate crosstalk impact on receiver performance on the system level including transmitter, receiver, package and PCB. In this paper, the impact of transmitter to receiver crosstalk induced at package balls and PCB vias is characterized using the internal eye-monitoring circuits. Critical factors causing this near-end crosstalk are identified and analyzed.**

Keywords—crosstalk, near-end crosstalk, transceiver crosstalk, ball crosstalk, via crosstalk

I. INTRODUCTION

As data rate and signal density increase for high speed serial link, crosstalk becomes an important factor determining the performance of the system. Crosstalk is induced between all kinds of interconnections on chip, package and PCB. More crosstalk analysis and cancellation techniques for horizontal interconnections have been investigated than vertical interconnections such as ball grid arrays (BGAs), socket pins and vias [1]-[4]. Even though length of vertical interconnections is shorter than horizontal interconnections, this coupling impact is not negligible due to difficulty of control. Unlike coupling of horizontal interconnections such as traces, there are not many techniques to reduce crosstalk on vertical interconnections connecting the package and PCB. Vertical coupling is not induced only at package balls but also socket pins and PCB vias connecting BGAs and PCB. To hook signals up to PCB and go down to the lower PCB layers, pads and vias are necessary. So, placement of pads and vias in PCB is also dependent on package ball assignment. Since ball pitch is fixed, increasing distance between signals is not feasible. Thus, the most frequent way is adding ground balls around important signals so that PCB also has shielding vias around

signals. Nevertheless, it is difficult to achieve perfect shielding on the package ball level. There are three kinds of crosstalk in high speed transceiver application, transmitter to transmitter, receiver to receiver and TX to RX. In this paper, TX to RX coupling is mainly studied and its impact on receiver performance is characterized, since TX to RX coupling impact is more significant than other cases. Some high speed protocols don't run TX and RX simultaneously so that probability of issues due to TX to RX coupling is not high. However, for FPGA, it is common that every TX and RX run individually with different protocols.

In Section II, source of coupling at package balls and PCB vias is summarized depending on wave propagation directions. Measurement setup and the eye monitoring circuit to characterize crosstalk impact are introduced in section III. By using the internal eye monitoring circuit and transceiver toolkit interface, TX to RX coupling impact on RX performance is analyzed depending on RX input signal conditions in section IV.

II. SIGNAL COUPLING AT VERTICAL INTERCONNECTIONS

Fig. 1 (a) shows a part of the package layout for ball grid arrays (BGAs). Each differential pair is surrounded by six ground balls. In many cases, these high speed signals are routed with strip lines on inner layers in PCB. Fig. 1 (b) depicts the perspective view of vertical interconnections including package vias, BGAs, package pads, PCB pads and PCB vias. Even though the signal balls are shielded by six ground balls, it is shown in Fig. 1 (a) that one signal of a differential pair faces diagonally another signal. Therefore, there is field interference between two signals, which induces a common mode noise. Although length of these vertical interconnections is short compared to package or PCB traces, coupling is not negligible depending on signal condition of victims and aggressors.

As shown in Fig. 1 (a) and Fig. 2, there are three coupling cases, TX to TX, RX to RX, and TX to RX. Each case has the exactly same physical dimension. However, coupling impact is different for each case since amount of crosstalk is not determined by only conductor's dimension but also by frequency contents of aggressors and sensitivity of victims.

978-1-4799-0708-3/13 $31.00 © 2013 IEEE

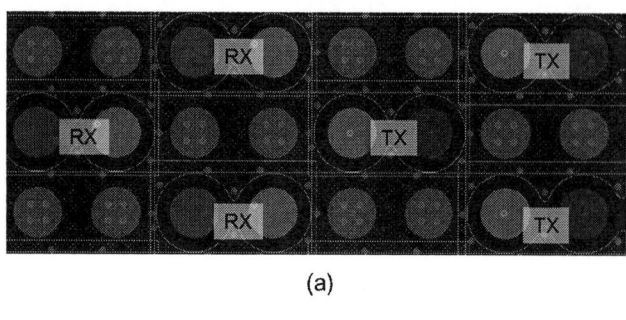

(a)

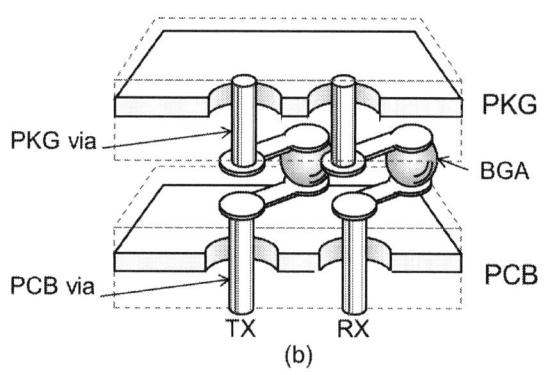

(b)

Fig. 1. (a) The package ball grid arrays (BGAs) on the bottom layer. TX and RX pairs are surroundedby six ground balls. Some of signal balls face each other diagonally. (b) The perspective view of vertical interconnections including package vias, BGAs, pads and PCB vias.

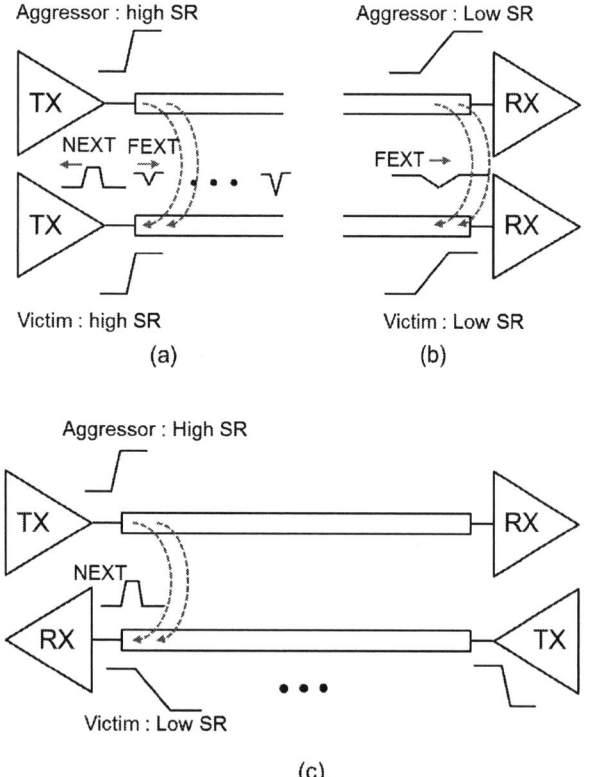

Fig. 2. (a) Transmitter to transmitter coupling (b) Receiver to receiver coupling (c) Transmitter to receiver coupling

For TX to TX coupling case (Fig. 2 (a)), edge of aggressor at TX is still sharp. It means the aggressor contains high frequency contents. What affects victim is far-end crosstalk (FEXT). Since the slew rate of victim TX is high, the sensitivity of victim to FEXT is not significant. Since the edge of the aggressor RX is smoothened through the lossy PCB traces or back plane channels before reaching the PCB vias, RX to RX coupling in Fig. 2(b) is not significant as well. Since the aggressor and victim's signals are aligned for both cases, crosstalk impact is appeared in jitter, which degrades timing margin budget, unless the victim's voltage swing is too small. Unlike RX to RX or TX to TX coupling, near-end crosstalk (NEXT) from TX to RX is the key factor affecting victim (RX)'s performance. Since the excited RX signal's rising time gets stretched by lossy channels, it is sensitive to noise. RX and TX signals are not aligned timing wise. Therefore, the coupling impact is shown in voltage and also timing margin loss depending on victim and aggressor's slew rate, victim's voltage swing and dimension of vertical interconnections.

III. MEASUREMENT SETUP FOR CHARACTERIZATION

There is no way to measure time domain jitter and s-parameter of TX to RX crosstalk. The internal eye-monitoring circuit enables to measure received eye [5][6]. Hence, crosstalk impact on the received signal is able to be observed. Fig. 3 presents measurement setup for TX to RX crosstalk impact on received signal. The aggressor signal is generated by the device (SVGX) whose output swing is controllable from 0V to 1.2V. TX swing is set with 1V and data pattern is PRBS 15 for the crosstalk measurement. For the reference case (no aggressor), output swing is set with 0. The output (12Gbps, PRBS9) of the pattern generator goes into RX through a 60cm SMA cable and 8 cm PCB traces. PCB via length is 1.5mm. To capture eye diagram of received signals, the internal eye monitoring circuit is utilized. The internal eye-monitoring circuit (Eye Mon in Fig. 3) consists of a phase interpolator (PI) and voltage slicer. One unit of the PI is 1/32UI and the unit step of the voltage slicer is 10mV.

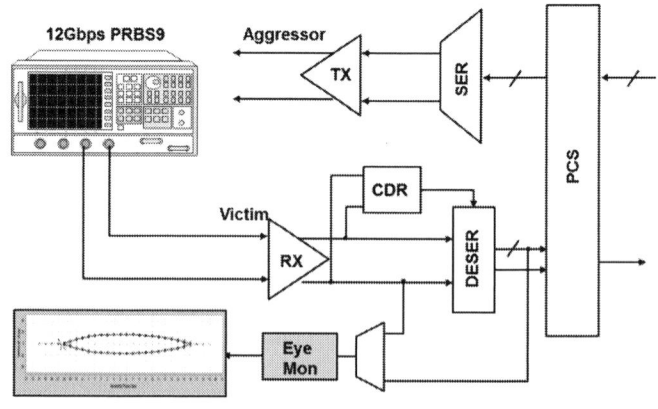

Fig. 3. Measurement setup for characterization of TX to RX crosstalk. TX path is composed with physical coding sub-layer (PCS), serializer (SER) and TX driver. The receiver consists of RX buffers, clock data recovery (CDR) and deserializer (DESER). It is able to monitor before CDR, after CDR and after DESER with the internal eye monitor (Eye Mon)

978-1-4799-0708-3/13 $31.00 © 2013 IEEE

IV. TX TO RX COUPLING IMPACT ON RECEIVED EYE

In this section, TX to RX coupling impact on received eye is presented depending on number of aggressors, victim signal swing, rising/falling time of aggressor, and length of vertical interconnections.

A. Number of Aggressors

TABLE I. shows TX to RX coupling impact on the received eye depending on number of aggressors. The input RX swing is 150mV in this measurement set. It is shown that crosstalk impact doesn't increase even if aggressors are more than two. As shown in Fig. 1(a), one RX victim has maximal two aggressors. Approximately, one TX has 0.04UI (unit interval)/10mV impact with the given package and PCB design (package vias, BGAs, pads and 1.5mm vias).

TABLE I.

TX aggressor	Eye Width	Eye Height
No	0.50 UI	50mV
1 TX (1.2V)	0.47 UI	40mV
2 TX (1.2V)	0.41 UI	30mV
3 TX (1.2V)	0.41 UI	30mV

B. Slew Rate of Victim

Since near-end crosstalk (NEXT) due to TX to RX coupling affects voltage margin, voltage swing of victim RX is a key factor of performance degradation. As increasing input swing of RX with 15ps t_r/t_f, its impact on the eye width is shown in Fig. 4 and TABLE II. Aggressor setup is identical with previous setup. It is clearly shown that the smaller swing is more sensitive to crosstalk. Near-end crosstalk mainly affects voltage margin. Therefore, the bigger signal gets degraded in eye height but not in eye width whereas the smaller signal is degraded in both eye height and width due to low slew rate.

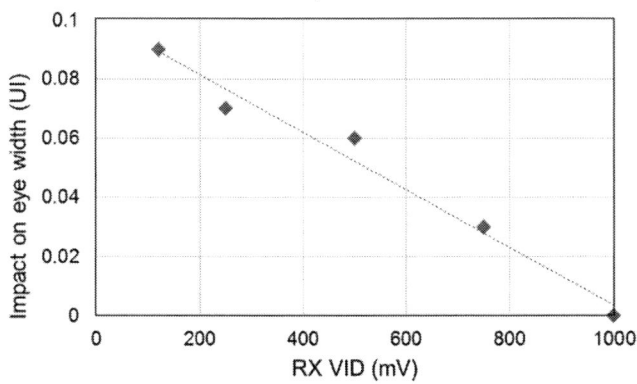

Fig. 4. Near-end crosstalk (NEXT) from TX to RX impacts on the received eye width vs. VID (voltage input differential which is the differential input voltage swing). The diamond points are data and the dotted line is a trend line to show linearity. Impact on eye width means difference of eye width with and without TX aggressors.

TABLE II.

Vin	120mV	250mV	500mV	750mV	1V
Δ Eye height	20mV	20mV	20mV	20mV	20mV
Δ Eye width	0.09UI	0.07UI	0.06UI	0.03UI	0UI

In this experience, all input conditions (from 120mV to 1V) are the same except for voltage swing. t_r/t_f is 15 ps. Smaller signal has lower slew rate, which caused higher sensitivity to crosstalk. Similarly, decreasing rising time reduces crosstalk impact as in TABLE III. These are all about RX slew rate.

TABLE III.

t_r / t_f of Victim	10ps	20ps	30ps
Δ Eye width	0.03UI	0.04UI	0.05UI

C. Rising and Falling Time of Aggressor

From previous experience, it is shown that slew rate of victim is important. It is similar for aggressors. Depending on rising (t_r) / falling (t_f) time, amount of crosstalk becomes different. Mostly, near-end crosstalk's shape is a long pulse (2 times of flight time) as shown in Fig. 5(a). And it is not dependent on t_r/t_f. Normally, NEXT is only depending on coupling ratio and length of interfered conductors.

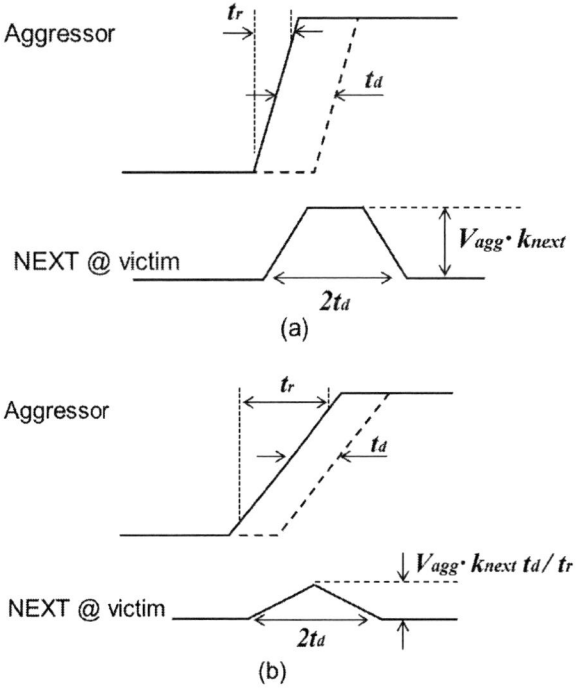

Fig. 5. Near-end crosstalk (NEXT) waveform (a) When t_r/t_f is shorter than flight time (t_d), the shape is a long retagular whose width is double of t_d. (b) When t_r/t_f is longer than t_d. The shape is a triangle. Amplitude is inverse proportional to t_r (or t_f).

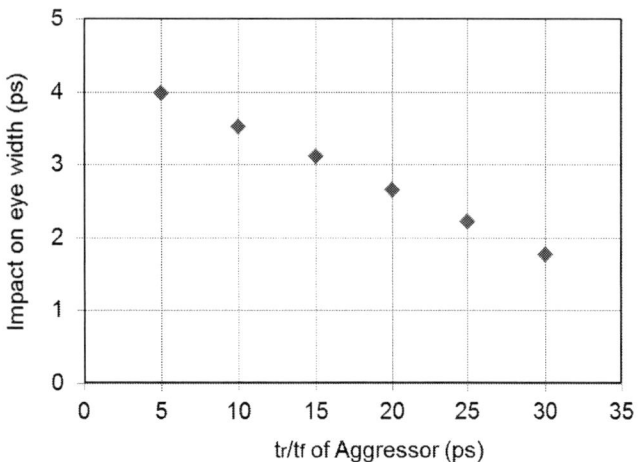

Fig. 6. Simulated TX to RX coupling impact on the received eye width vs. rising/falling time of aggressor.

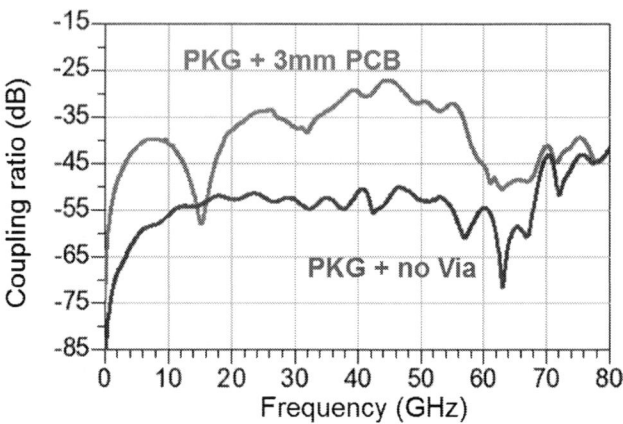

Fig. 7. Coupling ratio of NEXT with 3mm PCB vias and without PCB via.

But since length of vertical interconnection is very short compared to other horizontal interconnection, near-end crosstalk is not able to reach maximum, k_{NEXT} V_{agg}. Unlike typical NEXT waveform, it has a triangle shape as depicted in Fig. 5(b). The height of the vertex of the isosceles triangle is as below:

$$V_{NEXT\_MAX} = k_{NEXT} \cdot V_{agg} \cdot \frac{t_d}{t_r} \qquad (1)$$

k_{NEXT} is the coupling coefficient, V_{agg} is voltage swing of aggressor, t_d is flight time during coupled area (BGAs, pads and vias) and t_r is rising time of aggressor. Since this experience was not able to sweep rising time of aggressor, it has been done by channel simulation and summarized in Fig. 6. The input PRBS 9 signal's swing is 120mV. The aggressor is 1V with PRBS15. As described (1), crosstalk is inverse proportional to t_r of aggressor. It is also shown that its impact on eye width is inverse proportional to t_r.

D. Lengh of Coupled Interconnection

The distances between two balls, pads, or vias are pre-determined depending on the package design. But the length of the coupled area is highly depending on PCB design. If flight time, t_d, is long enough in the most cases of coupling between PCB or package traces, amplitude of NEXT is not affected by flight time as shown in Fig. 5(a), while FEXT is highly dependent on t_d. As presented in (1), crosstalk is proportional to t_d determined by length of coupled area. From eye-monitoring measurement, it is observed that 1.5mm via and 3mm via have 0.07UI difference.

V. CONCLUSION

As data rate increases, crosstalk is one of the major sources degrading system performance. Unlike horizontal coupling, vertical coupling doesn't have enough cancellation method.

Therefore, it is important to understand this vertical crosstalk impact on system performance. From a sort of characterization, it has been demonstrated that near-end crosstalk from transmitter to receiver impact is highly depending on input signal condition. Especially, crosstalk impact could be even worse for long reach application. Unlike horizontal coupling, it has been shown that near-end crosstalk at the short vertical interconnection is dependent on victims signal slew rate, dimension, aggressor signal rising time and flight time.

In order to reduce margin loss due to TX to RX coupling, there are two techniques. One is to reduce coupling itself and another is to cancel by using decision feedback equalizer (DFE). To reduce coupling without cost increase, it is feasible to change signal assignment or via locations so that parallel field interference is minimized by as changing field to orthogonal direction. And DFE improved sensitivity to crosstalk noise as recovering original signal. Effect by these two techniques will be followed up.

REFERENCES

[1] X.Chen, J.E.Schutt-Aine, and A.C.Cangellaris, "A strategy to optimize modal signaling over microstip lines with spacing variablity," *Signal and Power Integiry (SPI), 17th worksghop on*, France, May.2013, pp 1-4

[2] D.Chung, S. Sudhakaran, V.Satagopan, S.Hwang, "Rigorous breakdown of crosstalk in single-ended high speed memory interface," *DesignCon 2012*, Santa Clara, US, Feb. 2012

[3] X.Ye, "Intentional and un-intentional far end crosstalk cancellation in high speed differential link," *Electromagnetic Compatibility(EMC), IEEE international symposium on*, US, Aug.2011, pp791-796

[4] S.H.Hall and H.L.Heck, *"Advanced signal integrity for high speed digital designs,"* Wiley, 2009, pp177-193

[5] M.Shin, J.Shim, J.Kim, J.S.Pak, C.Hwang, C.Yoon, J.Kim, H.Kim, K.Park and Y.Kim, *EMC, IEEE international Sysmposium on*, US, Aug. 2008, pp1-6

[6] Application note, "Using the on-chip siganl quality monitoring circuitry feature in SIV transceivers," http://altera.com

Some Internal Crosstalk Reduction Schemes

Frédéric Broydé, Evelyne Clavelier
Tekcem
Maule, France
e-mail: fredbroyde@tekcem.com, eclavelier@tekcem.com

José E. Schutt-Aïné
University of Illinois at Urbana-Champaign
Urbana, Illinois, U.S.A.
email: jschutt@emlab.illinois.edu

Abstract — **We present and discuss different schemes which can be defined to mitigate internal crosstalk in transposed and untransposed uniform interconnections, and in non-uniform interconnections. We emphasize the unique characteristics of modal transmission schemes.**

I. Introduction

This paper is about a multichannel point-to-point electrical link, shown in Fig. 1, built in a multi-chip module (MCM) or PCB, providing $m \geq 2$ channels from a near-end interface and termination device (NIT) to a far-end interface and termination device (FIT). The link is linear and comprises a multiconductor interconnection having $n \geq m$ transmission conductors (TCs) and a reference conductor or ground conductor (GC). The link may also provide one or more channels from the FIT to the NIT.

We assume a close spacing of the TCs, which causes a significant *TC-to-TC coupling*, which collectively designates mutual capacitance between the TCs and mutual impedance between loops each comprising one of the TCs and the GC. Thus, some analog and/or digital processing of signals present on two or more TCs is necessary to obtain each channel with a sufficiently low *internal crosstalk*, that is to say a sufficiently low interaction with the other channels of the link.

We want to study different schemes which may be defined to mitigate internal crosstalk in transposed and untransposed uniform interconnections, and in non-uniform interconnections. Section II discusses uniformity and transposition. The sections III and IV present the internal crosstalk mitigation problem, and discuss possible solutions. Section V covers modal transmission schemes in transposed and untransposed interconnections.

II. Uniformity and Transposition

We use z to denote the curvilinear abscissa along the interconnection, which extends from $z = 0$ to $z = L$. We shall only consider frequency domain variables. In the framework of multiconductor transmission line (MTL) theory, at a given z, the interconnection is characterized by a per-unit-length (p.u.l.) impedance matrix denoted by $\mathbf{Z}'$, and a p.u.l. admittance matrix, denoted by $\mathbf{Y}'$. From the standpoint of the computation of wave propagation in the interconnection, it is legitimate to use values of $\mathbf{Z}'$ and $\mathbf{Y}'$ which are averaged over a length sufficiently smaller than the shortest wavelength of interest, denoted by λ. From the standpoint of measuring techniques, a sufficient length of the interconnection, compared to this wavelength, is necessary to obtain accurate measurements of $\mathbf{Z}'$ and $\mathbf{Y}'$, which are consequently averaged over the length of the sample under test.

Traditionally, *uniform* means independent of z. However, in this

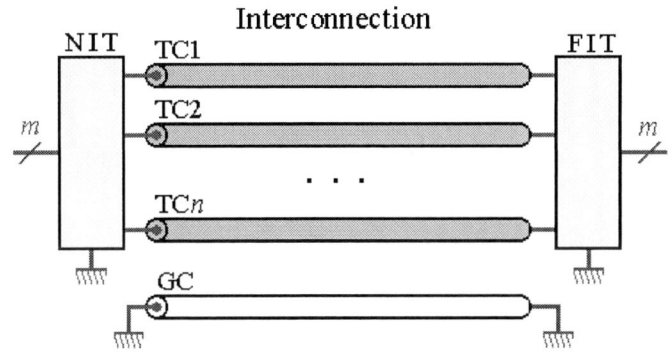

Fig. 1. A point-to-point link providing m channels, consisting of an interconnection, a near-end interface and termination device (NIT) and the far-end interface and termination device (FIT). The interconnection has n transmission conductors (TCs) and a reference conductor (GC).

paper, we shall define a uniform interconnection as being such that $\mathbf{Z}'$ and $\mathbf{Y}'$ are independent of z, after a suitable averaging over a reasonable length, e.g., $\lambda/100$, is taken into account.

At the beginning of the 20th century, engineers knew that an effective method of reducing crosstalk in telegraph and telephone transmission is the use of a separate circuit for each signal to be transmitted, the circuits being substantially perfectly balanced to each other by means of very frequent transposition of the TCs of each circuit [1]. In transposition, permutations of the positions of the TCs, at intervals along the interconnection, are used to simultaneously obtain a balanced interconnection, and a uniform interconnection in the bandwidth of interest.

A perfectly balanced interconnection comprises p pairs such that [2]: the TCs of the same pair have the same *averaged* p.u.l. impedance and p.u.l. admittance with respect to the reference conductor; and the excitation of any pair in differential mode induces no voltage and injects no current in any other conductor. Let us number the TCs of a p-pair interconnection in the following way: for $\alpha \in \{1, ..., p\}$, the TC number $2\alpha - 1$ is the 1st wire of the α-th pair, and the TC number 2α is the 2nd wire of the α-th pair. Let us use $\mathbf{X}'$ to denote any one of the *averaged* natural matrices $\mathbf{Z}'$ or $\mathbf{Y}'$. $\mathbf{X}'$ is of size $2p \times 2p$ and the interconnection is perfectly balanced if and only if

$$\forall \alpha \in \{1, ..., p\} \quad \forall \beta \in \{1, ..., p\}$$

we have

$$X'_{2\alpha-1\,2\alpha-1} = X'_{2\alpha\,2\alpha} \tag{1}$$

and $(\alpha \neq \beta) \Rightarrow \left(X'_{2\alpha-1\,2\beta-1} = X'_{2\alpha-1\,2\beta} = X'_{2\alpha\,2\beta-1} = X'_{2\alpha\,2\beta} \right)$

Particular properties are obtained with a super-balanced

978-1-4799-0708-3/13 $31.00 © 2013 IEEE

Fig. 2. A 2-pair super-balanced interconnection built in a PCB. The two colors used for the traces correspond to different layers.

interconnection, defined as a perfectly balanced interconnection in which any pair can be exchanged with any other pair without changing $\mathbf{Z}'$ or $\mathbf{Y}'$. The interconnection is super balanced if and only if there exist X_A, X_B and X_M such that [2]:

$$\forall \, \alpha \in \{1,...,p\} \quad \forall \, \beta \in \{1,...,p\}$$
$$X_{2\alpha-1,\,2\alpha-1} = X_{2\alpha,\,2\alpha} = X_A$$
$$X_{2\alpha-1,\,2\alpha} = X_{2\alpha,\,2\alpha-1} = X_B$$
$$(\alpha \neq \beta) \Rightarrow \begin{pmatrix} X_{2\alpha-1,\,2\beta-1} = X_{2\alpha-1,\,2\beta} \\ = X_{2\alpha,\,2\beta-1} = X_{2\alpha,\,2\beta} = X_M \end{pmatrix} \quad (2)$$

This corresponds to what Carson and Hoyt called "the ideal telephone transmission system" in 1927 [3, eq. (19)].

Transposition is a current technique used to obtain high-speed balanced interconnection, for instance in UTP category 5e twisted-pair cables for 1000BASE-T local area networks, which comprise 4 pairs and can be used up to 100 MHz over a distance of 100 m. It can also be used in a link built in a PCB or MCM [4] [5], for instance using a super-balanced interconnection such as the one shown in Fig. 2.

III. GENERAL FORMULATION OF INTERNAL CROSSTALK MITIGATION SCHEMES

Let us use $\mathbf{e}_S$ to denote the column vector of the open-circuit voltages applied by the NIT to the interconnection, for the m channels from the NIT to the FIT, and for a column vector of the input signals $\mathbf{x}_{IS}$ applied to these channels. $\mathbf{e}_S$ is of size $n \times 1$, and $\mathbf{x}_{IS}$ is of size $m \times 1$. A *premixing matrix* $\mathbf{A}_S$, of size $n \times m$ and rank m, defines linear combinations of signals performed in the transmitting circuit (TX-circuit) of the NIT, such that

$$\mathbf{e}_S = \mathbf{A}_S \, \mathbf{x}_{IS} \quad (3)$$

As said above, one or more channels from the FIT to the NIT may be present, in which case the link is bidirectional. Thus we need to consider a column vector of the open-circuit voltages applied by the FIT to the interconnection, denoted by $\mathbf{e}_L$, of size $n \times 1$, for a column vector of the input signals $\mathbf{x}_{IL}$, of size $q \times 1$, applied to these channels. A premixing matrix $\mathbf{A}_L$, of size $n \times q$ and rank q, defines linear combinations of signal performed in the TX-circuit of the FIT, such that

$$\mathbf{e}_L = \mathbf{A}_L \, \mathbf{x}_{IL} \quad (4)$$

The column vector of the voltages measured by the FIT is denoted by $\mathbf{v}_L$, and given by

$$\mathbf{v}_L = \mathbf{H}_{LS} \, \mathbf{e}_S + \mathbf{H}_{LL} \, \mathbf{e}_L + \mathbf{n}_L \quad (5)$$

where $\mathbf{H}_{LS}$ and $\mathbf{H}_{LL}$ are matrix transfer functions, both of size

$n \times n$, and $\mathbf{n}$ a noise vector. The FIT determines the column vector of the output signals $\mathbf{x}_{OL}$ of the m channels from the NIT to the FIT, $\mathbf{x}_{OL}$ being of size $m \times 1$ and given by

$$\mathbf{x}_{OL} = \mathbf{B}_L \, (\mathbf{v}_L - \mathbf{D}_L \, \mathbf{e}_L) \quad (6)$$

where $\mathbf{B}_L$ is a *demixing matrix*, of size $m \times n$, which defines linear combination of signals intended to recover the wanted signal sent by the NIT, and where $\mathbf{D}_L$ is a *duplexing matrix*, of size $n \times n$, whose purpose is the reduction (ideally, the cancellation) of the contribution of $\mathbf{x}_{IL}$ to $\mathbf{x}_{OL}$ to allow simultaneous bidirectional (full duplex) transmission. In a context where digital signal processing would be used to implement said linear combinations, $\mathbf{A}_S$ and $\mathbf{A}_L$ could each be referred to as a *coding matrix* or a *precoding matrix*, and $\mathbf{B}_L$ as a *decoding matrix*. We have:

$$\mathbf{x}_{OL} = \mathbf{B}_L \left(\mathbf{H}_{LS} \mathbf{A}_S \mathbf{x}_{IS} + \left[\mathbf{H}_{LL} - \mathbf{D}_L \right] \mathbf{A}_L \mathbf{x}_{IL} + \mathbf{n}_L \right) \quad (7)$$

which describes transmission from the NIT to the FIT. All variables in (7) may be frequency dependent. Thus, the mixing and demixing matrices can in principle provide preemphasis and deemphasis, respectively to obtain a compensation of the frequency dependent losses in the matrix transfer functions.

If the link provides only unidirectional (simplex), or alternate bidirectional (half-duplex) transmission, we may consider that $\mathbf{x}_{IL} = \mathbf{0}$ in (7). Here, the problem of internal crosstalk reduction consists in finding an interconnection structure (which determines $\mathbf{H}_{LS}$), a premixing matrix $\mathbf{A}_S$ and a demixing matrix $\mathbf{B}_S$ such that $\mathbf{A}_S$ and $\mathbf{B}_S$ each corresponds to a causal impulse response matrix, and such that, in the relevant frequency range, for any $\alpha \in \{1,...,m\}$

$$R'_\alpha \left(\mathbf{B}_L \mathbf{H}_{LS} \mathbf{A}_S \right) \leq \varepsilon_\alpha \left| \left[\mathbf{B}_L \mathbf{H}_{LS} \mathbf{A}_S \right]_{\alpha\,\alpha} \right| \quad (8)$$

where ε_α is an arbitrary positive real which defines the maximum allowed signal to crosstalk ratio, where $[\mathbf{M}]_{\alpha\beta}$ is the entry of the row α and column β of a matrix $\mathbf{M}$, and where, if $\mathbf{M}$ is of size $m \times m$, $R'_\alpha(\mathbf{M})$ is the αth deleted absolute row sum of $\mathbf{M}$, given by

$$R'_\alpha (\mathbf{M}) = \sum_{\substack{\beta=1 \\ \beta \neq \alpha}}^{m} \left| \left[\mathbf{M} \right]_{\alpha\,\beta} \right| \quad (9)$$

We see that (8) means that $\mathbf{B}_L \mathbf{H}_{LS} \mathbf{A}_S$ is sufficiently close to a diagonal matrix, and implies a requirement on the location of the eigenvalues of $\mathbf{B}_L \mathbf{H}_{LS} \mathbf{A}_S$, because each $R'_\alpha(\mathbf{B}_L \mathbf{H}_{LS} \mathbf{A}_S)$ is the radius of a Geršgorin disk [6, § 6.1]. If crosstalk cancellation is required, then $\varepsilon_\alpha = 0$ for any $\alpha \in \{1,...,m\}$, so that (8) means that $\mathbf{B}_L \mathbf{H}_{LS} \mathbf{A}_S$ is diagonal.

We are mostly interested in interconnections having a length L which is not very small compared to λ, because they exhibit a higher crosstalk when no crosstalk mitigation technique is used. In this case, $\mathbf{H}_{LS}$ corresponds to a causal time domain response which contains a propagation delay, so that $\mathbf{H}_{LS}^{-1}$ does not correspond to a causal time domain response. Thus, using $\mathbf{1}_n$ to denote the identity matrix of size $n \times n$, we see that, in the case $n = m$:

- $\mathbf{B}_L = \mathbf{1}_n$ and $\mathbf{A}_S = \mathbf{H}_{LS}^{-1}$ is not an acceptable solution of (8);
- $\mathbf{B}_L = \mathbf{H}_{LS}^{-1}$ and $\mathbf{A}_S = \mathbf{1}_n$ is not an acceptable solution of (8).

If the link provides full duplex transmission, internal crosstalk

978-1-4799-0708-3/13 $31.00 © 2013 IEEE

cancellation additionally requires that we find a duplexing matrix $\mathbf{D}_L$ corresponding to a causal impulse response matrix, such that, in the relevant frequency range, for any $\alpha \in \{1,...,m\}$

$$\sum_{\beta=1}^{m} \left| \left[(\mathbf{H}_{LL} - \mathbf{D}_L) \mathbf{A}_L \right]_{\alpha\,\beta} \right| \le \varepsilon_\alpha \left| \left[\mathbf{B}_L \mathbf{H}_{LS} \mathbf{A}_S \right]_{\alpha\,\alpha} \right| \qquad (10)$$

We see that (10) means that $(\mathbf{H}_{LL} - \mathbf{D}_L) \mathbf{A}_L$ is sufficiently close to a null matrix. If crosstalk cancellation is required, (10) means that $(\mathbf{H}_{LL} - \mathbf{D}_L) \mathbf{A}_L$ is a null matrix. In the case $q = n$, crosstalk cancellation is obtained if and only if $\mathbf{D}_L = \mathbf{H}_{LL}$, which of course corresponds to a causal impulse response matrix.

IV. CASE OF A UNIFORM INTERCONNECTION

We now assume that the interconnection is uniform in the meaning of § II. It may be transposed or untransposed. Here, we can compute $\mathbf{H}_{LS}$ and $\mathbf{H}_{LL}$ explicitly. Using [7, eq. 72] we get

$$\begin{aligned} \mathbf{H}_{LS} = \frac{1}{2} \left(\mathbf{1}_n + \mathbf{P}_L \right) e^{-L\mathbf{G}} \\ \times \left(\mathbf{1}_n - \mathbf{P}_S\, e^{-L\mathbf{G}}\, \mathbf{P}_L\, e^{-L\mathbf{G}} \right)^{-1} \left(\mathbf{1}_n - \mathbf{P}_S \right) \end{aligned} \qquad (11)$$

and

$$\begin{aligned} \mathbf{H}_{LL} = \frac{1}{2} \left(\mathbf{1}_n + e^{-L\mathbf{G}}\, \mathbf{P}_S\, e^{-L\mathbf{G}} \right) \\ \times \left(\mathbf{1}_n - \mathbf{P}_L\, e^{-L\mathbf{G}}\, \mathbf{P}_S\, e^{-L\mathbf{G}} \right)^{-1} \left(\mathbf{1}_n - \mathbf{P}_L \right) \end{aligned} \qquad (12)$$

where $\mathbf{G}$ is the lineic propagation matrix given by $\mathbf{G} = \sqrt{\mathbf{Z}'\mathbf{Y}'}$ and where $\mathbf{P}_S$ and $\mathbf{P}_L$ are the matrix of the voltage reflection coefficients at the near end and the matrix of the voltage reflection coefficients at the far end, respectively, given by

$$\mathbf{P}_S = \left(\mathbf{Z}_S - \mathbf{Z}_C \right) \left(\mathbf{Z}_S + \mathbf{Z}_C \right)^{-1} \qquad (13)$$

and

$$\mathbf{P}_L = \left(\mathbf{Z}_L - \mathbf{Z}_C \right) \left(\mathbf{Z}_L + \mathbf{Z}_C \right)^{-1} \qquad (14)$$

where $\mathbf{Z}_S$ and $\mathbf{Z}_L$ are the impedance matrices presented by the NIT and the FIT, respectively, to the interconnection, and where $\mathbf{Z}_C$ is the characteristic impedance matrix given by

$$\mathbf{Z}_C = \sqrt{\mathbf{Z}'\mathbf{Y}'}^{-1} \mathbf{Z}' \qquad (15)$$

The multiple reflection terms, $(\mathbf{1}_n - \mathbf{P}_S\, e^{-L\mathbf{G}}\, \mathbf{P}_L\, e^{-L\mathbf{G}})^{-1}$ in (11) and $(\mathbf{1}_n - \mathbf{P}_L\, e^{-L\mathbf{G}}\, \mathbf{P}_S\, e^{-L\mathbf{G}})^{-1}$ in (12), may cause an impulse response which lasts for a long time, and consequently increase dramatically the cost of the signal processing circuit used to implement $\mathbf{A}_S$, $\mathbf{B}_L$ or $\mathbf{D}_L$. It is therefore always advisable to require that

$$\left|\left|\left| \mathbf{P}_L \right|\right|\right|_\infty \quad \left|\left|\left| \mathbf{P}_S \right|\right|\right|_\infty \quad \left|\left|\left| e^{-L\mathbf{G}} \right|\right|\right|_\infty << 1 \qquad (16)$$

which is easily obtained if the NIR provides reflectionless matching, i.e. $\mathbf{P}_S = \mathbf{0}_{n\,n}$, or if the FIR provides reflectionless matching, i.e. $\mathbf{P}_L = \mathbf{0}_{n\,n}$, where $\mathbf{0}_{n\,n}$ is the null matrix of size $n \times n$.

V. MODAL SIGNALING

Let us now consider a uniform interconnection such that $\mathbf{Z}'\mathbf{Y}'$ is diagonalizable. The transition matrix from modal voltages to natural voltages, denoted by $\mathbf{S}$, is a solution of

$$\mathbf{S}^{-1} \mathbf{Z}'\, \mathbf{Y}'\, \mathbf{S} = \mathbf{\Gamma}^2 \qquad (17)$$

where

$$\mathbf{\Gamma} = \mathrm{diag}_n \left(\gamma_1, ..., \gamma_n \right) \qquad (18)$$

is the diagonal matrix of order n of the propagation constants.

The definition of $\mathbf{S}$ involves multiple choices. Let assume that such choices lead us to obtain $\mathbf{S}$ as a function of frequency. Since $e^{-L\mathbf{G}} = \mathbf{S}\, e^{-L\mathbf{\Gamma}}\, \mathbf{S}^{-1}$, we observe that, if $\mathbf{P}_S = \mathbf{0}_{n\,n}$ and $\mathbf{P}_L = \mathbf{0}_{n\,n}$, or $\mathbf{P}_S = -\mathbf{1}_n$ and $\mathbf{P}_L = \mathbf{0}_{n\,n}$, or $\mathbf{P}_S = \mathbf{0}_{n\,n}$ and $\mathbf{P}_L = \mathbf{1}_n$, then by (11) and (12) we may conclude that $\mathbf{S}^{-1} \mathbf{H}_{LS} \mathbf{S}$ and $\mathbf{S}^{-1} \mathbf{H}_{LL} \mathbf{S}$ are diagonal matrices. This indicates that we could consider using $\mathbf{A}_S = \mathbf{S}$ and $\mathbf{B}_L = \mathbf{S}^{-1}$ to solve (8). However, there is no guarantee that this solution is acceptable, because $\mathbf{S}$ and $\mathbf{S}^{-1}$ need not be frequency-domain descriptions of linear systems having a causal impulse response matrix. In fact, $\mathbf{S}$ and $\mathbf{S}^{-1}$ do not even need to be continuous functions of frequency.

This lead us to the general concept of modal signaling as defined in [8]. In a conventional modal signaling scheme:

■ the interconnection model used for the initial design of the link is a uniform MTL model, referred to as the underlying MTL model, which need not be a perfectly accurate model, for which the p.u.l impedance matrix and the p.u.l impedance matrix will be denoted by $\mathbf{Z}'_U$ and $\mathbf{Y}'_U$, respectively;

■ each of the m transmission channels from the NIT to the FIT is allocated to a modal electrical variable, that is a modal voltage or modal current of the underlying MTL model.

Assuming that $\mathbf{Z}'_U \mathbf{Y}'_U$ is diagonalizable, let us define, for the underlying MTL model, the transition matrix from modal voltages to natural voltages, denoted by $\mathbf{S}_U$ and the transition matrix from modal currents to natural currents, denoted by $\mathbf{T}_U$, as solutions of

$$\begin{cases} \mathbf{T}_U^{-1} \mathbf{Y}'_U \mathbf{Z}'_U\, \mathbf{T}_U = \mathbf{\Gamma}_U^2 \\ \mathbf{S}_U^{-1} \mathbf{Z}'_U\, \mathbf{Y}'_U\, \mathbf{S}_U = \mathbf{\Gamma}_U^2 \end{cases} \qquad (19)$$

where

$$\mathbf{\Gamma}_U = \mathrm{diag}_n \left(\gamma_{U1}, ..., \gamma_{Un} \right) \qquad (20)$$

is the diagonal matrix of order n of the propagation constants, $\mathbf{S}_U$ and $\mathbf{T}_U$ being additionally required to meet the necessary and sufficient condition for total decoupling [7, § 7] [9, § III]

$$\mathbf{S}_U = j\omega\, \mathbf{Y}'^{-1}_U\, \mathbf{T}_U\, \mathbf{c}_{UK} \qquad (21)$$

where $\mathbf{c}_{UK}$ is an arbitrary invertible diagonal matrix, possibly frequency-dependent, and having the dimensions of p.u.l. capacitance. The characteristic impedance matrix of the underlying MTL model, given by

$$\mathbf{Z}_{CU} = \sqrt{\mathbf{Z}'_U \mathbf{Y}'_U}^{-1} \mathbf{Z}'_U \qquad (22)$$

may be used to define the values of $\mathbf{Z}_S$ and $\mathbf{Z}_L$. Total decoupling entails that, for any $i \in \{1,...,n\}$, the propagation of the i-th modal voltage corresponds to the propagation of the i-th modal current, (in line with [7, eq. (48)] or [9, eq. (24)]). We may therefore assume that each of the m transmission channels is allocated to a modal voltage of the underlying MTL model, so that, if $m = n$,

$$\mathbf{A}_S = \left(\mathbf{Z}_S + \mathbf{Z}_{CU} \right) \mathbf{Z}_{CU}^{-1} \mathbf{S}_U\, \mathrm{diag}_n \left(\alpha_1, ..., \alpha_n \right) \qquad (23)$$

and

$$\mathbf{B}_L = \mathrm{diag}_n \left(\beta_1, ..., \beta_n \right) \mathbf{S}_U^{-1} \mathbf{Z}_{CU} \left(\mathbf{Z}_S + \mathbf{Z}_{CU} \right)^{-1} \qquad (24)$$

978-1-4799-0708-3/13 $31.00 © 2013 IEEE

where the α_i and the β_i are arbitrary nonzero possibly frequency dependent parameters. If $m < n$, $\mathbf{A}_S$ and $\mathbf{B}_L$ are submatrices of the matrices given by (23) and (24), respectively. A key aspect of modal signaling is that $\mathbf{A}_S$ and $\mathbf{B}_L$ are independent of L, if the α_i and the β_i are chosen to be independent of L. This simplifies the design of the NIT and FIT.

In a link using a transposed interconnection, the underlying MTL model typically assumes a symmetry which defines $\mathbf{S}_U$. For instance, if the underlying MTL model assumes a super-balanced interconnection, $\mathbf{S}_U$ may be chosen to be a frequency-independent real (FIR) matrix given by $\mathbf{S}_U = \mathbf{A}^{-1}$, where $\mathbf{A}$ is given by [2, eq. 9]. For instance, for the two-pair interconnection shown in Fig. 2, we may use

$$\mathbf{S}_U = \begin{pmatrix} 1/2 & 0 & 1/2 & 1 \\ -1/2 & 0 & 1/2 & 1 \\ 0 & 1/2 & -1/2 & 1 \\ 0 & -1/2 & -1/2 & 1 \end{pmatrix} \qquad (25)$$

Many designs based on a transposed interconnection only use the first p modes, referred to as differential modes, which correspond to the first two-columns in (25). The other p modes are, however, also available for transmission.

In the case of an untransposed interconnection, the underlying MTL model is typically chosen to be an approximate model such that $\mathbf{S}_U$ and $\mathbf{Z}_{CU}$ are FIR matrices, for instance a lossless model, or preferably a model which can accurately take high-frequency losses into account, such as the "fourth MTL" defined in [9, § IV]. If $\mathbf{S}_U$, $\mathbf{Z}_{CU}$ and $\mathbf{Z}_S$ are FIR, and if the α_i and the β_i correspond to causal time domain responses, then $\mathbf{A}_S$ and $\mathbf{B}_L$ correspond to causal time domain responses. Thus, the α_i and the β_i can be chosen to provide preemphasis and deemphasis.

The ZXtalk method [7, § 14] refers to a special case of modal signaling using an untransposed interconnection, in which a non-diagonal $\mathbf{Z}_S$ and/or a non-diagonal $\mathbf{Z}_L$ are used to approximate $\mathbf{Z}_{CU}$, in order to satisfy (16), so as to comply with (8). As an example, we consider a 4-channel point-to-point link in which the 20 mm-long interconnection is the multiconductor microstrip used in [9, § V], which presents resistive and dielectric losses. Here, $m = n = 4$. The link uses the ZXtalk method for simplex transmission, and only the FIT comprises a termination circuit. $\mathbf{A}_S$ and $\mathbf{B}_L$ are real and frequency independent, so that there is no equalization. The termination circuit is made of $2n - 1 = 7$ resistors. The signals at the far-end, computed with the lossy MTL model described in [9, § V], are shown in Fig. 3. We see that crosstalk is not canceled, because of the approximations made in the synthesis of the circuits of the link. However, compared to the single-ended link considered in [9, § V], a reduction of internal crosstalk of about 40 dB has been obtained above 10 GHz, and also a reduction of echo and linear distortions in the channels.

In the general ZXtalk method described above, the number of terms of the linear combinations performed in the NIT and FIT increases as n^2. This becomes a problem for large values of n. A special ZXtalk method [7, § 15] uses an interconnection for which we can use $\mathbf{S}_U = \mathbf{1}_n$, so that this problem is not present.

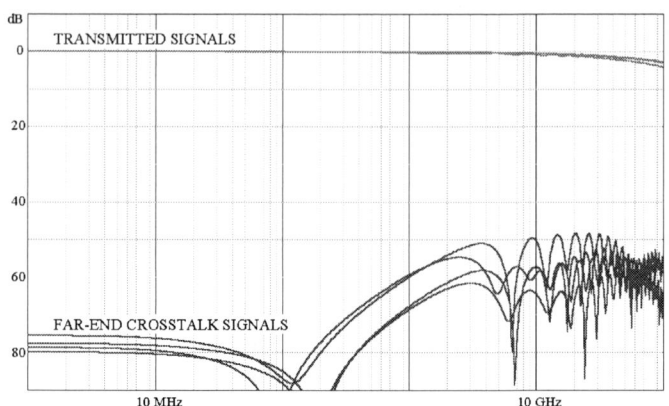

Fig. 3. Attenuations at the far-end when a signal is applied at the near-end of any channel in a ZXtalk link providing $m = 4$ channels. Sixteen curves are plotted, four for the transmitted signals and twelve for the far-end crosstalk signals.

VI. CONCLUSION

A link using an internal crosstalk reduction schemes must satisfy (8). We have stressed that, in addition, the premixing matrix $\mathbf{A}_S$ and a demixing matrix $\mathbf{B}_S$ complying with (8) must each corresponds to a causal time domain response. We have also noted that the cost of the analog or digital processing used to realize $\mathbf{A}_S$ and $\mathbf{B}_S$ is decreased when (16) is satisfied.

These considerations are used to support a view presented in [8]: modal signaling is a special case of noise subtraction where the signal processing requirements defined by (23) and (24) are light, and the ZXtalk technique is a special case of modal signaling in which the signal processing requirements are minimal because non-diagonal $\mathbf{Z}_S$ and/or $\mathbf{Z}_L$ are used to reduce reflections.

REFERENCES

[1] "Dr. Campbell's Memoranda of 1907 and 1912", *Bell System Tech. Jour.*, vol. 14, No. 4, Oct. 1935, pp. 553-572.

[2] F. Broydé, E. Clavelier, "Crosstalk in Balanced Interconnections Used for Differential Signal Transmission", *IEEE Trans. Circuits Syst. I: Regular Papers*, Vol. 54, No. 7, July 2007, pp. 1562-1572.

[3] J.R. Carson, R.S. Hoyt, "Propagation of Periodic Currents over a System of Parallel Wires", *Bell System Tech. Jour.*, vol. 6, No. 3, July 1927, pp. 495-545.

[4] D.G. Kam, H. Lee, J. Kim, J. Kim, "A New Twisted Differential Line Structure on High-Speed Printed Circuit Boards to Enhance Immunity to Crosstalk and External Noise, *IEEE Microwave Wireless Compon. Lett.*, vol. 13, No. 9, pp. 411-413, Sept. 2003.

[5] W.T. Beyene, A. Amirkhany, K. Kaviani, A. Abbasfar, "Design and Analysis of a High-Speed Channel for Coded Differential Signaling", *Proc. IEEE 21st Conference on Electrical Performance of Electronic Packaging and Systems, EPEPS 2012*, Tempe, pp. 7-10, Oct. 21-24, 2012.

[6] R.A. Horn, C.R. Johnson, *Matrix analysis*, Cambridge: Cambridge University Press, 1985.

[7] F. Broydé, E. Clavelier, *Tutorial on Echo and Crosstalk in Printed Circuit Boards and Multi-Chip Modules - Lecture Slides*, Second Edition, Excem, ISBN 978-2-909056-06-7, Feb. 2012.

[8] F. Broydé, E.Clavelier, "An Overview of Modal Transmission Schemes", *Proc. 17th IEEE Workshop on Signal and Power Integrity, SPI 2013*, May 2013, pp. 31-34.

[9] F. Broydé, E. Clavelier, "Multiconductor Transmission Line Models for Modal Transmission Schemes", *IEEE Trans. Components, Packaging and Manufacturing Technology*, Vol. 3, No. 2, pp. 306-314, Feb. 2013.

Crosstalk Mitigation in Dense Microstrip Wiring Using Stubby Lines

San K. Chhay, Richard K. Kunze, and Yunhui Chu
Intel Corporation
{san.k.chhay, richard.k.kunze, yunhui.chu}@intel.com

Abstract— **As the demand for low-cost electronic systems increases, reducing printed circuit board (PCB) layer count becomes highly desirable. But layer count reduction often limits the PCB routing layer options of high-speed input/output (I/O) signals, making it more difficult to meet the next-generation product's performance targets. For example, using densely wired microstrip routing on a PCB helps reduce layer count, but microstrip routing suffers from performance degradation due to increased far-end crosstalk (FEXT). Stub-alternated lines ("stubby lines") have been proposed to reduce or eliminate FEXT. In this paper simulation results are presented showing improved memory bus performance on densely routed microstrip channels implemented with stubby lines.**

Keywords—stubby line; microstrip; printed circuit board; crosstalk

I. INTRODUCTION

As the performance of computers continues to increase, one of the bottlenecks that hinders this trend is the data transfer rate of the signal channel between CPU and memory chips. For the widely used double data rate (DDR) memory bus, which is a single-ended, parallel bus, far-end crosstalk (FEXT) is often the limiting factor for achieving higher data transfer rates. Therefore, crosstalk mitigation may be employed to improve DDR channel performance and hence the speed of computer systems.

The adverse impact of FEXT on high-speed signaling, its expected worsening with higher data rates, as well as a variety of proposed mitigation approaches are described in [1]. The most common approach used to minimize the impact of crosstalk is through avoidance, i.e., at the expense of wiring density, e.g., by increasing the separation between traces or by increasing the separation of vias and/or adding ground vias. Newer approaches that preserve or even increase wiring density, like stub-alternated ("stubby") lines, have also been proposed [2]. In this work we examine the impact of stubby lines on DDR channel performance.

As seen in Figure 1, stubby microstrip wiring consists of short rectangular stubs added to the edges of the wire, orthogonal to the direction of propagation. These stubs effectively increase the mutual capacitance between the lines without significantly increasing the mutual inductance, and accordingly allow the designer to mitigate FEXT ([2]-[5]). The final difference between the mode velocities, together with the length of the stubby line section, determine the magnitude and polarity of the FEXT signal observed at the end of the line.

There are several design variables associated with stubby lines, e.g., line spacing, stub dimensions, number of stubs per unit length, and the separation of the stubs from each other and the adjacent line. The allowed range for these design variables is limited primarily by PCB manufacturing capability, which typically can provide a usefully large range of final FEXT values and wiring densities. Note that, in most practical cases, the trace width is also reduced from its normal nominal value when stubs are added in order to maintain the target impedance of the line.

While stubby lines can be used to simply compensate for FEXT that would otherwise be generated in a section of simple microstrip wiring, it can also be designed to compensate for crosstalk generated in other sections of the channel [5]. Effective use of the method, particularly for the latter case, requires careful channel analysis starting with a realistic assessment of the relative importance of crosstalk to the particular channel's margins. For example, stubby line wiring may not be appropriate for channels operating at slower transfer rates or when the wiring lengths are short.

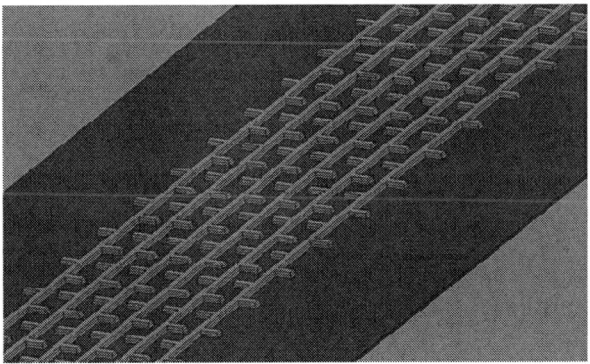

Figure 1: Stubby lines

For simplicity, the introductory discussion below is based on a single pair of traces that illustrate how stubby lines can mitigate FEXT. However, it still provides a good guideline for a general multi-conductor microstrip transmission line system since the major FEXT contribution is from the aggressor lines immediately adjacent to the victim line.

For a pair of traces, as a result of the difference between even and odd mode propagation delays, the FEXT pulse

978-1-4799-0708-3/13 $31.00 © 2013 IEEE

amplitude, $V_f(t)$, on a victim line can be described with the following formula[2]:

$$V_f(t) = \frac{t_f}{2}\left(\frac{C_m}{C_s} - \frac{L_m}{L_s}\right)\frac{dV_i(t-t_f)}{dt} \qquad (1)$$

Here t_f is the time of flight, V_i is the aggressor input voltage, C_m, C_s, L_m and L_s are the mutual capacitance, self-capacitance, mutual inductance, and self-inductance, respectively, per unit length. This equation is valid for times, t, before FEXT saturation occurs. Note that the FEXT amplitude is proportional to the difference between L_m/L_s and C_m/C_s; when the two are equal, FEXT is effectively 0. In such a case even and odd mode propagation delays are equal. The addition of stubs effectively increases C_m / C_s relative to L_m / L_s between circuit traces and thus modulates the crosstalk amplitude.

Microstrip wiring suffers from a disproportion between L_m / L_s and C_m / C_s caused by the inhomogeneous dielectric formed from the boundary between a board's dielectric material and air. This typically results in $C_m / C_s < L_m / L_s$ which, in turn, leads to a crosstalk pulse opposite in sign to that of the rate of change of the aggressor input voltage. Previous work [2-4] demonstrates various wiring methods to equalize the terms, L_m / L_s and C_m / C_s and thus avoid generating FEXT in a microstrip wiring section. One approach [5] focuses on adding enough stubs to increase C_m/C_s sufficiently to allow it to dominate over L_m / L_s, thus *reversing* the polarity of the far-end crosstalk signal. An example of this technique from a test board structure is shown in Figure 2. Here an approximately 0.8inch long segment of strongly coupled stubby line effectively cancels the far end crosstalk generated along the entire 8inch long coupled transmission line pair. Figure 3 is the lab measurement results showing the near perfect cancellation of FEXT for this pair of coupled lines.

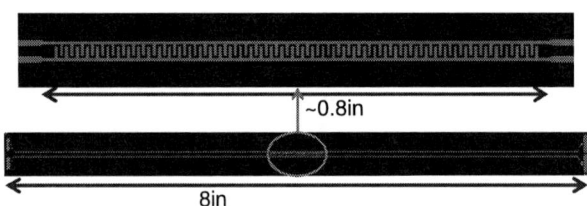

Figure 2: Test board layout for far-end crosstalk amplitude reduction using a short segment of strongly coupled stubby lines. Note that the crosstalk compensating stubby section can generally be placed anywhere along the line; here, it is placed near the center.

II. STUBBY LINE DESIGN METHODOLOGY

As mentioned in the previous section, there are two types of stubby line design, depending on the application. Type (I)

implements a segment of stubby line signal trace that generates zero/minimum FEXT. Such a design does not mitigate crosstalk generated elsewhere in the channel. Additionally, this type of design is scalable in length, that is, once the stub design parameters are established, it can be used for any routing length. Type (II) implements a segment of stubby line trace that minimizes or cancels out the FEXT of the entire channel by generating a FEXT signal of opposite polarity. The routing length of the stubby line segment needs to be optimized for the specific channel.

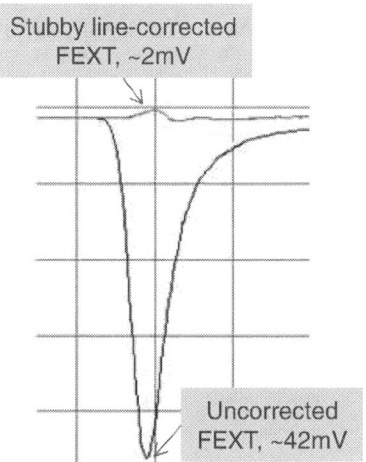

Figure 3: Measurement results showing net FEXT amplitude reduction due to stubby line compensation

The design flow for Type (I) is shown in Figure 4. The entire flow is focused on the segment that will be routed as stubby line. It should be noted that equation (1) is based on quasi-static assumption, so the actual performance of the stubby line over a broad frequency range needs to be validated. Therefore, time-domain simulation is necessary to validate the design and design retuning should be done as needed.

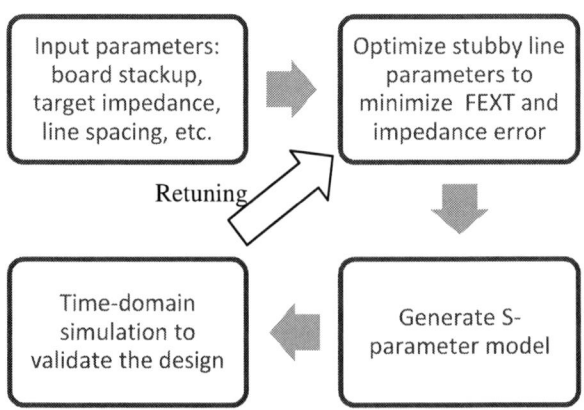

Figure 4: Design flow for Type (I) stubby line

The design flow for Type (II) is described in Figure 5. For this type of design, first the FEXT of the entire channel without stubby line segments is established to provide

978-1-4799-0708-3/13 $31.00 © 2013 IEEE

baseline results. The stubby line segments are then designed to generate the proper amount of FEXT of opposite polarity to cancel out the FEXT generated by the entire channel. The FEXT amplitude of opposite polarity also depends on the length of the stubby line, which is part of the optimization to match the baseline FEXT amount. Length optimization is usually preferred over retuning the stubby line design because the former is easier to do. However, sometimes retuning the stubby line design is required for impedance matching to the rest of the channel.

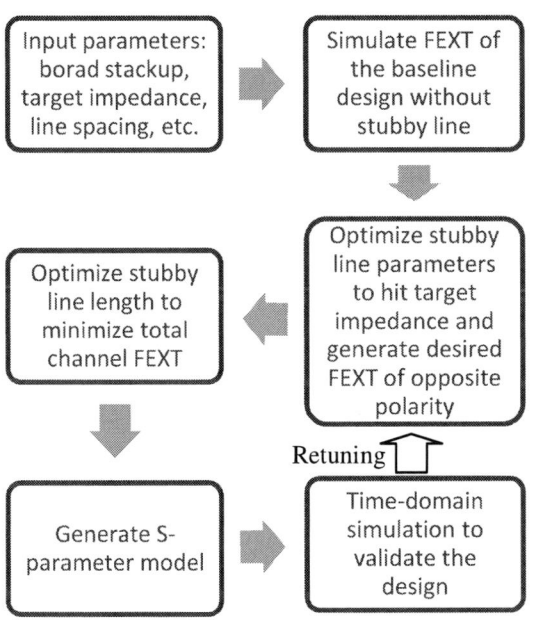

Figure 5: Design flow for stubby line that minimizes channel FEXT.

III. DDR BUS SIMULATION RESULTS

The signal channel of a typical memory interface is illustrated in Figure 6. The channel starts from the CPU package, followed by several segments of microstrip traces on the mother board including breakout, main routing, and DIMM (dual in-line memory module) field routing, which are modeled as non-ideal mutliconductor transmission lines, followed by the DIMM connectors and the DIMM card. For transmission line segments, 10-line models are used to take into account crosstalk within the byte lane and from the adjacent byte lane as well.

In the results below, two cases are compared: Case I is the regular microstrip routing and Case II is identical to Case I except that the open-field routing is composed of Type (I) stubby lines. To demonstrate the crosstalk reduction impact of the stubby lines, we first look at the time-domain FEXT voltage waveforms as shown in Figure 7. The waveform is measured on the victim line at the

receiving end while the excitation, which is a unit step function with finite slew rate, is at transmitting end of the aggressor line next to the victim.[1] Approximately 38mV reduction in p-p crosstalk voltage for the stubby line wiring (yellow) is achieved compared to the normal microstrip wiring (blue).

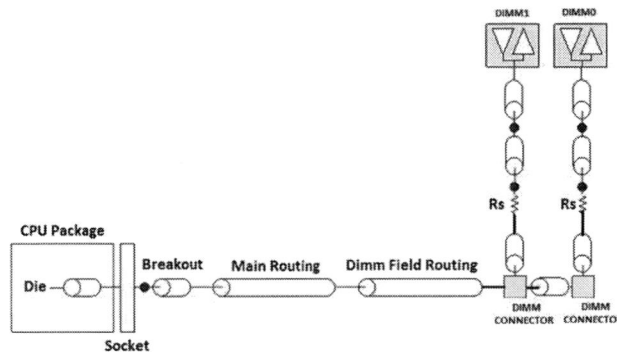

Figure 6: DDR channel model topology used in this study. Channel performance was compared for cases of the "Main Routing" microstrip constructed with and without stubby lines.

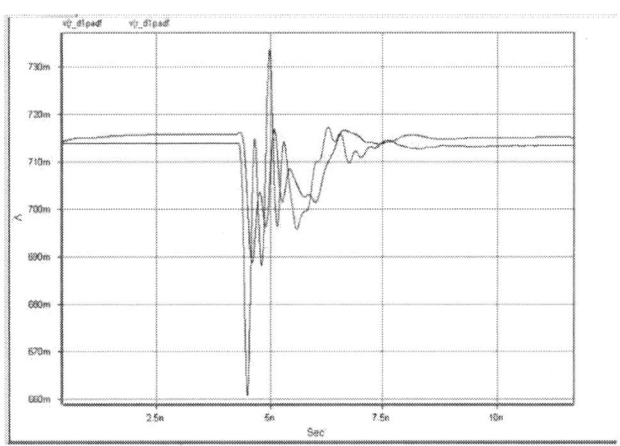

Figure 7: FEXT voltage waveforms at the receiving end. Blue: stubby line wiring; Yellow: normal microstrip wiring.

Next we compare the eye diagrams of the channel for the two cases. The eye diagrams in Figure 8 are obtained with fixed channel parameters, i.e., assuming all channel parameters, such as trace width, length, dielectric thickness in board stack-up, etc., take fixed values and there is zero manufacture/design variation. By using stubby line, the eye height (EH) is improved from 114mV to 140mV and the eye width (EW) from 294ps to 316ps. This data clearly demonstrates the improvement in overall channel performance just by replacing the open field normal microstrip routing with stubby line routing.

---

[1] The excitation and observation locations in the actual simulation are switched by utilizing the reciprocal principle.

Finally, we use one-million-case Monte Carlo simulation to take into account the manufacture/design variations for channel parameters and obtain the distribution of channel eye height and width. As shown in Figure 9, again we see obvious improvement from using stubby lines. For eye height, the mean value is improved from 35 to 66mV and standard deviation from 8.8 to 6.7mV. For eye width, the mean value is improved from 63ps to 86ps and standard deviation from 18ps to 11ps.

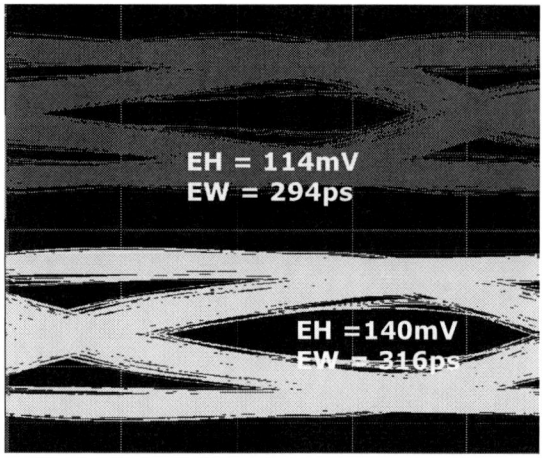

Figure 8: Eye diagram comparison. Red: normal microstrip wiring. Blue: stubby line wiring.

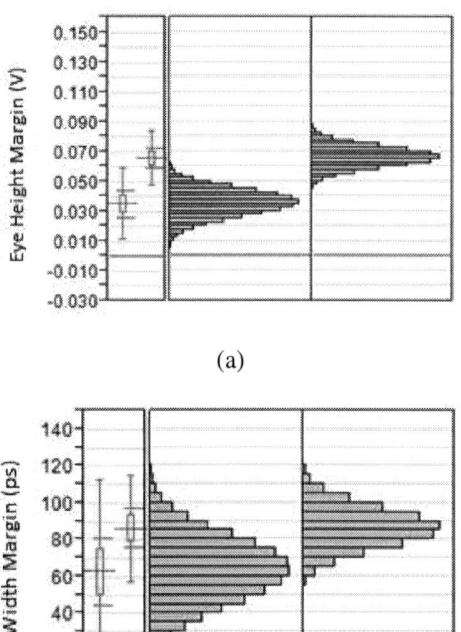

(a)

(b)

Figure 9: Eye height (a) and eye width (b) distributions. For both (a) and (b), the left-hand side is for normal microstrip routing and the right-hand side is for stubby line routing.

IV. SUMMARY

As demonstrated by both simulation and measurement, FEXT generated in closely spaced microstrip wiring can be effectively mitigated through the use of stubby lines. A comprehensive system level analysis predicts substantial channel performance improvement when simple microstrip wiring is replaced with stubby line structures.

REFERENCES

[1] B. Lee, X. Ye, R. Enriquez, K. Xiao, T. Ballou, and J. Johansson, "Design optimization for minimal crosstalk in differential interconnect," DesignCon 2012, February, 2012.

[2] S.-K. Lee, K. Lee, H.-J. Park, and J.-Y. Sim, "FEXT-eliminated stub-alternated microstrip line for multi-gigabit/second parallel links," Electronic Letters, February, 2008.

[3] I. Novak, B. Eged, and L. Hatvani "Measurement and simulation of crosstalk reduction by discrete discontinuities along coupled PCB traces," IEEE Trans. Instrumentation and Measurement, March/April, 1994.

[4] B. Eged, F. Mernyei, I Novak, and P. Bajor, "Reduction of far-end crosstalk on coupled microstrip PCB interconnects," IEEE IMTC/94, May, 1994.

[5] J. Abbott and R. Kunze, "Micro-Strip Crosstalk Compensation Using Stubs," U.S. patent application, June, 2009.

Simulation of the TSV-to-Device Coupling in 3D ICs for Short-Channel Strained Silicon Transistors

Krishnamurthy Yeleswarapu, Amit Ranjan Trivedi, and Saibal Mukhopadhyay

School of Electrical and Computer Engineering, Georgia Institute of Technology

Phone:+1.404.894.2688, Fax:+1.404.894.4641, E-mail: <krishnayeleswarapu,atrivedi31>@gatech.edu, saibal@ece.gatech.edu

Abstract—This paper analyzes Through-Silicon-Via (TSV)-to-device coupling due to the mechanical stress and the electrical field using three dimensional process and device simulation. The analysis considering 40nm and 28nm transistor demonstrates that TSV only has a minor impact on the current of short-channel devices and the effect diminishes with technology scaling.

I. INTRODUCTION

Three-dimensional (3D) integration of ICs using Through-Silicon-Vias (TSVs) has emerged as a promising technology. However, the effect of neighboring TSVs on the electrical parameters of a transistor, referred to as the TSV-to-device coupling, is a key concern for development of 3D ICs. The TSV-to-device coupling can originate from the mechanical stress caused by the mismatch in the coefficient of thermal expansion (CTE) between the TSV and surrounding material. The induced stress could impact the electrical characteristics (mobility and threshold voltage) of neighboring transistors (stress coupling) [1-7]. A second source of TSV-to-device coupling is the electric field from TSVs that can modulate the electrostatics of the transistors (field coupling) [8-11]. The TSV-to-device coupling is a key factor in deciding the Keep-Out-Zone (KOZ) around a TSV. A smaller coupling could reduce KOZ and hence, design footprint and wirelength [1].

The impact of TSV induced stress on transistors has been investigated and methods have been developed to minimize the effect [1-5]. However, recent experimental studies have observed that the TSV induced stress has a minimal impact on the characteristics of short-channel transistors [6-7]. The difference between earlier simulation and recent experimental results suggest the need for detailed analysis. Simulations have also been performed to characterize the electric coupling between TSV and devices [8-11].

This paper presents process and device co-simulation using Technology CAD (TCAD) considering TSV and transistors simultaneously. The paper studies the TSV-to-device coupling effects across technology nodes, and considers a single TSV as well as a regular TSV grid. The transistors and MOSFETs are simulated together to correctly capture the TSV-to-device coupling. For stress coupling, the coupled simulation accounts for all stress sources in the MOSFET channel including the engineered stress. This is an important difference from previous simulation as beyond 90nm nodes, stress induced by Shallow-Trench-Isolation (STI) and stress liners are used to enhance mobility. The second key difference is that our simulation considers both the stress and the electric field coupling. The simulations show that on, off, and linear

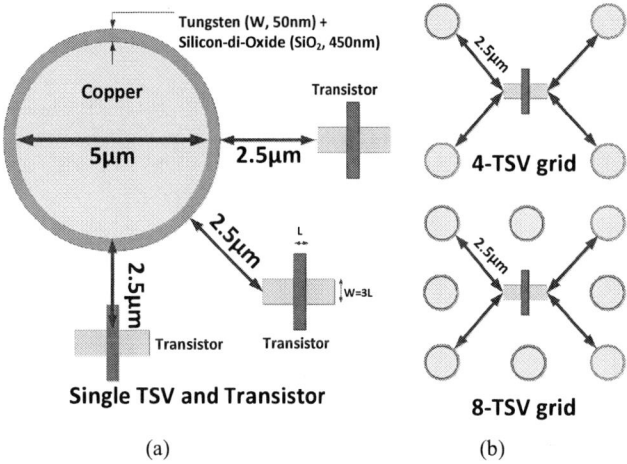

Fig. 1: The 3D TCAD simulation for TSV-to-device coupling: (a) Cartoon of the simulated system: NMOS/PMOS transistor with a TSV at 0°, 45° and 90°; (b) A 28 nm transistor surrounded by four and eight TSVs. The TSV is 5 μm in diameter, and is surrounded by 50 nm of W and 450 nm of SiO_2. The transistor channel has not been drawn to scale;

current of a 28nm transistor changes only by 3-4%, 7-8%, and 4-5%, respectively even when placed at the center of a grid of 8TSVs (diameter 6 μm). Moreover, the impact of the coupling reduces significantly with technology scaling.

II. SIMULATION METHODOLOGY

Three dimensional Technology CAD (TCAD) simulations are performed to analyze the effect of TSV induced stress on bulk-CMOS transistors. For understanding the impact of strain/stress on NMOS and PMOS transistors, we simulated three different orientations between TSV and the NMOS and PMOS device (Fig. 1). First a single TSV and transistor is simulated. Next, simulations are performed to quantify the change in currents of an NMOS/PMOS transistor when placed inside a grid of TSVs [6]. The multi-grid simulations are performed for two different cases: 4-TSV grid i.e. a transistor is surrounded by four TSVs, and 8-TSV grid i.e. a transistor is surrounded by eight TSVs (Fig. 1b).

The via-middle fabrication process steps are simulated through Sentaurus Process [12] to generate transistors and TSVs. The Via-middle process follows fabrication of the transistors followed by the TSVs and then the back-end-of-line (BEOL) designs. TSVs are generated following the Bosch DRIE process, where copper TSVs are coated with thin tungsten layer (~50nm), and thicker SiO_2 liner (~450nm). This arrangement reduces the CTE mismatch, and suppresses the TSV induced stress. The annealing temperature used is 250

978-1-4799-0708-3/13 $31.00 © 2013 IEEE

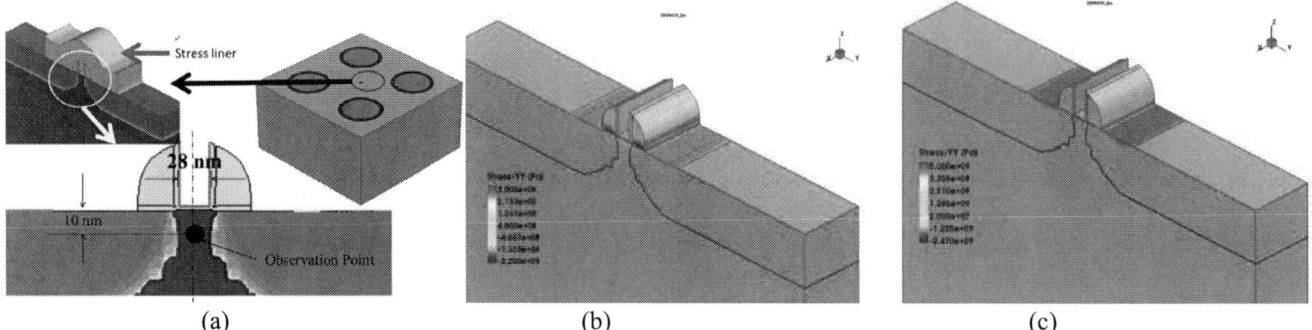

| | (a) | (b) | (c) |

Figure 2: The 3D TCAD simulation for TSV-to-device stress coupling: (a) the 3D simulation structure showing the TSVs and the transistors, (b) stress distribution for the 28 nm NMOS without any TSV; and (c) stress distribution for the 28 nm NMOS with a single TSV to the left. The tensile strain is higher at the edges of the channel due to the nitride being closer to it than in the region below the gate. The green and light yellow regions show the tensile stress being exerted by the stress liner. The stress at a depth of 10 nm below the surface is approx. 1.42 GPa, and the TSV adds around 140 MPa for a total of 1.56 GPa at the same point.

Table 1: Characteristics of the 28 nm NMOS and PMOS transistors obtained using TCAD simulation.

	28 nm NMOS	28 nm PMOS
I_{on}	749.7 µA/µm	464.8 µA/µm
I_{off}	2.994 nA/µm	3.008 nA/µm
I_{lin}	123.9 µA/µm	76.9 µA/µm
Sub V_T Slope	85 mV/decade	88 mV/decade

°C, and the temperature is then ramped down to 25 °C.

The transistors are calibrated to match the ITRS specifications for low operating power (LOP) technology [13]. High-k dielectric HfO_2 with SiO_2 as interfacial layer was utilized to enhance gate electrostatics [14]. Halos with retrograde doping profile are considered to control the short channel effects. The tensile and compressive nitride liners are used to engineer NFET and PFET channel stress, respectively. Nitride layers of thickness ~75nm are used, and the stress relaxation time is 100ps at a temperature of 600 °C.

Electrical characteristics of transistors are simulated through Sentaurus device [15]. TSVs are biased at logic '0' or logic '1' to account for the electrical field coupling between TSV and device. For an NMOS device, when the TSV is biased at 1V, the field originated from the TSV can cause (marginal) forward body-bias, resulting in reduced V_{th}. Similarly for PMOS device, TSV=0V results in marginal forward body bias. Hence, the V_{th} of NMOS and PMOS are expected to reduce with TSV=1V and 0V, respectively. Stress induced electron mobility modification is simulated through models in [15]. For hole transport, Intel's stress-induced hole mobility model [16] was used. The effect of stress on the threshold voltage is accounted for in the simulation, an increase in stress reduces the threshold voltage.

III. Simulation Results

Table 1 contains the key electrical characteristics of the baseline 28nm NMOS/PMOS without any neighboring TSVs. Fig. 2a shows the strain/stress distribution for the baseline 40nm NMOS transistor. The strain is greater at the edges of

Table 2: Percentage changes in I_{on}, I_{off} and I_{lin} for a 28 nm NMOS transistor with STI + stress liner due to a neighboring TSV.

	ΔI_{on} (%)	ΔI_{off} (%)	ΔI_{lin} (%)
TSV at 0° w.r.t. the NMOS channel			
TSV at 0V	1.42	3.73	2.62
TSV at 1V	2.42	5.37	3.74
TSV at 45° w.r.t. the NMOS channel			
TSV at 0V	2.11	4.64	3.31
TSV at 1V	3.05	6.37	4.40
TSV at 90° w.r.t. the NMOS channel			
TSV at 0V	0.93	2.86	1.77
TSV at 1V	1.37	4.18	2.69

Table 3: Percentage changes in I_{on}, I_{off} and I_{lin} for a 28 nm PMOS transistor with STI + stress liner due to a neighboring TSV.

	ΔI_{on} (%)	ΔI_{off} (%)	ΔI_{lin} (%)
TSV at 0° w.r.t. the PMOS channel			
TSV at 0V	1.76	5.28	3.34
TSV at 1V	1.24	3.51	2.35
TSV at 45° w.r.t. the PMOS channel			
TSV at 0V	2.96	8.41	5.71
TSV at 1V	2.42	6.26	4.68
TSV at 90° w.r.t. the PMOS channel			
TSV at 0V	2.00	5.81	3.82
TSV at 1V	1.46	4.05	3.01

the channel, where the nitride liner is very close to the source and drain regions, separated only by the oxide. The strain is comparatively lower in the center, where the thickness of the gate results in greater separation between nitride liner and channel. Fig. 2b shows the stress/strain pattern for the 28 nm NMOS transistor with a 6 µm diameter TSV at 2.5 µm to its left (Fig. 1a). The stress/strain pattern is similar to the one before, except a marginal increase in the tensile strain (by 100 – 200 MPa) in and around the channel. The analysis of the electrostatic potential at the center of the channel showed that with TSV at 1V, the substrate potential close to the channel increases marginally, resulting in a reduced threshold voltage.

978-1-4799-0708-3/13 $31.00 © 2013 IEEE

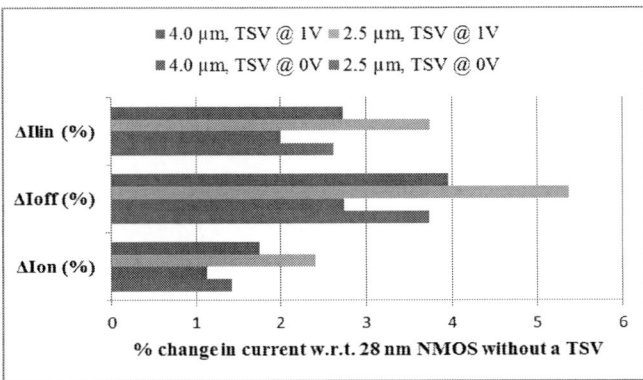

Figure 3: The changes in I_{on}, I_{off} and I_{lin} as a function of the TSV-to-device distance for an 28nm NMOS

A. Coupling between a transistor and single TSV

Table 2 summarizes the simulated effect of the TSV-to-device coupling for the 28nm NMOS. The percentage changes obtained are close to recent experimental results [3][4][5][17]. First consider the effect of stress at TSV=0V (i.e. no electric field coupling). With the TSV at 0V, the change in I_{on} was found to be within 2% for all three angles of the TSV to the transistor. This result is consistent with the observed increase (~150 MPa) in tensile strain at the center of the channel. The changes in I_{off} and I_{lin} are of a similar nature. The effect of additional TSV induced stress is higher for the linear current as mobility has a stronger impact on the linear current. The change in I_{off} is observed to be higher due to its exponential dependence on the threshold voltage. A higher strain reduces the threshold voltage. The change in current is highest when the TSV is at 45° to the transistor, and least when the TSV is

Table 4: Changes in I_{on}, I_{off} and I_{lin} for a 28 nm NMOS/PMOS transistor surrounded by 4 and 8 TSVs.

	TSV Grid	ΔI_{on}	ΔI_{off}	ΔI_{lin}
NMOS	4 TSVs @ 0V	2.8%	7.11%	4.88%
	8 TSVs @ 0V	3.77%	8.82%	5.14%
PMOS	4 TSVs @ 1V	-2.24%	-6.93%	-4.14%
	8 TSVs @ 1V	-2.63%	-8.02%	-4.81%

at 90° to the transistor, which is consistent with results obtained by West et al [7]. The difference in changes in current between the TSV at 0° and 90° is because of a greater change in electron effective mass when uniaxial strain is applied along the <110> direction (same as the channel), as predicted by empirical non-local pseudopotential theory [18]. For the 28 nm PMOS transistor with TSV=1V, the results (summarized in table 3) show a similar trend: I_{off} changes the most (4-6%), I_{on} changes the least (1-2%), and I_{lin} change is in between (3-4%). For the PMOS, the TSV at 90° results in the largest change in current. This is because TSV at 90° causes the largest change in the hole effective mass for same change in the stress.

The effect of combined stress and field coupling is evident from the results with TSV=1V for NMOS and TSV=0V for PMOS. With TSV=1V the field coupling marginally increases the body voltage for the NMOS device, resulting in a higher change in the current. The differential change due to TSV potential is comparable to the effect of the stress. A similar observation can be made for the PMOS device as well with the TSV at 0V. We assume that the TSV is placed inside the same n-well as the PMOS for this study.

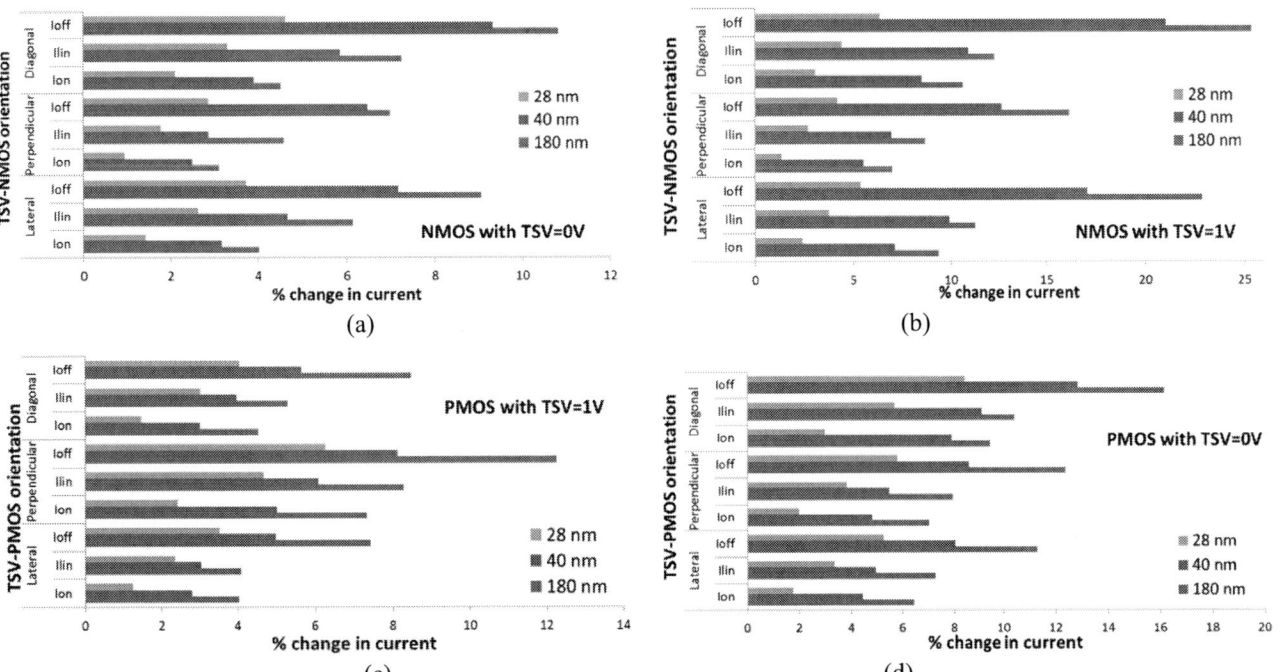

Figure 4: The effect of technology scaling on TSV-to-device coupling: (a) NMOS with TSV at 0V, (b) NMOS with TSV=1V, (c) PMOS with TSV=1V, and (d) PMOS with TSV=0V. The changes are computed with respect to the corresponding 180nm, 40nm and 28nm transistors, each with STI and nitride stress liner, but without any neighboring TSV.

We further study the effect of TSV voltage and TSV-to-device distance for the 28 nm NMOS and PMOS transistors. As expected, a higher TSV-to-device distance reduces the change in the current for both voltage levels of the TSVs. This is because both the field and the stress coupling reduce with an increase in distance. Figure 3 illustrates the observation for NMOS device and similar observation is made for PMOS.

B. Impact due to a grid of TSVs

We next study the effect of inserting a transistor inside an array of TSVs arranged as grid (Fig. 1b). The changes are computed with respect to a single 28 nm NMOS/PMOS transistor without a TSV and presented in Table 4. For the NMOS, the worst changes in on and off currents are ~ 4% and 9% respectively, which is only marginally higher than the change due to a single TSV. The results are presented assuming all TSV are at 0V for NMOS and at 1V for PMOS. The same is the case for the PMOS, with the worst on and off current changes being ~ 3% and 8% respectively. The reason for the smaller additional change when compared to the change caused by a single TSV is the destructive interference of stress by TSVs on opposite sides of the transistor, causing only a small change in the stress in the channel of the transistor due to the cumulative effect of all four/eight TSVs.

C. Effect of Technology Scaling

This section studies the effect of technology scaling on the TSV-to-device coupling. The 3D TCAD simulations are performed considering 40nm and 180nm NMOS/PMOS devices along with the 28nm device presented earlier. For all the cases the TSV diameter and the TSV-to-device distance is kept constant. This helps characterizing the effect of scaling the transistor technology only, decoupling it from scaling trends in TSVs. Fig. 4a shows the change in the I_{on}, I_{off} and I_{lin} for the NMOS transistors with the TSV at 0°, 45° and 90° considering TSV=0V. The figure clearly shows that the effect of TSV induced current change reduces appreciably with technology scaling. For example, in the 28nm device the maximum change in I_{on} is reduced to 2% compared to 4% in the 40nm device. The primary factor is the higher engineered stress in the channel of the device due to the thickness and extent of the nitride layer used to introduce tensile strain or compressive stress. The doping concentration and profile and the lower channel depth also contribute to the higher initial channel strain/stress. The effect of scaling is strongest in the change in I_{off}. Figure 4b shows the effect of scaling including the field coupling i.e. with TSV=1V for NMOS. The generic trend of Fig. 4b is repeated here as well. Fig. 5 shows the effect of scaling on the additional changes in the current between TSV=1V and TSV=0V conditions i.e. the changes due to the field coupling. The results indicate that the effect of the field coupling also reduces with technology scaling. This is attributed to the fact that the electric field inside channel (gate induced vertical field and source/drain induced lateral field) is higher at scaled technologies. Hence, perturbation due to additional TSV induced electric field is much smaller. Fig.

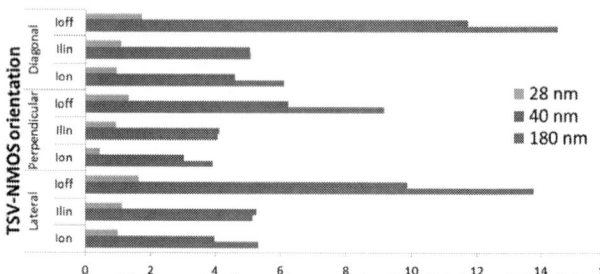

Figure 5: The difference in the percentage change in NMOS currents between TSV=0V and TSV=1V due to the effect of field coupling.

4(c) and 4(d) shows the effect of scaling on TSV-to-PMOS coupling for various orientations. The change in I_{on}, I_{off} and I_{lin} reduces with scaling for the PMOS transistors as well (Fig. 4c and 4d). As in case of the NMOS, the difference between the changes due to TSV=0V and TSV=1V reduces as well. .

IV. CONCLUSION

This paper studies the effect of stress and field coupling between TSVs and devices through 3D process-device co-simulation. TSVs introduce additional tensile strain in the transistor channel. However, the magnitude of the induced strain is much smaller compared to the tensile (compressive) nitride liner used to improve NMOS (PMOS) performance in nanometer technologies. Likewise the TSV induced electric field also modifies the electrostatics inside the transistor channel and modifies its electrical characteristics. However, the TSV induced field is much smaller than the internal electric field in the channel, particularly in the nanometer technologies. Hence, the TSV-to-device coupling has only a marginal impact on the electrical characteristics on neighboring transistors in scaled technologies. The analysis suggests that with scaling the transistors can be placed much closer to a TSV, providing opportunities for optimization of the Keep-Out-Zone and hence, design performance.

Acknowledgement: This work is based on material supported in part by National Science Foundation (CCF-0917000 and CNS-1218745).

REFERENCES

[1] K. Athikulwongse et al, Intl. Conf. on Computer Aided Design, 2010,
[2] A. Mercha,.; et al, IEE Intl. Electron Devices Meeting, 2010.
[3] C. Okoro, et al, Intl. Interconnect Technology Conference, 2008.
[4] C. McDonough et al, IEEE Intl. Reliability Physics Symposium, 2011.
[5] C. McDonough et al, IEEE International Interconnect Technology Conference and Materials for Advanced Metallization, 2011.
[6] H.B. Chang et al, Symposium on VLSI Technology (VLSIT), 2012.
[7] J. West et al, Symposium on VLSI Technology (VLSIT), 2012.
[8] H. Khan, et. al. IEEE Intl. Cong. On 3D System Integration, 2009.
[9] J. Cho, et al. IEEE Conf. on Electrical Performance of Electronic Packaging and Systems, 2009.
[10] C. Xu et. al. IEEE Trans. on Electron Devices, vol. 58, no. 11, 2011.
[11] A. Trivedi, et. al. IEEE Electron Device Letters, vol. 32, no. 8, 2011.
[12] Sentaurus Process User Guide, Version E-2010.12, December 2010
[13] http://www.itrs.net/reports.html
[14] R. Chau et al, IEEE Electron Device Letters, vol.25, no.6, 2004
[15] Sentaurus Device User Guide, Version E-2010.12, December 2010
[16] B. Obradovic et al, Intl. Work. on Computational Electronics, 2004
[17] H. Chaabouni, et al. IEEE Electron Devices Meeting (IEDM), 2010.
[18] E. Ungersboeck et al, Intl. Work. on Computational Electronics, 2006

Efficient Adaptive Mesh Refinement for MoM-based Package-Board 3D Full-wave Extraction

Arkaprovo Das, Rahit R. Nair and Dipanjan Gope

Department of Electrical Communication Engineering,
Indian Institute of Science,
Bangalore 560012, India
Email: arka_reek@yahoo.co.in, rahit.r.nair@gmail.com and dipanjan.gope@gmail.com

Abstract—The accurate solution of 3D full-wave Method of Moments (MoM) on an arbitrary mesh of a package-board structure does not guarantee accuracy, since the discretizations may not be fine enough to capture rapid spatial changes in the solution variable. At the same time, uniform over-meshing on the entire structure generates large number of solution variables and therefore requires an unnecessarily large matrix solution. In this work, a suitable refinement criterion for MoM based electromagnetic package-board extraction is proposed and the advantages of the adaptive strategy are demonstrated from both accuracy and speed perspectives.

Keywords—Method of Moments, Signal Integrity, Adpative Mesh Refinement

I. INTRODUCTION

The rapid proliferation of connected sensors have ushered in a new era of designs with multiple technologies, frequencies and functionalities. To address the system-level requirements of such a design, technology trends like System-in-Package (SiP) and System-on-Chip (SoC) are becoming increasingly relevant. To ensure the fidelity of chip-to-chip communication and system-level performance, package-board electrical performance parameters like signal integrity (SI), power integrity (PI), electromagnetic interference (EMI) are critical. Increasing pin-counts to satisfy functionality requirements and decreasing layer-counts to maintain cost-effectiveness renders a 3D full wave electromagnetic solution necessary for accurate system modeling.

To solve a 3D package-board problem with full-wave accuracy, three different computational techniques are widely used: Finite Element Method (FEM), Finite Difference Time Domain (FDTD) technique and Method of Moments. While the former two are volume based methods, the latter is a surface based formulation. FEM uses tetrahedrons and FDTD uses grid-based cubical Yee cells for discretization. MoM on the other hand, uses triangular mesh elements to accurately discretize the surface of the geometry. Due to lesser number of mesh elements MoM has an advantage of a smaller matrix size. However, due to Green's Function interactions, the MoM matrix is dense and its solution presents a time and memory challenge. Therefore, depending on fast solver algorithms like Fast Multipole Method (FMM) [1], MoM is widely used for any 3D electromagnetic application.

The solution of Maxwell's equations on an arbitrary mesh of the structure under consideration might not yield accurate results. To obtain a converged solution the mesh needs to be refined at appropriate locations of high spatial variations, such that the solution remains invariant with any further mesh refinement. Increasing the refinement with uniform sized mesh elements over the entire geometry results in the structure being unnecessarily over-meshed and consequently leads to increased solution time. Therefore, it is time-memory cost efficient if the geometry is refined in such a way so that the physics of the problem can be captured with minimum number of optimally placed and sized mesh elements- a method known as adaptive mesh refinement. While existing literature covers adaptive mesh refinement for FEM [2] and FDTD [3] based methods, research work in integral equation based counterpart for full-wave problems is relatively untouched. For MoM based methods involving static cases, local error estimates between potentials at adjacent nodes on the geometry has been taken as refinement criterion in [4]. Adaptive mesh generation scheme for 2D magnetostatic integral equations has been covered in [5], a fuzzy logic based scheme for non-linear 3D magnetostatic cases in [6], and a framework to permit easy hp-refinement using Generalized Method of Moment (GMM) is presented in [7].

In this work, different refinement indicators are studied in an adaptive refinement environment for 3D Electric Field Integral Equation (EFIE) full-wave MoM based package-board extraction. Consequently, the most suitable refinement indicator is identified and the advantages of the adaptive strategy are demonstrated from both accuracy and speed perspectives. The paper is organized as follows. The 3D full-wave MoM formulation using Rao-Wilton-Glisson (RWG) basis function [8] is explained in section 2. Section 3 explains the adaptive refinement environment built on top of an open-source mesher. Section 4 details the study of different refinement criterion. The relative advantages of the adaptive scheme with appropriate refinement criterion in terms of time and memory usage benefits are studied in section 5. In section 6, numerical examples are provided to validate the presented scheme.

II. 3D FULL-WAVE MoM FORMULATION USING RWG BASIS

To solve a 3D full-wave problem using MoM, the equation that is solved for is given below:

$$E_i = j\omega A + \nabla \phi \qquad (1)$$

where E_i is the incident electric field on the given geometry. Equation (1) is the Electric Field Integral Equation (EFIE). The vector and scalar potentials are given by:

$$\mathbf{A(r)} = \frac{\mu}{4\pi} \int_S \frac{e^{-jk|\mathbf{r}-\mathbf{r'}|} \mathbf{J(r')}}{|\mathbf{r}-\mathbf{r'}|} ds' \tag{2}$$

$$\Phi(\mathbf{r}) = \frac{1}{4\pi\varepsilon} \int_S \frac{e^{-jk|\mathbf{r}-\mathbf{r'}|} \rho(\mathbf{r'})}{|\mathbf{r}-\mathbf{r'}|} ds' \tag{3}$$

The current density on the entire surface is modeled with RWG basis functions [8] and the coefficients are computed from a MoM matrix solution.

III. ADAPTIVE MESH REFINEMENT ENVIRONMENT

The adaptive meshing platform is implemented as a wrapper on top of open source mesh generator Gmsh[9]. The refinement algorithm works by way of feedback from solver.

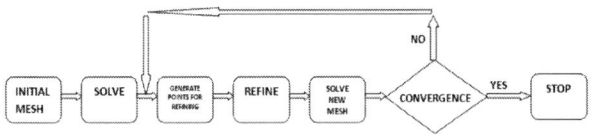

Figure 1: Steps involved in adaptive mesh refinement

As shown in Fig. 1, the initial mesh is taken in by the solver and from the solution, a set of points are chosen for refinement based on the refinement indicator. A background mesh density file (Gmsh .pos file) is created using a finer desired mesh density for the refinement points and the original mesh density for the others. Fig. 2 shows an example of adaptive meshing. The original mesh is shown in Fig. 2a. The background mesh with refinement factors is shown in Fig. 2b and the corresponding mesh generated after one adaptive meshing pass is shown in Fig. 2c.

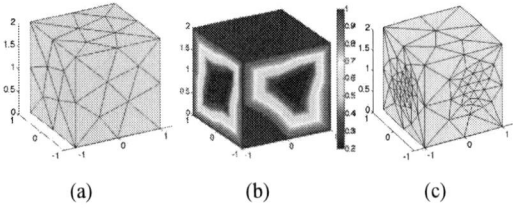

Figure 2: (a) Mesh of a cube (b) background mesh used for adaptive meshing (c) mesh after adaptive meshing according to the background mesh

IV. REFINEMENT INDICATOR STUDY

Refinement Indicator can be defined as an electromagnetic parameter based on which the points for further refinement can be chosen such that the converged solution is reached with least mesh-count.

A. Choice of refinement indicator

In the present work, the suitability of different electromagnetic parameters as an appropriate refinement indicator is compared and analyzed. In order to capture the geometry effects and distinctly separate it from the high frequency effects, an electrically small microstrip line 0.1λ long at the frequency of operation, with air as the dielectric substrate and a ground plane is taken. The line is shorted at one end, and excited with a lumped port at the other. The entire problem is solved using in-house MoM EFIE solver with RWG basis function. Fig. 3 shows the comparative logarithmic variations of different electromagnetic parameters for the

structure. 30% points are printed based on descending order of magnitudes of the respective parameter, and are shown in the figures as black dots. These are the points that are considered for finer refinement in the next pass.

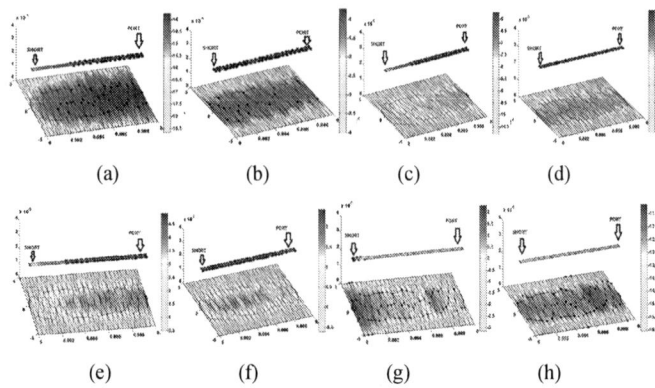

Figure 3:Mesh plot showing variations of (a) Charge (b) Current (c) Charge density (d) Current density (e) Electric field (f) Magnetic field (g) Current - Potential ratio (h) Charge - Potential ratio on the microstrip structure in logarithmic scale

- Charge: The charge distribution on the entire structure is evaluated using divergence property of RWG basis in conjunction with the charge-current continuity equation. As evident from Fig. 3a, charge variation is in accordance with the conventional transmission line theory for a sub quarter wavelength structure, where it is high towards the ports and zero near the shorted end. Therefore, charge captures the physics on one part of the geometry nearer to the port, while the region near the short gets neglected for further refinement.

- Current: The normal component of current across an RWG edge is numerically equal to the coefficient related to that edge found by solving the system of equations obtained by formulating EFIE. The current behaves complimentary to charge which is high towards the shorted end, and gets reduced to a lower value at the ports for structure size less than quarter wavelength. However, it tends to spread during its return path through a wider ground and as a result, most of the points on the signal trace are printed first for adaptive refinement while the priority of points on the ground plane becomes secondary (Fig. 3b).

- Charge or current density: Owing to its smaller width, the charge density (Fig. 3c) or current density (Fig. 3d) on the signal trace is much higher than that on the ground. Therefore, the points on the ground plane are ignored. In fact charge or current density in a triangle, unlike charge or current, is independent of the size of the triangle and therefore will continue refining at the same position on successive adaptive passes. Therefore they cannot be considered as adaptive indicators.

- Electric field: The electric field (Fig. 3e) is evaluated using the relation :

$$E_s = -j\omega A - \nabla\emptyset \tag{4}$$

where E_s is the scattered electric field anywhere on and outside the structure. Similarity with Fig. 3c explains that electric field follows a similar variation as charge density, which makes it unsuitable for being a good refinement indicator. Dimensionally, a physical intuition can be attributed to the integral form of Gauss' law, where the Electric field follows charge density.

- Magnetic Field: The magnetic field (Fig. 3f) on other hand follows a variation similar to electric surface current density. This is due to the fact that the tangential component of magnetic field on a PEC is given by electric surface current density, in accordance with the boundary relation:

$$\hat{n} \, X \, \vec{H} = \vec{J}_s \qquad (5)$$

In addition, H-field has units of A/m, suggesting it dimensionally follows current density.

- Current-potential ratio (I/V): The plot of surface current to potential ratio on the geometry is depicted in Fig. 3g. This represents inductive admittance in the form of $\frac{1}{j\omega L}$.

Intuitively, higher the I/V, the less inductive impedance is faced for current flow at that point, making it a preferred path and therefore a prime candidate for refinement.

- Charge-potential ratio (Q/V): Charge to potential ratio on the other hand captures the relative effect of charge over potential at a point. Physically it represents the electric admittance in the form of $j\omega C$. Therefore, higher the Q/V, the least impedance is faced from a capacitive point of view. (Fig. 3h).

- Combination of I/V and Q/V: The EFIE system comprises of both scalar and vector potentials which are related to capacitive and inductive components respectively. The resultant current flow is tuned by their combined effect. Therefore, I/V and Q/V together capture the overall admittance and serves as an excellent refinement indicator. This is validated in the next section with convergence results.

B. Convergence Results

In this section a comparative convergence study is presented for 2 different cases - microstrip line, and microstrip multiple bends with a ground plane. In Fig. 4a, the convergence results for the 0.1λ long microstrip line is shown. The relative Z-parameter error with final converged value is shown in logarithmic scale. Fig. 4b shows the regions of dense mesh refinement for a higher pass number, with combination of I/V and Q/V as refinement indicator.

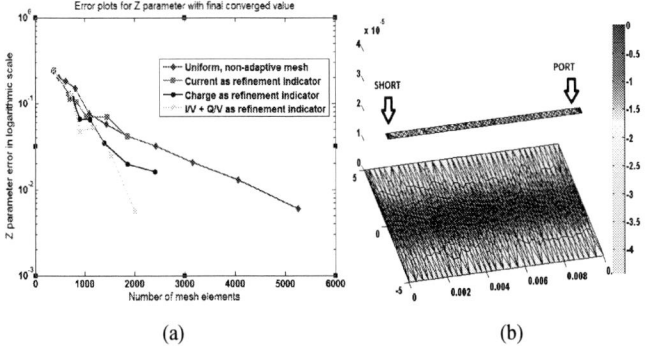

(a)   (b)

Figure 4: (a) Convergence result for microstrip line (b) Mesh plot after 4th adaptive pass with combination of I/V and Q/V as refinement

The convergence plot above clearly suggests that with a combination of I/V and Q/V as the refinement indicator, the convergence towards the actual solution is faster than the rest. The mesh plot reveals that the refinement is on intended regions for faster convergence. Fig. 5 shows the same for the microstrip with bend structure. The convergence threshold in both the examples is set to 1% norm by norm relative error of the Z matrix.

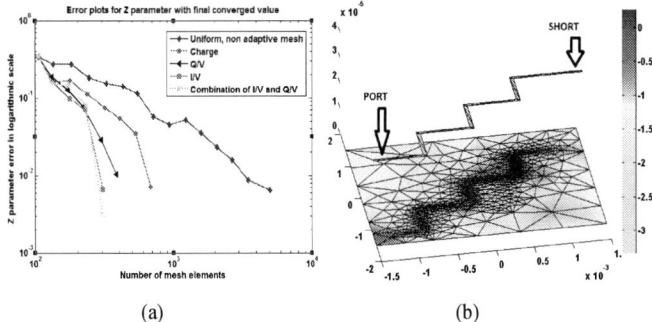

(a)   (b)

Figure 5: (a) Convergence result for microstrip multiple bends (b) Mesh plot after few adaptive passes with combination of I/V and Q/V as refinement indicator

C. Multiple ports and multiple frequency sweep

The concepts explained using 1-port network can also be generalized to incorporate multiple ports. Since a typical MoM solution is based on extracting Y parameters, each port or RHS solution resembles the situation presented in the earlier studies. Therefore for multiple ports, all RHSs solutions are summed up in a single vector which is used to extract refinement points based on the combination of I/V and Q/V.

For problems involving frequency sweep over a range of frequencies, the refinement indicator is first evaluated for the centre frequency of the entire frequency range, and is then used for successive refinement of the geometry.

V. COST ESTIMATES

For a conventional MoM based solver, the solution time is dominated by LU decomposition of the MoM matrix, where the order of complexity is $O(N^3)$, N being the number of unknowns. The memory requirement in this case is $O(N^2)$. On the other hand, fast solver algorithms like Fast Multipole Method (FMM) reduces the order of complexity for both solution time and memory requirement to $O(N \log N)$.

For conventional solvers, the solution time speed up for adaptive mesh refinement over uniform non-adaptive meshes can be given by:

$$Speed\ Up = \frac{\sum_{i=1}^{P_1}(N_i)_{na}^3}{\sum_{j=1}^{P_2}(N_j)_{a}^3} \qquad (6)$$

where P_1 is the number of passes required for non-adaptive solution to converge, P_2 is the number of passes till convergence for an adaptively refined solution, N_i or N_j being the number of unknowns at i-th or j-th pass respectively, and the subscripts 'na' and 'a' denote non-adaptive and adaptive respectively. Similarly, the memory savings can be given by the following relation:

$$Memory\ savings = \frac{[max\{(N_i)_{na}\}]^2}{\left[max\left\{(N_j)_{a}\right\}\right]^2}, \forall\ i = 1\ to\ P1, j = 1\ to\ P2 \qquad (7)$$

For fast solvers, using similar analysis, the solution time speed up and memory savings for fast-solver solutions is given by the following equation:

$$Speed\ up = \frac{\sum_{i=1}^{P_1}(N_i \log N_i)_{na}}{\sum_{j=1}^{P_2}(N_j \log N_j)_a} \ \& Memory\ savings = \frac{max\{(N_i \log N_i)_{na}\}}{max\left\{(N_j \log N_j)_a\right\}}, \forall\, i = 1\ to\ P1, j = 1\ to\ P2 \quad (8)$$

Using (6)-(8) the time-memory saving results are calculated and compared for microstrip trace in Table I. The adaptive refinement results are calculated using combination of I/V and Q/V as refinement indicator.

TABLE I.   TIME-MEMORY SAVINGS RESULTS FOR MICROSTRIP LINE: ADAPTIVE REFINEMENT VS. UNIFORM REFINEMENT

Profile	Conventional solver	Fast solver
Solution time Speed up	85.22 times	5.44 times
Peak Memory Savings	18.73 times	4.32 times

VI.   NUMERICAL RESULTS FOR PACKAGE GEOMETRY

In this section numerical results are presented to explain the performance of adaptive mesh refinement for a package geometry. Fig. 6a shows the top view of a simplified package structure, while its uniform mesh is shown in Fig. 6b.

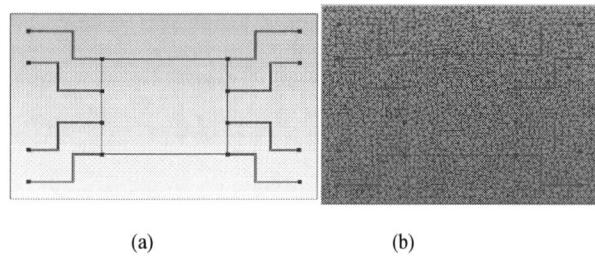

(a)   (b)

Figure 6: (a) Top view of simplified package (b) Uniform mesh elements on the package structure

Fig. 7 explains the convergence results and mesh plot for the above structure using adaptive mesh refinement. Here, one of the terminals is excited with a lumped port while the rest are grounded. Combination of I/V and Q/V is selected as the refinement indicator.

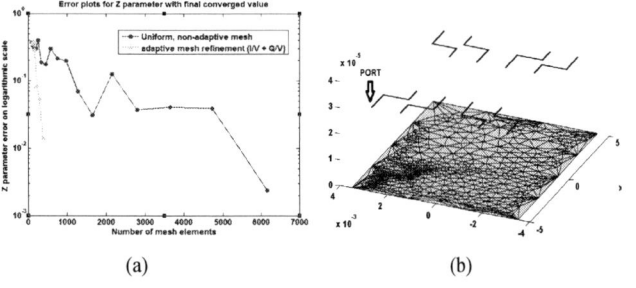

(a)   (b)

Figure 7: (a) Convergence results (b) Mesh plot showing adaptive refinement for a higher pass (angular view)

Fig. 7b reveals that the refinement is adaptively done on appropriate regions which results in much faster convergence.

In the next exercise, an 8x8 Y-parameter matrix is extracted, by successively exciting the same geometry at 8 different terminals with lumped port excitation while the rest are shorted. The convergence plots and meshes for this multiple port excitation problem is shown below (Fig. 8).

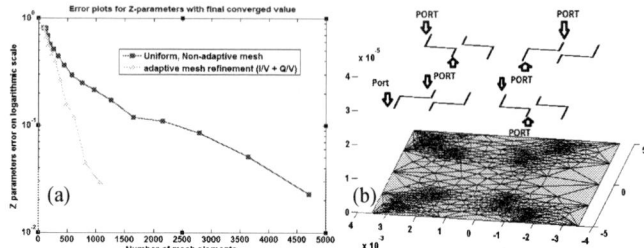

(a)   (b)

Figure 8: (a) Convergence results for multiple-port excitation (b) Mesh plot showing adaptive refinement for a higher pass (angular view)

It takes 16 passes and 4711 mesh elements to reach convergence with uniform refinement as against 10 passes and 1088 mesh elements for adaptive refinement scheme.

A comparison of magnitudes of Z11 and Z43 are shown over a multi-frequency sweep (Fig. 9). It explains that the initial mesh results may be highly erroneous, while the adaptive and uniform refined meshes at convergence yield similar but correct results. However, as explained throughout the paper, fastest solution can be achieved by adaptive refinement.

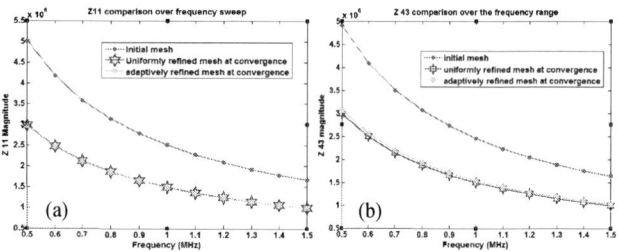

(a)   (b)

Figure 9: comparison for 3 different meshes (a) Z11 (b) Z43

ACKNOWLEDGEMENTS

The authors thank Angiometrix Inc. for support and Raghavan Subramaniyan for discussions related to this work.

REFERENCES

[1] L. Greengard, V. Rokhlin, "A fast algorithm for particle simulations," *J. Computational Physics*, vol. 73, pp. 325-348, Dec. 1987.

[2] F.J.C. Meyer and D.B. Davidson, "Adaptive-mesh refinement of finite-element solutions for two-dimensional electromagnetic problems", *IEEE AP Magazine*, vol. 38, pp. 77-83, 1996.

[3] Yaxun Liu and C.D. Sarris, "Numerical Error Analysis and Control in a Dynamically Adaptive Mesh Refinement (AMR) - FDTD Technique", *IEEE MTT-S International Microwave Symposium Digest*, 2006.

[4] Dong-Kuk Park , Chang-Hoi Ahn , Soo-Young Lee and Jung-Woong Ra, "Adaptive mesh refinement for boundary element method and its application to stripline analysis", *IEEE Trans. on Magnetics*, 1990.

[5] J. Ikaheimo, K. Forsman and L Kettunen, "Adaptive mesh generation in 2D magnetostatic integral formulations", *IEEE Transactions on Magnetics*, vol. 33, pp. 1736-1739, 1997

[6] W.Hafla, A. Buchau, A.Bardakcioglu, C. Scheiblich, W.M Rucker, "Fuzzy Logic Adaptive Mesh Refinement for 3D Nonlinear Magnetostatic Problems Using Integral Equation Method", *12th Biennial IEEE Conference on Electromagnetic Field Computation*, 2006

[7] N.V.Nair, B. Shanker, "An hp-refinement scheme for surface integral equations using the generalized method of moments", *IEEE APSURSI*, 2012.

[8] S.M Rao, D.Wilton, A.W Glisson, "Electromagnetic scattering by surfaces of arbitrary shape", *IEEE Trans. on Antenna and Propagation*, vol. 30, 1982.

[9] C. Geuzaine , Jean-François Remacle, "Gmsh: a three-dimensional finite element mesh generator with built-in pre- and post-processing facilities",*International Journal for Numerical Methods in Engg.,2009*.

A Novel EBG Structure with Super-Wideband Suppression of Simultaneous Switching Noise in High Speed Circuits

Jai Narayan Tripathi, Jayanta Mukherjee,
and Prakash R. Apte
Department of Electrical Engineering
Indian Institute of Technology Bombay
Mumbai, India 400076
Email: {jai,jayanta,apte}@ee.iitb.ac.in

Raj Kumar Nagpal, Nitin Kumar Chhabra,
and Rakesh Malik
TRnD, STMicroelectronics Pvt. Ltd.,
Greater Noida, Uttar Pradesh,
India 201308
Email: {rajkumar.nagpal,nitin.chhabra,rakesh.malik}@st.com

Abstract—**A novel uniplanar electromagnetic band-gap structure to maintain power integrity by suppressing simultaneous switching noise (SSN) is presented. The EBG structure with stopband from 750 MHz to 5.10 GHz is designed, fabricated and validated using network analyzer. Simulation results are verified by measurements and compared with the earlier published structures. Suppression of resonant cavity modes of power plane by EBG structure is also shown. The adoption of EBG structure in power deliver network is recommended to reduce the high frequency noise coupling between neighboring devices. These structures further help in better EMI/EMC compliance of the product by attenuating the propagation of high frequency noise between devices. The EBG structure usage can be on board, package or at die level.**

Keywords—*Electromagnetic Band-Gap (EBG) structure, Simultaneous Switching Noise (SSN), Power Integrity (PI), Cavity Modes.*

I. INTRODUCTION

In high speed systems, the size and threshold voltage of switching devices is reducing and at the same time, switching speed is increasing, hence suppression of Simultaneous Switching Noise (SSN) is becoming very significant for proper operation of the device. SSN can affect the system performance by causing the false switching of logic circuits. To overcome the problem of SSN, either decoupling capacitors are used or vias are added around the 'hot spots' to confine this noise [1][2]. The decoupling capacitors are effective up to few hundreds of MHz only [3]. Embedded capacitors can be used up to some GHz but they need expensive dielectric material [4]. Electromagnetic Band- Gap (EBG) structures are becoming popular as an alternate to decoupling capacitors as they can effectively reduce SSN in high speed systems. EBG structures are periodic structure with physical dimensions and topologies, designed to potentially block the EM wave propagation in a particular frequency band [5][6]. EBG structures are cost effective and easy to design.

This paper presents a uniplanar EBG power/ground plane for stop band attenuation in band of 1-5 GHz. Uniplanar structures have advantage over mushroom structures that they don't need an extra metal layer and thus are easy to design and have lower cost. The proposed power plane is compared with

the earlier published uniplanar structures of same dimensions, same number of patches and same dielectric material [7]-[10]. The structure (chosen size 90 x 90 mm^2) is a square having 9 cells (patches) with same dimensions. The proposed unit cell structure has triangle structure on side a, which is mirrored. Side b is a rectangle structure mirrored again. The use of such interconnects resulted in better performance than traditional EBG structures having similar bridges at all the sides [7]-[10]. Interconnect structure between the cells is designed keeping in view the dc current requirement coming from maximum current which this structure should be able to sustain, as well as the minimum inductance which this structure should offer to isolate desired noise coupling between the cells.

II. STRUCTURE DESIGN AND SSN SUPPRESSION

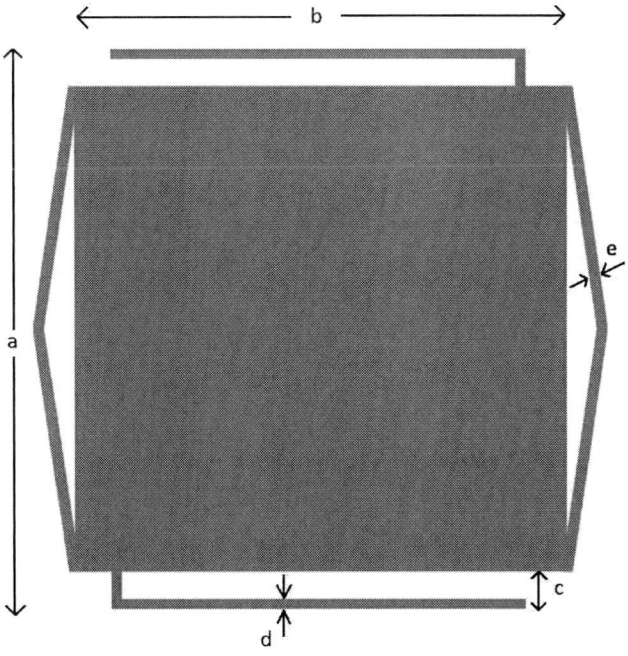

Fig. 1. Single cell of EBG Structure with dimensions

Fig. 1 shows a single cell with dimensions where $a = 30$

mm, $b = 28\ mm$, $c = 2\ mm$, $d = 0.5\ mm$, $e = 0.5\ mm$. There is a central square which is having strips attached to it wherein a is the dimension of the square cell, b is the length of the sides of central square of the cell. c is the difference between a and b. d and e are the width of strips. All the strips are 0.5 mm wide. Fig. 2 shows the complete design of the power plane using EBG structure. This structure is having different shapes of bridges at vertical sides and horizontal sides as shown in fig. 1. Compared to the traditional EBG structures having similar bridges at all the sides [7]-[10], this structure increases the effective inductance between two adjacent cells resulting in broader stop-band.

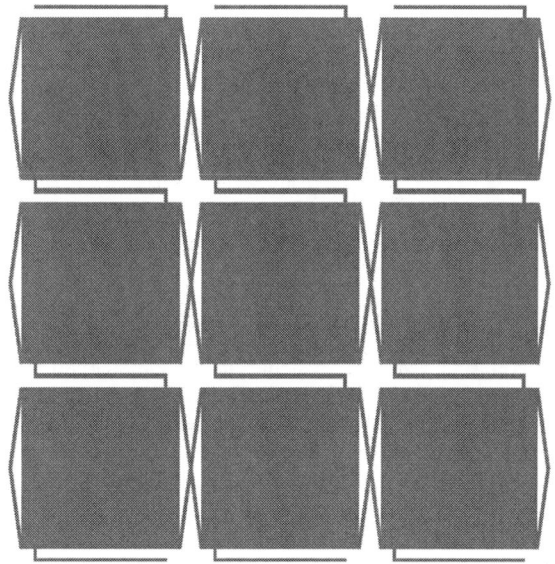

Fig. 2. Complete 9 cell EBG Structure for Power Plane

III. POWER INTEGRITY

To maintain power integrity (PI) in a high speed system the target impedance of the power delivery network (PDN) should be below the system required target impedance in the desired frequency range [3]. The EBG structure are designed to attenuate the power plane spectral noise propagation on power delivery network (PDN).

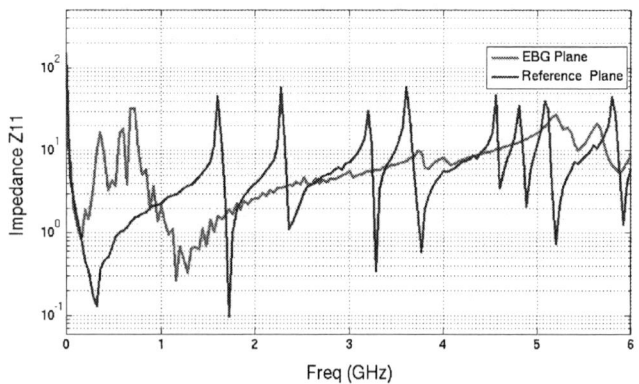

Fig. 3. Self Impedance (Z_{11}) of reference plane and EBG plane

In power planes, there are resonance modes due to cavity effects which need to be damped in order to control the

propagation of noise. The EBG structure unit cell are designed and inductive coupled to modify the self impedance (Z_{11}) of the system (fig 3).

Fig. 3 shows the self-impedance profiles of EBG plane and reference plane which is having a simple rectangular shape. The cavity modes can be observed at various frequencies for reference board which have been eliminated in EBG plane. Fig. 4 shows the measured and simulated $|S_{21}|$ for the EBG board where port 1 is at (46 mm, 45 mm) and port 2 is at (74 mm, 74 mm). There is a stop-band ranging 4.35 GHZ (from 700 MHz to 5.1 GHz).

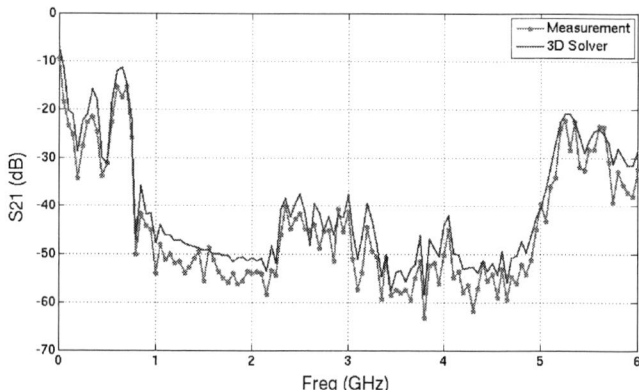

Fig. 4. S_{21} parameters of board (measured and simulated)

IV. EQUIVALENT CIRCUIT AND CURRENT DISTRIBUTION ANALYSIS

The chosen EBG structure is approximated to first order equivalent electrical model. The noise suppressing capability of proposed structure in PDN can be explained using its lumped equivalent RLC model. Figure 5 depicts the equivalent model for a single cell of proposed EBG power plane structure. The values of R, L, and C in the lumped equivalent model are derived from conventional transmission line method and insertion loss filter design approximations. The RLC model is iteratively optimized to produce the desired S_{21} performance. The elemental values of R, L and C parameters are tuned to a wide band width of 4.35 GHz. After optimization the values of R, L, C components are found as $R_1 = 2.53\ m\Omega$, $L_1 = 21.10 pH$, $C_1 = 57.32\ pF$, $R_2 = 692.80\ m\Omega$, $L_2 = 1.92\ nH$, $C_2 = 2.60\ pF$, $R_3 = 0.001\ m\Omega$, $L_3 = 95.0\ pH$, and $C_3 = 3.10\ pF$. The s-parameters obtained form the RLC equivalent model can be seen in fig. 6.

The optimized EBG structure is fabricated, validated using 3D CAD tool and measured using VNA. The Full wave simulation in ADS is carried out for verification of the lumped model approximation of complete EBG ground plane. The suppression of noise can also be seen by insertion loss between adjacent ports on the proposed structure. The stopband is 4.35 GHz when the insertion loss is observed below -30 dB, as shown in figure 4.

Figure 7 shows current distribution at 4 GHz. There are three subfigures showing current at three different ports. It can be seen from figure, that the magnitude of surface current reduces as it propagates from one port to another. Similar nature of current distribution is also observed for other frequencies in

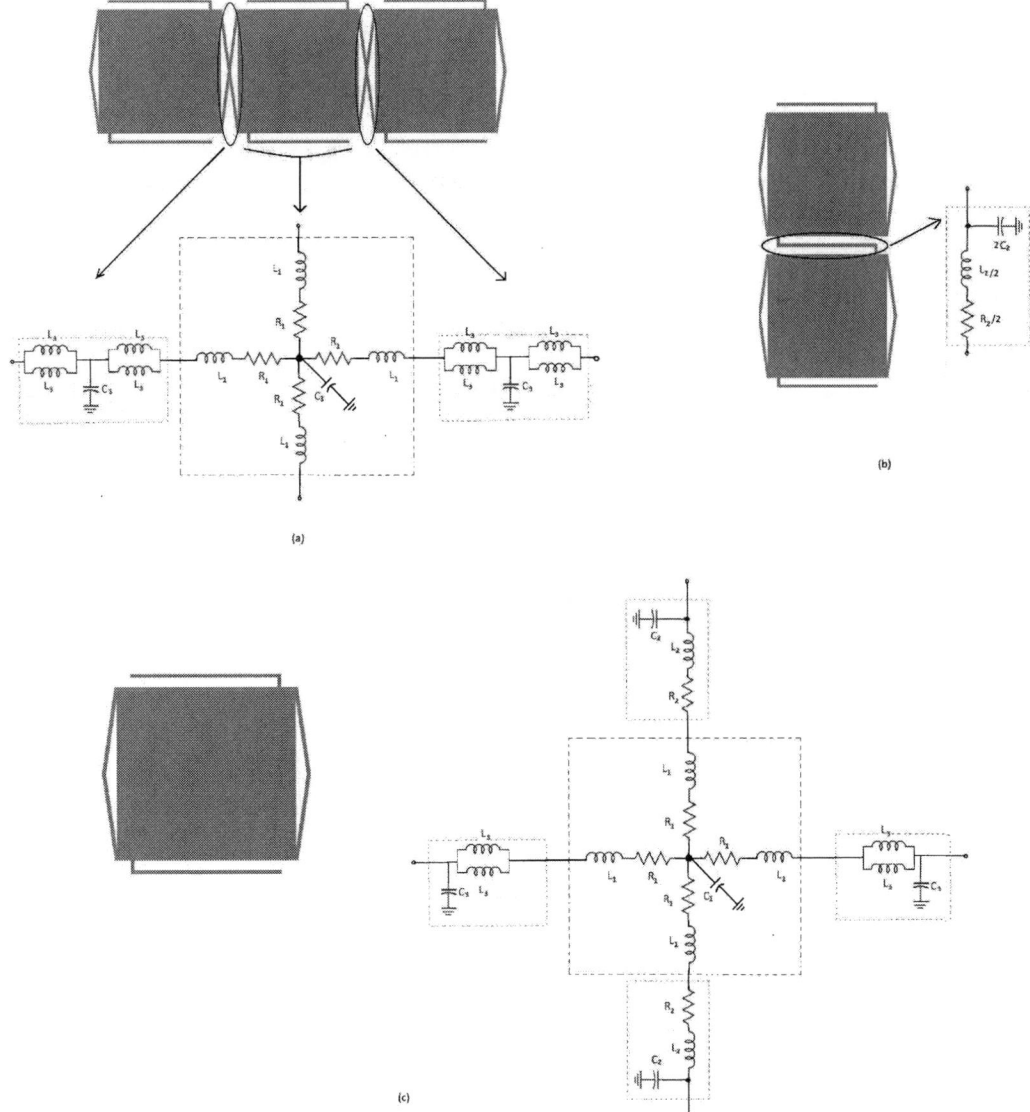

Fig. 5. Equivalent lumped model for proposed EBG ground plane: (a) Row wise (b) Column wise (c) single cell approximation.

the band of 750 MHz- 5.10 GHz. This nature of surface current distribution indicates that PDN noise is reduced drastically when the proposed structure is used, as it provides band gap for 750 MHz - 5.10 GHz band (since PDN Noise is primarily dominant distributed below 6 GHz [7]).

This is compared with the other published structures in table I. For the purpose of proper comparison, the dimensions and port locations are taken exactly the same. The methodology is as follow.

1) S_{21} performance requirement for EBG structure is given as specifications.
2) Select the cell size which is optimum for given board dimension and bandwidth requirement.
3) Find first order RLC equivalent model of single cell using the transmission line theory

4) Make array and evaluate S_{21}, iterate and optimize till S_{21} performance is met with performance margin
5) With optimized RLC value, modify and finalize the cell geometry.
6) Validate the cell array using 3D EM tool for S_{21} performance.
7) Fabricate the EBG structure and do measurement for correlation with 3D EM S_{21} performance

V. RESULTS

Table I shows the results obtained by the present structure with the previously published structures of the same dimensions and same dielectric material. The significant improvement in the stopband can be observed. Compared to the traditional EBG structures having similar bridges at all the sides [7]-[10], this structure shows improvement in stopband

978-1-4799-0708-3/13 $31.00 © 2013 IEEE

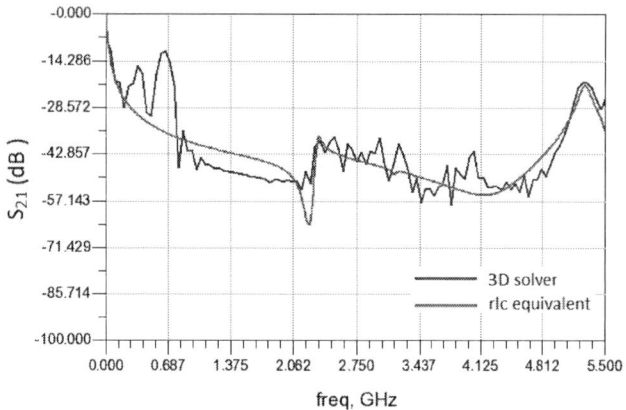

Fig. 6. S-parameters obtained from equivalent rlc model

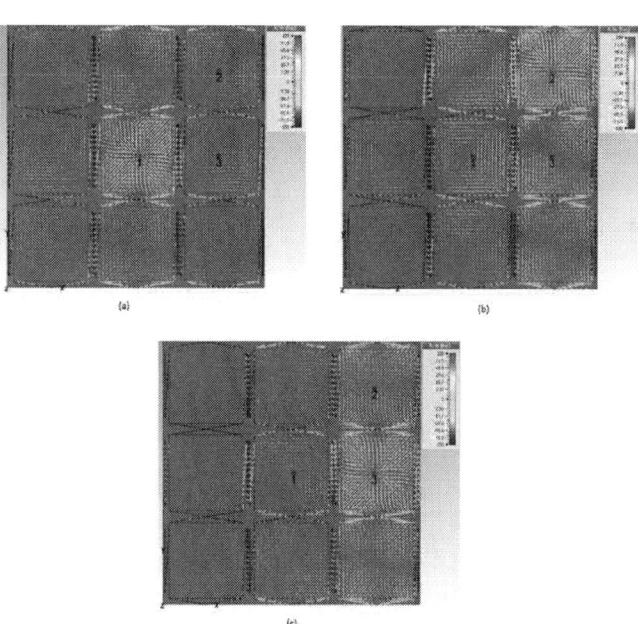

Fig. 7. Current distribution at various ports at 4 GHz

by increasing the effective inductance between two adjacent cells.

TABLE I. COMPARISON OF STRUCTURE WITH OTHER STRUCTURES

S. No.	Publication	Board Dimensions	Dielectric Height & Material	Stop-Band (below -30 dB)
1.	T. -L. Wu et al [8]	90 x 90 mm²	0.4 mm / FR4	3 GHz
2.	D. B. Lin et al [10]	90 x 90 mm²	0.4 mm / FR4	3.08 GHz
3.	T. -L. Wu et al [7]	90 x 90 mm²	0.4 mm / FR4	4 GHz
4.	Present Work	90 x 90 mm²	0.4 mm / FR4	4.35 GHz

VI. CONCLUSION

A novel, cost-effective, uniplanar EGB power plane is proposed in this paper with super-wideband suppression of PDN noise from 750 MHz to 5.10 GHz. Performance is compared with previously published conventional structures of exactly same dimensions and dielectric material. Excellent agreement between simulated and measured results are found.

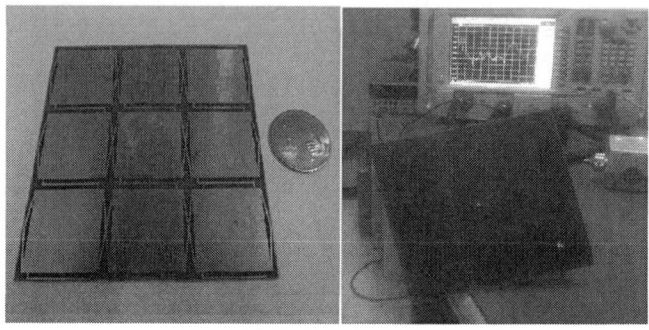

Fig. 8. Fabricated EBG board and measurement set-up

REFERENCES

[1] G. Chen, K. L. Melde, "A Cavity Resonance Suppression in Power Delivery Systems Using Electromagnetic Band Gap Structures," IEEE Trans. Advanced Packaging, vol. 29, no. 1, pp. 21-30, Feb. 2006.

[2] J. N. Tripathi, R. K. Nagpal, N. K. Chhabra, R. Malik, J. Mukherjee, "Maintaining Power Integrity by damping the cavity-mode anti-resonances' peaks on a power plane by Particle Swarm Optimization", 2012 13th International Symposium on Quality Electronic Design, ISQED'12, pp. 525-528, March 2012, Santa Clara, USA.

[3] M. Swaminathan, A. Ege Engin, *Power Integrity Modeling and Design for Semiconductors and Systems*, Prentice Hall, 2008.

[4] F de Paulis, M. H. Nisanci, A. Orlandi, "Practical EBG Application to Multilayer PCB : Impact on Power Integrity," IEEE Electromagnetic Compatibility Magazine, Vol. 1, No. 3, pp. 60-65, 2012.

[5] R. Shelby, D. Smith, S. Shultz, "Experimental verifications of a negative index of refraction," Science, vol. 292, pp. 77-79, Apr. 6, 2001.

[6] D. Sievenpiper, L. Zhang, R. F. Jimnenez Broas, N. G. Alexopolous, E. E. Yablonovitch, "High-impedance electromagnetic surfaces with a forbidden frequency band," IEEE Tran. Microw. Theory Tech., vol. 47, no. 11, pp. 2059-2-74, Nov. 1999.

[7] T. -L. Wu, C. -C. Wang, Y. -H. Lin, T. -K. Wang, and G. Chang, "A novel power plane with super-wideband elimination of ground bounce noise on high speed circuits," IEEE Microw. Wireless Comp. Letters, Vol. 15, No. 3, 174176, Mar. 2005.

[8] T. -L. Wu, Y. H. Lin, and S. T. Chen, "A novel power planes with low radiation and broadband suppression of ground bounce noise using photonic bandgap structures," IEEE Microw. Wireless Compon. Lett., vol. 14, no. 7, pp. 337339, Jul. 2004.

[9] Y. J. Kim, K. B. Yang, and Y. S. Kim, "Wideband simultaneous switching noise suppression in mobile phones using miniaturized electromagnetic bandgap structures," Journal of Electromagnetic Waves and Applications, Vol. 23, Nos. 14-15, 1929-1938, 2009.

[10] D. -B. Lin, K. -C. Hung, C. -T. Wu, and C. -S. Chang , "A Serpent Bridge Electromagnetic Bandgap Structure for Suppressing Simultaneous Switching Noise," Journal of Electromagnetic Waves and Applications, Volume 23, Issue 2-3, pp. 213 - 220, 2009.

[11] W. -H. Chen, H. Zhang, and J. Wang, "A New Uniplanar Electromagnetic Bandgap Power Plane With Broadband Suppression of Simultaneously Switching Noise", Progress In Electromagnetics Research M, Vol. 1, 95-99, 2008.

[12] J. N. Tripathi, R. K. Nagpal, N. K. Chhabra, R. Malik, J. Mukherjee, and P. R. Apte, "Power Integrity Analysis and Discrete Optimization of Decoupling Capacitors on High Speed Power Planes by Particle Swarm Optimization", International Symposium on Quality Electronic Design 2013, Mar. 4 - 6, 2013, Santa Clara, USA.

[13] D. M. Pozar, Microwave Engineering, 2nd edition, Wiley Publication, 1998.

Application of Qualitative Imaging Methods to Electrical Performance-Aware Package Board Design

Nikita Ambasana, Dipanjan Gope
Department of Electrical Communication Engineering
Indian Institute of Science
Bangalore, India
ambasananikita@gmail.com, dipanjan.gope@gmail.com

Arun Chandrasekhar
Intel Corporation
Bangalore, India
arun.chandrasekhar@intel.com

Abstract— **Package-board co-design plays a crucial role in determining the performance of high-speed systems. Although there exist several commercial solutions for electromagnetic analysis and verification, lack of Computer Aided Design (CAD) tools for SI aware design and synthesis lead to longer design cycles and non-optimal package-board interconnect geometries. In this work, the functional similarities between package-board design and radio-frequency (RF) imaging are explored. Consequently, qualitative methods common to the imaging community, like Tikhonov Regularization (TR) and Landweber method are applied to solve multi-objective, multi-variable package design problems. In addition, a new hierarchical iterative piecewise linear algorithm is developed as a wrapper over LBP for an efficient solution in the design space.**

Keywords— *Crosstalk, Interconnect Design, Linear Back Projection*

I. INTRODUCTION

With the increasing popularity of system-centric design and Systems in Package (SiP) and Systems on Chip (SoC) technologies, the fidelity of signaling across package-board interconnects becomes crucial. Conventionally, the electrical analysis of package-board structures was done in parts, using SPICE models for several sections and lumped element models for others. However, with the increasing bitrates, the dimensions of most package-board interconnects are approaching and exceeding the wavelength of operation, so the lumped element analysis has to be replaced with full wave Electromagnetic (EM) modeling [1]. Several EM tools based on 2D, 2.5D and 3D modeling are commercially available and the field of fast EM solvers is well researched and mature. With low cost packages having less layer count, 3D solvers are becoming mandatory for better accuracy. However, both, in requirement of computational time as well as memory, 3D solvers are expensive. Hence, their use for simulation is feasible for verification purposes, but for design they prove to be time consuming under the current design methodologies which require several forward solution iterations.

The package design problem is becoming more relevant in the electronics industry because of higher density of I/O pins, increasing bit rates and reducing form factors. The performance specifications for commercial cost-effective applications are becoming more stringent [2]. Some methods currently used for package-board design are genetic algorithms, convex optimization based methods, designer intuition with trial and error, parametric sweeps and monte-carlo simulations. The authors in [3] have given a framework for package interconnect design using response surface methodology for optimization. In [4] a simulation study based method to reduce crosstalk using low k dielectric, adding guard trace and adding ground gap is explained. In [5] a methodology to improve the electronic package design process by performing multi-disciplinary design and optimization using Genetic Algorithm (GA) is given. A hybrid method combining Artificial Neural Networks (ANN) and GA is applied in [6] for optimization of package geometry variables using electrical modeling of interconnects for cost functions and weight determination. In [7] several schemes for mitigating differential crosstalk are outlined. An adjoint-variable approach to frequency-domain design sensitivity analysis is proposed for the optimization of high-frequency structures with full-wave EM solvers in [8]. The work done in [8] pertains to antenna design but the idea of sensitivity based approach to design gives a new perspective towards the package design problem.

The main objective of the work presented here is to develop a design tool that generates the design values of multiple specified design parameters for supplied many objectives, such that the design cycle time which is dominated by the runtime of the underlying 3D field solver is minimized.

II. TOMOGRAPHIC RECONSTRUCTION VS PACKAGE DESIGN

Package-board interconnect design essentially involves finding a set of values for design variables in a bounded design space which satisfy the given specifications. Like in image reconstruction through optical, electrical or electro-magnetic measurements popularly termed as tomography, the design of packages is a multi-objective, multi-variable problem where the objectives, analogous to the measurements in tomography, are functions of the design variables (substrate material, geometry etc.) analogous to the spatial material property distribution in tomography. Consider a nominal package design where C_i are the objectives, x_k are the design variables and f_i are the transformation functions. Then,

$$C_i = f_i(x_{00} + \Delta x_0, x_{10} + \Delta x_1, \dots); \quad C_{i0} = f_i(x_{00}, x_{10}, \dots) \quad (1)$$

$$C_i = f_i(x_{00}, x_{10}) + \frac{\partial f_i(x_{00}, x_{10}, \dots)}{\partial x_0} \Delta x_0 + \frac{\partial f_i(x_{00}, x_{10}, \dots)}{\partial x_1} \Delta x_1 + \dots \quad (2)$$

978-1-4799-0708-3/13 $31.00 © 2013 IEEE

Ignoring the higher order terms in Δx_k equations (1) and (2) can be written as a system of linear equations,

$$C_i - C_{i0} = \sum_k \frac{\partial f_i(x_{00}, x_{10}, \dots)}{\partial x_k} \Delta x_k \; ; i \in [1, n], k \in [1, m] \quad (3)$$

where $\frac{\partial f}{\partial x}$ represents the sensitivity matrix. Thus the non-linear continuous forward problem is now expressed in its linearized and discrete form as:

$$\lambda = Sg \quad (4)$$

where, λ is the vector of normalized objectives, S is the normalized sensitivity matrix and g is the vector of normalized variables. The tomographic reconstruction problem also frames an identical set of equations where λ represents the measured values in terms of fields for Radio Frequency (RF) imaging, or capacitance for the Electrical Capacitance Tomography (ECT) and g represents the material properties of the object-under-investigation. Consequently the structure parameters are obtained from measured results by solving the inverse problem

$$g = S^{-1}\lambda \quad (5)$$

The matrix S is under-determined, thus S^{-1} does not exist and there is no unique solution to this problem. Also, it is ill-conditioned, therefore regularization methods are often used to obtain the solution. Linear Back Projection (LBP) is one of the methods commonly used in the process and is given by:

$$g = S^T \lambda \quad (6)$$

Equation (6) obtains the material parameters of the structure under consideration. Commonly the results are represented as pixel values of an image, thus the term imaging. LBP yields poor quality images, but it is widely used because of its simplicity and low cost. Alternative qualitative methods involve regularization tools and iterative methods involving Newton-Raphson or Steepest-Descent algorithms to find the solutions that minimize errors to the measured data. Based on the standard Tikonov Regularization (TR) procedure, the solution of equation (4) can be expressed as

$$\hat{g} = (S^T S + \mu I)^{-1} S^T \lambda \quad (7)$$

Where, $\hat{g}$ is the approximation to g, μ is the regularization constant and I is an identity matrix. Applying Newton-Raphson approach to find a root in the space, iteratively, we get,

$$\hat{g}_{k+1} = \hat{g}_k - (S^T S + \mu I)^{-1} S^T (S\hat{g}_k - \lambda) \quad (8)$$

The initial guess g_0 can be obtained by either TR or LBP. Landweber (LW) iterative method is a variation of the steepest gradient descent method, widely used in optimization theory. It can be expressed as,

$$\hat{g}_{k+1} = \hat{g}_k - \alpha S^T (S\hat{g}_k - \lambda) \quad (9)$$

If, $\epsilon_k = (S\hat{g}_k - \lambda)$

$$\epsilon_k = \begin{cases} \max(\hat{g}_k), & \text{if } \epsilon_k > \max(\hat{g}_k) \\ \epsilon_k \text{ if } \min(\hat{g}_k) < \epsilon_k < \max(\hat{g}_k) \\ \min(\hat{g}_k) & \text{if } \epsilon_k < \min(\hat{g}_k) \end{cases}$$

where α is the relaxation factor.

III. APPLICATION OF QUALITATIVE IMAGING TECHNIQUES IN PACKAGE DESIGN

The analogy between the image reconstruction problem and package design model has been discussed in the previous section. In a typical package interconnect design, the specifications are usually to achieve return loss and crosstalk less than a particular limit. Therefore, unlike the image reconstruction method, in package design, instead of equality the target could be an inequality. Two different approaches are presented here. The first is a direct application of regularization based iterative approaches. The second is a binary search process used as a wrapper over LBP also referred to here as Hierarchical Search LBP or HSLBP. The latter attempts to capture the non-linearity in the system response by applying a piecewise-linear model.

A. Iterative Imaging Techniques (TR,LW)

The iterative techniques expressed in section II are meant to find one single solution point in the solution space. Hence, modifications have been made to the existing techniques to search a region or boundary points rather than a particular solution. For an explanation of the method, consider the ultimate objective to be O_o and objectives at each iteration to be O_k. Let γ be a constant. Let μ and α respectively be the Tikhonov and Landweber constants. Let D_0 be the design space corners, the design space corners indicate the extremities of the design variable variations. Let F_0 be the objectives calculated by the forward solution at the design corners D_0. Let S be the sensitivity matrix calculated by applying equation (3) using Finite Differences. Let λ_k be the set of normalized objectives and g_k be the obtained normalized design variables at each iteration. Assume desired goal is to achieve $F_k \leq O_0$. Let M be the maximum number of iterations

Algorithm 1:
  Step 1: Set $M, O_o, D_0, \mu, \alpha, \gamma$
  Step 2: Set $k = 0; D_k = D_0; O_k = O_0$
  Step 3: Compute F_0 by forward solution;
      Compute S, λ_0 from D_0, O_0
  Step 4: Compute $g_0 = (S^T S + \mu I)^{-1} S^T \lambda_0$ [For TR]
      $g_0 = S^T \lambda_0$ [For LW]
  Step 5: If $k < M$
      Compute D_{k+1} from g_k;
      Simulate D_{k+1} to get F_{k+1}
      $k = k + 1$
      Else
      Go to Step 7
  Step 6: If $F_k \leq O_0$
      Go to Step 7
      Else
      $g_{k+1} = g_k - (S^T S + \mu I)^{-1} S^T (S\hat{g}_k - \lambda_k)$ [For TR]
      $g_{k+1} = g_k - \alpha S^T (S\hat{g}_k - \lambda_k)$ [For LW]
      $O_k = O_k + \gamma O_0; \lambda_k$ changes accordingly
      Go to Step 5
  Step 7: Design variables D_k
      Exit

Fig.1. (a) and (b) show the movement of the solution points with iterations for the LW and TR methods. The example is a structure of two coplanar transmission lines. The spacing between the transmission lines and the dielectric constant are

978-1-4799-0708-3/13 $31.00 © 2013 IEEE

the variables and constitute the design space shown by cyan colored dots and the objectives are the return loss and the Near End Cross Talk (NEXT). The solution region is the colored black dots in the upper right corner of the space; the blue triangles show the actual solutions.

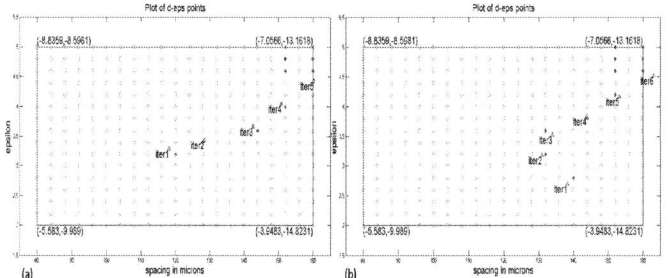

FIGURE 1. MOVEMENT OF SOLUTIONS IN (A) LW (B) TR METHODS

B. Hierarchical Search LBP (HSLBP)

The HSLBP method is a piece-wise approximation technique that starts with the entire space as the linear model and at each iteration, attempts to search for a solution in a smaller space formed on certain criterion.

Algorithm 2:

> *Step 1: Set $M, O_o, D_0, \mu, \alpha, \gamma$*
> *Step 2: Set $k = 0; D_k = D_0; O_k = O_0$*
> *Step 3: Compute F_0 by forward solution;*
>   *Compute S, λ_0 from D_0, O_0*
> *Step 4: Compute $g_0 = S^T \lambda_0$*
> *Step 5: If $k < M$*
>   *Compute D_{k+1} from g_k;*
>   *Simulate D_{k+1} to get F_{k+1}*
>   *$k = k + 1$*
>   *Else*
>   *Go to Step 7*
> *Step 6: If $F_k \le O_0$*
>   *Go to Step 7*
>   *Else*
>   *Find new D_k s.t O_0 lies between F_k and F_0*
>   *$O_k = O_k + \gamma O_0; \lambda_k, S_k$ change accordingly*
>   *Compute $g_k = S_k^T \lambda_k$*
>   *Go to Step 5*
> *Step 7: Design variables D_k*
>   *Exit*

Fig. 2 shows the movement of solution points and corners selection in the HSLBP method and Fig. 3 shows the |S13| curves at consecutive iterations.

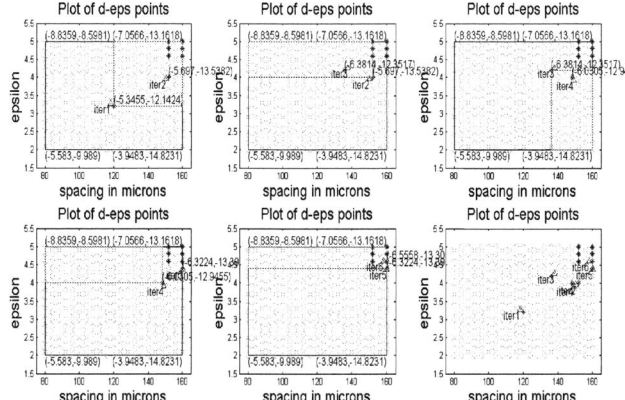

FIGURE 2. DETERMINATION OF CORNERS & SOLUTIONS IN HSLBP METHOD

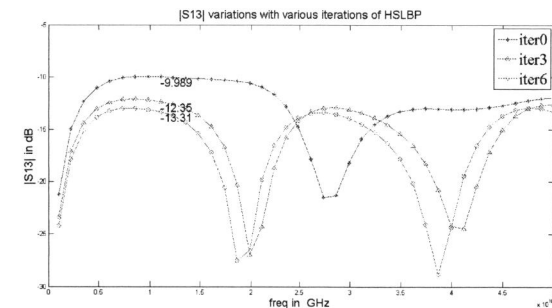

FIGURE 3. |S13| VARIATIONS WITH ITERATIONS

IV. NUMERICAL EXPERIMENTS

The experiment presented here is on an eight port structure with four transmission lines, of equal width. The design variables are the spacing between the four lines (d1,d2,d3), the dielectric constant of the substrate (ϵ), the loss tangent of substrate (tan δ) and the thickness of the substrate (t), the objectives were the NEXT for line 1 with respect to lines 2,3 and 4. Fig. 4 shows the structure.

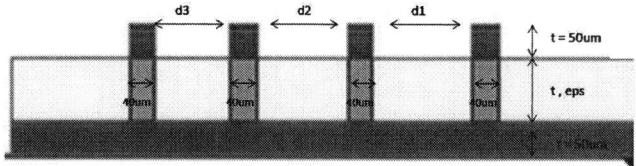

FIGURE 4. STRUCTURE FOR EXPERIMENT

Table I gives the nominal design values.

TABLE I. NOMINAL DESIGN VALUES FOR EXPERIMENT

Variable	Values
Transmission line width	40um
Transmission line length	3750um
Transmission line thickness	50um
Transmission line spacing(s)	80um
Dielectric thickness	100um
Dielectric constant	2
Dielectric tan δ	0
Ground plane thickness	50um

Table II gives the detailed design values obtained for target objectives for a single test case for different methods. Table III gives the percentage of cases passing the tests in a range of iterations, giving an idea of the convergence behavior of the three methods. The maximum number of iterations for each method was set to 25 and the values of $\mu = 1$, $\alpha = 1$ and $\gamma = 0.01$ were used.

TABLE III. COMPARATIVE STUDY OVER 2744 TEST CASES

Range No. Iterations	% Cases Pass HSLBP	% Cases Pass TR	% Cases Pass LW
1-5	55.07%	42.31 %	44.06 %
6-10	8.97%	2.62 %	1.17 %
11-15	10.28%	0.87 %	0.11 %
16-20	2.84%	0.0036 %	0.0036 %
21-25	0.62%	0	0

978-1-4799-0708-3/13 $31.00 © 2013 IEEE

Table II. Single test case design values for experiment

Method	S13 desired (db)	S13 Got (db)	S15 desired (db)	S15 Got (db)	S17 desired (db)	S17 Got (db)	Variables						No. Iter
							d1(um)	d2(um)	d3(um)	T(um)	tan δ	eps	
HSLBP	-13.09	-16.95	-29.11	-30.11	-37.67	-37.72	80	80	120	100	0.0015	2	20
TR	-13.09	-17.09	-29.11	-31.32	-37.67	-37.76	80	100	140	125	0	3	7
LW	-13.09	-17.09	-29.11	-31.32	-37.67	-37.76	80	100	140	125	0	3	3
Commercial Tool (SNLP)	-13.09	-16.6	-29.11	-31.6	-37.67	-37.7	114.08	120.95	138.93	127.31	0.000654	3.1481	25

Table IV gives the number of iterations taken by the three methods and a Sequential Non-Linear Programming (SNLP) based Commercial Tool (CT) method for optimization. Time for a given method is No. Iterations times Forward Solve Time.

TABLE IV. COMPARATIVE STUDY OF PERFORMANCE

Case No.	S13 desired (db)	S15 desired (db)	S17 desired (db)	HSLB P No. Iter	TR No. Iter	LW No. Iter	CT (SNLP) No. Iter
1	-16.1	-28.8	-36.6	6	**25**	6	14
2	-16.1	-30.3	-33.0	10	**25**	**25**	11
3	-16.1	-30.3	-36.6	10	**25**	**25**	11
4	-15.1	-28.8	-33.0	3	3	**25**	11
5	-15.1	-28.8	-36.6	14	3	3	11
6	-15.1	-30.3	-34.8	1	4	1	11
7	-15.1	-30.3	-36.6	7	4	**25**	11
8	-14.7	-28.8	-36.6	6	3	3	2
9	-14.7	-29.7	-36.6	1	3	1	2
10	-14.7	-30.3	-36.6	1	3	1	10

Bold indicates target was not achieved in Max. Iterations

From Table III and IV it can be concluded that the HSLBP algorithm works best for finding a boundary in a design space that satisfies the targeted goals. Both in number of iterations taken to achieve a solution as well as the percentage of cases in which a solution is found, the HSLBP method shows better results. Fig. 5 (a), (b) and (c) shows the |S13|, |S15| and |S17| curves for the nominal design and design variables achieved for each of the methods mentioned in Table II.

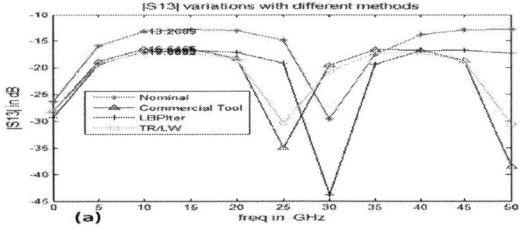

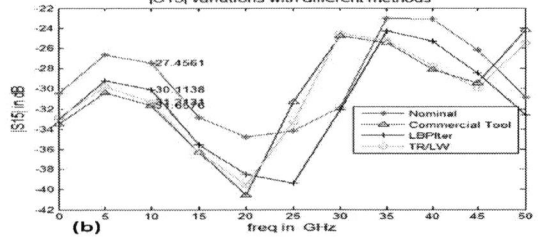

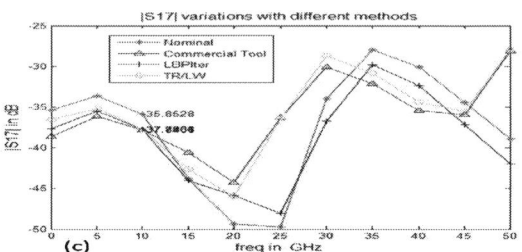

FIGURE 5. (A) |S13|, (B) |S15|, (C) |S17| CURVES FOR DIFFERENT METHODS

REFERENCES

[1] W.T.,Hao Shi , Feng, J., Xingchao Yuan Beyene, "Electromagnetic modeling methodologies and design challenges of packages for 6.4-12.8 Gbps chip-to-chip interconnects," in *Antennas and Propagation Society International Symposium*, 2004, pp. 3325-3328.

[2] Xiang G. , Brunet P. Sheach K., "Modeling of flip chip bump patterns to minimize crosstalk on a BU-BGA package design," in *Microelectronics and Packaging Conference*, Rimini, 2009, pp. 1-4.

[3] Vikram Jandhyala and Henning Braunisch Arun V. Sathanur, "A Hierarchical Simulation Flow for Return-Loss Optimization of Microprocessor Package Vertical Interconnects," *IEEE Transactions on Advanced Packaging*, vol. 33, no. 4, pp. 1021-1034, November 2010.

[4] Jian Song, Fengman Liu, Haifei Xiang, Wei Gao, Lixi Wan Haidong Wang, "Crosstalk Analysis and Optimization of High-Speed Interconnections," in *International Conference on Electronic Packaging Technology and High Density Packaging*, 2011, pp. 963-966.

[5] Hamid Hadim Tohru Suwa, "Multidisciplinary electronic package design and optimization methodology based on genetic algorithm," in *IEEE Transactions on Advanced Packaging*, vol. 30, 2007, pp. 402-411.

[6] Subramanian G.V., Raju S. Kumar V.A. Kumar N.S., "Hybrid modeling and optimization of VLSI interconnects for signal integrity using neuro-Genetic algorithm," in *Electromagnetic Interference and Compatability(INCEMIC)*, 2006, pp. 120-126.

[7] Xiaoning Ye,Raul Enriquez,Kai Xiao,Ted Ballou,Jimmy A Johansson, Beomtaek Lee, "Design Optimization for Minimal Crosstalk in Differential Interconnect," in *DesignCon* , 2012, pp. 1-20.

[8] Snezana Glavic,Mohamed H. Bakr,John W. Bandler Natalia K. Georgieva, "Feasible Adjoint Sensitivity Technique for EM Design Optimization," *IEEE Transactions on Microwave Theory and Techniques*, vol. 50, no. 12, pp. 2751-2759, December 2002.

[9] Lihui Peng W. Q. Yang, "Image reconstruction algorithms for electrical capacitance tomography," *Measurement Science and Technology*, vol. 14, pp. R1-R13, 2003.

[10] Guy Demoment, "Image reconstruction and restoration: Overview of common estimation structures and problems," *IEEE Transactions on acoustics speech and signal processing*, vol. 37, no. 12, pp. 2024-2037, December 1989.

[11] Donald R. Wilton, Allen W. Glisson Sadsiva M.Rao, "Electromagnetic Scattering by Surfaces of Arbitrary Shape," in *IEEE Transactions on Antennas and Propagation*, 1982, pp. 409-419.

Characterization of TSVs by Cascaded Daisy Chains

Yi-Chen Wu, Kai-Bin Wu, Kang-Yun Yang, Ting-Yi Huang, and Ruey-Beei Wu

Department of Electrical Engineering and Graduate Institute of Communication Engineering,

National Taiwan University, Taipei, Taiwan, 10617, R.O.C.

E-mail: rbwu@ew.ee.ntu.edu.tw

Abstract—The Characterization of stacked TSVs is difficult because of expensive experiments and low feasibility. In addition, the calibration of transmission lines connecting TSVs and coupling among TSVs is the main obstacle when horizontally connected TSVs are measured. Therefore, the equivalent transmission matrix of the coupling effect between two adjacent TSVs is derived, and a set of test structures is proposed to extract characteristics of a signal TSV. Under weak coupling conditions, the effect caused by transmission lines and coupling among signals can be eliminated by a calibration mechanism. Hence, extraction results can be used to predict the electrical behavior of stacked TSVs, and experiments are inexpensive and feasible.

Keywords-characterization, calibration, coupling, equivalent transmission matrix

I. INTRODUCTION

Optimizing the performance of integrated circuits (ICs) in a limited space has become an important issue as electronic devices continue to decrease in size. One of the most promising technologies that aim to solve this issue is called three dimensional integrated circuits (3D-ICs), which stacks ICs vertically. Compared to traditional techniques such as wire-bonding and microbumps, TSVs provide shorter transmission paths, higher input/output (I/O) density and lower cost.

However, the characterization of stacked TSVs is a difficult problem. Most of the time, 3-D full wave simulators and circuit simulators are used to extract TSVs' characteristics [1]. As the size decreases and frequency increases, simulators may fail to take all possible effects into consideration, especially when the manufacture uncertainty is taken into account. Therefore, it is also inevitable to design test structures to achieve the goal. A single TSV is too small to measure, so test structures are usually composed of two or more signal TSVs connected by metal lines in the redistribution layer (RDL) or in the interposer. Nevertheless, effects of metal lines cannot be calibrated from this kind test structure [2]-[3]. In addition, though the relation between TSVs and metal lines are investigated, the coupling effect among TSVs is not eliminated [4]-[5]. Hence, the main drawback of most structures is about the failure to calibrate the metal lines and coupling of TSVs.

The double-sided probing system was developed to do the calibration of vertical interconnects [6]. This system can characterize a single TSV after de-embedding the connection part. However, the application is limited because not every process is suitable to use the system. Besides the special probing system, the daisy chain, which means horizontally connected TSVs, was proposed to characterize TSVs. However, it is only used to detect discontinuity in the TSV

channels [7]-[8]. As a result, there still lacks practical and accurate methods to characterize stacked TSVs.

In order to extract the characteristics of stacked TSVs, a set of test structures between two metal layers is designed. In addition, coupling among TSVs is investigated, and the equivalent transmission matrix of coupling between two adjacent TSVs is obtained. By using the equivalent transmission matrix to express the coupling effect, the calibration of transmission lines and coupling can be easily done. After extraction, it is inexpensive and highly feasible to model stacked TSVs as cascaded single TSVs.

II. COUPLING EFFECT AND EQUIVALNET TRANSMISSION MATRIX AMONG SIGNALS

Since the lossy silicon substrate can be modeled as a capacitor in parallel with a conductor, coupling among TSVs has three main parts: the mutual inductance, the mutual capacitance and the conductance. As the number of TSVs increases, coupling among signals becomes more complicated and unpredictable. Therefore, in daisy chains, the calibration of coupling is essential to extract the characteristics of TSVs. Hence, different aspects of the coupling effect should be examined. In the setting, there are 4 signals (S) and 8 grounds (G) as shown in Fig. 1, and the dielectric constant of oxide liner, silicon subtract, and the conductivity of silicon subtract are 3.2, 11.9, and 10 S/m respectively. Other electrical parameters are shown in Table 1.

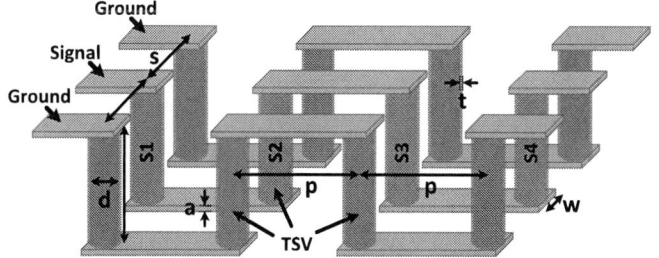

Fig. 1. Test structure for extraction of coupling effects

Only considering the vertical TSV in quasi-static field solver (horizontal RDL metal is ignored), Ansys Q3D [9], in Fig. 1, the mutual inductance between two adjacent signals (L_{12} and L_{23}) is more than eight times larger than the mutual inductance between two non-adjacent signals (L_{13} and L_{14}) as depicted in Fig. 2. In addition, the mutual capacitance caused by the silicon substrate among TSVs is shown in Fig. 3. The mutual capacitance between two adjacent signals (C_{12} and C_{23}) is several ten times larger than the mutual capacitance between

978-1-4799-0708-3/13 $31.00 © 2013 IEEE

two non-adjacent signals (C_{13} and C_{14}). Moreover, the conductance has the relation with the mutual capacitance as follows:

$$G_{Si} = \frac{\sigma_{Si}}{\varepsilon_{Si}} C_{Si} \qquad (1)$$

Table 1. Process parameters

Symbol	Description	Values
p	Pitch	200μm
s	Spacing	120μm
t	Thickness of oxide liners	2μm
d	Diameter of TSVs	56μm
h	Height of TSVs	120μm
w	Width of transmission lines	90μm
a	Thickness of transmission lines	12μm
$\varepsilon_{r,liner}$	Permittivity of oxide liners	3.2
$\varepsilon_{r,silicon}$	Permittivity of oxide dielectric	11.9
$\sigma_{silicon}$	Conductivity of silicon substrate	10 S/m

When the mutual capacitance between two non-adjacent TSVs is much smaller, the conductance will also be smaller. Hence, the impedance of shunt C_{Si} and G_{Si} between them is much larger than it of two adjacent signals, and the channel can be seen as open. As a result, when used to characterize TSVs, the test structure can be considered only having coupling between two adjacent signals.

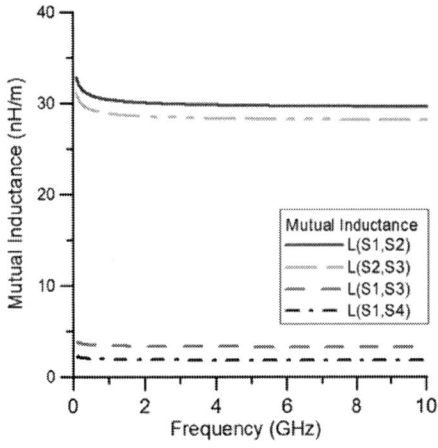

Fig. 2. Mutual inductance of 12 TSVs with 4 signals (nH/m)

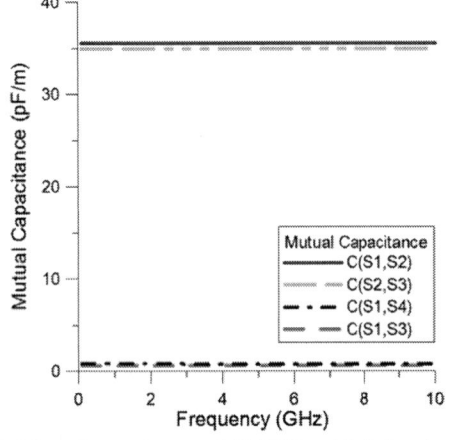

Fig. 3. Mutual capacitance of 12 TSVs with 4 signals (pF/m)

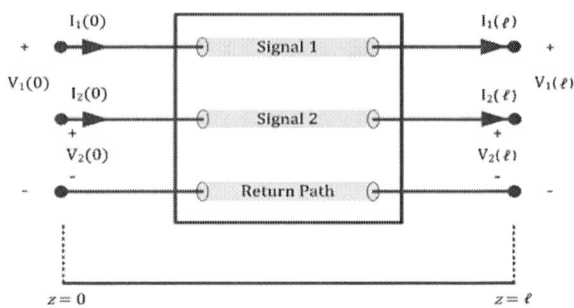

Fig. 4. Two-wire transmission lines system

In order to take the coupling effect into consideration, how to formulize it as an equivalent transmission matrix is an important issue. Firstly, two adjacent signal TSVs can be seen as a two-wire transmission lines system as illustrated in Fig. 4. As the per-unit-length impedance and admittance matrices are

$$\mathbf{Z} = \begin{bmatrix} Z_{11} & Z_m \\ Z_m & Z_{11} \end{bmatrix} \qquad (2)$$

$$\mathbf{Y} = \begin{bmatrix} Y_{11} & Y_m \\ Y_m & Y_{11} \end{bmatrix} \qquad (3)$$

the following is the state-transition matrix of it when ℓ is much smaller than the wavelength [10].

$$\begin{bmatrix} V_1(0) \\ V_2(0) \\ I_1(0) \\ I_2(0) \end{bmatrix} = \begin{bmatrix} 1 & 0 & Z_{11}\ell & Z_m\ell \\ 0 & 1 & Z_m\ell & Z_{11}\ell \\ Y_{11}\ell & Y_m\ell & 1 & 0 \\ Y_m\ell & Y_{11}\ell & 0 & 1 \end{bmatrix} \begin{bmatrix} V_1(\ell) \\ V_2(\ell) \\ I_1(\ell) \\ I_2(\ell) \end{bmatrix} \qquad (4)$$

If a transmission line, which is d in length, is much shorter than the wavelength, its transmission matrix can be represented by the per-unit-length impedance Z_{TL} and per-unit-length admittance Y_{TL} of the transmission line.

$$\begin{bmatrix} V_2(\ell) \\ I_2(\ell) \end{bmatrix} = \begin{bmatrix} 1 & Z_{TL}d \\ Y_{TL}d & 1 \end{bmatrix} \begin{bmatrix} V_1(\ell) \\ -I_1(\ell) \end{bmatrix} \qquad (5)$$

Hence, when signals 1 and 2 are connected by the transmission line at $z = \ell$, $V_2(\ell)$ and $I_2(\ell)$ are substituted by (5). In addition, both ℓ and d are much smaller than the wavelength, so higher-order items can be neglected. The result is shown as

$$\begin{bmatrix} V_2(0) \\ I_2(0) \end{bmatrix} =$$

$$\begin{bmatrix} 1 & Z_{TL}d - 2(Z_m - Z_1)\ell \\ 2(Y_{11} + Y_m)\ell + Y_{TL}d & 1 \end{bmatrix} \begin{bmatrix} V_1(0) \\ -I_1(0) \end{bmatrix} \qquad (6)$$

Fig. 5. Assumption of the equivalent transmission matrix

Secondly, under the condition that both ℓ and d are small, it is assumed that the coupling effect between two adjacent signals is equivalent to a transmission matrix $\hat{M}_c$. When cascaded in series with signals and the transmission line, $\hat{M}_c$ is in the middle of the transmission line as shown in Fig. 5. Moreover, $\hat{M}_{TSV}$ and $\hat{M}_{TX/2}$, in the same form as (5), mean the transmission matrix of a signal TSV and a half of the transmission line respectively. After the comparison between Fig. 4 and Fig. 5, $\hat{M}_c$ can be derived as below

$$\hat{M}_c = \begin{bmatrix} 1 & -2\,Z_m\,\ell \\ 2\,Y_m\,\ell & 1 \end{bmatrix} \tag{7}$$

To sum up, when used to characterize TSVs, the structure in Fig. 1 can be seen only having coupling between two adjacent signals. In addition, this coupling effect is modeled as an equivalent transmission matrix as (7). Based on the analysis, section III is about structures to extract characteristics of TSVs.

III. DESIGN OF TEST STRUCTURES

In this paper, a set of test structures is designed, and there are two main strengths of the proposed method. One is about the feasibility. For the cost of stacking TSVs is high, it is necessary to design test structures within two metal layers. The other is about the calibration mechanism. In this method, the effects caused by transmission lines and coupling can be eliminated. As depicted in Fig. 6, the first sub-structure is composed of 4 signals and 8 grounds, which are connected horizontally by metal lines. Parameters are listed in Table 1. As ports are set as follows, [M$_{full}$] is its transmission matrix.

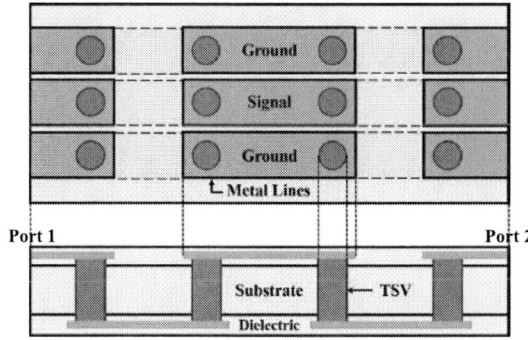

Fig. 6. First sub-structure

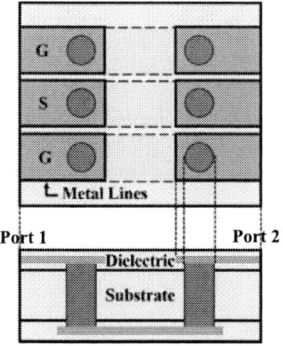
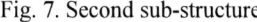

Fig. 7. Second sub-structure

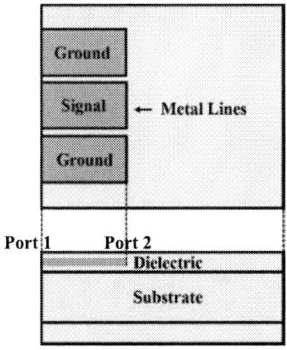

Fig. 8. Third sub-structure

Moreover, the second sub-structure is half of the first one (Fig. 7), and the third sub-structure is about metal lines of the most left part of the first one (Fig. 8). Ports are set as depicted, and their transmission matrices are [M$_{half}$] and [M$_{TX}$].

As a result, two relations can be derived as (8) and (9). [M$_{TSV}$] means the transmission matrix of a signal TSV, and [M$_c$] is the coupling effect between two adjacent signals.

$$[M_{full}] = [M_{half}] \times [M_c] \times [M_{half}] \tag{8}$$

$$[M_{half}] = [M_{TX}] \times [M_{TSV}] \times [M_{TX}] \times [M_c] \times [M_{TX}] \times [M_{TSV}] \times [M_{TX}] \tag{9}$$

From (8), [M$_c$] can be derived as below

$$[M_c] = [M_{half}]^{-1} \times [M_{full}] \times [M_{half}]^{-1} \tag{10}$$

If it is assumed that [Y] and [U] are shown as

$$[Y] = [M_{TX}] \times [M_c] \times [M_{TX}] \tag{11}$$

$$[U] = [M_{TSV}] \times [Y] \tag{12}$$

then (9) can be reformed as follows

$$[M_{TX}]^{-1} \times [M_1] \times [M_{TX}]^{-1} = [U]^2 \times [Y]^{-1} \tag{13}$$

Hence, a set of [P] and [λ] can be obtained from (13), and [U] can be solved using eigendecomposition.

$$[U] = [P] \times [\lambda] \times [P]^{-1} \tag{14}$$

and [M$_{TSV}$] can be derived as below.

$$[M_{TSV}] = [P] \times [\lambda] \times [P]^{-1} \times [Y]^{-1} \tag{15}$$

IV. RESULTS AND DISCUSSION

As a verification, the calculation results from the proposed method are compared to the extraction results from full-wave simulations. As shown in Fig. 9, there is high consistency in both return loss and insertion loss. In addition, the variation between them is very small, as evident from Table 2.

In applications, the effect caused by connections such as bumps cannot be neglected. Therefore, by adjusting the length of transmission lines in test structures, it can be easily taken into consideration. For example, bumps are added into the set of test structures mentioned above, with 20μm in height and 80μm in diameter. After shortening transmission lines to meet the requirements, the results of a signal TSV with effects of connections can be extracted and depicted as Fig. 10. Moreover, the variation is also very small, as evident in Table 3.

Table 2. Variation between calculation results and extraction results.

Items		Values
$\lvert S_{11} \rvert$ @ 10GHz	Extraction	-27.8dB
	Calculation	-28.5dB
$\lvert S_{21} \rvert$ @ 10GHz	Extraction	-0.312dB
	Calculation	-0.327dB

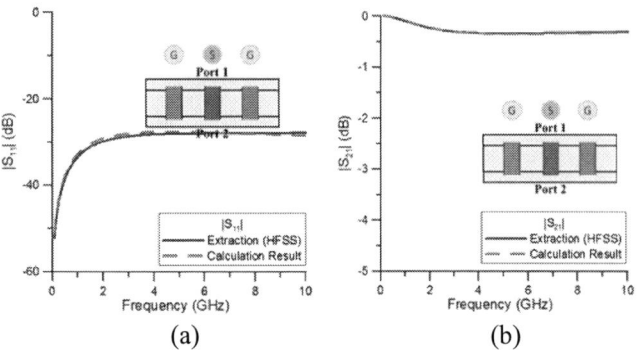

Fig. 9. Comparison of calculation results and extraction results. (a) Return loss. (b) Insertion loss.

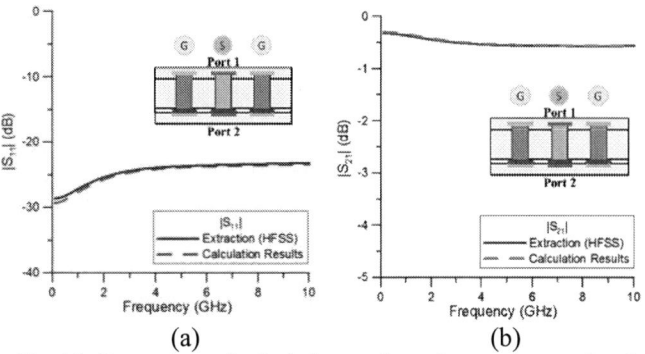

Fig. 10. Comparison of calculation results and extraction results of TSVs with effects of connections. (a) Return loss. (b) Insertion loss.

Table 3. Variation between calculation results and extraction results of TSVs with effects of connections.

Items		Values
$\lvert S_{11} \rvert$ @ 10GHz	Extraction	-23.2dB
	Calculation	-23.4dB
$\lvert S_{21} \rvert$ @ 10GHz	Extraction	-0.571dB
	Calculation	-0.564dB

From the results and discussion, it is proved that TSVs can be characterized by the proposed structures and calculations. Moreover, the accuracy of it is high enough to predict the electrical behavior of stacked TSVs. Hence, the characterization of TSVs can be done by the proposed method.

V. CONCLUSIONS

Because the characterization of stacked TSVs is not easy, an inexpensive and feasible method is proposed to achieve the goal. First of all, coupling among TSVs is examined, and the equivalent transmission matrix of coupling between two adjacent signals is derived. Secondly, a set of test structures is designed. The effects of transmission lines and coupling among TSVs can be eliminated by a calibration mechanism. In addition, the characteristics of TSVs can be extracted from simple calculations. At the end, this method is verified by comparing the calculation results with the extraction results from full-wave simulations. Moreover, the accuracy remains high when the effects caused by connections are also investigated. Therefore, stacked TSVs can easily be characterized by the method proposed in this paper.

ACKNOWLEDGEMENT

This work was supported in part by the National Science Council, Republic of China, under Grant NSC 100-2221-E-002-223-MY3, National Taiwan University under 101R89083, and NSC 101-2218-E-002 -010.

REFERENCES

[1] J. S. Pak, C. Ryu, and J. Kim, "Electrical characterization of through silicon via (TSV) depending on structural and material parameters Based on 3D full wave simulation," in *Int'l. Conf. Electron. Mat. Packag.*, Daejeon, Korea, Nov. 11–19, 2007.

[2] N. Kim, D. Wu, D. Kim, A. Rahman, and P. Wu, "Interposer design optimization for high frequency signal transmission in passive and active interposer using through silicon via (TSV)," in *IEEE Electron. Compon. Technol. Conf.*, Lake Buena Vista, Florida, USA, May 31–June 3, 2011, pp.1160–1167.

[3] T. G. Lim, Y. M. Khoo, C. S. Selvanayagam, D. S. W. Ho, R. Li, X. Zhang, G. Shan, and X. Y. Zhong, "Through silicon via interposer for millimetre wave applications," in *IEEE Electron. Compon. Technol. Conf.*, Lake Buena Vista, Florida, USA, May 31–June 3, 2011, pp.577–582.

[4] H. Kim, J. Cho, M. Kim, K. Kim, J. Lee, H. Lee, K. Park, K. Choi, H.-C. Bae, J. Kim, and J. Kim, "Measurement and analysis of a high-speed TSV channel," *IEEE Trans. Compon., Packag., Manuf. Technol.*, vol. 2, no. 10, pp.1672–1685, Oct. 2012.

[5] J. Kim, J. S. Pak, J. Cho, E. Song, J. Cho, H. Kim, T. Song, J. Lee, H. Lee, K. Park, S. Yang, M.-S. Suh, K.-Y. Byun, and J. Kim, "High-frequency scalable electrical model and analysis of a through silicon via (TSV)," *IEEE Trans. Compon., Packag., Manuf. Technol.*, vol. 1, no. 2, pp.181–195, Feb. 2011.

[6] K.-C. Lu, Y.-C. Lin, T.-S. Horng, S.-M. Wu, C.-C. Wang, C.-T. Chiu, and C.-P. Hung, "Vertical interconnect measurement techniques based on double-sided probing system and short-open-load-reciprocal calibration," in *IEEE Electron. Compon. Technol. Conf.*, Lake Buena Vista, Florida, USA, May 31–June 3, 2011, pp.2130–2133.

[7] J. Kim, D. Jung, J. Cho, J. S. Pak, J. M. Yook, J. C. Kim, and J. Kim, "High-frequency measurements of TSV failures," in *IEEE Electron. Compon. Technol. Conf.*, San Diego, California, USA, May 29–June 3, 2012, pp. 298–303.

[8] D. H. Jung, J. Kim, H. Kim, J. J. Kim, J. Kim, J. S. Pak, J.-M. Yook, and J. C. Kim, "Frequency and time domain measurement of through-silicon via (TSV) failure," in *IEEE Electrical Performance Electron. Packag. Syst.*, Tempe, Arizona, USA, Oct. 21–24, 2012, pp. 331–334.

[9] Q3D Extractor, Ansys Corporation, Canonsburg, PA, USA [Online]. (http://www.ansys.com)

[10] C. R. Paul, *Analysis of Multiconductor Transmission Lines*. New York: Wiley, 1994, chs. 7.

Design and Verification of SMT MMIC Package using a 20 GHz LNA, a 40 GHz LNA and a 40GHz Digital Attenuator

Inkwon Ju* , In-bok Yom* and Keun Kwan Ryu**
Satellite & Wireless Convergence Research Department*
Electronics and Telecommunications Research Institute*
Department of Electronic Engineering**
Hanbat National University**
Daejeon, South Korea
juinkwon@etri.re.kr

Abstract— A surface mounting (SMT) low temperature cofired ceramic (LTCC) MMIC package was developed using new vertical transition consist of a trough line, a slab line, and shielded multilayer coplanar waveguides (SMCPWs) for DC to 50 GHz band applications. A 20 GHz LNA, a 40 GHz LNA and a 40 GHz 5-bit digital attenuator were packaged to verify the performances of the developed LTCC SMT MMIC package. The packaged 20 GHz LNA has less than 0.3 dB gain degradation, compared with the on-wafer measurement. The packaged 40 GHz LNA has some differences with on-wafer measurement due to the size mismatch with the MMIC package. The packaged 40 GHz attenuator exhibits a negligible degradation of the attenuation accuracy, compared to the on-wafer results.

Keywords— *Integrated circuit packaging, interconnections, ceramics, surface mounting.*

I. INTRODUCTION

A microwave monolithic integrated circuit (MMIC) is used in a transmitter and a receiver of various types of radio systems such as portable communication systems, military communication systems, satellite communication systems, and the like.[1] The MMIC is manufactured through a semiconductor manufacturing process and is formed in an unpacked bare chip.[2] Therefore, the MMIC needs an appropriate package.

A highly efficient package for MMIC of microwaves and millimeter waves must satisfy mechanical, electrical, and environmental requirements. In an aspect of mechanical and environmental requirements, the package must provide a function of protecting an internal circuit from surrounding environments. In an aspect of electrical requirements, the package must provide a minimum insertion loss and a high isolation between terminals and also provide a function of shielding electromagnetic waves in order to reduce electromagnetic interference (EMI). Also, another important electrical requirement is to not allow the package to give an adverse effect on a circuit performance or to make parasitic cavity resonance deteriorate the circuit performance.

Generally, an internal circuit mounted onto a surface mount package may be electrically connected with an external circuit by vertical interconnection and also may need a low-loss vertical interconnection in order to completely transfer signals of a microwave band and a millimeter wave band.

Recently, several research activities have been reported for MMIC packages. For surface mounting MMIC package, high performance vertical transitions are required. The conventional vertical transition mainly used a via-hole [3], and the via-hole mainly acts on an inductance and generates a discontinuity.[4] The vertical transition which uses the coaxial line in order to remove the discontinuity of via-hole was reported [5]. The vertical transition consist of three via-holes which are similar to the CPW structure [6] and the five wire lines vertical transition were reported [7].

We developed a SMT LTCC MMIC package using new vertical transition consist of the trough line, the slab line, and SMCPWs. The trough line and the slab line were employed in order to minimize the discontinuity of the conventional via-hole vertical transition. The SMCPWs were used to reduce the radiation loss and crosstalk of the transition between the multiple transmission lines.

To verify the proposed SMT MMIC packages, a 20 GHz LNA, a 40 GHz LNA and a DC to 40 GHz 5-bit digital attenuator were packaged and tested. The performances of the packaged MMICs were compared with the on-wafer measurement results and the conventional metal package.

II. DESIGN OF SMT MMIC PACKAGE

The structure of the proposed SMT LTCC MMIC package is shown in Fig. 1 and consists of the two vertical transitions, cavity for MMIC, DC and ground pads. The outer dimensions of the MMIC package are 5.0 mm x 5.0 mm x 1.2 mm with a package cover. RF pad size in the bottom is 220 μm x 200 μm and DC pad size is 250 μm x 250 μm. The package consists of a six layers LTCC substrate and another six layers LTCC cover. And the LTCC package is surface mounted on the RO4003C substrate of 8 mil thickness.

978-1-4799-0708-3/13 $31.00 © 2013 IEEE

We adopt the trough line and the slab line for the structure of the vertical transition. The trough line and the slab line are facilitated to arrange the input-output transmission lines because one side or two sides of the ground are opened. The round wire of the slab line and the trough line was made with the signal via in the multi-layered MMIC package. And the ground is constructed of ground planes and ground via holes.

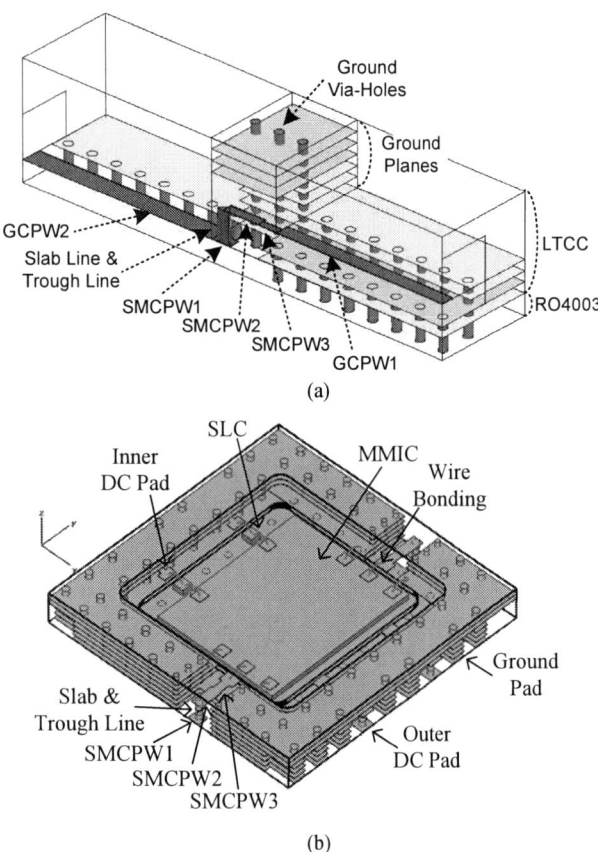

(a)

(b)

Fig. 1 The proposed LTCC SMT MMIC package structure (a) 3-D view of symmetric plane for single vertical transition. (b) The packaged MMIC in the SMT MMIC package.

The characteristic impedances of the trough line and the slab line are approximately given by the Eq. (1) and (2) respectively [8].

$$Z_0 = \frac{138}{\sqrt{\varepsilon_r}} \log\left(1.17 \times \frac{s1}{d2}\right)(\Omega) \qquad (1)$$

$$Z_0 = \frac{138}{\sqrt{\varepsilon_r}} \log\left(\frac{4 \times s1}{\pi \times d2}\right)(\Omega) \qquad (2)$$

In order to reduce the radiation loss and crosstalk, the SMCPWs were used for the proposed vertical transition as shown in Fig. 1. Conventionally, the complicated structure like a SMCPW was analysed with the conformal mapping technique [9] or the full-wave analysis [10].

We used the analytical method of the SMCPW structure using the circuit library of the linear simulator ADS in order to get the initial design values of HFSS and reduce simulation time. The characteristics of SMCPW cannot be analysed through the CPW, the microstrip line, or the strip line library

which is generally used as transmission line in ADS. But there are two kinds of the library to model the multi-layer transmission line in ADS. One is the printed circuit board (PCB) library and the other is the multilayer library. The PCB library is based on a quasi-static analysis in an enclosed region with stratified layers of a single dielectric. The multilayer library is based on method of moments and Green's function method and it handles arbitrary dielectric layers and arbitrary metal thickness.

The proposed vertical transition is constructed of the multiple transmission lines of the SMCPW1-trough line-slab line- SMCPW2-SMCPW3. For the design of the vertical transition using ADS, the trough line and the slab line were replaced with the strip line. The ground plane space of the strip line is 700 μm and this is the same as the s1 of the slab line and the trough line. The strip line has the 50 ohm characteristic impedance when the conductor thickness and the line width of the strip line are 100 μm. Fig. 2 shows the schematic of the single vertical transition by ADS circuit libraries.

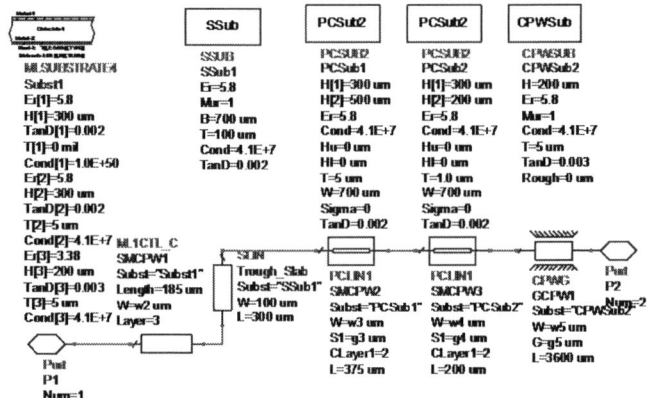

Fig. 2. The schematic of the single vertical transition by ADS circuit libraries.

III.   VERIFICATION OF THE SMT MMIC PACKAGE

To verify the proposed SMT MMIC package, the packaging and measurements of the MMICs are required. A 20 GHz LNA [11], a 40 GHz LNA [12], and a DC to 40 GHz 5-bit digital attenuator [13] were selected to verify the performances of the MMIC package. The 20 GHz LNA, the 40 GHz LNA, and the DC to 40 GHz 5-bit digital attenuator were produced with NGST (Northrop Grumman Space Technology) 0.15 um GaAs pHEMT process.

The wire bonding between RF signal pads were made three 0.7 mil gold wires to reduce the bonding inductance. And the ball bonding was used in the RF ground pads. The MMICs were bonded on SMT MMIC package by conductive epoxy substitute for eutectic because of the rough surface of SMT MMIC package.

The 20 GHz LNA is a two-stage amplifier with balanced structure and has normally 1.6 dB NF and 18 dB gain at the bias conditions of 2.3 V and 60 mA. Fig. 3 shows the photo of the packaged 20 GHz LNA in the SMT MMIC package.

978-1-4799-0708-3/13 $31.00 © 2013 IEEE

As shown in Fig. 4 the packaged 20 GHz LNA has less than 0.3 dB gain degradation, compared with the on-wafer measurement. The gain ripple is flattened by using of the single layer capacitors (SLC) in the SMT MMIC package than on-wafer results. The return loss has some degradation, but still less than -20 dB in the interesting frequency band.

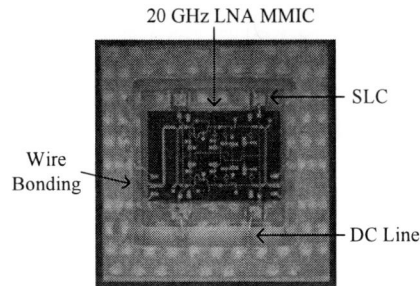

Fig. 3. The packaged 20 GHz LNA in the SMT MMIC package.

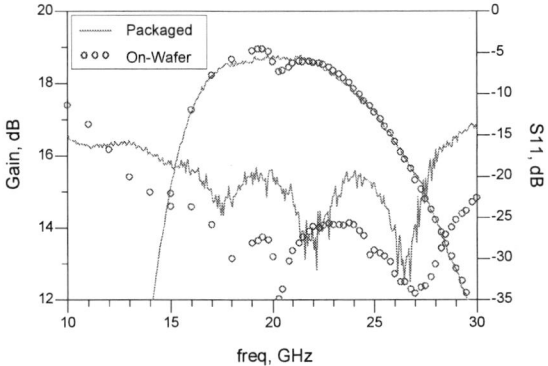

Fig. 4. Measurement results of on-wafer and the packaged 20 GHz LNA.

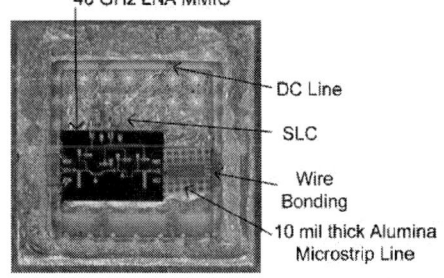

Fig. 5. The packaged 40 GHz LNA in the SMT MMIC package.

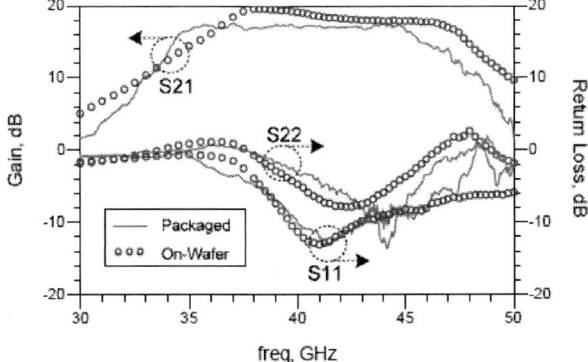

Fig. 6. Measurement results of on-wafer and the packaged 40 GHz LNA.

The 40 GHz LNA is a two-stage amplifier and has normally 2.8 dB NF and 18 dB gain at the bias conditions of 2.15 V and 15 mA. Fig. 5 shows the photo of the packaged 40 GHz LNA in the proposed SMT MMIC package.

A 1 mm long microstrip line of 10 mil thick alumina substrate was used behind LNA output port because the 40 GHz LNA length is 2.2 mm, whereas the length of cavity for MMIC is 3.2 mm. Only signal pad bonding with relatively long bonding wire were made between the 1 mm long microstrip line and others interconnection pad due to thickness difference. The MMIC GaAs substrate thickness is 4 mils. The only signal pad bonding with long bonding wire induce high inductance than ground-signal-ground pads bonding method with short bonding wire.

As shown in Fig. 6, the packaged 40 GHz LNA has 1 to 3 dB gain degradation and about 2 GHz operating frequency shift, compared with the on-wafer measurement. If a suitable package were used, the gain degradation and the frequency shift would be decreased.

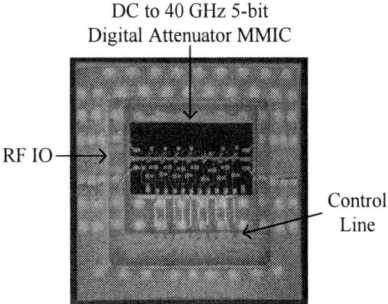

Fig. 7. The packaged DC to 40 GHz 5-bit digital attenuator in the SMT MMIC package

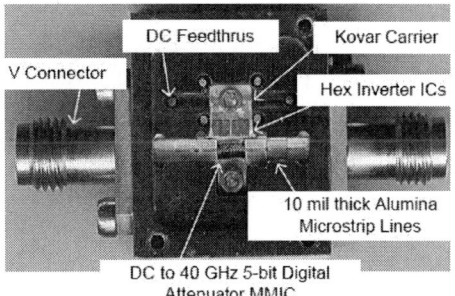

Fig. 8. The packaged DC to 40 GHz 5-bit digital attenuator in the conventional metal carrier and the metal package

The DC to 40 GHz 5-bit digital attenuator has 1 dB resolution, 23 dB dynamic ranges and high attenuation accuracy over all attenuation range and full 40 GHz bandwidth. The input and output return losses of the attenuator are better than 14 dB over all attenuation states.

Fig. 7 shows the photo of the packaged DC to 40 GHz 5-bit digital attenuator in the proposed SMT MMIC package. And, Fig. 8 shows a conventional metal carrier and a metal package for the digital attenuator. The conventional metal package consists of a Kovar carrier, mounting screws, DC feedthrus, alumina substrate microstrip lines, and V-connecters.

978-1-4799-0708-3/13 $31.00 © 2013 IEEE

As shown in Fig. 9, the proposed SMT MMIC packaged attenuator exhibits a negligible degradation of the attenuation accuracy compared to the on-wafer results, whereas the metal package shows high degradation of performance and cavity resonance around 24 GHz.

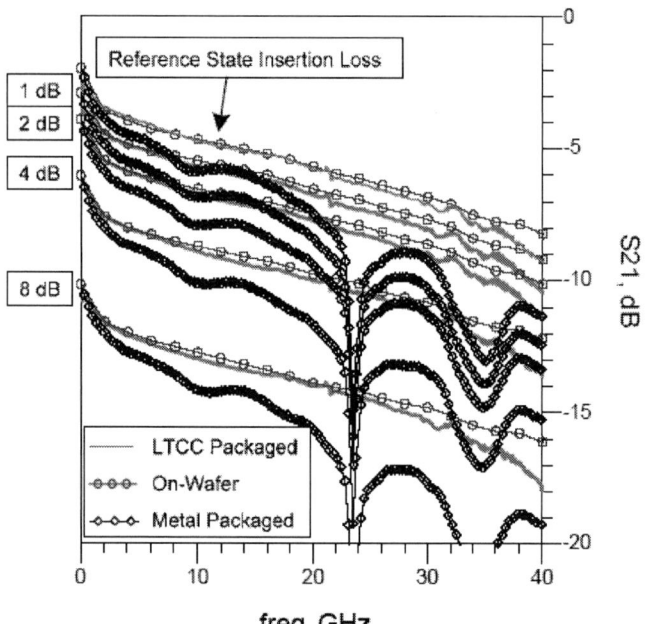

Fig. 9. Measurement results of on-wafer, metal package and the proposed SMT MMIC package for the 5-bit digital attenuator MMIC.

IV. CONCLUSION

We designed the SMT MMIC package using the trough line, the slab line, and the SMCPWs. The performances of the proposed SMT MMIC packages were verified by the packaging and measurements of the 20 GHz LNA, the 20 GHz LNA and the DC to 40 GHz 5-bit digital attenuator. The proposed SMT MMIC package has not only a merit of low cost but also an advantage of high performance for millimeter-wave MMICs compare to the conventional metal package.

ACKNOWLEDGMENT

This research was funded by the MSIP(Ministry of Science, ICT & Future Planning), Korea in the ICT R&D Program 2013.

REFERENCES

[1] J.-C. Jeong, D. Shin, I. Ju and I.-B. Yom, "An S-Band Multifunction Chip with a Simple Interface for Active Phased Array Base Station Antennas," in *ETRI Jounal*, Vol.35, No.3, June 2013, pp. 378-385.

[2] J.-C. Jeong, D.-P. Jang and I.-B. Yom, "A Ka-Band 6-W High Power MMIC Amplifier with High Linearity for VSAT Applications," in *ETRI Jounal*, Vol.35, No.3, June 2013, pp. 546-549.

[3] Y.-H. Suh, D. Richardson, A. Dadello, S. Mahon and J. Harvey, "A Low-Cost High Performance GaAs MMIC Package using Air-Cavity Ceramic Quad Flat Non-Leaded Package up to 40 GHz," in *35th European Microwave Conference 2005 - Conference Proceedings 3, art. no. 1610233, pp. 1491-1494*.

[4] T. Tischler, M. Rudolph, A. Kilk and W. Heinrich, "Via Arrays for Grounding in Multilayer Packaging-Frequency Limits and Design Rules," in *IEEE MTT-S Digest*, 2003, pp. 1147~1150.

[5] R. Valois, D. Baillargeat, S. Verdeyme, M. Lahti and T. Jaakola, "High Performances of Shielded LTCC Vertical Transitions From DC up to 50 GHz," in *IEEE Transaction on Microwave Theory and Techniques*, Vol.53, No.6, June 2005, pp. 2026-2032.

[6] J. Heyen, T. V. Kerssenbrock, A. Chernyakov, P. Heide and A.F. Jacob, "Novel LTCC/BGA Modules for Highly Integrated Millimeter-Wave Transceivers," in *IEEE Transaction on Microwave Theory and Techniques*, Vol.51, No.12, Dec 2003, pp. 2589-2596.

[7] F. J. Schmuckle, A. Jentzsch, W. Heinrich, J. Butz and M. Spinnler, "LTCC as MCM Substrate: Design of Strip-Line Structures and Flip-Chip Interconnects," in *IEEE MTT-S Digest*, 2001, pp. 1903~1906.

[8] Liao, Samuel Y., "Microwave Circuits Analysis and Amplifier Design," *Prentice-Hall*, 1987, pp.32~33.

[9] S. Gevorgian, L.J.P. Linner and E.L. Kollberg, "CAD Models for Shielded Multilayered CPW," in *IEEE Transaction on Microwave Theory and Techniques*, Vol.43, No.4, Apr 1995, pp. 772-779.

[10] S. S. Bedair and I. Wolff, "Fast, Accurate and Simple Approximate Analytic Formulas for Calculating the Parameters of Supported Coplanar Waveguides for (M)MIC's," in *IEEE Transaction on Microwave Theory and Techniques*, Vol.40, No.1, Jan 1992, pp. 41-48.C. J. Kaufman, Rocky Mountain Research Lab., Boulder, CO, private communication, May 1995.

[11] Inkwon Ju, In-Bok Yom, "High Step Accuracy MMIC Channel Amplifier using Nonlinear Temperature Compensation for Ka-band OBS Satellite Transponder," in *24th AIAA International Communications Satellite Systems Conference*, ICSSC 2, pp. 863-869.

[12] Byung-Jun Jang, In-Bok Yom, and Seong-Pal Lee, "Millimeter Wave MMIC Low Noise Amplifiers Using a 0.15um Commercial pHEMT Process," in *ETRI Journal*, Vol. 24, No. 3, June 2002, pp.190-196.

[13] Inkwon Ju; Youn-Sub Noh; In-Bok Yom, "Ultra broadband DC to 40 GHz 5-bit pHEMT MMIC digital attenuator," in *35th European Microwave Conference 2005 - Conference Proceedings 2, art. no. 1610096, pp.995-998*.

978-1-4799-0708-3/13 $31.00 © 2013 IEEE

A Novel Common-Mode Filter for Multiple Differential Pairs with Low Crosstalk and Low Mode Conversion Level

Chi-Hsuan Cheng
Graduate Institute of Communication Engineering
National Taiwan University
Taipei, Taiwan
f99942024@ntu.edu.tw

Tzong-Lin Wu
Graduate Institute of Communication Engineering
National Taiwan University
Taipei, Taiwan
wtl@cc.ee.ntu.edu.tw

Abstract—**In this paper, a two-pair common-mode filter is proposed to solve the crosstalk and mode conversion problem of conventional ones in multi-pair form, such as USB3.0 and PCI-Express II. Coupling of the two differential pairs is obviously reduced by a shielding ground plane between them. Moreover, mode conversion problem resulted from asymmetry is also solved. Some simulations have been done to prove the good signal integrity over gigahertz bands brought by the proposed structure, including low loss and distortion of differential mode, a deep transmission zero to suppress common-mode noise, and lower crosstalk and mode conversion level.**

Keywords—common-mode filter, crosstalk, mode conversion, signal integrity (SI)

I. INTRODUCTION

In recent years, differential transmission lines have been playing an important role in high-speed digital circuits due to their high immunity to noise and electromagnetic interference (EMI). Ideally, a differential signaling pair has symmetric traces and propagates balanced differential signals. Nonetheless, in practical circuits, imbalance will arise in the course of the unavoidable discontinuities such as bend, asymmetric routing, patterned ground plane, incoherent waveform and fabrication variation. These imbalances will induce unwanted common-mode noise which will couple to the shields, grounds, or I/O cables and form radio frequency interference (RFI) or EMI problems [1], [2].

Besides, multiple pairs of differential signaling lines are used for higher data rate and better audio / video quality in latest interfaces such as USB 3.0 and PCI-Express II. As the number of differential lines increase, it is more difficult to route the traces symmetrically, which implies that problem resulted from common-mode noise will be more serious. Furthermore, crosstalk between differential pairs will also reduce signaling quality, which is considered as signal integrity (SI) problem.

Since common-mode noise is inevitable in practical differential signaling lines, it is important to design good common-mode filter to keep the transmission quality. Recently, some common-node filters using patterned-ground structure are

proposed [3]. These designs are low-cost and wideband suppressive, but they are too space-occupying and not suitable for multilayer structure, and the patterned-ground is also a main channel of crosstalk. Some common-mode filters with broadband suppression are also developed for multilayer printed circuit board (PCB) or low-temperature co-fired ceramic (LTCC) [4], [5] , but crosstalk problems among pairs cannot be avoided when applied to multiple pair designs.

In this paper, a two-pair common-mode filter applying one-cell mushroom structure with suppression over GHz frequency range is proposed. Lower differential-mode crosstalk level can be also achieved with some shielding planes, which implies the improvement of signal integrity. In addition, with this placement, lines are fully symmetric for both two pairs, and as a result, mode conversion will not occur when signal passes through this structure. Furthermore, this structure is also low loss and low distortion for differential-mode signal, which can also be proved in eye diagram.

II. STRUCTURE AND EQUIVALENT CIRCUIT MODEL

A. Anslysis of Mushroom-Like Common-Mode Filter

A common-mode filter should not only suppress common-mode noise but also act as an all-pass filter for differential-mode signal. Traditional mushroom-like structure can easily meet these two requirements, which can be observed by applying even- and odd-mode analysis. As Fig. 1 shows, when differential-mode signal passes through, the symmetric plane will acts as a perfect electric conductor (PEC). Under this circumstance, the traces behave like normal transmission lines, and as a result, have no influence on the signal. When it comes to common-mode noise, the symmetric plane will act as a perfect magnetic conductor (PMC). In consequence, the shape of mushroom is kept. When the structure resonances in some frequency bands, the wave on the traces will be guided to the ground and cannot transmit anymore. Hence, common-mode suppression can be achieved.

However, if two filters are put nearby, three problems that greatly degrade the performance of filter will occur. The first one is the incensement of crosstalk level, especially that of

978-1-4799-0708-3/13 $31.00 © 2013 IEEE

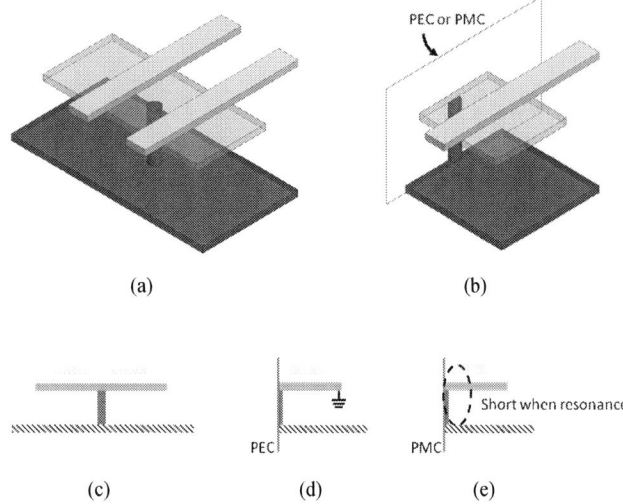

(a)   (b)

(c)   (d)   (e)

Fig. 1. Analysis of mushroom-like common-mode filter. (a) 3D virw of the whole structure. (b) Mode analysis by applying PEC or PMC on the symmetric plane. (c) Front view of the whole structure. (d) Equivalence as normal transmission line for odd (or differential) mode. (e) Equivalence as a shunt resonator for even (or common) mode.

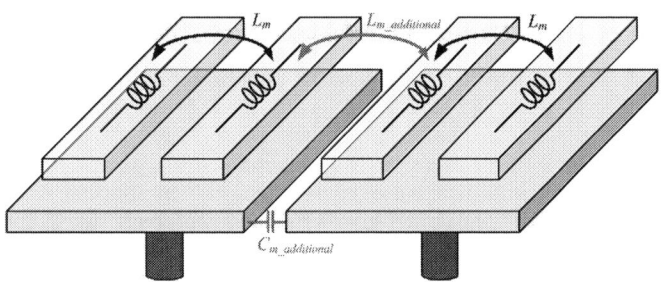

Fig. 2. Additional coupling between pairs which will change the symmetry of filter.

differential mode. The second problem is that mode conversion become obvious since the structure is not wholly symmetric anymore. The third problem, which results from the coupling between pairs, will change the filter characteristic such as stopband of common-mode and make it more difficult to model or design.

B. Proposed Structure and Equivalent Circuit Model

In order to solve the previous problems, shielding and symmetric structure is essential, and based on this concept, a two-pair common-mode filter is proposed. As illustrated in Fig. 3, two mushroom-like structures are placed in opposite side and mirror symmetrical to the ground plane. Each part includes a differential pair which is shielded to another with the ground plane. Besides, horizontal symmetry can also be wholly achieved, so it is expected to bring out good performance. The common-mode stopband of each pair can be predicted with the even-mode equivalent circuit model, as shown in Fig. 4. It is composed of two transmission lines and a inductance which is shunt to ground. One transmission line represents the signal trace and the other is for the patch of mushroom. The return path of the first line is shorted with the signal line of the second

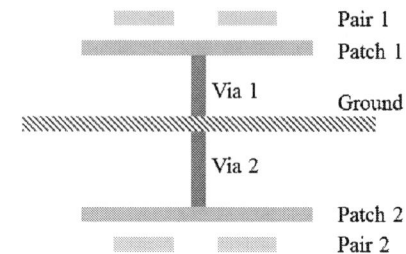

Fig. 3. Proposed two-pair common-mode filter in concept.

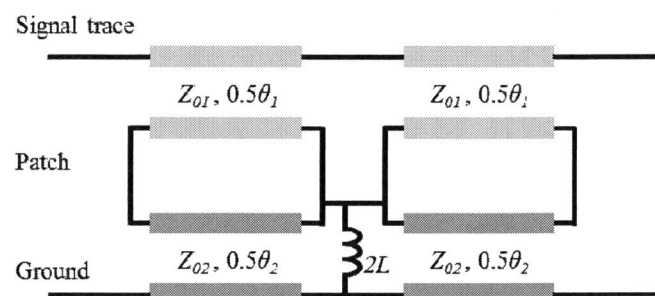

Fig. 4. Equivalent circuit model for the even-mode signal of the proposed common-mode filter.

one since they stand for the same conductor, that is, the patch. At the center of the second transmission line, an inductance is shunted to ground to model the characteristic of via. The value of inductance is twice as much as that extracted with full-wave simulator due to the PMC boundary condition.

C. Implemtation and Extraction of Circuit Element

This filter is implemented in a 6-layer printed circuit board with substrate of FR4. Since there is only through-hole via available in this processing and larger value of inductance is needed for lower stopband of common mode, some modification is essential. As shown in Fig. 5, the upper traces and patch are set in layer 1 and 2 respectively, while the lower ones are set in layer 6 and 5. The ground plane inside the filter is in layer 3, while the outside ones are on layer 2 and 5, which is mostly used in 6-layer PCB. The through-hole vias connected to the patches is grounded with the vias outside the filter after a short routing in layer 4, which can increase the equivalent value of inductance. Besides, they are shifted a little in the opposite away from the center such that the electrical behaviors of the two pairs can be designed separately.

The relevant dimensions are shown in the following: Width of signal lines and routing in layer 4 $w = 0.18$ mm; separation of signal lines $s = 1.2$mm; length and width of the patch $(l, W) = (11, 1.6$ mm); distance between patch and ground plane $d = 0.2$ mm; dimension and pitch of the through-hole vias $(D, p) = (0.3, 0.8$ mm); and the substrate thickness $(h_1\text{-}h_5) = (0.1, 0.2, 0.4, 0.2, 0.1$ mm). Regarding the electrical characteristics of the FR4 substrate, the dielectric constant $\varepsilon_r = 4$ and the loss tangent $\tan\delta = 0.02$. The corresponding extracted values in the equivalent circuit at 5 GHz are as follow: Impedance of signal trace and patch $(Z_{01}, Z_{02}) = (56, 36 \ \Omega)$; Electrical length $(\theta_1, \theta_2) = (117°, 123°)$; Total equivalent inductance $L = 0.44$ nH.

978-1-4799-0708-3/13 $31.00 © 2013 IEEE

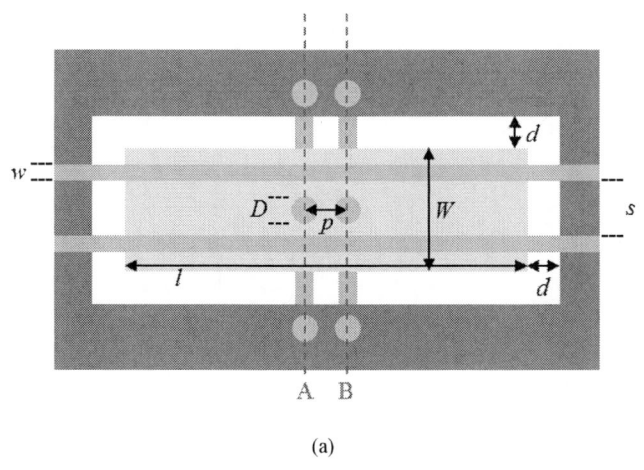

(a)

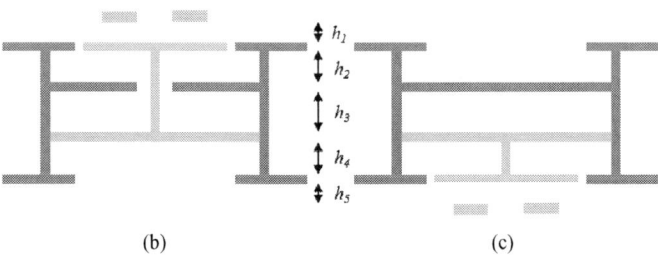

(b)   (c)

Fig. 5. Configuration of proposed common-mode filter in 6-layer PCB. (a) Top view where the ground plane in layer 3 is not shown. (b) Side view: pair 1 and ground at cut A where through-hole via is presented as blind via. (c) Side view: pair 2 and ground ay cut B where through-hole via is presented as blind via.

III. SIMULATION RESULTS

Some simulation have been done to testify the performance of proposed structure. With the circuit simulation of equivalent model, it is estimated that there will be a transmission zero of common mode signal at 2.85 GHz, and as illustrated in Fig. 6, the result agrees well to that of full-wave simulation tool CST. Insertion loss and group delay of differential-mode transmission are shown in Fig. 7 and as good as expected since the odd-mode impedance is designed to be 50 Ω and there is no discontinuity on the traces. With the aids of shielding ground plane, the differential-mode crosstalk level, as shown in Fig. 8, is less than -60dB from DC up to 6 GHz, which is much smaller than the reference board. Moreover, in Fig. 9, due to full symmetry, the mode conversion of proposed solution board is obvious reduced (more than 40dB) compared to the reference board.

Eye diagram is a useful and helpful measure to access the channel quality for digital circuits. A differential $2^{11}-1$ pseudo-random bit sequence at 5 Gbps with voltage level of 1 V, rise time of 35 ps, and fall time of 35 ps is excited into the solution board and an ideal differential pair respectively to obtain the eye diagram in CST, as shown in Fig. 10. The eye diagram quality is compared in terms of eye width and eye height, which are both summarized in Table I. From this, good signaling quality is proved once more.

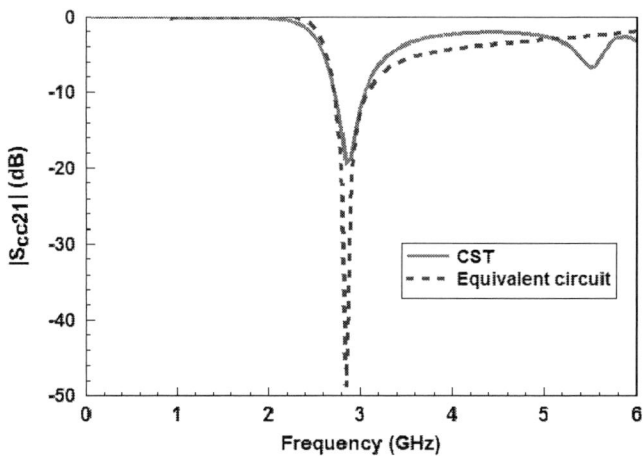

Fig. 6. Verification of the equivalent circuit model by full-wave simulation CST.

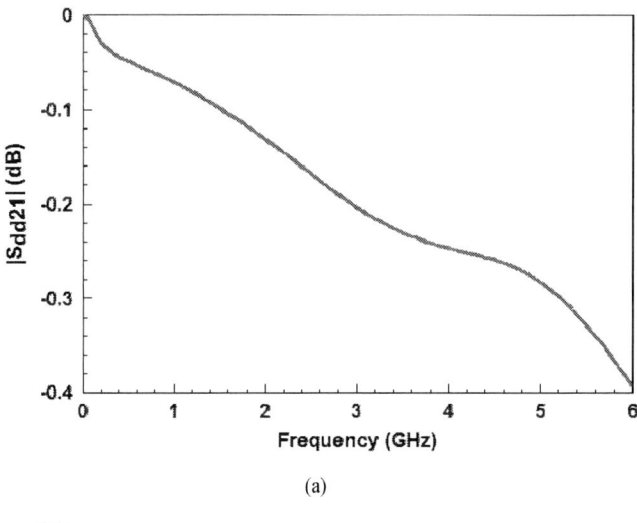

(a)

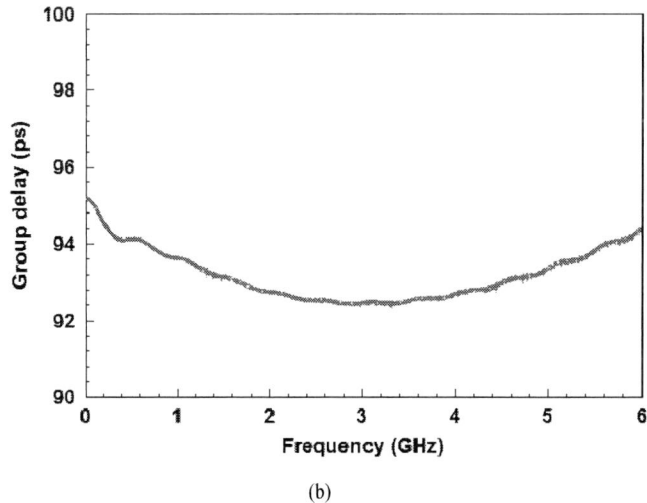

(b)

Fig. 7. Low-distortion charcteristic of proposed structure. (a) Insertion of differential mode. (b) Group delay of differential mode.

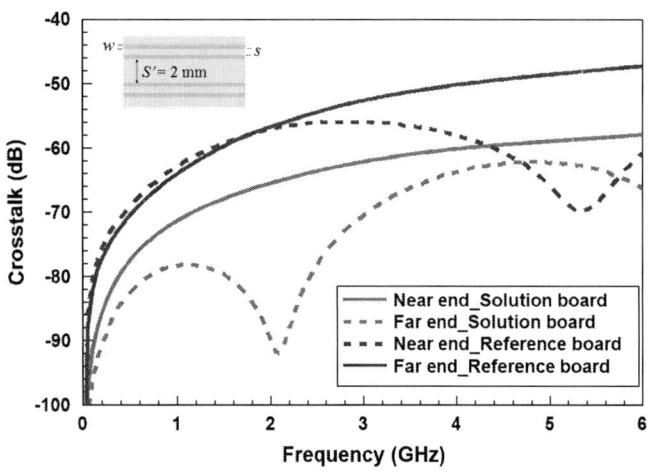

Fig. 8. Differential mode crosstalk of solution board compared to reference board.

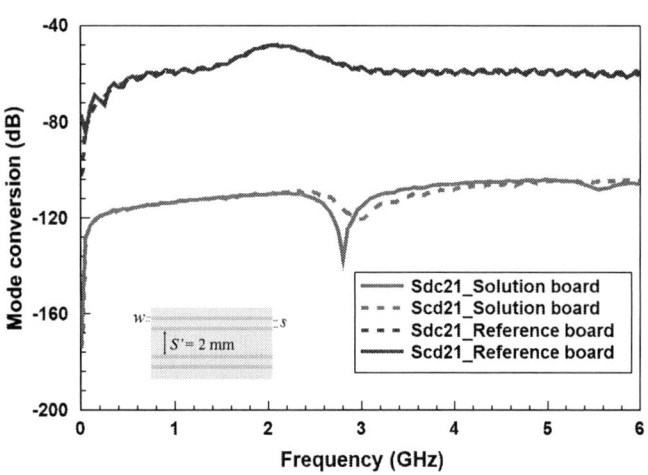

Fig. 9. Mode conversion of solution board compared to reference board.

TABLE I
EYE PARAMETER SUMMARY

	Eye width	Eye height
Solution board	197 ps	966 mV
Reference board	198 ps	981 mV

IV. CONCLUSION

A new structure that can suppress common-mode noise of two differential pairs at the same time while keeping the signaling quality is proposed. By placing the two pairs in different layers and a shielding planes between them, crosstalk level is obviously lower than that by conventional placing. In addition, because of the fully symmetric structure, the mode conversion problem can also be solved. Furthermore, since the coupling of the two pairs is removed, the common-mode stopband will not shift and become easier to design. Some simulations have been done with CST to prove the good performance of signal transmission and noise suppression.

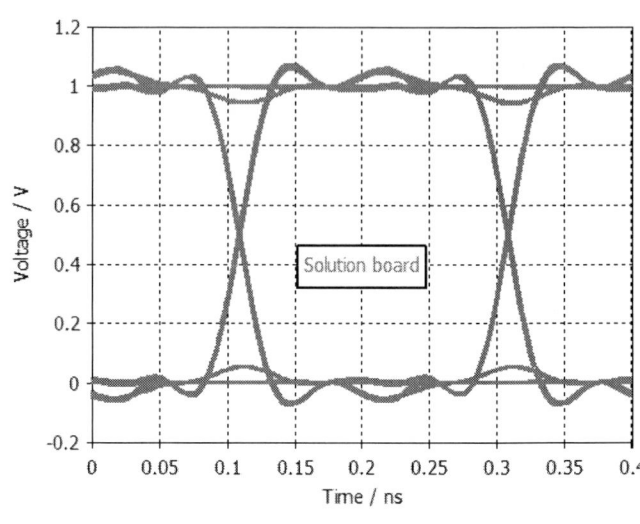

(a)

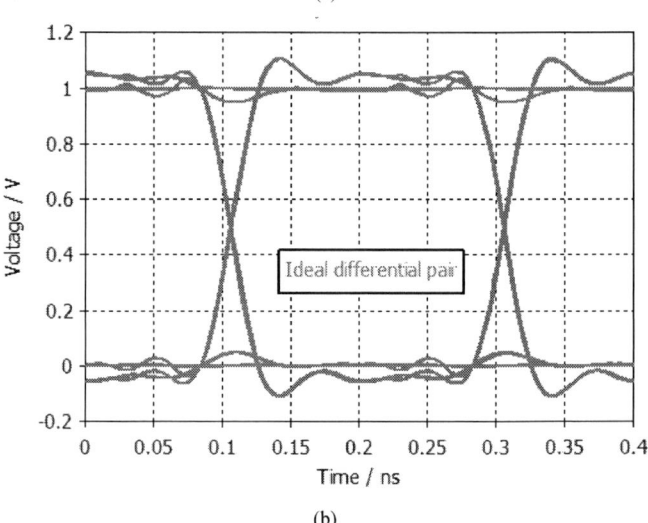

(b)

Fig. 10. Eye diagram of (a) Solution baord (b) Ideal differential pair.

REFERENCES

[1] B. Archambeault, S. Connor, and J. Diepenbrock, "EMI emissions from mismatch in high-speed differential signal trace and cables," in *Proc. IEEE Int. Symp. Trans. Electromagn. Compat.*, Hawaii, July 2007.

[2] D. M. Hockanson, J. L. Drewniak, T. H. Hubing, T. P. Van Doren, F. Sha, and M. Wilhelm, "Investigation of fundamental EMI source mechanisms driving common-mode radiation from printed circuit boards with attached cables," *IEEE Trans. Electromagn. Compat.*, vol. 38, no. 4, pp. 557–566, Nov. 1996.

[3] W.-T. Liu, C.-H. Tsai, T.-W. Han, T.-L. Wu, "An embedded common mode suppression filter for GHz differential signals using Periodic Defected Ground Plane", *IEEE Microwave and Wireless Components Letters*, vol. 18, no4, pp. 248-250, Apr, 2008.

[4] C.-Y. Hsiao, C.-H. Tsai, C.-N. Chiu, T.-L. Wu, "Radiation Suppression for Cable-Attached Packages Utilizing a Compact Embedded Common-Mode Filter," *IEEE. Trans. Compon. Packag. Manuf. Technol.*, vol. 2, no. 10, pp. 1696-1703, OCT, 2012.

[5] B. C. Tseng and L. K. Wu, "Design of miniaturized common-mode filter by multilayer low-temperature co-fired ceramic," *IEEE Trans. Electromagn. Compat.*, vol. 46, no. 4, pp. 571–579, Nov. 2004.

Tests for Time Domain EM Solvers for Stability and Towards Passivity

G. Antonini [1], A. E. Ruehli [2], D. Romano [1], L. Jiang [3]

[1] *Dipartimento di Ingegneria Industriale e dell'Informazione e di Economia*
Universita' degli Studi dell'Aquila, Via G. Gronchi 18, L'Aquila, Italy,
giulio.antonini@univaq.it, daniele.romano.vis@gmail.com

[2] *EMC Lab Missouri Univ. Science and Technology, Rolla, MO, USA, albert.ruehli@gmail.com*

[3] *Department of Electrical/Electronic Engineering, University of Hong Kong, Hong Kong, ljiang@eee.hku.hk*

Abstract—**Stability issues for time domain integral equation based EM solvers have been a problem for a many years. In this paper, we consider ways to test for stability as well as other criteria towards passivity. The tests are designed to monitor the EM solver while the time domain solution is progressing. Hence, the solution can be terminated and improved with a passivization technique when a violation of the criterion has been detected. Also, the tests help to understand the cause of instabilities and results can be used for passivization.**

I. INTRODUCTION

It has been known for many years that the time domain solutions for integral equation solvers are unstable for some problems. This is an issue even today, e.g., [1]. For this reason, frequency domain solutions are sometimes preferred for integral equation solvers in spite of the fact that time domain solutions can be faster than the ones for frequency domain. Recently, it also became clear that a large class of problems have additional stronger requirements. Several different solutions in the time domain are frequently connected in series to model a channel of connected components. Hence, stability in the time domain for a single component is no longer sufficient since it does not guarantee stability of the connected systems and/or circuits.

A class of techniques has been developed recently for the frequency domain to improve the solution properties since each one of the series connected frequency solutions should also be passive. This insures the that the Fourier transformed time solutions are passive, which is *enforced* after the EM solution has been obtained, e.g., [2]. The approach is called *passivity enforcement*. In this paper, we pursue a different approach.

Rather than making corrections as a post-processing step, we want to test at the EM solver level while the solution is progressing in the time domain. Importantly, a solution can be time domain stable, but not passive. Time domain stability issues have been considered early on for PEEC solvers [3]. Passivity has recently been considered for the frequency domain [4]. However, the tests in the time domain are different.

In this paper, we consider issues for the testing in the time domain by applying different monitoring techniques while the solution is in progress. It is desirable to stop the solution once a problem is detected such that stabilization techniques can be applied. The testing of desirable properties like *passivity* is more difficult in the time domain than the frequency domain.

In general, the size of the system matrices for practical problems is too large to perform many internal test operations. Also, the additional compute time for the test should be minimal. Fortunately, stability is easy to monitor since for an unstable solution measurable voltages and/or currents can be observed. Instability results in an exponential growth of the solution

$$x(t) = x(0)e^{\alpha t} \tag{1}$$

where $x(0)$ may be extremely small. However, since $\alpha > 0$ the exponential will eventually dominate for $t > T_0$ where T_0 is hard to predict. This is clearly an easily monitored situation. Therefore, time domain stability has the advantage that eventually at each node in the PEEC model can be used to monitor the solution.

We also consider additional observation techniques which are aimed at tests towards time domain passivity. The conditions for time domain passivity are stated in [2]. One of them is the computation of the energy supplied at the ports by the sources in the form

$$\int_{-\infty}^{t} v(\tau) \, i(\tau) d\tau > 0 \tag{2}$$

The problem with these tests is that, in principle, it is required that (2) is tested for any $v(t)$ ant $i(t)$. Clearly, these quantities are related to the excitation source used which is related to the specific solution at hand. Hence, applying the test to a specific solution does not prove that the PEEC model is universally passive. However, we aim at utilizing some of the PEEC solver specific properties to limit possible test waveforms. For example, it is clear that slow waveforms will not excite the high frequencies which lead to stability and passivity issues.

Scattering parameters or reflection coefficients are very useful for the passivity tests since it separates the incident

978-1-4799-0708-3/13 $31.00 © 2013 IEEE

power from the reflected power [2]

$$\int_{-\infty}^{t} [a^2(\tau) - b^2(\tau)]d\tau > 0 \qquad (3)$$

such that the incident power $a^2(\tau)$ is larger than the reflected power $b^2(\tau)$. We present a circuit variable version of (3) in Section II.

It is important to understand that other issues differentiate passivity in a time domain solutions from the frequency domain. Basically, for frequency domain solutions we can limited the highest frequency up to which the solution is considered. Of course, this does not guarantee passivity above this frequency. However, solutions which are not strictly passive [5] can be sufficient to assume passivity up to a maximum frequency.

It is clear that the quality of the numerical EM solution gets worse with higher frequencies. High frequencies can play a key role in the stability and passivity preservation. The input signals spectrum often includes very high frequencies which excite very high frequency resonances in the EM problem where the solution is not stable and not passive. Another issue which we do not consider in this short paper is the impact of the numerical damping due to the time domain integration method employed.

Passivization techniques have been used to remove some of these problem issues [4]. In general PEEC code are used for the solution of interconnect and package, e.g, [6], [5]. We use two PEEC solvers for our experiments.

Another issue we cannot consider is a multi-port situation. We observe that the passivity requirements in [2] for a multi-port basically requires a test for each of the port of the EM system under investigation. However, even for a one port system, the terminations at each of the ports is of importance. The passivity of the system depends on whether the ports which are used in the application must be passive. However, it is sufficient to observe time instabilities by monitoring the voltage on a single node or port in the PEEC circuit due to the exponential growth (3).

II. EM SOLVER AND MONITORING CIRCUIT

A PEEC formulation is used with an MNA (Modified Nodal Analysis) descriptor equation solver. All PEEC models have the common circuit node at infinity, which is called ground in a SPICE type circuit solver. Also, the partial inductances and potential coefficient couple all elements even if they are not sharing a common dc path. In this case, the transparency or observability is restricted to higher frequencies if the dc path is not fully connected. However, it is well known that passivity issues occur mostly at high frequencies. Again, instability results in an unstable time solution at all nodes in the PEEC circuit. For this work, we use PEEC solvers for both the time as well as the frequency domain using similar partial element models in both domains. The formulation of a PEEC solver is well known today e.g., [7]. Of course, we exclusively will use full-wave (FW)PEEC models.

The (FW)PEEC model has the form of (4) which also includes potential independent voltage sources $\mathbf{V_i}$

$$\begin{bmatrix} s\mathbf{P_d}^{-1} & (\hat{\mathbf{I}}+\mathbf{M})\mathbf{A}_\ell \\ \mathbf{A}_\ell^{\mathbf{T}} & -(s\mathbf{L_p}+\mathbf{R}) \end{bmatrix} \begin{bmatrix} \mathbf{\Phi_n} \\ \mathbf{I}_\ell \end{bmatrix} = \begin{bmatrix} (\hat{\mathbf{I}}+\mathbf{M})\mathbf{A_i}\mathbf{I_i} \\ \mathbf{N}\,\mathbf{V_i} \end{bmatrix} \quad (4)$$

and where unknowns are $\mathbf{\Phi_n}$ the node potentials and $\mathbf{I}_\ell$ is the set of currents for the partial inductances. Also $\mathbf{P_d}$ is the diagonal part of the partial coefficient of potential matrix while $\mathbf{M}$ has the off-diagonal terms and $\hat{\mathbf{I}}$ is a unit

matrix. $\mathbf{A}_\ell$ is the matrix KCL for the inductive currents and $\mathbf{A_i}$ for the inputs. $\mathbf{L_p}$ are the partial inductances and $\mathbf{R}$ the series resistances. $\mathbf{I_i}$ are potential input currents and $\mathbf{V_i}$ potential voltage source in series to the inductive branches. In this formulation, the matrix $\mathbf{N}$ selects which loop has potential voltage sources $\mathbf{V_i}$ in series to the loops. For the time domain, we replace the Laplace variable $s \to d/dt$.

A. Time domain reflection coefficient

Here, we use for the single port reflection coefficient in the time domain [8] to derive a power-oriented version. We pick up a node in the EM PEEC circuit where we attach the monitor circuit shown in Fig. 1. In PEEC, we can connect

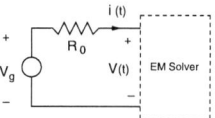

Fig. 1. Monitor circuit for properties of EM solution.

the return connection to the ground node at infinity such that any one of the nodes can be monitored. To derive the instantaneous power reflection coefficient for the scalar case we use [8]

$$\rho(t) = \frac{v_r(t)}{v_i(t)} = \frac{v(t) - R_0 i(t)}{v(t) + R_0 i(t)}. \qquad (5)$$

Multiplying and dividing (5) by $i(t)$ we finally obtain

$$\rho(t) = \frac{v(t)\,i(t) - R_0\,i(t)^2}{v(t)\,i(t) + R_0\,i(t)^2}. \qquad (6)$$

We test the passivity rule, e.g., [2] in the form of (6) for the instantaneous power produced. By integrating over time we also consider the energy relation. Since all initial conditions in the PEEC model are zero, we start the integration at $t = 0$. Here, we do not consider the multi-port situation.

III. DISCRETIZATION AND PASSIVIZATION

If an exact solution of Maxwell's equation could be obtained, then it would clearly be passive. All methods are based on a necessary discretization of the geometry and a differential or integral equation solution based on the Method of Weighted Residuals (MWR). Unfortunately, refinement of the discrete solution including the mesh would be too costly and result in excessive compute times. Practical solutions use reasonably fine mesh subdivisions. However, discretization clearly can lead to the loss of passivity and stability.

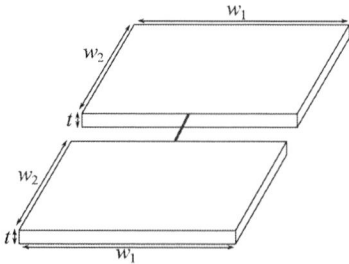

Fig. 2. Small dipole antenna test problem.

We will call the frequency for which the meshing is sufficient f_{max} which, in the time domain, is given by the highest input spectral component. A key observation gave us insight into a fundamental aspect of the problem at hand. Even if we have 20 cells per wavelength at f_{max} at a frequency $10 f_{max}$, we only have 2 cells/wavelength which leads to a very poor representation of the phase and a non-analytic solution.

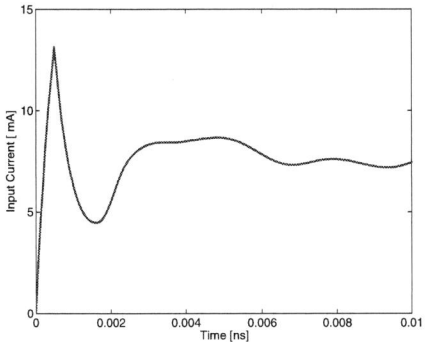

Fig. 3. Input current response for antenna.

Several *passivization* techniques exist for PEEC solutions for (FW)PEEC models. Briefly, some are: (a) resistive damping; (b) macromodels for partial elements; (c) subdivision or improvement of the integral coefficients; and (d) resistive partial elements at very high frequencies. Some of the techniques are considered in [3], [9], [6], [4]. The solution in the time domain is much more revealing since an unstable solution will result in a totally unacceptable solution while the issues may stay hidden in the frequency domain. Clearly, time integration methods have no place in the frequency domain while they play a key role in the time domain.

IV. NUMERICAL EXPERIMENTS

To explore the issues, we use two different examples with different properties.

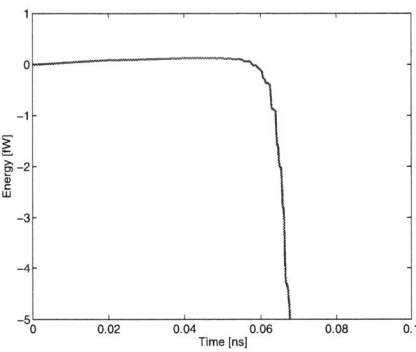

Fig. 4. Energy integral result for antenna.

The first example is for a loss-less dipole patch antenna shown in Fig. 2 in which $w_1 = w_2 = 4.5$ mm. The thickness is $t = 50$ μm and the spacing between them is 500 μm. Each side has 5 cells of 0.375 mm and two edge half-cells. The line between the two patches is the monitor in Fig. 1. Clearly, the applied waveforms must have a very high frequency content, else the check cannot detect the problem. We used a 0.5ps ramp function. To get a sufficient resolution, we used a time step of 0.05 ps. Only a

selected set of results can be included in the paper. Figure 3 shows the system input current $I(t)$. Also, the energy integral (2) turns negative as is evident from Fig. 4 indicating non-passivity. The transient reflection coefficient in (5) or (6) are shown in Fig. 5 with non-stable and passive result.

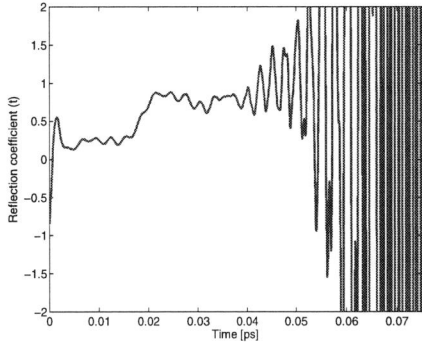

Fig. 5. Ramp Transient reflection coefficient for antenna.

As the last time domain result for the antenna problem, we give the energy integral in Fig. 6. The last result for the antenna problem is the frequency impedance $Z(f)$ which has negative real parts.

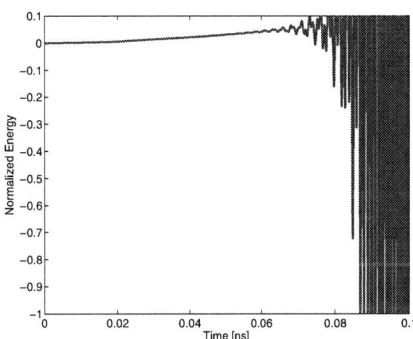

Fig. 6. Normalized energy scattering integral for antenna.

This is shown in Fig 7. All these results point to the fact that the antenna solution is neither stable nor passive.

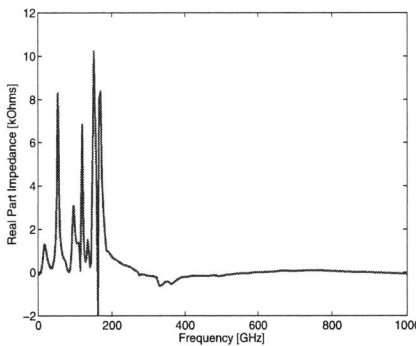

Fig. 7. Real part of frequency domain domain for antenna.

As a second example, a spiral inductor has been modeled. The geometry is shown in Fig. 8. The width and the thickness of the copper conductors are $w = 15$ μm and

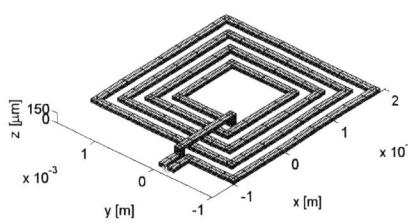

Fig. 8. Geometry of the spiral inductor example.

$t = 10$ μm, respectively. The horizontal spacing is 400 μm, the vertical one is 200 μm.

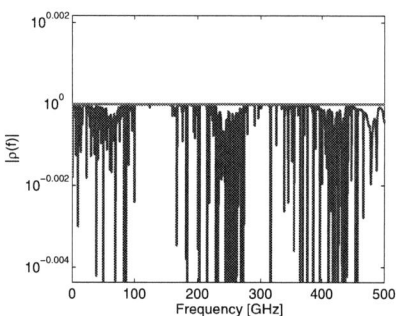

Fig. 9. Frequency domain reflection for spiral.

For the spiral example in Fig. 8, both the frequency response in Fig. 9 as well as the transient response Fig. 10 show that the reflection coefficient ρ stay below 1 showing passivity in both the time and frequency domain. This is further confirmed by the positive real part of the input impedance $Re[Z(f)] > 0$ shown in Fig. 11 which is necessary also for stability.

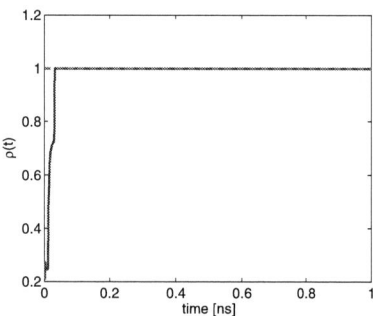

Fig. 10. Transient reflection coefficient for spiral.

In the last Fig. 12, we show that the time solution is stable even for the small time step 0.3 ps and a fast current ramp rise time of 1 ps. The fact that the solution is stable is consistent with the results in the last three figures.

V. ACKNOWLEDGMENT

A. Ruehli acknowledges insightful discussions with M. Nakhla and R. Achar which impacted this work.

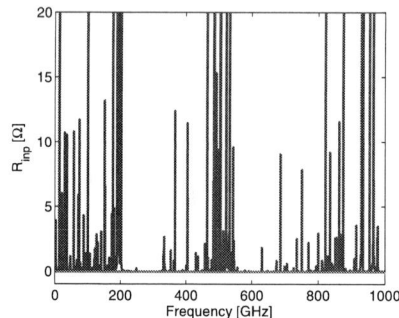

Fig. 11. Real part of the input impedance $Z(f)$ for spiral.

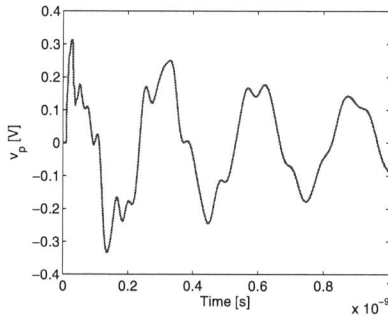

Fig. 12. Time domain response to current step for spiral.

VI. CONCLUSIONS

We show in this paper, that we can monitor the stability in the time domain to interrupt the solution. We also consider additional tests which are oriented towards passivity. Examples are given to illustrate the process for a time domain PEEC solution.

REFERENCES

[1] B. H. Jung, T. K. Sarkar, S. W. Ting, Y. Zhang, Z. Mei, Z Ji, M. Yuan, A. De, M. Salazar-Palma, and S. M. Rao. *Time and frequency domain solutions of EM problems using integral equations and a hybrid methodology.* John Wiley and Sons, New York, 2010.

[2] P. Triverio, S. Grivet-Talocia, M. S. Nakhla, F. G. Canavero, and R. Achar. Stability, causality and passivity in electrical interconnect models. *IEEE Transactions on Advanced Packaging*, 2007.

[3] J. Garrett, A. Ruehli, , and C. R. Paul. Accuracy and stability improvements of integral equation models using the partial element equivalent circuit PEEC approach. *IEEE Transactions on Antennas and Propagation*, 46(12):1824–1832, December 1998.

[4] A. E. Ruehli, G. Antonini, and Lijang L. Passivization of EM PEEC solutions in the frequency and time domain. In *ICEAA Int. Conf on EM in Adv. Applications*, Torino, Italy, September 2013.

[5] A. E. Ruehli and A. C. Cangellaris. Progress in the methodologies for the electrical modeling of interconnect and electronic packages. *Proceedings of the IEEE*, 89(5):740–771, May 2001.

[6] J. Nitsch, F. Gronwald, and G. Wollenberg. *Radiating nonuniform transmission-line systems and the partial element equivalent circuit method.* John Wiley and Sons, New York, 2009.

[7] A. E. Ruehli, G. Antonini, J. Esch, J. Ekman, A. Mayo, and A. Orlandi. Non-orthogonal PEEC formulation for time and frequency domain EM and circuit modeling. In *Proc. of the IEEE Int. Symp. on Electromagnetic Compatibility*, volume 45, pages 167–176, May 2003.

[8] J. E. Schutt-Aine and R. Mittra. Scattering parameter transient analysis of transmission lines loaded with nonlinear terminations. *IEEE Transactions on Microwave Theory and Techniques*, 36(3):529–535, March 1988.

[9] G. Antonini, D. Deschrijver and T. Dhaene. Broadband macromodels for retarded Partial Element Equivalent Circuit rPEEC method. *IEEE Transactions on Electromagnetic Compatibility*, 49(1):34–48, February 2007.

AUTHOR INDEX

Achar, Ramachandra 137
Ahmadloo, Majid.......... 29
Ambasana, Nikita 247
Antonini, Giulio 115
Antonini, G. 263
Antoulas, Athanasios 133
Apte, Prakash R. 243
Aronsson, Jonatan 119
Asai, Hideki.......... 47
Aydiner, Alaeddin 51
Bae, Bumhee 91
Baek, Hyunho 103
Bandinu, Michelangelo 125
Bang, Kyongmo 165
Bautista, Harold 215
Benner, Peter 115
Beyene, Wendem 169
Beyreuther, Anne 65
Bhatheja, Kushagra 151
Biondi, Alessandro 99
Brenner, Pietro 129
Broyde, Frederic 227
Butt, Khalid 119
Canavero, Flavio.......... 99
Cehn, Guang 211
Cha, Seungyong 215
Chand, Kundan 155
Chandrasekar, Karthik.......... 79
Chandrasekhar, Arun 247
Chandrasekhar, Janani 223
Chang, Xin 39
Chen, Duo.......... 219
Chen, Kevin 165
Cheng, Chi-Hsuan 259
Cheng, Chung-Kuan 87
Chernobryvko, Mykola 15
Chew, Weng Cho 203
Chhabra, Nitin 151
Chhabra, Nitin Kumar.......... 243
Chhay, San 231
Chiang, ChunTong 57
Chinea, Alessandro.......... 125
Cho, Minki 199
Choi, Baekkyu 215
Choo, Wernshin 211
Chu, Yunhui 231

Clavelier, Evelyne 227
Cnagellaris, Andreas 11
Comberiate, Thomas 25
Commens, Matthey 35
Coutts, Ryan.......... 87
Crisp, Richard 165
Dahl, David.......... 65
Das, Arkaprovo 239
De Luca, Giovanni 115
De Zutter, Daniel.......... 15, 99
Delino, Julius 103
Dia, Yun.......... 61
Duan, Xiaomin 65
Eisenstadt, William 103
Elad, Danny.......... 7
Eo, Yungseon 207
Fang, Jiayuan.......... 173
Gope, Dipanjan 239, 247
Grivet-Talocia, Stefano 125, 129
Gu, Xiaoxiong 73
Gundurao, Anil.......... 3
Ha, Myunghyun 51
Hein, Matthian 179
Huang, Ting-Yi 251
Humbla, Stefan 179
Jenkins, Keith 73
Jiang, Jenny 223
Jiang, L. 263
Jiao, Dan.......... 219
Jiayuan, Fang.......... 187
Jin, Jianming.......... 107
Jinping, Zhang 187
Ju, Inkwon 255
Jung, Daniel H. 91
Kabir, Muhammad 141
Kaleem, Saqib 179
Kaminski, Noam 7
Khazaka, Roni.......... 141
Kim, Sukjin 91
Kim, Hyewon.......... 207
Kim, Sungjoo 159
Kim, Ga Won 215
Kim, Dongsu 183
Kim, Junghoon J. 91
Kim, Jungho.......... 91
Kim, Kee Sup 69
Kollipara, Ravi 83

978-1-4799-0708-3/13 $31.00 © 2013 IEEE

Kong, Sunkyu 91
Kung, Jaeha 199
Kunze, Richard 231
Lan, Hai 83
Lang, Klaus-Dieter 65
Lee, Seung-Cheol 35
Lee, Junghyun 207
Lee, Byunghyun 69
Lee, Sang Min 69
Lefteriu, Sanda 133
Liu, Yang 87
Liu, Hui 155
Lu, Tianjian 107
Luo, Chong 173
Ma, Xiao 11
Madden, Chris 83
Malik, Rakesh 151
Malik, Rakesh 243
Manfredi, Paolo 99
Martin, Aaron 147
Melde, Kathleen L. 21
Min, Max (Sunghwan) 215
Moon, Seong Jae 159
Mori, Hirouaki 7
Mukherjee, Jayanta 243
Mukhopadhyay, Saibal 199
Nagpal, Rajkumar 151
Nagpal, Raj Kumar 243
Nair, Rahit R. 239
Nakhla, Michel S. 137
Natu, Nitish 69
Ndip, Ivan 65
Nguyen, Cuong 223
Nouri, Behzad 137
Ochoa, Juan 11
Oh, Dan 79, 155, 211, 223
Okamoto, Keishi 7
Okhmatovski, Vladimir 119
Olivadese, Salvatore 129
Park, Sung Joon 69
Park, Se Hoon 183
Park, Jong Chul 183
Ping, Liu 187
Polstyanko, Sergey 35
Pytel, Steven 35
Qian, Zhi-Guo 203
Qian, Zhiguo 51
Qing, He 219

Razmadze, Alex 79
Ren, Jihong 83
Rentsch, Sven 179
Romano, D. 263
Ruehli, A. E. 263
Ryu, Jong-In 183
Ryu, Woong Hwan 159
Ryu, Keun Kwan 255
Ryu, Woong Hwan 69
Saibal, Mukhopadhyay 235
Sangmin, Lee 159
Schuster, Christian 65
Schutt-Aine, Jose 25, 227
Schutt-Aine, Jose E. 203
Scogna, Ciccomancini Antonio 57
Sekine, Tadatoshi 47
Seler, Ernst 43
Shi, Hong 155
Shim, Yujeong 223
Shumakher, Evgeny 7
Skibin, Stanislav 191
Soldo, Denis 35
Stevanovic, Ivica 191
Sun, Zhuowen 165
Sun, Shishuang 211, 223
Swaminathan, Madhavan 51, 69
Toriyama, Kazushige 7
Tran, Dinh 215
Tretiakov, Yuri 223
Tripathi, Jai Narayan 151, 243
Trivedi, Amit 235
Triverio, Piero 111
Tsang, Leung 39
Ubolli, Andrea 125
Vande Ginste, Dries 15, 99
Vargas, Marcos A. 21
Vennam, Prakash 35
Wang, Xiaoqing 147
Wei Wu, Jun 203
Weigel, Robert 43
Wojnowski, Maciej 43
Wu, Yi-Chen 251
Wu, Kai-Bin 251
Wu, Tzong-Lin 259
Wu, Ruey-Beei 251
Wunsch, Bernhard 191
Yalamanchili, Sudhakar 199
Yan, Zhuo 83
Yang, Melinda 3
Yang, Kang-Yun 251

Yeleswarapu, Krishnamurthy 235
Yi, Ming 51
Yi, Yang 195
Yom, In-bok 255
Yook, Jong Min 183
You, Eileen 3
You, Se-ho 215
Yuan, Weiliang 159

Zhang, Jingping 173
Zhang, Jianmin 223
Zhou, Yaping 195
Zhou, Nian 215
Zhu, Jianfang 219
Zhang 87

9781479907083